Integral Approaches to Tribo-testing in Mechanical Engineering

Integral Approaches to Tribo-testing in Mechanical Engineering

Dr. H. Prashad

Consultant,
Centre for Tribology Incorporated, (CETR),
Campbell, CA, USA.

BSP BS Publications

CRC is an imprint of the Taylor & Francis Group,
an informa business

Distributed in India, Pakistan, Nepal, Myanmar (Burma), Bhutan, Bangladesh and Sri Lanka by **BS Publications**

Distributed in the rest of the world by
CRC Press LLC, Taylor and Francis Group,
6000 Broken Sound Packway, NW, Suite 300, Boca Raton, FL 33487, USA

First published in India by
BS Publications, 2009

ISBN 13: 9781439806081
ISBN 10: 143980608X

Printed in India by
Sanat Printers, Kundli

Published by
BS Publications
4-4-309, Giriraj Lane, Sultan Bazar, Hyderabad - 500 095 (A.P.) India.

Dedicated to

"The Almighty whose Blessings, Radiance, Inspiration, Silent Guidance Graced me towards Enlightenment,* Ultimate *Truth and Self-realization"

Dr. H. Prashad

Contents

Section - II
Lubricants

Chapter - 2

Latest Tribology Testing of Lubricating Oils and Effect of Nano-additives on Oils

Chapter - 3

Diagnosis of Deterioration of Lithium Greases Used in Rolling-Element Bearings by X-Ray Diffractrometry

Chapter - 5

A Combinatorial Approach to Elucidating Tribochemical Mechanisms

Section - III
Bearing under Electric Current

Chapter - 6

A State-of-the-Art Review of Bearings and Lubricants in Electrical Environment

Chapter - 7

Causes of Shaft Voltages in Rotating Machines

Chapter - 8

Magnetic Flux Density Distribution on Track Surfaces of Rolling-Element Bearings

Chapter - 9

Unusual Causes of Rolling-Element Bearings Failure

Chapter - 10

Localized Electrical Current in Rolling-Element Bearings

Chapter - 11

Rolling-Element Bearings Under the Influence of an Electric Current

Section - IV

Bio-tribology

Chapter - 12

Tribometrology of Skin

Section - V
Tribology in Engines

Chapter - 19
Investigations of Low Friction Ring Pack for Gasoline Engines

Chapter - 20
Investigations into Characterization and Simulation of Cylinder Linear Surface Finishes

Section - VI

Films and Coatings

Chapter - 24

Integrated Multi-Sensor Tribo-SPM (Scanning Probe Microscope) Testing for Nano/Macro Tribo-Metrology

Chapter - 25

Evaluation of the Bond Strength of Thermal Sprayed Coatings

Chapter - 26

Friction and Wear of a Rubber Coating in Fretting

Section - VIII
Indentation, Scratch and Mapping

Section - IX
CMP-Chemico Mechanical Processing

Section - X
Cosmetics

Preface

This book on "Integral Approaches to Tribo-testing in Mechanical Engineering" brings out the solution of various complex technical problems projected to be of tribology and mechanical engineering concerning hard and soft coatings, adhesion, scratch, wear, tribological studies in bioengineering, cylinder liners, gasoline engines, micro/nano studies in lubricants and additives including effect of electric current and influence of magnetic flux density on bearing surfaces. Also, diagnosis and investigations are reported of such cases those are closely associated with the design, system, quality, malfunctioning, maltreatment, material, operational and various unforeseen causes related to performance of the machines besides the significance of Tribology in multifold applications.

The different investigations, their analysis and solutions are brought out in this book are unique in nature, but the symptoms of the problems and related failure were projected to be closely revolved around testing, data analysis, diagnosis of mechanical engineering components, and tribology. However, detailed diagnosis after investigations/testing and repeated trials revealed the origin of the causes and sources of problems and thereby initiation of failures, which were dormant in nature. The solutions to most of these problems disclosed/opened a new era and direction, which is focused on development and analysis in the field of tribology and mechanical engineering through nano/micro studies/testing and investigations to tackle multi-dimensional complex issues.

The various chapters of this book deal with the individual unique problems and their solutions particularly pertaining to the comprehensive tribo-mechanical testing of hard coatings, multi-sensing advance testing of thin and thick coatings for adhesion and delamination, scratch, wear and adhesion testing of LCD display coatings, bond strength of thermal sprayed coatings, mapping of thin overcoats including multi-sensor tribo-scanning microscope testing for nano/micro tribo-metrology. Methods of characteristics evaluation of electrical contact properties of carbon nanotube coated surfaces have been included. Magnetization of bearings leading to premature failure due to unrecognized flow of a localized electric current in the bearings causing flutings by the induction effects on the track surfaces and various unforeseen causes leading to the failure. Also, details of theoretical and experimental analysis of magnetic flux density distribution as a diagnostic tool for critical failures are included. Furthermore, techniques have been brought out in a few chapters of the book for the tribo-metrology of the skin and in depth studies in bioengineering covering micro/nano scale mechanical and tribological characterization for orthopedic applications and studies related to friction coefficient pertaining to bovine knee articular cartilage.

The friction and wear characteristics of rubber coating in fretting and precision measurements of experimental properties of visco-elastic properties of industrial polymers have been analyzed for trouble free performance evaluation in various industrial equipments. Furthermore, a few chapters have been devoted to the characterization and simulations of cylinder liner surface finishes and study of low friction ring pack for gasoline engines along with an approach to elucidating tribo-chemical mechanism.

The life estimation of turbine oils highlighting a methodology and criterion for their acceptance in power plants equipments for the reliable operation in the system being very significant for the maintenance engineers has been included besides an efficient non-conventional integral approach using multi-sensing technology for testing of lubricating oils, greases and effect of nano additives on lubricants. Also, chapters on lubricants are included, which have a potential to analyze their performance in a roller bearing by X-ray diffractrometry, and variation and recovery of resistivity of lubricants in the bearings under different conditions of operation.

In short, this book on "Integral Approaches of Tribo-testing in Mechanical Engineering " deals with the unique multifaceted problems and their solutions and in depth analysis/investigations of nano/micro studies of different coatings, bioengineering, polymers, cosmetics, chemico-mechanical processing, indentation, scratch and mapping, lubricants and additives, system problems including reliability assessment and new testing and diagnosis approaches in various dimensions in 10 Sections with 37 Chapters. In the net shell book presents original typical investigations and studies as the summarized professional experience of various experts in the area of mechanical engineering and industrial tribology. The work presented in various Chapters may prove to be useful for engineers and technologists of different industries, students, research engineers/scientists, academicians and others who grapple with the complex problems of testing of engineering components and tribology for multifaceted research for future developments.

The author is extremely thankful to the large number of professionals/ engineers from Centre for Tribology Incorporated (CETR), CA, USA and various experts and organizations who have used CETR tool for advanced research and brought very useful industrial data in publications besides specialists from BHEL Corporate R & D Division, Hyderabad, India, those who have participated in different capacities in solving/rendering assistance to identify tricky and multi-dimensional complex tribology and mechanical engineering problems closely linked with system, testing, design and operation.

Thanks are due to Dr. Norm Gitis, President of CETR-Centre for Tribology Inc.CA, USA, for giving opportunity/permission to publish studies/ investigations carried out at CETR and publications by world-wide various other organizations/institutions using CETR tool for in-depth involved systems, testing and in solving complex unique problems concerning tribology and mechanical engineering.

Furthermore, without silent sacrifice of my wife, Darshan, and my beloved children Rajeev & Shwetlana, Poojan and Gauri, and little Kriya who has co-operated in my work with her sweet smile without any interference at the time when they need my involvements for their needs, this task would not have been completed in the time frame.

Finally, my gratitude is due to Almighty, whose blessings, radiance, continued inspiration, pouring of booming energy and silent guidance that has nurtured me from time to time to follow the path of untiring research for the ultimate progress.

May 2009

Dr. Har Prashad

Consultant (CETR)

E-mail: hprashad@cetr.com

About the Author

Dr. Har Prashad is presently consultant of Centre for Tribology Incorporated, (CETR), Campbell, CA, USA. He is retired Senior Deputy General Manager (Tribology) from the Bharat Heavy Electricals Limited, Corporate Research and Development Division, Hyderabad, India. He obtained an M.E. (Hons) degree in Mechanical Engineering in 1970, and subsequently, Ph.D. in Tribology. He worked with the Indian Institute of Petroleum, Dehra Dun, and with the Design Bureau at Bokaro Steel Ltd., Dahnbad, Bihar, before joining BHEL in 1974.

At BHEL, he was associated with setting up of the Tribology Laboratory at the Corporate R&D Division, Hyderabad, particularly design and development of various bearing test rigs for hydrodynamic and rolling-element bearings. He has done substantial design/development work on magnetic, dry and other special types of bearings. He has developed energy saving double decker high-precision rolling-element bearings, and established the performance of these bearings both theoretically and experimentally.

His areas of interest include diagnostic monitoring, failure analysis and bearing performance evaluation. He has done significant original work to establish the behaviour of different bearings and lubricants under the influence of electrical current. He has established electrical analogy for dynamic analysis of bearings. Assessment of flow of current through rolling-element bearings by study of magnetic flux density distribution on the bearing surfaces, theoretical evaluation of corrugation pattern, bearing life estimation, resistivity and recouping of resistivity phenomenon in lubricants, theory that explains the causation, morphology and rate of formation of electrical current damage are some of the outstanding original contributions of Dr.Prashad especially for the engineers engaged in the design or operation of heavy rotating electrical machinery.

Dr. Prashad has published more than 115 papers in both national and international journals, particularly in ASME-Journal of Tribology, STLE-Tribology Transactions, WEAR, Lubrication Science, STLE-Lubrication Engineering, Tribotest, Tribology International, IE (I) and BHEL journals and delivered many invited talks. He has patents to his credit. He is recipient of the Corps of Electrical and Mechanical Engineering Award for his paper, "Condition Monitoring of antifriction bearings" of Institution of Engineers (India). He was also awarded with Corps of Engineers Medal Award – 1998 along with other various awards for his contributions and publications. From time to time he was awarded merit certificates from Institution of Engineers for his publications. He was convener and organising secretary of Second

International Conference on Industrial Tribology, held at Hyderabad during December 1999.He was the member of technical papers review panel of Third International Industrial Tribology Conference to be held in Jamshedpur, India, during November 2001.

He is the author of two books. The book on "Tribology in Electrical Environments" is published from United Kingdom by Elsevier Publishers in December 2005. The other book on 'Solving Tribology Problems of Rotating Machines" is published by Woodhead Publishers, U.K, in February 2006.

He has served as Head, Tribology Department, at BHEL, Corporate R and D, Hyderabad. He is on the review panel for technical publications in the journals of Wear, Tribology International, Lubrication Basics, Tribo test, International Journal of Comadem, Journal of Tribology Society of India, Institution of Engineers (India), and assessment committee of the awards for National Research and Development Corporation of India. He is a Fellow of Institution of Engineers (India) and Member of Society of Tribologists and Lubrication Engineers, U.S.A. He is life member of Tribology Society of India and presently he is the executive committee member of the Society. He was joint secretary of Tribology Society of India during 1997-99, and was the editor of Tribology Newsletter during this period. For his achievements, his name has been included in Marquis Who's Who in the world published in 2001 from USA. Also, his name is included in International Directory of Distinguished Leadership-2001, published by American Biographical Institute. His bio-data is published in the directory of "Contemporary Who's Who-2004" by the same institute.

Contact details:

Dr. H. Prashad

1-2-319/A, Gagan Mahal,

Domal Guda,

302 Central View Apartments,

Hyderabad-500029

India

e-mail: har.prashad@gmail.com

(Ex. Senior Deputy General Manager (Tribology), Bharat Heavy Electricals Limited (BHEL), Corporate Research and Development Division, Hyderabad, India.)

Section - I

Tribology

1
Tribology, its Significance and Multifold Applications

1.1 Introduction

In the mechanical engineering nomenclature various new terms with the broad engineering background have been introduced, notably Tero-Technology, Group Technology and Tribology. Tero-technology is based on the Greek word "tere in", which literally means: to hold, to care for.

Tribology is defined as *the science of interacting surfaces in relative motion.* The word *tribology* comes from the Greek *tribos*, meaning rubbing. The entire related phenomenon in the mechanical engineering covering friction, wear and lubrication have been brought under the engineering head "TRIBOLOGY". Thus, tribology broadly covers and embraces the fields of physics, metallurgy, material science, chemistry, mathematics, mechanical and chemical engineering, fluid dynamics and very closely interacts with the field of computational science and engineering.

In any machine there are lots of components parts that operate by interacting with each other together. Some examples are bearings, gears, cams and tappets, tyres, brakes, and piston rings. All of these components have two surfaces, which come into contact, support a load, and move with respect to each other. Sometimes it is desirable to have low *friction,* to save energy, or high friction, as in the case of brakes. Usually to minimize the wear, the components are lubricated.

The study of friction, wear, lubrication and contact mechanics is all-important parts of tribology. Related aspects are *surface engineering* (the modification of a component's surface to improve its function, for example by applying a surface coating), *surface roughness*, and *rolling contact fatigue* (where repeated contacts cause fatigue to occur).

Tribology, in fact attempts to investigate process of mechanics in detail along with the complex effects of energy and material dissipation. It includes not only the work of engineers who use interacting surfaces for the transmission of motion but also of forces, work etc. The importance of tribological studies in present day technologies can be judged by the fact that over 8000 papers are published every year in this field.

One can very well understand that in plants where high temperature, dust and a rather aggressive environment in general are encountered the "war on wear', which is what tribology is all about, has to be a part and parcel of human life. The extent of a successful plant operation in present day could hinge, to a great extend, on the success of tribology.

With advancement of industrialization, mechanization and automation have increased considerably. This has lead directly and consequently to expensive breakdown and mechanical failure in industry. Efforts to prevent such maintenance breakdowns and mechanical failures due to various reasons particularly that of friction and wear have lead to the fundamentals of energy conservation through Tribology in industries i.e. "the fundamentals of interacting surfaces in relative motion". In spite of best efforts problems and challenges in Tribology persists in machines used in power sector equipments including Steam and Gas Turbines, Hydro Turbines, Nuclear Turbines, Turbo generators, and Hydro generators, etc.

1.1.1 *When Two Surfaces are pressed together*

Surfaces may look smooth, but on a microscopic scale they are rough. When two surfaces are pressed together, contact is made at the peaks of the roughness or *asperities*. The real area of contact can be much less than the apparent or nominal area. At the points of intimate contact, adhesion, or even local welding, can take place. If it is required to slide one surface over the other then it is needed to apply a force to break those junctions.

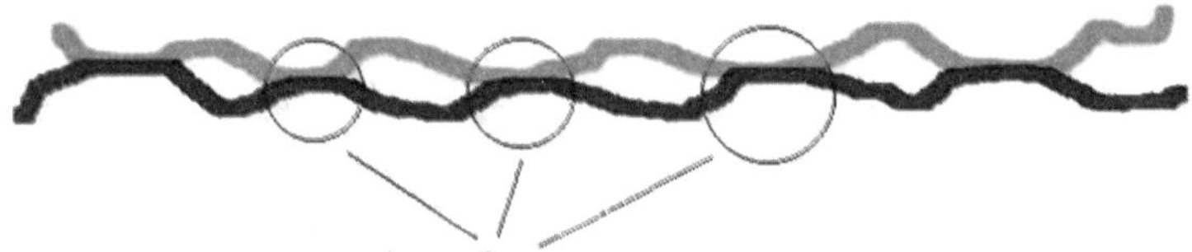

1.1.2 *The Force of Friction*

The friction force is the resistance encountered when one body moves relative to another body with which it is in contact. The static friction force is, how hard it is required to push object to make it move, whilst the dynamic friction force is how hard it is pushed to keep it moving. The ratio of the frictional force F to the normal force P is called the coefficient of friction.

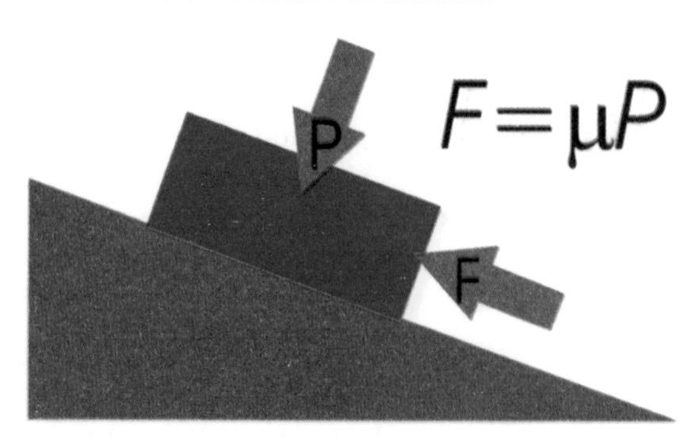

Usually low friction (in a car engine for example) is required so as not to waste excessive energy getting it moving. But in some cases high friction is needed, particularly in brakes for example. Friction is also important for car tires to grip the road and between shoes and the ground for walking.

1.1.3 *Keeping the Surfaces apart - Lubrication*

If a layer of oil is put between the surfaces then these can be separated and easily slide one over another with reduced friction and wear. Mineral oils are the most common lubricants, but other low shear strength materials are also used; for example graphite, PTFE, and soft metals like lead or gold.

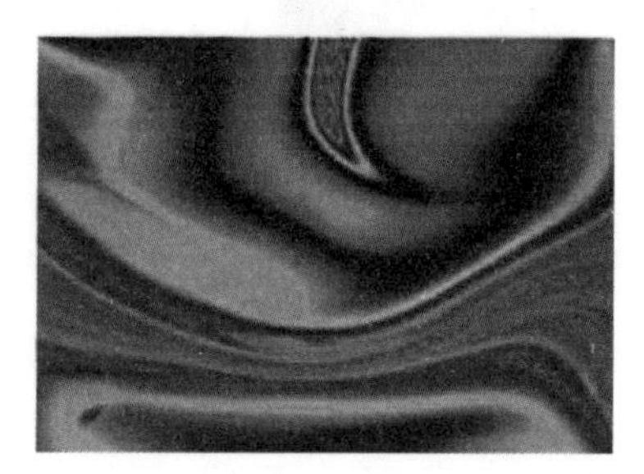

The selection of the best lubricant and understanding the mechanism by which it acts to separate surfaces in a bearing or other machine component is a major area for study in tribology.

1.1.4 *When Surfaces Wear Out*

If one surface is slid over another then the asperities come into contact and there is a possibility of wear out. The breaking of the entire little junction can cause material removal (called *adhesive* wear). Or the asperities of a hard surface can plough grooves in a soft surface (called *abrasive* wear).

Wear is usually unwelcome; it leads to increased clearances between moving components, increased mechanical loading and may cause even fatigue. But in grinding and polishing process the generation of high wear rates is desirable.

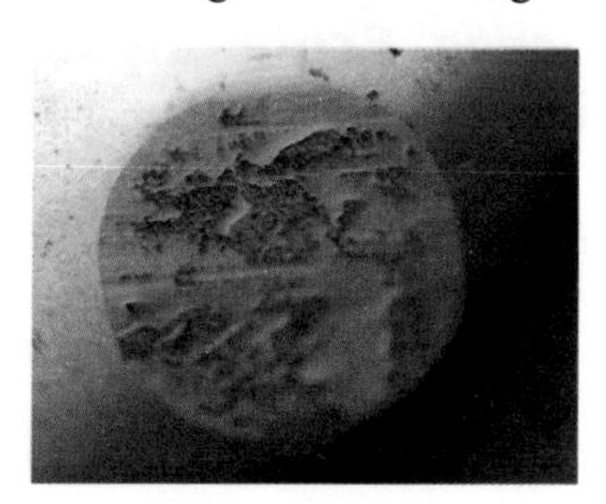

Apart from adhesive and abrasive wear, there are other mechanisms whereby material can be removed from a surface. *Erosive* wear occurs when particles (or even water droplet) strike a surface and break off a bit of the material. Hard particles can become trapped in contacts and cause material to be removed from one or both of the surfaces. One of the main reasons for frequent change of car engine oil is that it becomes contaminated with hard debris particles that can wear out the engine components.

1.1.5 *Stress and Strain at the Contact*

The design of rolling bearings and gears is such that the load is supported on a small area. This leads to high stresses over small areas of the components. This can cause high friction, wear, and contact fatigue. *Contact mechanics* is therefore an important part of tribology.

The analysis of contact stress is generally difficult. Simple component geometries can be analyzed using hand calculations. More complex component shapes frequently require analysis by numerical methods.

1.2 Causes Leading to Tribological Failures

Industrial equipments in power sector or in any allied industry in operation may break down or fail due to the various reasons. However, some of the tribological reasons are broadly classified as:

Environment : Causing corrosion, creep, hydrogen embrittment

Material : Improper material causing fatigue, and inadequate friction and wear characteristics of the used material

Abuse : Notably overloading of equipment, neglect of lubricant and improper selection of lubricant.

Wearing out : As a result of incorrect/malfunctioning of equipment

The above causes lead to the unscheduled breakdown of the equipments causing loss of man-hours, money, energy and such untimely breakdown can be avoided by the correct tribological practices.

1.3 Tribology Losses, Economic Aspects and Energy Saving due to Tribology in Industries

A committee of the 'Organization for Economic Co-operation and Development', as it is understood today, first introduced tribology, as a new concept around 1967. It concerns with the study of friction, wear and lubrication of interacting surfaces in relative motion. The famous Jost Report 'Lubrication (Tribology) - Education and Research: a Report on the Present Position and Industry Needs', Department of Education and Science, was the first all pervading and well researched document submitted to the Government of U. K. In 1966 this report focused on the enormous loss to the British Industry on account of friction and wear. In the year 1979, a similar report was published in West Germany, which estimated the loss in Germany to be of 10 billion DM per year. Tribology related losses for the U.S. industry was estimated in the 'Final

Report to the Congress of the United States' by the National Commission on Materials Policy in the year 1979. A summary of these reports is presented in Table 1.1. Details of the Jost report of economics aspects and energy saving due to tribology is given in Table 1.2. A loss of approximately 100 billion dollars per year in US industry was estimated. In all these efforts, it was felt that 25% to 50% of the losses could be avoided by the systematic application of tribological practices.

Table 1.1 Economic losses due to tribology (friction and wear)

Study Report	National Loss per year	Savings potential per year	Source
UK 1966	£ 2 billion	£ 515 million (About 25% of total loss)	Jost report [1]
FRG 1976	DM 10 billion	DM 5 billion (50% of total loss)	Research report, T76-36, BMFT [2]
USA 1979	$ 100 billion	-	Final report to congress

Similar detailed studies have not been conducted for Indian industries to provide information regarding tribology related losses and the potential savings. However, according to certain estimates, the loss of material in Indian industries due to causes like abrasion, corrosion and other wear processes is about Rs. 20,000 million per year. In view of the published studies furnished earlier it can be safely concluded that a very substantial amount of such wear related losses could be saved by systematic application of tribology practices. In 1997-98, the total energy consumed in India was 241 MTOE (million tonnes of oil equivalent). According to the American report, about 11% of the total consumed energy can be saved through tribology. Thus, the potential saving in this respect would amount to about Rs. 65,000 million per year (assuming crude oil price at $ 15 per barrel) for the Indian economy. All these savings are possible by applying the existing knowledge without significant investment on fresh R & D efforts.

Table 1.2 Economic Aspects and Energy Saving due to Tribology (British Industry Experience)

Particulars	Amount Saved (£ Million)
Reduction in energy conservation through lower friction	20
Reduction in man power	10
Saving in lubricant costs	10
Saving in maintenance & replacement costs	230
Saving in loss consequential upon breakdowns	115
Saving in investment due to higher utilization & greater mechanical efficiency of machinery	22
Savings in investment through increased life of machinery	100
Total	**515**

To resolve the present energy crisis and increase of energy consumption which is asymptotically increasing and getting twofold every decade, tribology plays a very significant role in overall energy scenario. Directly and consequentially Tribology guides industry to save energy by the following industrial tribological developments.

1. By developing lubricants, which allows longer drain periods;
2. by proper selection of lubricants for a given application;
3. by reducing storage and handling losses of lubricants;
4. by recycling of the used lubricants;
5. by avoiding unnecessary leakage through seals joints and fittings (In general 20 to 30% oil is lost from poor quality fittings);
6. by correct design of bearings to minimize power loss;
7. by developing new cost saving materials with better wear and friction characteristics.

It has been highlighted that about 40% of the World's energy generation is spent by various means in friction and wear. That is why Tribolgy bears a vital relationship to the universal energy problem of the industrialized part of the world.

Besides the above, as early as 1966, i.e. in the beginning of the tribology movement in United Kingdom, it was stated that in case of 500 MW turbines sets, then being built, on bearings not designed to correct tribological principles, the friction loss could be up to 5 MW. At least 35-40% of this friction would not be there, had the bearings been designed on the correct tribological basis. Furthermore, the additional steam required to overcome such unnecessary friction led to an equally unnecessary capital expenditure on generating equipments of over one million pound that could have been saved.

1.4 Systems Approach for Accurate and Quick Solution in Steel Plants

1.4.1 Formation of Tribological System

Generally speaking, two prominent tribo-systems, namely, lubricated and non-lubricated, have been identified as they require different considerations. Non-lubricated tribo-systems are predominantly present in the iron and steelmaking groups where wear processes occur under direct material to material contact initiated by impact, abrasion and fluid erosion, at ambient or elevated temperatures. Obviously, for the solutions of such problems the most important tribological factors to be considered are M (material) and A (design, manufacture and assembly) and they are improved for the best results.

Lubricated tribo-systems are predominantly present in the rolling mills and auxiliary groups. Therefore, wear problems related to lubricated surfaces are considered. After adequate improvements of M and A factors, fullest improvement of L (lubrication) and C (service condition) factors are taken up for the best results.

1.4.2 *System Description*

The following methodology is adopted to achieve the benefit of a system approach in arriving quickly at accurate solutions quickly. In applying this approach effectively, the system boundary is first defined judiciously. A bigger boundary would require unnecessary extra analysis whereas a smaller one would omit analysis of some relevant parameters affecting the performance of the system. The following constitute the essential aspects of the system:

Input variables : Quantum and type of velocity, quantum and type of load, temperature, duration of load, and velocity and distance of travel, etc.

Useful output : Guidance of motion, work, information, material, force, velocity etc.

Loss input : Friction, friction power loss, rise in temperature, noise and vibration, wear (loss of material), etc.

Elements, a : (1) Moving part, (2) Stationary part, (3) Lubricant, (4) Atmosphere

Properties, p : (1) Volume properties: geometry, chemical composition, metallurgical structure, elastic modulus, hardness, mechanical strength, density, thermal conductivity, etc. (2) Surface properties: surface roughness and surface composition.

(3) Lubricant: system dependent and system independent. (4) Atmosphere: chemical composition, pressure and temperature etc.

Relation, r : This indicates the tribological interactions between the elements of a tribo-system, such as contact, friction, lubrication and wear processes.

While adopting the systems approach to tribological problems, either in lubricated or in unlubricated situations, all the relevant information with the numerical values wherever possible is collected. This would include:

- Magnitude and direction of forces and velocities.
- The type of force (like unidirectional, reciprocating or rotating).
- The type of velocity (like sliding or rolling).
- The working temperature.
- Variation of force, velocity and temperature with respect to time.
- Duration of applied force and velocity.
- Displacement of tribo-element under load and velocity.
- Chemical composition of the interacting surfaces.
- Properties of lubricant (like viscosity, type of base oil and additive systems).
- Environment and its constituents (like air, fumes, vapour pressure and Temperature).

- Volume properties of the interacting surfaces (like geometry, dimensions, volume, weight, hardness, toughness, mechanical strength, thermal conductivity etc).
- Surface properties (like surface roughness in microns, surface layer composition and the thickness).
- Mode of lubrication (like boundary, hydrodynamic, elastohydrodynamic etc).
- Method of application of lubricant and its rate of flow.
- Friction values.
- Wear types and rates.
- Clearances and tolerances of interacting surfaces and so on.

This information is extremely important in having a look at the problem in its entirety. The identification of the loss output variables like "type of wear"; "rate of wear" and "severity of wear" is normally the first step, followed by finding the "location of wear". The identification of "location of wear" immediately focuses on the troubled component needing attention. Similarly, the identification of the type of wear defines the line of action to reach the optimum solution in a scientific way without going through the long route of hit and miss solution. The other support information regarding input variables, system structure and output variables bring the actual area of deficiencies like "M", "A", "L" or "C" into focus. For lubricated tribological systems, ferrography and spectrometric oil analysis are very helpful in identifying the rate and mode of wear as well as the location of wearing component of the system. For un-lubricated systems, such information is obtained by the physical and laboratory analysis of the failed components.

Once the deficient factors for a tribological component are precisely determined, the follow-up action need to be initiated for improvement. The follow-up action is required to be taken in the in-house manufacturing shops for improving the aspects A, M and C and in the production plants for improving aspects L and C. The improvement actions are discussed with the component and equipment suppliers with a view to bring about the desired results. Based on these principles and techniques many improvements in tribology can be achieved.

1.5 Energy Conservation by Tribology in Power Sector—Technical Prospects

Various power generating machines ranging from 1.5 MW to 500MW have been designed and installed in our country. Considering the potential importance of these machines in terms of speeds, mass, and physical size, the tribological solutions are extremely significant and closely related with the energy conservation during power generation and maintenance.

Equipments, which need close monitoring during design and operation for tribological aspects have been studied and certain important aspects are discussed.

1.5.1 *Steam Turbine Sets*

- The high-pressure casing of turbo sets is allowed to slide on the pedestal and low-pressure casing on the foundation to provide differential movements. These supports and guides are subjected to a combination of sliding motion, pressure and differential temperature induced stresses. These create tribological problems like misalignment, due to unsymmetrical friction forces.
- Rubbing or malfunctioning of bearing occurs due to wear causing change in internal clearance between stator and rotor. The degree of wear depends on the choice of material pairs, lubrication, pressure and speed and has been resolved judiciously accordingly.
- By the use of hard material coating on valve spindles of governing systems, various tribological problems have been solved.
- Tribological problems of turbo generators have been reduced with proper selection and assembly of laminations of the core. This minimizes relative rubbing between adjacent laminations consequentially leading in saving of energy and increase in life of laminations.

1.5.2 *Rotor*

- In the steam turbines, erosion in the blades of LP (low Pressure) stages is caused by water droplets causing impingement on the tip of the leading edge of the last stage blades. Flame hardening and hard coating are in use to reduce the erosion.

  The most popular design however is to braze a shield made of cobalt based alloy or tool steel to protect the blade from erosive wear—is a tribological solution for the extension of blade life.
- Wear is the common phenomenon in the end winding and slip ring brushes. Cyclic loading due to self-weight causes wear in the end windings and results in fretting. Selection of proper insulation material has been adapted for appropriate solution to overcome remedy.
- Slip ring transmits current through brushes to the rotor windings. These results in brush wear and temperature rise due to high surface speeds. Tribological measures using graphite impregnated with synthetic lubricant has been developed to overcome this problem in a optimum way.

1.5.3 *Bearings and Instabilities*

- Science of Tribology has developed various bearings required to be used depending on load, speed, clearance and stability criterion. Partial arc – cylindrical bearings, two-lobe, three-lobe, four-lobe, tilting-pad and offset bearings are commonly used.

  The precise optimum design of the bearing by tribological consideration has led to run the bearings in super-laminar regime to reduce the power losses to the bare minimum. Energy efficient bearing is one of the outstanding contributions to the industry by Tribologists.
- Most of turbo machines having flexible rotors operate above first and second critical speeds. These systems are subjected to oil whirl instabilities and parametric instabilities. To tackle rotor dynamic problems, tribology has developed various techniques to determine precisely dynamic coefficients of the bearings so as to avoid the bearing instabilities to reduce power losses and make the bearings energy-efficient.
- A new approach using electrical analogy was developed to determine dynamic coefficients of bearings and discussed in the book on "Tribology in Electrical Environmental" published by Elsevier, 2006.

1.5.4 *Seals*

- Steam seal problem mainly occurs in HP turbine where the pressure drop is more. The gland arc has been constructed by the tribological principles in six segments and spring backed so that during misalignment and shaft run out, hard contact is prevented and life of seals is extended.
- If the shaft is whirling in synchrony with running speed, the temperature at the contact point rises as compared to the other side of the shaft and produces thermal bending. This may cause catastrophic failure. Proper seal design by optimum tribological design is the only solution for such remedies.

1.5.5 *Hydro Sets*

- Thrust bearings, guide bearings, brakes and seals are the important tribological components in hydro-sets, which governs primarily the functional performance and are the source of energy conservation in hydro sets

1.5.6 *Thrust Bearings*

- Thrust bearings in hydro sets support the dead weight of the rotor and the hydraulic thrust, which is in the order of few hundred tons to 4000 tons in a large machine with a surface speed of 30-40 m/s.
- High load causes distortion. High bearing temperature causes thermal distortion, which lead to an increase in the gap between runner and bearing at the edge. As a result, oil film in these areas does not contribute

to load carrying capacity. This reduces film thickness at the central zone because of non-uniform distribution of load on pad causing higher load at the central zone. This results in metal-to-metal contact causing bearing failure.

- Hence by the help of tribology, analysis of both thermal and elastic distortion, which is very important for successful operation and reduction of power losses in the bearings, is carried out.
- The recent developments of thrust bearings are PLEF as bearing material and development of bimetallic bearings. Bimetallic bearings consist of segments made into two separate parts. The top part is babbitted bronze and lower part made of steel, contains machined grooves for flow of oil beneath the pad surfaces. This tribological design results in good cooling of bearing and reduction of operating temperature and consequentially the power loss.
- PTFE base bearing is another break-through in the area of tribology towards the energy conservation. PTFE has a coefficient of friction 0.07 as against 0.3 for conventionally used Babbitt material.

  Thus, during start/stop regime, much smaller torque is required for rotation and heat generated is reduced substantially. Furthermore, jacking up operation as used in conventional bearings is eliminated. This further adds to the saving of energy by using PTFE bearings.

1.5.7 *Guide Bearings*

- The vertical rotors are inherently unstable. Hence, guide bearings used for both hydro turbine and generator are of tilting-pad type. These bearing are very stable in operation. However, some times fretting are observed between bolt guides bearing pad. This results in increase in the bearing clearance and subsequently higher vibration levels and power loss.

  This problem is tackled by Tribology by providing hard coatings on the mating surfaces.

1.5.8 *Brakes*

Brakes are used for stopping the hydro set. Brakes are hydraulically operated and act against the collar made of mild steel or medium carbon steel. The brakes minimize the distance travelled by the rotor in absence of adequate oil film. Thus, wear of bearing is reduced and life is not affected by stopping the machine.

Adequate care is taken by tribological means in selecting the contact pressure so that the heat generated does not cause plastic thermal deformation and damage to the brake track.

1.5.9 *Seals*

The hydro turbine seals used for preventing entry of water into the guide bearing housing are made of fiber-reinforced plastic material with asbestos fibers and resins as ingredient. Seal is designed and seal material is selected to consider the performance characteristics and frictional losses during operation.

1.5.10 *Erosion*

The hard quartz particles in water erode the guide vanes and turbine blades and life is drastically reduced. Applying hard and tough coating on the blades of chromium and nickel has solved the problem. The same tribological technique is extended for other areas wherever erosion problem persists.

1.6 Energy Efficient in Rolling Element Bearings

- Rolling-element bearings widely used in rotating machines have different life span depending on load and speed, and in general failure of these bearings are caused by surface stress.
- Tribology has immense contribution to increase the load carrying capacity of these types of bearings by development of much harder material without changing the configuration of the bearings.
- Load carrying capacity has also been increased through surface coatings and improvement of lubricant characteristics.
- Double Decker High Precision Rolling-Element Bearing developed has low power loss under identical operating conditions as compared to conventional bearings, is the contribution by tribological efforts.

1.7 Tribology in Medical Sciences

- Tribology is based on the natural laws, is equally applicable to other spheres besides that of Mechanical Engineering.
- The Tribology centers along with Medical Centers have has bridged the interface between Tribology and medical engineering, where there are many unresolved problems encompassing the concept of interacting surface in relative motion.

The followings are a few significant contribution of Tribology to Medical Sciences:

- Development of lubricating fluids, which are compatible with human body and posses rheological characteristics similar to those of synovial fluids.
- Design and development of total end prostheses replacement of hip, shoulder, knee, elbow and finger joints.

- Design and development of artificial heart valve required to operate 40 million cycles per annum for many years without any wear and with full reliability----is a remarkable contribution of tribology to the humanity
- Use of Teflon coatings on the surface of cooking vessels reduces the consumption of oil in preparation of foodstuff there by resulting in improved health by minimizing the various heart related ailments.
- Copper based stainless steel utensils, combines with the needs of external aesthetics, and controls the adulteration of foodstuff, besides high heat conductivity resulting in fuel economics and fast preparation of food.
- Even in the area of dentistry, the choice of appropriate implants for replacement or simply wear out by repetitive use, the principles of tribology is applied for correct selection of material so as to control friction in the medium in which it operates---may be in the mouth or in the joints.

1.8 Multifold Benefits

1.8.1 *In Steel Plants:*

- Unplanned maintenance downtime has come down by 29% over the last five years by rigorous tribological practices.
- Specific power rate kWh/t of steel in Tata Steel has come down by 10% during the last five years.
- Specific Lubricant Consumption (kg/t of steel) has come down by 31% during the past Five years.
- The conveyor belt replacement rate has come down by 16% in the last five years.
- The cost of maintenance material has been brought down by 8.4% of the Production Cost.

1.8.2 *Human Life and Welfare*

- Appropriate coating material and thickness can be optimized by tribological studies in the various products/fields of science and technology applicable to human welfare.
- Evaluation of Skin creams, Contact lenses, Shaving blades and many other Biological studies are the significant contributions of Tribology for human Welfare.

1.8.3 *Glamour and Cosmetics*

It is obvious that war against wear is a continuous process. Ladies have been known to fight aging to look young by spending large sums of money on “cosmetics”. In a way, the application of these cosmetic principles in industry is tribology, which alters the skin texture by altering the skin surface temporarily.

1.9 Scope for Research in Energy Conservation by Tribology

- In the area of wear to avoid under and over design of machinery.
- Design criterion for identifying the type of wear – abrasive, adhesive, corrosive and fatigue.
- Mathematical wear model.
- Tribology in electrical environment.
- Improvement in energy efficient bearings.
- Optimisation of magnetic bearings.
- Development of non-conventional lubricants.

1.10 Tribological Know–How–Gap

There has been a great advancement in the area of Tribology still, following know-how-gaps can be highlighted.

- Lubricant stability at higher temperature is presently a know-how-gap and information available is relatively poor.
- Lubricants working at subzero temperature.
- Design of various types of bearings are required to be modernized keeping in view optimum performance.
- Bearing materials having less wear rate under higher load and with less coefficient of friction under various regimes of operation.
- Preventive maintenance and "signature analysis" technique to reduce the breakdown time of machinery and other hazards to increase the life of bearings.
- Bearings and lubricants working in electrical environment i.e. Tribology in electrical fields.
- Bearings for higher load and speed characteristics
- Phenomenon of adhesion, physico absorption and chemico absorption.

1.11 Challenges to Tribology in the Twenty First Century

The Tribology is entering in the area of nano from macro to micro. This brings the following challenges for its further advancement.

- Understanding of tribological phenomenon at atomic and sub atomic level
- To connect phenomenon of lubrication, friction and wear and their integrated effect on a tribological system
- To know the boundary of tribological interaction between a successful application and a catastrophic failure

- Study the run-in surfaces survival form chemical and mechanical attacks of contaminants without changing the surfaces from the well run-in state
- Examine how materials below the surfaces survive the mechanical stresses and diffusion of chemicals
- How can the wear rate of the surfaces be maintained so low during its total life time without changing geometry and load/stresses
- Most of the successful tribological applications where both lubrication and wear are present are based on experience than on theoretical knowledge
- In short Tribology in 21st century be expected to develop into a science where the entire life cycle of a machines is considered and optimized starting from manufacturing, productivity scrapping and recycling of its components.

1.12 Conclusions

From the above studies, the following conclusions are drawn for the future work in the area of Tribology:

- For energy conservation by tribology, lot of work still require to be done in the area of wear phenomenon to avoid under and over design of machinery.
- Design criterion are required to be developed which relates and identify the type of wear i.e. abrasive, adhesive, corrosive and fatigue wear with operating conditions i.e. load, speed, temperature, environment and material properties.
- Tribology in electrical environment needs to be developed further which governs wear of machinery by the surface phenomenon of physico and chemico absorption, magnetization, charge accumulation etc.
- Development of energy efficient magnetic bearings and bearing less machines are the outstanding contribution of Tribology in electrical environment, however much more still required to be done.
- In short, a multi disciplinary approach is required to achieve success in the area of Tribology in uplifting the industry and in energy conservation apart from its contribution to Medical Sciences, large Machines and Allied fields

Section - II

Lubricants

2

Latest Tribology Testing of Lubricating oils and Effect of Nano-additives on Oils

2.1 A General Review

There are several solutions suggested to develop an unified approach to tribological testing of oils. For e.g. evaluation of the coefficient of friction and wear, it is applied to monitor friction force, normal load, and in-situ wear dynamics, as well as to clearly distinguish the running-in and stable-wear components. Tests on lab-bench testers are obtained to start from getting a Stribeck curve in a wide range of loads and speeds. During the tests, monitoring additional parameters like contact acoustic emission and contact electrical resistance are highly beneficial for sensitive and comprehensive characterization of oils, greases and study of the effects of nano-additives on lubricants [2.1].

2.2 Introduction

There are numerous techniques known to study tribological properties of lubricants. A common problem in tribological literature, however, is the wide scatter of test results obtained by different authors on different testers for different applications.

After detailed studies, several solutions were recommended to overcome data fluctuations, and also developed the unified approach to the tribological testing of lubricants, namely:

- To evaluate friction and wear characteristics on the lab-bench tester,
- it is important to start from obtaining a Stribeck curve in a wide range of loads and speeds, so that to clarify whether the test conditions correspond to boundary, mixed or hydrodynamic regime,
- and to modify them to ensure the regime fully corresponding to the one in the target practical application
- and finally not to perform "blind" tests in the unknown lubrication regime and thus neglecting the dramatic friction and wear changes at different zones of the Stribeck curve.

Furthermore, when evaluating the coefficient of friction, it is important to measure both its friction force and normal load components (instead of assuming a static normal load and so neglecting flatness of test samples and run out of the motion).

For wear evaluation, it is important to monitor its actual dynamics during the test and clearly distinguish its running-in and stable components (instead of taking into account only the post-test final wear depth and so neglecting the fact that most of it could be due to the initial test period).

Additional parameters of contact acoustic emission and contact electrical resistance are beneficial for more sensitive characterization of lubricants for monitoring bonding and debonding processes, formation and destruction of tribo-films, initiation and propagation of wear tracks.

2.3 Experimental Apparatus

The tests were performed on the Universal Tribometer, which can be configured either as a micro-tribometer UMT-2 or a macro-tribometer UMT-3 developed and marketed by CETR. It can accommodate test samples of various shapes and dimensions (up to 150 mm in both lateral and vertical directions) and perform all common lubricant tests, including ball/pin-on-disc, disc-on-disc, block-on-ring, nut-in-screw, drill-in-hole, mill-on-block, shaft-in-bushing, 4-ball, etc. Easily interchangeable compatible rotary and linear drives allow for almost any combination of rotary and linear motions (including fast oscillations) of test specimens. Oils can be applied either continuously with a peristaltic pump or in the beginning of the test into corrosion-resistant containers.

The applied loads are servo-controlled (with a closed-loop feedback) from 1 mN to 1.2 kN. The speeds (from 0.001 to 5,000 rpm), and temperatures from – 20 to + 1100 deg. Centigrade, as well accelerations and positions are also precisely controlled. The loads and speeds can be programmed to have different values on different areas of the test specimens consistently during the test, allowing for wear evaluation and comparison on those areas as if they were obtained after a whole battery of tests. In case of start-stop cyclical tests, both the load-unload and acceleration-deceleration profiles can be either programmed or downloaded from the real machine, allowing for the full simulation of target application.

During testing, up to 16 parameters are monitored, displayed and stored in the computer for further analysis: up to three forces (in all 3 axes), three torques (around all 3 axes), wear depth, temperature, humidity, electrical contact resistance, high-frequency contact acoustic emission, audible noise, etc. An integrated digital optical microscope allows for both precise samples positioning and imaging of the wear track.

2.4 Experimental Procedures

Prior to friction and wear testing, it is crucial to find out what lubrication regime is targeted (boundary, mixed or hydrodynamic) so as to perform the tests in the correct regime. All tribological parameters differ dramatically in different regimes. These regimes, in turn, depend on the lubricant thickness and viscosity, surface roughness, speed, load, temperature, etc. Therefore, comparison of oils by their friction and wear characteristics has to be done within the same lubricating regimes. For example, oils with viscosity modifiers may show advantages in the hydrodynamic regime and can be easily change/switch over from the mixed to hydrodynamic regimes, but there are no advantages in boundary lubrication. Lubricants with wear reducing additives may exhibit better characteristics in boundary lubrication, but no advantages in a hydrodynamic regime.

The best way to compare frictional properties of lubricating oils is to obtain their Stribeck curves and determine their friction levels in all three regimes. Automatic developed procedure on CETR, UMT includes testing of lubricants at different speeds and loads, each ranging by 3 orders of magnitude, so that the total speed-to-load ratio ranges by 6 orders of magnitude; it allows for detection of all three lubrication regimes. The test on each level takes 1 minute; the software calculates average friction coefficient during the last 15 seconds and automatically plots the Stribeck curve.

After the Stribeck curves are obtained and all three regimes are determined, the test conditions are chosen that correspond to the target application lubrication regime. In these conditions, a wear test is run with simultaneous monitoring of normal load, friction force and coefficient COF, wear depth W and rate, contact electrical resistance ER and acoustic emission AE. Humidity and temperature are monitored, if needed – controlled within 5 to 100 %RH and –20 to +1100 °C, as the machine is fully equipped with this facility.

Furthermore, conventional base oil was tested with and without carbon nano-particles of high lubricity[2.3, 2.4]. For the nano-additives two common types of fullerenes were used, spherical (bucky-balls) and cylindrical (nano-tubes). The structure of fullerenes is composed of a sheet of linked hexagonal rings (like in graphite), but with additional pentagonal rings that prevent the sheet from being planar. Total of five lubricants were tested (the base oil pure and with each of the two nano-additives at two concentrations, 3% and 8% by weight).

Three different tests have been performed, each repeated five times:

- friction measurements at five sliding-to-rolling levels,
- sliding friction measurements at five temperatures,
- sliding friction tests at five speed-to-load levels.

2.5 Results and Discussion

Comparative tests of oils 1 and 2 were performed. Quick testing on 10 speed and load levels with monitoring the COF and ER versus the speed-to-load ratio produced two different Stribeck curves (Fig. 2.1). A higher-viscosity oil 1 showed higher COF (solid lines) and ER (broken lines) data, friction and near-zero ER in boundary lubrication at very low speed/load ratio (as it did not form protective layers separating the surfaces). Also, slightly higher friction in the hydrodynamic lubrication at high speed-load ratio, but lower friction and more dramatic ER increase in the mixed lubrication regime is indicated (as its higher viscosity allowed for a hydrodynamic effect at lower speed/load levels than for the lower-viscosity oil 2).

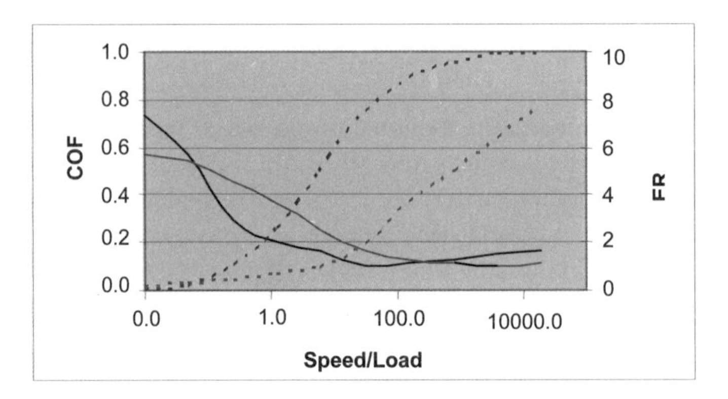

Fig. 2.1 Stribeck curves for oils 1 (blue) and 2 (red):

Based on the oil target application, a speed/ratio of 6 was chosen for a durability test. The durability results are shown in Figure 2.2. Oil 1 failed, its friction and wear increased sharply after 20 minutes, with ER dropping and AE increasing about 10 minutes earlier. Oil 2 performed well and formed a protective film, reflected in an ER increase and AE decrease after about 20 minutes, followed by reduction in both friction and wear rate.

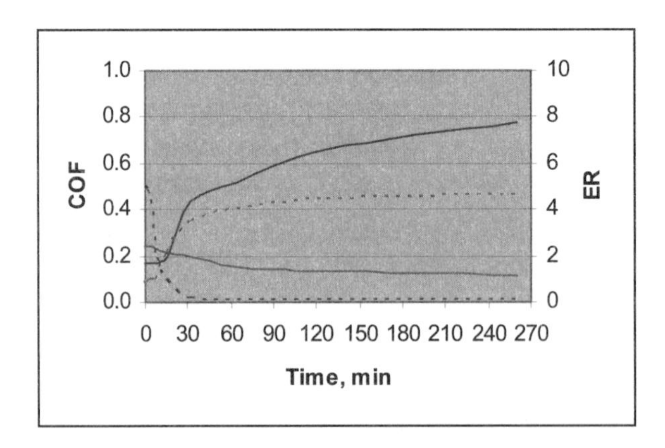

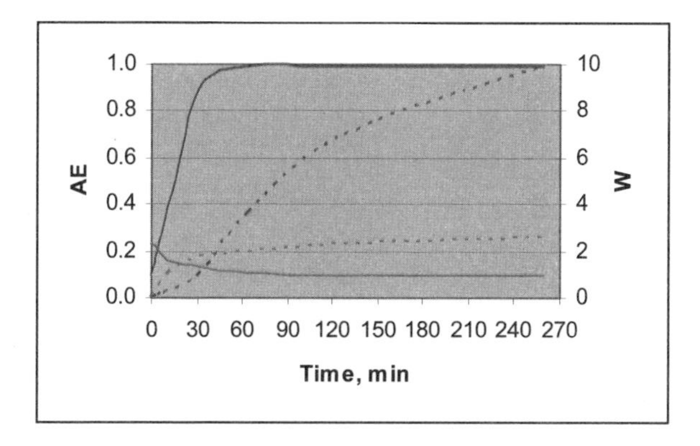

Fig. 2.2 Durability test for oils 1 (blue) and 2 (red): COF and AE - solid lines, ER and W – broken lines.

Three summary plots below show (Figure 2.3) the effect of nano-additives with lubricating oils. Plots clearly bring out that both the bucky-balls and nano-tubes have reduced sliding friction at normal temperatures in both in boundary and mixed lubrication, similarly to earlier observations[2.4]. Both types of fullerenes did not show any substantial effect on friction in rolling, hydrodynamic lubrication, and at elevated temperatures. In all the tests, the effects of bucky-balls and nano-tubes were similar[2.5].

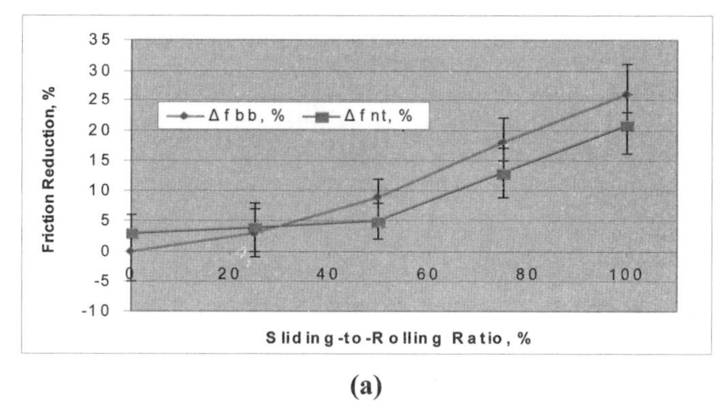

(a)

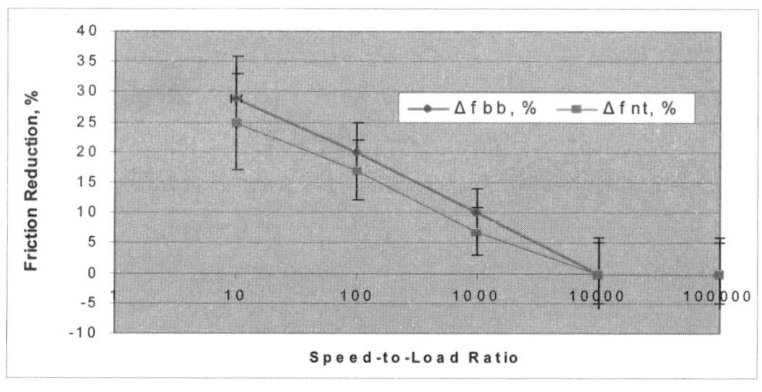

(b)

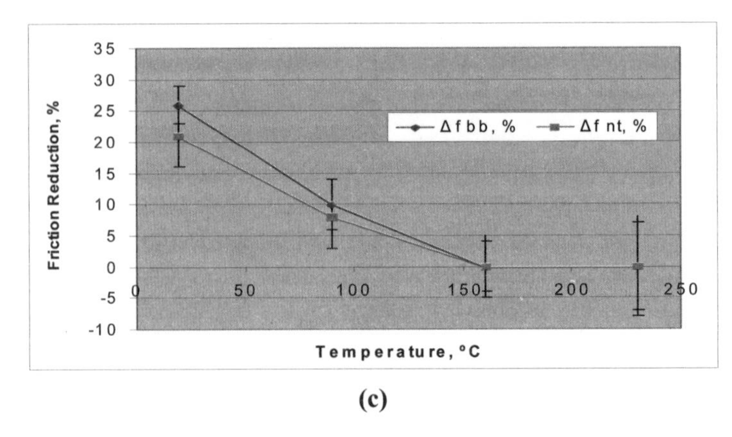

(c)

Fig. 2.3 Friction reduction of fullerenes versus rolling-to-sliding (a), speed-to-load (b), temperature (c)

Furthermore, it is intended to test fullerenes and inorganic fullerene-like additives, produced and added with various methods.

2.6 Conclusions

The multi-sensing technology is effective for tribology testing of lubricants and lubricants with nano-additives. For comprehensive frictional characterization, lubricants have to be tested and compared in all three lubrication regimes. For durability characterization, wear testing should be performed in the application-specific lubrication regime. The UMT tester with multiple sensors, automated test procedures and data presentation is suitable for sensitive and practical lubricants tests and lubricants with nano-additives.

References

[2.1] Gitis V. Norm, "Effective Triblogy Testing of Lubricating Oils" Proceedings of WTC 2005, World Tribology Congress III, WTC2005-63078

[2.2] Gitis V.Norm, "Tribology Testing of Lubricating Oils with Nano-Additives"

[2.3] X. Li, W. Yang, *Nanotechnology*, **18**, (2007), 1-7.

[2.4] B. Gupta, B. Bhushan; *Lubrication Engineering*, **50**, (1994), 524-528.

[2.5] Gitis V. Norm, Kuiry Suresh, Prashad, H. "Latest Tribology Testing of Lubricating Oils and Effect of Nano-additives on Oils", ISFL, 6th International Symposium on Fuels and Lubricants, march 9-12,2008, New Delhi.

3

Diagnosis of Deterioration of Lithium Greases Used in Rolling-Element Bearings by X-Ray Diffractrometry

3.1 A General Review

In this chapter, diagnosis of deterioration of lithium greases used in rolling-element bearings by X-ray diffractometry (XRD), and analysis by X-ray fluorescence spectrometry (XRFS) and atomic absorption spectrophotometry (AAS) is reported. These techniques give reproducible reliable data to establish the severity of deterioration of greases recovered from the active zone of the bearings. XRD appears to be suitable to quickly diagnose the chemical changes that occurred in the soap residue of the greases, available even in small quantity, compared to the other analytical and performance evaluation techniques.

The analysis gives diagnosis of the fresh lithium greases, chemical compositions and formation of new compounds in the greases after being used in the rolling-element bearings operated under the influence of electrical fields, and also under pure rolling friction without the effect of electrical fields on the roller bearing test rig. The deteriorated greases recovered from various motor bearings have also been analyzed and results are compared with that of the greases from the test bearings.

The X-ray diffraction analysis shows that the chemical composition of the soap residue of the fresh lithium grease is lithium stearate ($C_{18}\ H_{35}\ Li\ O_2$), which does not change in the bearings operated without the effect of electrical fields under pure rolling friction. However, lithium iron oxide ($Li_5\ Fe\ O_2$) peaks have been detected after prolong operation of the bearings. On the contrary, under the influence of electrical fields the chemical composition of the greases is changed to lithium palmitate ($C_{16}H_{31}Li\ O_2$), and peaks of gamma lithium iron oxide (γ-$LiFeO_2$) and lithium zinc silicate ($Li_{3.6}Zn_{0.2}Si\ O_2$) have been detected. The surface of the bearings is found corroded after operation under electrical fields. Atomic absorption spectrophotometry analysis has shown considerable increase of lithium percentage as lithium carbonate and lithium hydroxide in the used greases in the acquous solution compared to the fresh grease.

The investigations given in this chapter, along with the study of damaged/corrugated bearing surfaces, have potential to diagnose the cause of bearing failure under the influence of electrical fields, and also, to establish the severity of deterioration of the lithium greases used in the bearings.

3.2 Introduction

It is often necessary to analyse a grease for quality, developmental work, trouble-shooting and machinery-health diagnostic studies. The amount and depth of analytical work required will, of course, depend on desired objectives. This can range from a simple to a great deal of sophisticated, expensive, and time-consuming analysis. There are various analytical techniques available; physical tests, infra-red spectroscopy, mass spectrometry, emission and atomic absorption spectrophotometry, X-ray fluorescence spectrometry (XRFS), and X-ray diffraction (XRD) and various other performance evaluation tests[3.1].

Despite the wide use of lubricating greases, most of the studies are concerned with their structure, manufacturing, processing, mechanical testing and evaluation[3.2, 3.3]. A few papers deal with the effect of soaps and base oils on the frictional behaviour of greases[3.4, 3.5]. Greases, because of their high consistency and lack of mobility have a tendency to work in starved conditions. This diminishes the film thickness of a grease compared to flood lubrication, which has been reported by various authors by experimental and the theoretical studies[3.6, 3.7, 3.8]. The abilities that a lubricating grease must possess are to form a film of lubricant, to prevent corrosion, to resist temperature rise without becoming unduly softened in the process, to withstand shear without breakdown of the structure.

It has also been studied that greases used in the bearings are subjected to a continuous shear rate and they suffer degradation with which the effective viscosity diminishes. The process would gradually lead to a final stage where greases having lost their original structure would appear like a suspension of spherical particles of metallic soap in to the base oil[3.9]. However, greases, as bodies of rheological complex behaviour, have not been assessed for their behaviour in the rolling-element bearings operating under the influence of electrical fields and how fast deterioration and degradation of the greases would initiate. Behaviour of grease constituents under electrical fields may be quite complex. Film of metal-soap having strong bounding nature to the crystal lattice of the metal and its high breaking strength may also add to the complexity. Polar-chain molecules of a grease may tend to orient towards the field while non-polar hydrocarbons would behave indifferently. Contamination of a grease may also change its bahaviour.

The greases have an effective yield stress and they resist the deformation elastically until the shear stress greater than the yield stress is applied. Quite a few investigations have been reported related to the electro-physical and electrochemical phenomena in friction, lubrication, contact interaction and fretting corrosion[3.10, 3.11]. The understanding of viscous properties of grease in relation to shear rate, temperature and pressure is still not complete.

Although the above factors have an influence on the behaviour and deterioration of the greases, but it is not easy to quantify them, because friction between surfaces of a bearing in a lubricating media is accompanied by complicated physico-chemical processes taking place between the plastically deformed surface layers and surrounding medium. As a result of this interaction, secondary structure of lubricating medium may form which may eliminate the adhesion of conjugate surfaces. The composition of the products obtained by interacting surfaces with the lubricating medium under friction depends on the medium, bearing surfaces and conditions of interaction[3.12]. Investigations of these changes may give valuable information to diagnose the mechanism of the medium-metal interaction during friction.

The chapter presents net effect on the change of structure of greases used in rolling element bearings after being operated separately under the influence of electric fields and under pure rolling friction, and also to detect newly formed compounds in the medium. XRD technique has been applied in this work to diagnose the changes in chemical composition of the soap residues of the used greases apart from the analysis by XRFS and atomic absorption spectrophotometry techniques.

3.3 Experimental Facilities

3.3.1 *Bearing Test Machine and Test Procedure*

The bearing test machine shown in Fig. 3.1, Ref.[3.13, 3.14] was used for experimental investigations. In the present studies, bearings type NU 326 lubricated with lithium base grease having a resistivity of 10^7 ohm-cm have been tested at 1100 rpm under 1000 kgf and 500 kgf of radial loads acting at two different directions (90° to each other) for a duration of 250 hours by passing 50 A (a.c.). During the test period voltage across the bearing varied from 1.12 to 2.3 V between points A and B (Fig 3.1 of Ref. [3.13, 3.14]). Corrugated patterns on bearing surfaces and effect of operating parameters on the threshold voltages and impedance response have also been studied[3.13, 3.14]. Similarly, NU 326 bearings were tested under identical conditions, without the passage of current under pure pure rolling friction.

3.3.2 *X-Ray Diffraction (XRD) Unit*

A Philips PW 1140 X-ray diffraction unit, complete with generator and gonimeter, was used for analysis of soap residues of different greases. Instrument parameters are listed in Table 3.1. By X-ray diffraction, symmetry and regularity in arrangement of atoms was made visible when a monochromatic X-ray beam irradiates soap residues. In this way unknown compound could be identified since XRD shows the typical lines for each compound separately, satisfying Bragg's law.

$$n\lambda = 2\, d \,\text{Sin}\, \theta \qquad (3.1)$$

Where

n = order of diffraction

λ = wave length

θ = angle

d = inner planner distance (^{o}A)

The particular advantage of diffraction analysis is that it discloses the presence of substance as that substance actually exists in the sample, and not in terms of its constituent chemical elements.

3.3.3 *X-Ray Fluorescence Spectrometry (XRFS)*

This is an analytical tool for rapid quantitative and qualitative determination of elements present in a sample. When a beam of X-rays is directed onto the surface of the specimen, secondary fluorescent radiation is emitted by the specimen, which contains wave- length characteristics of each element present. The secondary radiation is directed onto an analysing crystal separating out the wave- lengths, which are recorded by a detector. The whole arrangement is in the form of a vertical gonimeter. Since the wave- length is a function of angular position of the crystal and gonimeter, the elements in the specimen can be identified. A Philips model PW 1140 with channel control PW 1390 has been used for the elements present in the soap residues of the greases.

3.4 Experimental Investigations

3.4.1 *Grease Application to Bearings and Sampling Methodology*

A permissible measured quantity of grease was filled in inner and outer bearing caps (about 250 gms each), and a small quantity in the working surfaces of the rollers as per the standard procedure of bearing lubrication. Periodically, after a few hours of operation, varying between 41 and 250 hours, bearing caps and outer race with rollers were dismantled and a small quantity of grease (5-7 gms) was collected using sterilized glass rod from the working surfaces of the rollers for analysis, and then the bearings were reassembled without addition of fresh grease. Thus, the deterioration of the grease, once filled, was studied for a complete test duration without creating starvation of lubricant on the roller surfaces. Similarly, grease samples from different motor bearings, collected from the working surfaces of the rollers (after about 6000 hours of operation) were also analyzed.

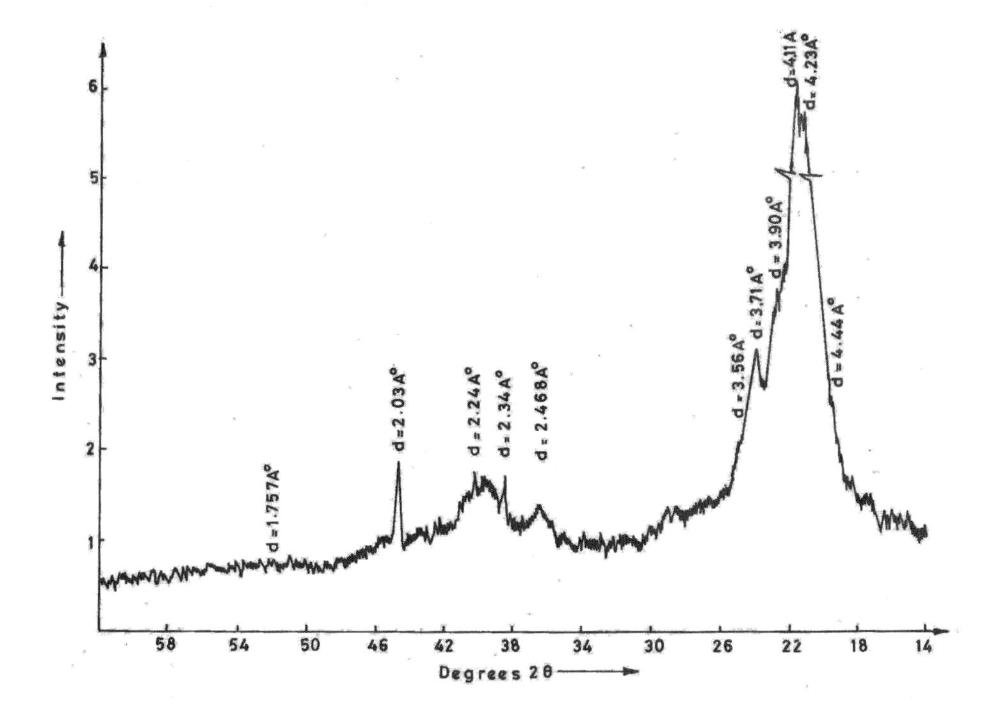

Fig. 3.1 X-Ray Diffraction pattern of soap residue of fresh grease

Table 3.1 X-Ray Diffractometry Instrument Parameters

Model	– Philips PW 1140
Generator	– 3 Kw, 100 Kv Max., 100 mA Max.
Horizontal Gonimeter	– PW 1380 Philips
Detector	– Xenon filled proportional Counter
Channel Control	– PW 1390 with recorder PW 8203 (single pen recorder)
Debye Scherrer Camera	– 2 nos. (114mm)
Powder Sample	– 5 gms fine powder
Solid Sample	– 1.5 cm x 1 cm x 0.5 cm approx. (with smooth surface)

3.4.2 *Investigation by X-Ray Diffractometry (XRD)*

Soap and oil contents of fresh grease and the grease samples from the test bearings and from motor bearings have been separated by dissolving them in a petroleum ether. The soap residues, thus obtained from different grease samples, have been subjected to XRD investigations. Fig. 3.1 shows XRD analysis of the soap residue of fresh grease. Figures 3.2 and 3.3 indicate XRD analysis of the soap residues of grease samples from the bearings after operation under electrical fields and under pure rolling friction on the bearing test machine for a duration of 250 hours. Similarly, Figs. 3.4 and 3.5 illustrate analysis of the grease samples from a.c. motor bearing I and II, respectively.

XRD plots of different samples, thus obtained, have been compared with standard cards to diagnose their chemical composition. Tables 3.2 and 3.3 show the experimental matching peaks d(A°) of the soap residues with those of the standard cards and their chemical compositions. The standard cards have been chosen from the reference standards[3.15] to match closely with the minimum three peaks of the experimental XRD plots (Figs. 3.1 to 3.5).

3.4.3 *Analysis by Atomic Absorption Spectrophotometry (AAS)*

To determine the change in the composition of the grease samples, the percentage of lithium metal present as lithium hydroxide and lithium carbonate were determined by an atomic absorption spectrophotometer of Pye Unicam-make. The percentage of lithium in an acquous solution of a fixed quantity of different samples were compared with that of the fresh grease. Table 3.4 indicates the lithium percentage and relative change in its percentage compared to the fresh grease.

The presence of metals Fe, Zn, Li, Ni, Cr, Si and their percentage in the ash extracted from different grease samples have also been analysed by AAS.

3.5 Experimental Results

3.5.1 *Physico-Chemical Properties, Functional Performance Characteristics and 'Recouping' Property of the Grease*

Functional performance characteristics and physicochemical properties of the fresh grease were analysed. The detailed analysis of the grease properties used in NU 326 bearings is given in Ref.[3.16]. The grease exhibits viscosity of 99.92 centistokes at 40°C and 9.68 centistokes at 100°C. It shows low temperature torque as 50 gm-cm and unworked consistency as 236 and drop point as 185°C. Oil separation of grease is 0.44 percent at 130°C. The average time taken to initiate wear scar on balls of a 4-ball-wear tester at 2000 rpm at 100 kgf load is 7.7 min. The range of variation of resistivity of grease with respect to time (from 1 to 90 min) and potential drop (from 10 to 500 V) is 0.63×10^7 to 115×10^7 ohm-cm. The grease shows increase in resistivity at different rates–the maximum being 23.67 times at 50 V[3.16].

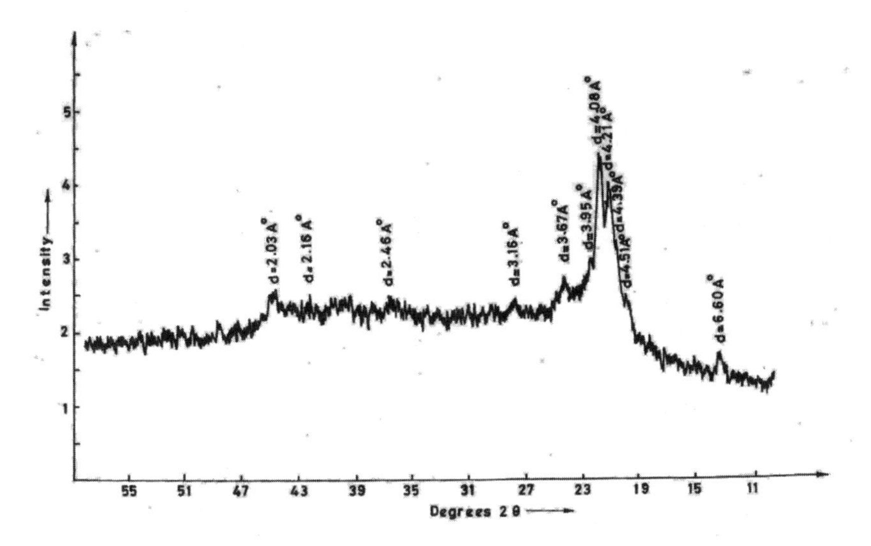

Fig. 3.2 X-Ray diffraction pattern of soap residue of grease from NU 326 bearing after operation under electrical fields for 250 hrs

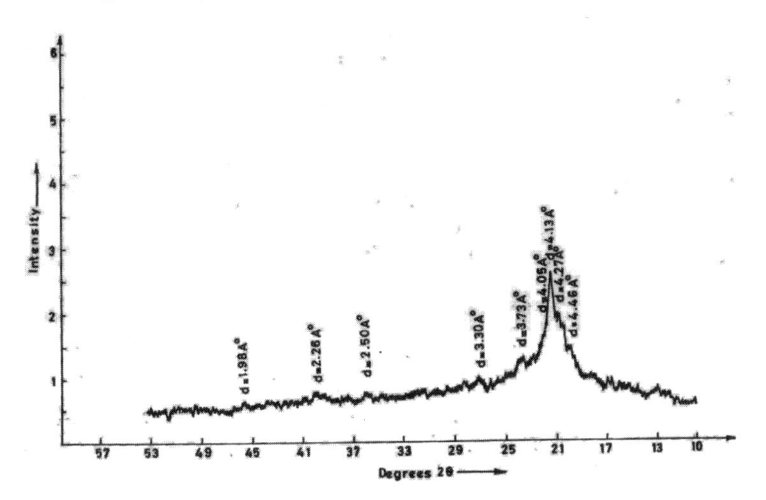

Fig. 3.3 X-ray diffraction pattern of soap residue of grease from NU 326 bearing after operation without electrical fields for 250 hrs

In various grease samples taken at different stages from the working surfaces of the rollers, no significant changes were detected in the drop points. However, penetration and viscosity could not be determined because of the very small quantity of the available samples. Also, no significant changes were determined in the total acid and basic numbers of the oil residues of the grease samples after 250 hrs of operation in comparison to the fresh sample. Furthermore, IR spectra of the oil residues at different stages show C-H bending at 1370 cm^{-1} and 1460 cm^{-1},which suggest that the oil in the greases has not been significantly affected[3.16].

It is also shown in Ref.[3.16] that the resistivity of grease increases under electrical fields, and on disconnecting the applied voltage and keeping the system idle, the resistivity of grease tends to approach the original value. If the grease is re-exposed to the same potential drop, its resistivity again increases, but with different rates. The 'recouping' property is found to be non-uniform.

3.5.2 *Bearings Condition After Test*

The bearings NU 326, after testing separately, under electrical fields and under pure rolling friction for a duration of 250 hrs, were examined. The inner race of the bearing tested under the electrical field was found to be fully corroded with corrugations[3.14], whereas corrosion and corrugations were not observed on the bearing tested under pure rolling friction without the passage of current.

The bearing I, NU 230, used in the motor was found to be corroded after 6000 hrs of operation but was not corrugated. On the contrary, the bearing II, NU 230, used in another motor was fully corroded and corrugated.

3.6 Results and Discussions

3.6.1 *XRD and XRFS Analysis of Greases from NU 326 Bearings*

From the XRD analysis as illustrated in Table 3.2 and Fig. 3.1, it is established that chemical composition of the fresh grease–before filling in the bearings –is lithium stearate ($C_{18} H_{35} Li O_2$). After operation of the bearing under electrical fields for a duration of 41 hrs, the chemical composition of the grease undergoes changes from lithium stearate (intermediate changes) and faint corrosion on the bearing surfaces indicates decomposition of the grease which is also supported by a four times increase in the relative percentage of free lithium (Table 3.4).

After 250 hours of operation, the chemical composition of the fresh grease in the bearing changes to lithium palmitate ($C_{16}H_{31} LiO_2$). Also peaks of gamma lithium iron oxide (γ-LiFe O_2) and lithium zinc silicate ($Li_{3.6} Zn_{0.2} Si O_2$) are found (Table 3.2, Fig. 3.2). Besides this the bearing surfaces were found to be fully corroded(3.14). On the contrary, no change in the chemical composition of the grease was detected after 41 hrs of operation of the NU 326 bearing under pure rolling friction without passage of current. However, after 250 hrs of operation the formation of lithium iron oxide (Li_5FeO_2) was detected, but chemical composition still remained as lithium stearate (Table 3.3). Moreover no corrosion on the bearing surfaces was found, which suggests that the grease was not decomposed. However, crystalline structure of the fresh grease (Fig. 3.1) changes to amorphous structure under pure rolling friction as well as in a bearing operating under electrical fields (Figs. 3.2 to 3.5).

Table 3.2 X-Ray Diffraction (XRD) Analysis of Soap Residues of Different Grease Samples

Particulars	Fresh Grease		Grease From NU 326 Bearing (After Operation Under Electrical Fields)				
	Matching Standard Card (4-0352)	Experimental Analysis (XRD)	Matching Standard Cards			After 41 hrs	After 250 hrs
			4-0385	17-937	24-684	Experimental Analysis (XRD)	Experimental Analysis (XRD)
Matching Peaks (dA)			4.53				4.51
			4.34				4.39
	4.46	4.44					
	4.23	4.23	4.22			4.23	4.21
						4.20	
	4.11	4.11	4.12				4.12
					4.03	4.08	4.08
	3.97	3.97	3.97		3.97		3.95
		3.90				3.90	
		3.71	3.77			3.75	
				3.67			3.67
		3.56	3.54				
					3.18	3.16	3.16

Table 3.2 *Contd...*

	Fresh Grease		Grease From NU 326 Bearing (After Operation Under Electrical Fields)				
	Matching Standard Card (4-0352)	Experimental Analysis (XRD)	Matching Standard Cards			After 41 hrs	After 250 hrs
			4-0385	17-937	24-684	Experimental Analysis (XRD)	Experimental Analysis (XRD)
	2.46	2.46	2.46	2.363	3.18	3.16	3.16
				2.184			2.16
	2.02	2.03	2.06	2.025		2.04	2.03
	1.75	1.75		1.771		1.78	
				1.602			
						1.54	
	1.45	1.43		1.481			
Chemical Composition	Lithium Stearate ($C_{18}H_{35}LiO_2$)	$C_{18}H_{35}LiO_2$	Lithium Palmitate ($C_{16}H_{31}LiO_2$)	Gamma Lithium Iron Oxide γ-$LiFeO_2$	Lithium Zinc Silicate $Li_{3.6}Zn_{0.2}SiO_2$	Transitional (unidentified)	i) $C_{16}H_{31}LiO_2$ ii) γ-$LiFeO_2$ iii) $Li_{3.6}Zn_{0.2}SiO_2$
		Fig. 3.1					Fig. 3.2

This may be noted that during interaction of surfaces under rolling friction, individual components of the lubricating medium react with bearing metal to form a layer of secondary structures, which may get destroyed by normal and tangential stresses. That is why the number of components taking part in the reaction or formed in the process of friction will be changed. Thus, the exact final compound in the grease will not be easy to identify (as in case of the grease from the bearing after 41 hrs of operation under electrical fields). And it also might be difficult to identify the exact matching composition of the formed compounds in presence of a number of components of the lubricating media and overlapping/close existence of peaks of various compounds (Tables 3.3).

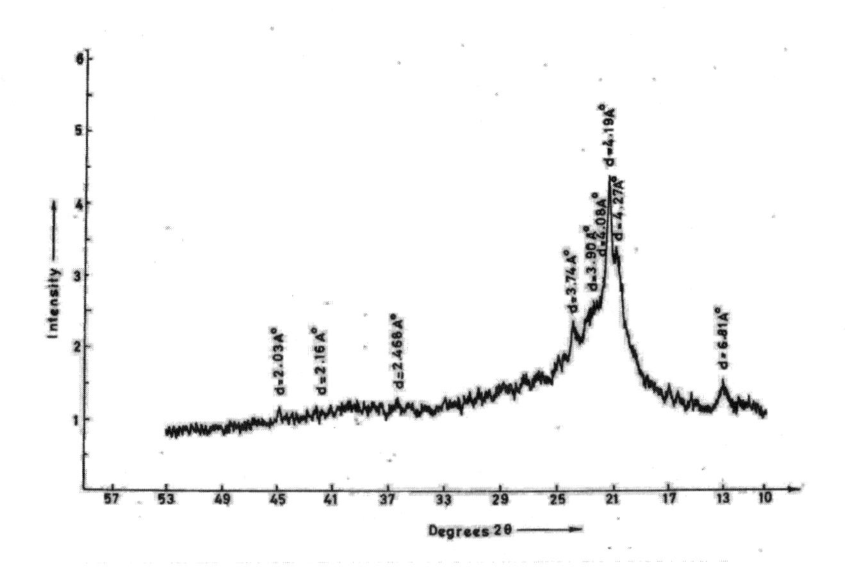

Fig. 3.4 X-ray diffraction pattern of soap residue of grease from NU 230 motor bearing I

The role of oxygen is very significant in contact interacting surfaces. It has been found that oxygen takes an active part in the processes of friction. Contact stresses, temperature, composition of lubricating medium, and impurities greatly influence the rate of oxygen consumption during friction. The increase of contact stresses leads to a corresponding increase in the rate of oxygen consumption[3.12]. That is why oxides of the metals i.e. lithium iron oxide is formed easily in rolling friction as shown in Tables 3.2 and 3.3.

The decomposition of the zinc additive i.e. zinc dithiosphosphate or zinc dialkyl dithiosphosphate under the influence of electrical fields has lead to the formation of lithium zinc silicate in presence of high relative percentage of free lithium and silica impurity in the grease under high temperature in the asperity contacts (Table 3.4). On the contrary, lithium zinc silicate compound has not been detected in the grease from the bearing operated without the influence of electrical field (Table 3.3).

The formation of lithium iron oxide and lithium zinc silicate also contributed to medium-metal interaction, rate of chemical reaction, affinity of lubricating medium components for the metal, availability of the free metal and the temperature. Besides this, reactions having fast rates will be more energetic and carried out to a higher extent. The rate of reaction also determines the end composition of interacting products and their characteristics.

The detection of Fe, Zn, Ni, Li, Cr, Si by XRFS and AAS in the grease samples also suggests the formation of lithium iron oxide and lithium zinc silicate in presence of oxygen.

3.6.2 *XRD Analysis of Greases from Motor Bearings*

XRD analysis of the grease from bearings I and II of the a.c. motors indicates that the original chemical composition of lithium stearate of fresh grease is not changed in the bearing I, but it is changed to lithium palmitate in the bearing II (Table 3.3). Furthermore, corrosion and corrugation pattern on the surfaces of bearing II are found, and peaks of gamma lithium iron oxide and lithium zinc silicate are detected. However, peaks of lithium iron oxide has been detected from the grease of bearing I (Fig. 3.4). Comparison of Fig. 3.5 with that of Fig. 3.2 indicates that the motor bearing II was exposed to electrical fields, and comparison of Fig. 3.4 with that of Fig. 3.3 suggests that motor bearing I was not exposed to electrical current.

3.6.3 *Analysis of Data from Atomic Absorption Spectrophotometry (AAS)*

The AAS analysis illustrates that change in relative percentage of free lithium is from 4 to 5.2 within a duration of 41 and 250 hrs after operation under electrical fields, whereas without the influence of electrical field this change is only from 2.48 to 2.88 (Table 3.4). This indicates that the change in the chemical composition of the greases is more significant under the influence of electrical fields than in pure rolling friction. The corrosion of bearing surfaces also justify this conclusion[3.14]. The relative percentage of free lithium would have been much higher than 5.2 if the bearing surfaces would not have been corroded.

The absence of corrosion on the surfaces of the bearing (tested without current) and the low value of relative percentage of lithium (maximum 2.96) in the grease show the effect of pure rolling friction. However, the bearing surfaces would have been corroded after prolonged operation due to the significant change in the composition of the grease. Relative percentage of free lithium (2.1), corrosion of bearing surfaces and XRD analysis of the grease from motor bearing I suggest the effect of rolling friction after 6000 hours of operation (Fig. 3.4). Whereas higher relative percentage of free lithium (3.8), XRD analysis of the grease from motor bearing II and complete corrosion of the surfaces indicate the exposure of current to the bearing (Table 3.4, Fig. 3.5).

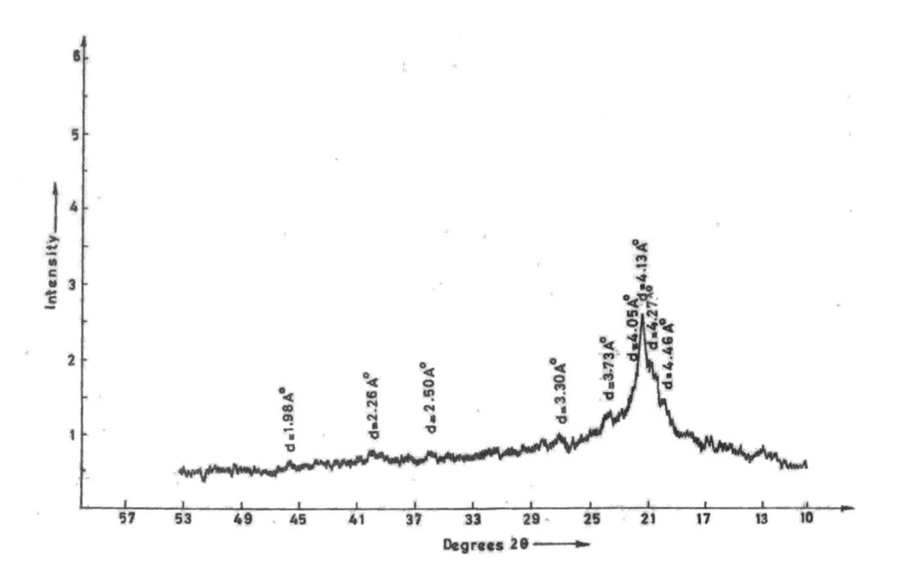

Fig 3.5 X-ray diffraction pattern of soap residue of grease from NU 230 motor bearing II.

3.6.4 *Decomposition of Greases and Bearing Failure*

The capability of a lubricant to carry the load at the point of contact of the friction surfaces in a bearing depends on the degree of lubricant molecule orientation and by the bond of the molecules to the bearing interacting surfaces[3.17]. When this bond is disturbed – under the influence of electrical current, rolling friction or shear – the lubricating film is destroyed. Under these conditions the lubricating grease looses its property to resist high pressures and capability to separate friction surfaces. This leads to changes in the structure of the grease, change of higher hydrocarbon molecules to low hydrocarbon molecules, increase in percentage of free lithium in the lubricating medium and formation of other compounds as shown in Tables 3.2, 3.3. The 'recouping' of the resistivity of the grease under electrical fields suggests the stretching of the molecules and the zero 'recouping' indicates the change of molecules of higher hydrocarbons to low hydrocarbons[3.16]. This finally leads to contact of mechanical surfaces, which causes a sharp increase of the coefficient of friction and heating of the bearing surfaces. The outer race temperature of the bearing – operated under electrical fields – as high as 90°C has been detected. And the bearing surfaces under high contact pressure were found to be corrugated[3.14].

The chemical process leading to the formation of lithium iron oxide, lithium hydro-oxide, lithium carbonate, lithium zinc silicate and low hydrocarbon molecules is explained as under:

$$R\,COO\,Li \rightarrow Li^{+} + R\,COO^{-}$$

$$R\,COO\,Li \rightarrow Li^{+} + R_1\,COO^{-}$$

(R is $C_{17}\,H_{35}$ Or $C_{15}\,H_{31}$, and R_1 is $C_{15}\,H_{31}$)

$$H_2O \rightarrow H^{+} + OH^{-}$$

$$R\,COO^{-} \rightarrow CO_2 + \text{Organic Compound}$$

$$CO_2 + H_2O \rightarrow H_2\,CO_3$$

$$H_2\,CO_3 \leftrightarrows 2H^{+} + CO_3^{=}$$

$$2\,Li^{+} + CO_3^{=} \rightarrow Li_2CO_3$$

$$Li^{+} + OH^{-} \rightarrow Li\,OH$$

$$Li^{+} + Fe^{+} + O_2 \rightarrow \gamma\text{-}Li\,Fi\,O_2$$

$$Li^{+} + Fe^{+} + O_2 \rightarrow Li_5\,Fi\,O_2$$

$$Li^{+} + Zn^{+} + Si\,O_2 \rightarrow Li_{3.6}\,Zn_{0.2}\,SiO_2$$

The formation of lithium hydroxide (Li OH) and lithium carbonate ($Li_2\,CO_3$) make the dielectric alkyline in nature and corrode the bearing surfaces. It also increases the percentage of lithium in the acqueous solution (Table 3.4). The percentage increase in lithium content is correlated to the corresponding decrease in carboxylic group as indicated in infrared spectra in contrast to the fresh grease and the soap residue[3.16].

It may be noted that the process leading to the damage of the bearing is usually caused by the presence of oxygen and water. Corrosion on the bearing surfaces develops as an array of pits, which often lowers the fatigue life of a bearing causing corrugations on the surfaces under the influence of electrical fields[3.14, 3.16]. Similar corrugation patterns are observed on the motor bearing II.

In view of the above, the mechanism of a bearing failure under the influence of electrical current is explained as:

When current passes through a bearing, silent discharge occurs through the inner race, rolling-elements and outer race which leads to electrochemical decomposition of the low resistivity lithium grease (10^7 ohm-cm) and subsequently corrosion of the bearing surfaces, which finally leads to increased wear and high local temperatures on asperities, before pitting starts on the surfaces and bearings fail. This effect is accompanied by an increase in the lithium content of the grease and change in its chemical composition accompanied by the formation of lithium iron oxide and lithium zinc silicate compounds.

Table 3.3 X-Ray Deffraction (XRD) Analysis of Soap Residue of Different Grease Samples

Particulars	Grease from NU 326 bearing (After operation without electrical fields)			Grease From Motor Bearing I	Grease from Motor Bearing II
	Matching Standard Cards	After 41 hours	After 250 hours	After 6000 hours	After 6000 hours
Matching Peaks (dA)	**24-623**	**XRD Analysis**	**XRD Analysis**	**XRD Analysis**	**XRD Analysis**
			4.46		4.39
		4.26	4.27	4.27	4.26
		4.13	4.13	4.19	
	4.12	4.05	4.05	4.08	4.07
					3.96
		3.91	3.93	3.90	3.90
					3.67
	3.77	3.71	3.73	3.74	
			3.30		3.16
	2.65		2.5	2.5	2.46
		2.25	2.26	2.16	2.24
		2.03		2.03	2.03
			1.98		
		1.757			
		1.43			
Chemical Composition	Li_5FeO_2 Lithium Iron Oxide	$C_{18}H_{35}LiO_2$	i) $C_{18}H_{35}LiO_2$ ii) Li_5FeO_2	i) $C_{18}H_{35}LiO_2$ ii) Li_5FeO_2	$C_{18}H_{31}O_2$ γ-$LiFeO_2$, $Li_{3.6}Zn_{0.2}SiO_2$
			Fig. 3.3	Fig. 3.4	Fig. 3.5

Table 3.4 Atomic Absorption Spectrophotometry Analysis of Different Grease Samples

Test Details	Fresh Grease in NU 326 Bearings	Duration (hours) After Operation under Electrical Fields in NU 326 Bearing				Duration After Operation without Electrical Fields in NU 326 Bearing				Fresh Grease in Motor Bearings	From Motor Bearing I	From Motor Bearing II
		41	101	151	250	41	110	205	250		After about 6000 hours of operation	
Lithium percentage as lithium hydro-oxide and lithium carbonate in acquious solution of grease	0.0125	0.05	0.06	0.061	0.065	0.031	0.037	0.036	0.035	0.0067	0.014	0.026
Increase in lithium percentage as compared to fresh grease sample		4.0	4.8	4.9	5.2	2.48	2.96	2.88	2.80		2.1	3.8

On the contrary, a bearing under pure rolling friction does not corrode the surfaces in the initial stages, and also change in lithium content of the grease is much lower. Furthermore, no change in chemical composition takes place. However, formation of lithium iron oxide, corrosion of the surfaces and change of crystalline structure of the grease to amorphous leads to bearing failure after prolong operation.

3.7 Conclusions

The above study draws the following conclusions[3.18]:

1. X-ray diffraction anlaysis indicates that the structure of the fresh lithium grease is lithium stearate ($C_{18}\ H_{35}\ Li\ O_2$), which changes to lithium palmitate ($C_{16}\ H_{31}\ Li\ O_2$); the lower fraction of hydrocarbons after operation of a bearing under electrical fields. Where as the original structure of the grease is not changed under pure rolling friction.
2. After prolonged operation, crystalline structure of a fresh lithium grease in a bearing changes to amorphous structure.
3. When the bearing is operated under electrical fields, the lithium grease gets decomposed and gamma lithium iron oxide (γ-$Li\ Fe\ O_2$) and lithium zinc silicate ($Li_{3.6}\ Zn_{0.2}\ SiO_2$) are formed in presence of Zn, Fe and $Si\ O_2$.
4. Under pure rolling friction, lithium iron oxide ($Li_5\ Fe\ O_2$) is formed in the presence of free Li and Fe in the grease.
5. Percentage increases of free lithium in acqueous solution is about two fold in the grease from a bearing operated under electrical fields than under pure rolling friction.
6. During bearing operation free lithium percentage in the grease increases and then the formation of lithium iron oxide takes place.
7. The formation of lithium hydro oxide and lithium carbonate make the dielectric alkyline and corrode the bearing surfaces, which finally leads to increased wear and failure of a bearing.

The present studies of XRD, XRFS and AAS of greases, along with the study of bearing surfaces, have a potential to diagnose the bearing damage under the influence of electrical fields, and also to establish the deterioration of the lithium greases used in the bearings.

References

[3. 1] Staton, G. M., “Grease Analysis A Modern Multitechnique Approach,” Presented at the NLGI 43rd annual meeting in St. Louis, Missouri, October, 1975.

[3.2] Langborne, P.L., "Grease Lubrication: A Review of Recent British papers", *Proc. Inst. Mech. Eng.*, **184,** Part 3F, p 82.

[3.3] Barnett, R.S., "Review of recent USA Publications on Lubricating Grease", Proc. Inst. Mech. Eng., **184,** Part 3F, p 87. Also in Wear, **16,** p 87 (1970).

[3.4] Godfrey, F., "Friction of Greases and Grease Components during Boundary Lubrication", *ASLE Trans.*, **7,** p 24 (1964).

[3.5] Horth, A.C. Sproule, L. W., and Pattenden, W. C., "Friction Reduction with Greases," *NLGI Spokeman*, **32,** p 155 (1968).

[3.6] Lander, W., "Hydrodynamic Lubrication of Proximate Cylindrical Surfaces of Large Relative Curvature", *Proc. Inst. Mech. Engr.*, **180,** Part 3B, p 101 (1965).

[3.7] Dowson, D. and Whitaker, A. V., "The Isothermal Lubrication of Cylinders", *Trans. ASLE*, **8,** p 224 (1965).

[3.8] Dowson, D. and Whomes, T. L., "Side Leakage Factors for a Rigid Cylinder Lubricated by an Iso-viscous Fluid", *Proc. Inst. Mech. Eng.*, **181,** p 165 (1966).

[3.9] Polacios, J. M., "Elastohydrodynamic Films of Lithium Greases", Macanique, Materiaux, Electricite (GAMI) **365-366,** 176, (1980).

[3.10] Pastnikov, S. N., "Electrophysical and Electrochemical Phenomena in Friction, Cutting and Lubrication", VNR, 1978).

[3.11] Matveyevsky, R.M., Markov, A.A., and Buyanovsky, I. A., Proc. of Conf. Physico-chemical Mechanics of Contact Interaction and Fretting Corrosion (in Russian), p 46, Kiev (1973).

[3.12] Aksyenoy, A. F., Beliansky, V. P., and Shepel, A. Y., "On some Aspects of the Interaction Mechanism between Hydrocarbon Liquid and Metals during Friction", 3rd Inter. Trib. Cong. Eurotrib 81, Lubricant and their Applications, **III,** pp 7-16 (1981).

[3.13] Prashad, H., "Effect of Operating Parameters on the Threshold voltages and Impedance Response on Non-Insulated Rolling Element Bearings under the Action of Electrical Currents," *Wear*, **117,** 2, pp 230-240 (1987).

[3.14] Prashad, H., "Investigations of Corrugated Pattern on the Surfaces of Roller Bearings Operated under the Influence of Electrical Fields," ASME/ASLE, Trib. Conf., San Antonio, Texas Oct. 5-Oct. 8 (1987).

[3.15] Powder Diffraction File, Search Manuals, Organic/Inorganic compounds, Publication SMO-27, published by JC PDS, International Centre for Diffraction Data, 1601 Parklane, Swarthmore, PA 19081, USA (1977).

[3.16] Prashad, H., "Experimental Study on Influence of Electrical Fields on Behaviour of Greases in Statically Bounded Conditions and when Used in Non-Insulated Bearings," *BHEL Jour.*, **7**, 3, pp 18-34 (1986).

[3.17] Gersimov, M., Novacheva–Ranheva, A., Staynov, S. T., "Study of Fatty and Oily Base Effect on the Lubricating peoperties of Lithium Lubricants," 3rd Inter. Trib. Cong. Eurotrib 81, Lubricant and their Application, **III,** pp 102–113 (1981).

[3.18] Prashad,H.,"Diagnosis of Deterioration of Lithium Greases Used in Rolling-Element Bearings by X-ray Diffractrometry" Tribology Transactions,Vol. 32, 2, pp 205-214 (1989).

4 Acceptance Criterion for Turbine Oils

4.1 A General Review

The function of turbine oil is to reduce friction and wear in bearings, to serve as a coolant and sealant, to act as hydraulic medium in governor and to protect metallic surfaces from rusting and corrosion. Oils have a tendency to get oxidized during usage specially in the presence of heat, water, air and certain metallic impurities which act as catalysts for oxidation.

Oxidation of oil causes formation of acid products, resulting in the increase of Total Acid Number (TAN) and decrease of Rotating Bomb Oxidation Test (RBOT) values. Oil manufacturers have correlated RBOT, percentage depletion of anti-oxidant and acid values with Turbine Oil Oxidation Stability Test (TOST) values and proposed a certain definite RBOT value as rejection criterion for turbine oil. Whereas, there is no definite methodology and criterion for acceptance has been stream lined.

In this study, RBOT values of different turbine oils have been determined for fresh oils after TOST of different durations in the laboratory and for the used oils received from different sites. Also, an attempt is made to establish a methodology to correlate kinetics of oxidation of oils using RBOT & TOST values. The quality of oils as well as acceptance/rejection criterion to determine life of the turbine oil, based on the experimental results, has been discussed.

4.2 Introduction

Modern turbine oil should have a very high level of functional properties and the oil has to be formulated in such a way that these properties are maintained at adequately high level during usage in the system. Furthermore, turbine oil should have good demulsibility, viscosity, and resistance to air entrainment and foaming as well as corrosion and rust protection characteristics. Oils have a tendency to get oxidized during usage. When oil undergoes oxidation, its ability of water shedding gets lowered and permanent water emulsion is formed. Emulsified oil cannot provide adequate oil films in bearings and in extreme cases cause scoring on the surfaces of bearings and gear teeth. Also, excessive foaming can interfere with heat removing capability and promote further oxidation.

Oxidation of oil causes formation of acid products resulting in the increase of Total Acid Number (TAN) and decrease of Rotating Bomb Oxidation Test (RBOT) values[4.1]. Turbine Oil Oxidation Stability Test (TOST) and RBOT values were used to establish quality of an oil as well as acceptance/rejection criterion to determine residual life of turbine oils[4.2 & 4.3]

A kinetic approach was also adopted to arrive at a comparative evaluation of turbine oils. The oils were degraded in the laboratory in RBOT apparatus at different temperatures, followed by RBOT life estimation, as per ASTM-D 2272.

Laboratory tests of fresh turbine oils ('A','B' and 'C') of three different suppliers were carried out before and after ageing under TOST conditions of different durations. Under both these conditions, the measurement of properties of RBOT, TAN and EA (Elemental Analysis) were carried out to establish the deterioration of the oils. The correlations of TOST, TAN, RBOT values and elemental analysis have been used for analysis to evaluate the quality of the turbine oils investigated.

4.3 Experimental Investigations

4.3.1 *Turbine Oil Stability Test (TOST)*

For TOST, ASTM - D 943 method has been used. TOST ageing was carried out for oils 'A','B' and 'C' received from the different suppliers. TOST ageing was carried out up to 2000 hr and change in properties like TAN, RBOT and EA were established to study the severity of the deterioration. TOST basically signifies the oxidation stability of turbine oils using different crudes, effect of processing, blending and additive packages.

4.3.2 *Rotating Bomb Oxidation Test (RBOT)*

RBOT ageing of fresh oils samples ('A','B','C') of the different suppliers was carried out at 140°C, 150°C, and 160°C for kinetic studies (as per ASTM – 2272). A plot of log of RBOT value versus inverse of temperature (1/T) in degree absolute was generated and is shown in Fig.4.1.

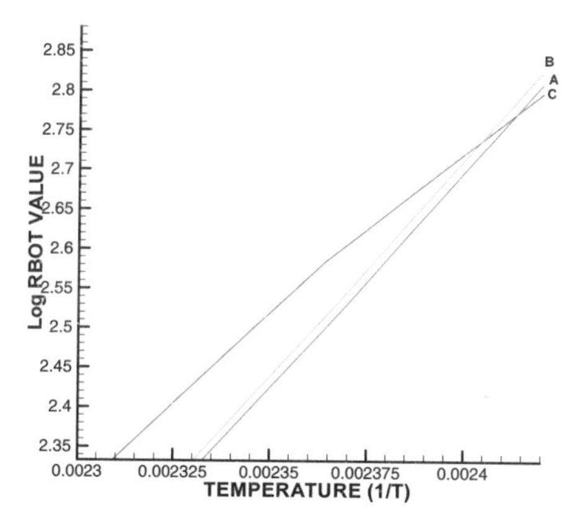

Fig. 4.1 Variation of the Log of RBOT values versus inverse of absolute temperature for the different oils

Activation energies were calculated from the slope of the plot by Arrhenious - equation. Furthermore, RBOT life was derived from the graph at 90°C for all the three samples. Thus, RBOT measurements were carried out on fresh, aged and service oils received from different sites.

4.3.3 *Four-Ball Tests*

As these turbine oils are recommended to be used for geared turbines, the oils are expected to possess load carrying capacity, which is measured by FZG. To establish these characteristics and also to distinguish between the characteristics of extreme pressure and anti-wear additives, four - balls EP (extreme pressure) and four - balls wear tests, as specified in IS 8406 and revised in 1993, were carried out.

4.4 Data Deduction

General properties of the laboratory tests of fresh oil samples 'A','B' and 'C' have been determined as per ASTM and found to be within the specified limits (Annexure –1). Data diagnosed by RBOT and TAN values of fresh oils and after TOST ageing for 1500 and 2000 hrs (as per ASTM - D943) are given in Table 4.1. Table 4.2 gives data of RBOT at different temperatures (140°C, 150°C, and 160°C) for all the fresh oils. Behavior and service life of each oil, supplier - wise, has been determined and brought out in Table 4.3. Data of activation energies, as calculated from log time versus inverse of temperature as depicted in Fig.4.1 (kinetic studies), is shown in Table 4.4 along with RBOT life of turbine oils at 90°C.

4.5 Results and Discussion

4.5.1 *Analysis and Life Estimation of Oil 'A'*

RBOT values of TOST ageing as per ASTM D 943 of fresh oil 'A' after 2000 hrs and 1500 hrs as determined experimentally, are 140 minutes and 157 minutes respectively. The higher RBOT value of 157 minutes at 1500 hrs indicates that the oil got less oxidised as compared to 2000 hr having RBOT of 140 minutes. The change in RBOT values from 317 minutes of fresh oil to 157 minutes after 1500 hrs of TOST shows rapid oxidation of the oil as compared to that from 1500 hrs to 2000 hrs where the change in RBOT value is only from 157 minutes to 140 minutes. This indicates the initial fast ageing followed by slow ageing. The initial ageing of 53% by RBOT in 1500 hrs reduces to 5% in this span of 1500 hrs to 2000 hrs of use. The degradation of the oil is also confirmed by the increase in the values of Fe and Cu as determined by the elemental analysis (Table 4.1).

The fresh oil 'A' indicates RBOT value of 317 minutes at 150°C (Tables 4.1 & 4.3). Its 25% cut-off value is 79 minutes, which is generally considered to be the oil reclamation/rejection limit as per ASTM - D 4378 (Tables 4.3).

Table 4.1 RBOT and tan values, and elemental analysis levels of the fresh turbine oils and after tost ageing for 1500 and 2000 hours (as per astm - d943)

	OIL 'C'			OIL 'B'			OIL 'A'		
	TAN (mg KOH /gm)	RBOT (minutes)	Elemental Analysis Ppm (Fe,Cu,Zn)	TAN (mg KOH/gm)	RBOT (minutes)	Elemental Analysis Ppm (Fe,Cu,Zn)	TAN (mg KOH/gm)	RBOT (minutes)	Elemental Analysis, Ppm (Fe,Cu, Zn)
Fresh Oil	0.04	385	35, <2, <2	0.1	327	38, <2, <2	0.1	317	28, <2, <2
Ageing for 1500 Hrs	0.16	210	36, <2, <2	0.23	220	96,18. <2	0.3	157	110, 100,2
Ageing for 2000 Hrs	--	72	--	--	172	--	--	140	--

Table 4.2 Kinetic study of rbot life of the fresh turbine oils at different temperatures (as per astm –d– 2272) rbot values (in minutes)

OIL	140° C	150° C	160° C
'C'	660	385	215
'B'	680	327	168
'A'	655	317	165

Table 4.3 Service life comparison of the turbine oils

Turbine Oil Used in the Machine	Site Name	Fresh Oil RBOT (minutes)	The 25% Cut - off Value (minutes)	RBOT of TOST Aged Oil after 2000 hrs (minutes)	Service Oil RBOT (minutes)	Machine Running Time at the Time of Sample Withdrawal
'C'	Z	385	96	72	340	3 Years
'B'	Y	327	82	172	245	5 Years
'A'	X	317	79	140	215	2 Years

Table 4.4 RBOT life and activation energy of the turbine oils

OIL	ACTIVATION ENERGY Δ E (k.Cal.)	RBOT LIFE OF OIL at Working Temperature of 90° C (minutes)
'C'	21.6	18750
'B'	24.6	41336
'A'	24.5	39738

Ratio of RBOT life of oil 'B' to oil 'A' is 1.04
Ratio of RBOT life of oil 'B' to oil 'C' is 2.20

The RBOT value reaches the cut - off limits of 79 minutes after 2650 hrs of TOST ageing as per graphic evaluation. Also, oil 'A' sample collected from site 'X' after 2 years of service has shown RBOT value of 215 minutes (Table 4.3). This is well above the 25% cut-off value. This shows that the oil has not reached its reclamation or its rejection limits as per RBOT property. From the RBOT values of aged oil samples taken from the service intermittently from site 'X', it is estimated that the 25% cut - off value of 79 minutes will be reached approximately after 6 years (extrapolated graphical estimation).

I.e. 2650 hrs of TOST ageing = 6 year of residual service life of 'A' oil approximately.

So, the net life of fresh oil 'A' is estimated as 8 yrs approximately.

4.5.2 *Analysis and Life Estimation of Oil 'B'*

The fresh oil 'B' indicates RBOT value of 327 minutes at 150°C (Table 4.2). Its 25% cut off value is 82 minutes (rejection limit as per ASTM).

RBOT value of TOST ageing of fresh oil 'B' after 2000 hrs has been determined as 172 minutes (Table 4.1 & 4.3), which is above the rejection limit. RBOT value reaches the cut-off limit of 82 minutes after 3450 hrs of TOST ageing (graphical value). Furthermore, the higher RBOT value of 220 minutes at 1500 hrs as determined experimentally indicates that the oil is gradually oxidised from its RBOT value of 327 minutes (fresh oil) to 172 minutes (after 2000 hrs). This can be assessed from the fact that up to 1500 hrs of TOST ageing, RBOT is reduced by 33%, followed by 15% in the span of 1500 hrs to 2000 hrs. The change of ppm levels in elemental analysis indicates the deterioration of the oil (Table 4.1).

Also, oil 'B' sample collected from site 'Y' after 5 yrs of service has shown RBOT value of 245 minutes (Table 4.3). This value is well above the 25% cut - off limit and it shows that oil has not reached its reclamation /rejection limit with respect to RBOT property. From the RBOT values of aged oil samples taken from service intermittently from the site 'Y', it has been estimated that the 25% cut - off value of 82 minutes will be reached approximately after 10 years (extrapolated graphical value).

I.e. 3450 hrs of TOST ageing = 10 years of residual service life of oil 'B' approximately.

So, the net life of oil fresh 'B' is estimated as 15 years.

4.5.3 *Analysis and Life Estimation of Oil 'C'*

RBOT value of fresh oil 'C' has been determined as 385 minutes at 150°C as shown in Table 4.1&4.3. Its 25% cut - off value is 96 minutes.

RBOT value of TOST ageing of fresh oil 'C' after 2000 hrs has been determined as 72 minutes (as shown in Table 4.1). The significant change in RBOT value of fresh oil i.e. 385 minutes to 210 minutes after 1500 hrs and 72 minutes at 2000 hrs shows that the oil 'C' deteriorates and gets oxidized at a higher pace with the higher duration of use. This can be assessed from the fact that up to 1500 hrs of TOST ageing, RBOT is reduced by 45% followed by 36% in the span of 1500 hrs to 2000 hrs. This shows that it may not give higher life, but for medium life this oil is adequate. No significant change in ppm levels is indicated in elemental analysis. This shows that inhibitors used in the oil are quite stable up to 1500 hrs of TOST. It has been determined graphically that RBOT value of oil 'C' reaches the cut - off limit of 95 minutes after 1850 hrs of TOST ageing.

Also, oil 'C' sample collected from site 'Z' after 3 years of service has shown RBOT value of 340 minutes (Table 4.3). This value was above 25% cut - off limit like oils 'A' and 'B', and it shows that the oil 'C' has not reached its reclamation/rejection limit with respect to RBOT property in this span of use.

But for higher duration of use, it may deteriorate very fast as indicated in laboratory investigations of RBOT value of 72 minutes after 2000 hrs of TOST as against 210 minutes after 1500 hrs of TOST (Table 4.1). It is further confirmed from the RBOT value of the aged samples taken from service intermittently (from site Z). It has been estimated that the 25% cut - off value of 96 minutes will be reached approximately after 3 years (extrapolated graphic value).

I.e. 1850 hrs of TOST ageing = 3 years of residual service life of oil 'C' approximately.

So, the net life of fresh oil 'C' is estimated as 6 years.

4.5.4 *Four-Ball Test Results*

It was observed that the results of weld load test of all the three oils did not conform the IS specification No. 8406, 1993, because the quantity of EP additive used in the oils was not adequate. However, mean Hertz load value was found to be within the range. Also, four-ball wear test indicated that the scar diameter is within the specified limit.

Besides the above, the kinetic studies conducted at IIT, New Delhi, on the above three oils indicated the same order of the quality as found in investigations carried out at BHEL, Corporate R &D Division.

4.5.5 *Overall Comparison of the Oils*

Fresh oils 'A','B' and 'C' has almost identical characteristics and meets the general specifications. However, oils 'A' and 'B' have higher TAN values as compared to oil 'C'. TAN of fresh oil 'C' is 0.04 mg g^{-1} KOH/gm in comparison to 0.1 mg g^{-1} KOH/gm for oils 'A' and 'B' against the normal specified standard value of 0.2 mg g^{-1} KOH/gm. This shows that ratio of TAN value of oils 'A' and 'B' to that of oil 'C' is 2.5 times. This results in slow increase of TAN value of oil 'C' during TOST ageing up to 1500 hrs. Of test as indicated in Table 4.1. The ratio of TAN value of oil 'B' to oil 'C' and oil 'A' to oil 'C' after 1500 hrs of TOST ageing is 1.44 and 1.89, respectively as against 2.5 for fresh samples of these oils. RBOT values of all the fresh oils were found to be more or less identical. The ratios of RBOT of fresh oils 'B' and 'C', and oils 'A' and 'C' vary between 0.86 and 0.83, respectively. After 1500 hrs of TOST, RBOT ratios of oils 'B' and 'C', and oils 'A' and 'C' change to 1.05 and 0.77, respectively. But after 2000 hrs of TOST, the deterioration of oil 'C' is very significant. The ratios of RBOT after 2000 hrs of TOST for oil 'B' and 'C', and for oils 'A' and 'C', change to 2.4 and 1.94, respectively, whereas this ratio for the oil 'B' to 'A' is 1.23. From this, it is evident that the performance of oil 'C' deteriorates much faster after long use as compared to oils 'A' and 'B', and the loses its oil inhibition oxidation properties. However, the lesser value of TAN makes the oil 'C' stable against oxidation up to 1500 hrs of TOST ageing (Table 4.1).

Similar conclusion can be drawn by the kinetic study of RBOT life of fresh turbine oils at different temperatures. At 140°C and 150°C, there is insignificant change in the RBOT values of these oils: The values change only from 655 minutes to 680 minutes at 140°C and 317 minutes to 385 minutes at 150°C but at 160°C RBOT values for oil 'C' is 215 min as against 168 min for oil 'B' and 165 min for oil 'A' (Table 4.2). The higher difference in RBOT value with increase in temperature indicates that oil 'C' is deteriorating at a higher rate as compared to 'A' and 'B' at higher temperature of use and will have less life comparatively. From the activation energy, it is further confirmed that oil 'C' is inferior to that of oils 'A' and 'B' because activation energy of oil 'B' and 'A' are 24.6 Kcal and 24.5 Kcal respectively as against 21.6 Kcal for oil 'C' as shown in Table 4.4.

Furthermore, results of kinetic studies at 90°C of RBOT working temperature indicates that RBOT life of oil 'B' is 2.20 times that of oil 'C' and 1.04 times that of oil 'A' (Table 4.4).

4.6 Conclusions and Recommendation

Followings are concluded based on the studies conducted[4.4]:

1. The life of oil 'B' is 2.20 times that of oil 'C' and 1.04 times that of oil 'A'.
2. The deterioration of oil 'B' is gradual and uniform with time as against that of oils 'A' and 'C'.
3. Lesser TAN value of oil 'C' make it stable for medium duration of use. For higher duration of use, the deterioration is very fast.
4. A gradual and uniform deterioration of oil by RBOT not exceeding 33% and 15% may be allowed as a criterion after TOST ageing of 1500 hrs and 2000 hrs, respectively. This is required to achieve the life span of 15 years of turbine oils at actual site use
5. From the TOST and RBOT values, the service life of the oils can be assessed
6. Based on the TOST ageing up to 1500 and 2000 hrs as monitored by RBOT, TAN and RBOT kinetic studies, the quality of the tested oils is found to be as per following order:

   (1) Oil 'B', (2) Oil 'A' and (3) Oil 'C'

In short, turbine oil 'B' is found to be superior to the oils 'A' and 'C'. This may be due to the blending of various additive packages. However, it may be necessary to monitor each batch of fresh oils for properties to assess the quality of the oils since batch-to-batch variation was observed in the studies.

References

[4.1] Goel, P.K., Jaya Prakash, K.C, Srivastava S.P, "Oxidation Stability of Steam Turbine Oil and Laboratory Methods of Evaluation", ASLE, Vol.2,pp 89-95,1984.

[4.2] Warne, T.M., Vienna, P.C., "High Temperature Oxidation Testing of Lubricating Oil", ASLE Vol. 40, pp 211-217, 1984

[4.3] Herderm, D.,Vienna, P.C, "Control of Turbine Oil Degradation during Use", ASLE, vol. 37, pp 67-71, 1980.

[4.4] Murthy T.S.R., Prashad,H.,Jagga,C.R., "Life Estimation of Turbine Oils – A Methodology Criterion for Acceptance", Petrotech 2003, January 9-12, New Delhi.

Annexure - I

Properties of Turbine Oils

Sl No.	Property	Oil 'A'	Oil 'B'	Oil 'C'
1.	Density at 15° C	0.870	0.865	0.868
2.	KV at 40° C	48.140	46.400	46.500
	KV at 100° C	7.100	7.120	7.500
3.	Viscosity index	110	110	110
4.	T A N	0.1	0.1	0.04
5.	Pour point	-6	-6	-6
6.	Demulsibility			
a)	Free water (ml) collected during the test (starting with 45ml)	41	41	39
b)	% water in oil after 5 hr test	0.35	0.3	0.4
c)	%Left after centrifuging	Nil	Nil	Nil
7.	Foaming characteristics			
	At 24° C	Nil	Nil	Nil
	At 93.5° C	Nil	Nil	Nil
	At 24° C after the test at 93.5° C	Nil	Nil	Nil
8.	Air release value at 50° C	1 min 8 s	1 min 8s	1min 20 s
9.	Four ball EP test			
a)	Weld load (kg)	150	150	150
b)	Mean Hertz load (MHL) (kg)	39	46	43

5
A Combinatorial Approach to Elucidating Tribochemical Mechanisms

5.1 A General Review

A new type of combinatorial tribological experiment is presented, which explores a series of tribological conditions, such as load and relative velocity, spatially separated as a ''library'' on one single sample. As an example, a library displaying the results of tribological testing of an additive under a series of different loads has been prepared and analyzed. The tribological information acquired during the testing has been correlated with spectroscopic information from the tribologically stressed surface. The use of imaging and small-area X-ray photoelectron spectroscopy has allowed the identification of the different tribologically stressed areas and the acquisition of detailed spectroscopic information. The composition and the thickness of the tribofilm were found to be dependent on the applied load. The use of the combinatorial approach shows the potential to greatly facilitate rapid characterization of new lubricant additives.

5.2 Introduction

The combinatorial synthetic approach, together with high-throughput screening of compounds, has been applied in pharmaceutical chemistry since the early eighties. A large number of molecules are synthesized in parallel and subsequently probed for chemical, physical and medicinal properties. This approach was first brought into materials science in the field of high temperature superconductors[5.1], where a spatially addressable library of potential candidates for high-T_c was fabricated and tested and it has been adapted to different fields in materials science, such as semiconductors[5.2] or metallurgy[5.3]. The advantage of the combinatorial synthetic approach in both chemistry and materials science is the rapid production of many substances of varying compositions, which are subsequently analyzed in a massively parallel way. This methodology speeds up the search for new substances with the desired properties.

In tribology, and especially in tribo-chemistry, one has to deal with a large parameter space, since the friction, wear and tribo-chemical reactions of a given tribological system have been shown to depend strongly on the applied conditions (e.g. relative velocity, contact pressure, sliding time, temperature)[5.4]. To understand the reactions that can occur in this system it is necessary to explore a significant portion of the parameter space. To date, this has been a very time-consuming process, involving an enormous number of experiments, suggesting that a variant of the combinatorial analytical approach could also be useful in a tribological context.

A wide variety of surface-analytical methods have been used to investigate tribo-chemical reaction products. X-ray photoelectron spectroscopy (XPS), scanning Auger microscopy (SAM), time-of-flight secondary ion mass spectroscopy (ToF-SIMS) and X-ray absorption near-edge structure (XANES) are frequently used methods for the chemical characterization of tribo-chemical reaction films. Most of these methods can combine imaging and spectroscopy. While scanning Auger electron microscopy (SAM) has been used to map elemental distribution of a contact area[5.5], the use of imaging XPS (i-XPS) has been shown recently to be of value in investigating the distribution of chemical species according to their chemical state[5.6, 5.7]. So far, the imaging possibilities of these methods were used on relatively simple systems, allowing the contact and non-contact regions to be distinguished. By carrying out a combinatorial experiment, one can take further advantage of the imaging capability. In this type of experiment a parameter library is built, applying various tribological conditions (as a function of the lateral position on the disk) on a single sample, which is subsequently analyzed by an imaging surface analytical technique. The tribological information (coefficient of friction, wear) and the spectroscopic results can afterwards be mapped onto the parameter library (see Fig. 5.1). The correlation of this data should provide insight into the application range of a lubricant additive system and help uncover mechanistic reaction pathways.

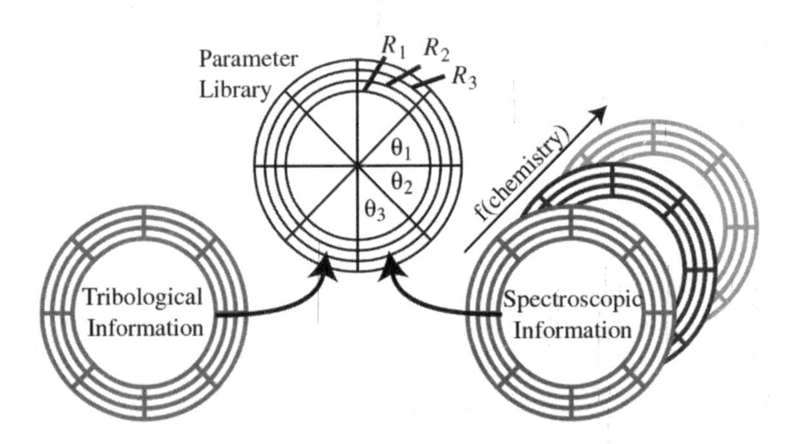

Fig. 5.1 General principle of the combinatorial approach: a parameter library is produced by varying the tribological stress on a single sample depending on the lateral e.g. radial (R) and angular (θ) position on the disk. The tribological information acquired during or after the test (coefficient of friction, wear) and the spectroscopic information (chemistry of surface film) from surface-analytical examination of the sample can then be mapped onto this parameter library and correlated.

Very little work has been done applying a combinatorial approach to tribological problems. Green and Lee used an AFM with chemically patterned cantilevers and tip arrays to probe adhesive forces between carboxylic acid, alcohol and methyl groups[5.8]. This approach reveals tribological information on the nanometer scale, whereas the tribological load scanner introduced by Hogmark uses a macroscopic, crossed cylinder configuration[5.9]. In the latter setup, two elongated cylinders repeatedly slide across each other with varying load in such a manner that each point along the sliding track of both cylinders experiences a unique load. This test was used in the evaluation of hard coatings.

A model system (di-isopropyl zinc dithiophosphate (i-ZnDTP) in decane) has been chosen as a lubricant additive system in the present study for two reasons: firstly, the tribological system was to be kept as well defined as possible for the first experiments with this new approach, and secondly because ZnDTP is a well studied system, which readily allows comparison of our own results with the literature. The formation of tribo films from ZnDTP has been studied extensively and has been described by several authors[5.4, 5.10, 5.11, 5.12].

5.3 Experimental

Steel disks (AISI 52100) were polished to a final surface roughness (R_a) below 10nm using silicon carbide paper and diamond paste. The samples were ultrasonically cleaned in ethanol and analyzed for surface contamination by XPS immediately prior to tribo-testing. Commercial 4mm ball-bearings (AISI 52100) were used as a counter-face.

A 1 wt.% solution of di-isopropyl zinc dithiophosphate (i-ZnDTP) in decane was used as a lubricant. To dissolve the additive in the solvent, the solution was stirred at 60 °C for 30 min. The tribo-tests were carried out at room temperature (24 $\pm$ 0:5 °C) and the relative humidity of the air was recorded during each test and found to be always between 22 and 38%. The sample was fully immersed in the lubricant during testing. A CETR 2 tribometer (Center for Tribology, Inc., Campbell, CA, USA) was used to run the tests using a ball-on disk geometry. This tribometer allows the independent programming of normal load, velocity, radius and duration. Prior to the actual tests, running-in of the ball was performed at 5N load with a speed of 31.4 mm/min for 2 h, outside the region to be used for XPS analysis and in the presence of the test lubricant. The running-in was performed in order to create a flat spot on the bottom of the ball. This flat spot was then placed in contact with the surface for the subsequent runs and defined the apparent contact area.

In Fig. 5.2, a schematic of the combinatorial tribotest is presented. The experiment consisted of producing five concentric tribologically stressed annuli on a single sample; in all cases a speed of 31.4 mm/s was used. In each annulus a different load (ranging from 0.05 to 5N) was applied. Each annulus consisted of 11 overlapping wear tracks, with radii differing from each other by 25 μm (see insert of Fig. 5.2). The width of each wear track was defined by the flat spot produced during the running-in. It was found to have a diameter of approximately 150 μm: The 11 wear tracks created together an annular test region spanning more than 250 μm. Before the XPS measurements the sample was ultrasonically cleaned in cyclo-hexane. On the tribologically stressed sample, an O(1s) map was collected, which evidences the different wear regions. Within each tested area a small-area XPS analysis was carried out. The analyzed area consisted of a spot of 120μm diameter, and thus completely inside the tribologically stressed area.

The XPS analyses were performed on a PHI 5700 system with an Omni Focus IV lens system. Spectroscopic maps were acquired using the imaging capabilities of the lens system. The analyzed spot (diameter 120 μm) was electrostatically rastered over the sample (typically 64 × 64 pixels, 2 × 2mm). For each pixel a full spectrum of the selected energy region was acquired. The typical size of an XPS map with respect to the tested areas is displayed in Fig. 5.2. The acquired spectroscopic maps were processed with PHI Multipak (V6.0) software.

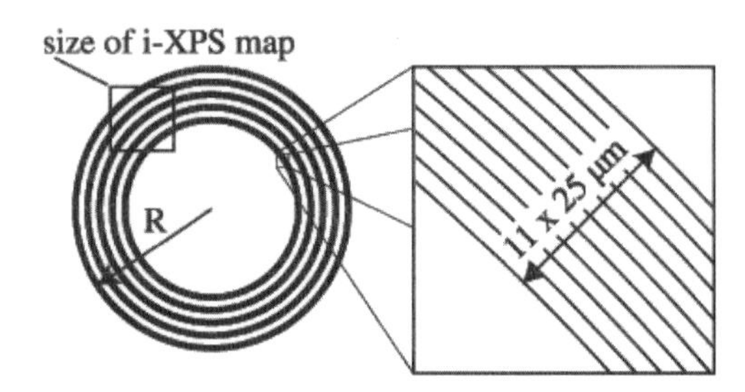

Fig. 5.2 Schematic of the tribotest: the tests were performed in 5 concentric annuli consisting of 11 single tracks, each with a radius differing by 25µm: The wear tracks are partially overlapping due to the width of the apparent contact area, which is defined by the flat spot (@150µm) produced on the ball during the running-in of the tribotest. For each annulus, a different load (L) was applied (R = 3.5mm, L = 0.05N; R = 4mm, L = 0:1 N; R = 4.5mm, L = 0.5 N; R = 5mm, L = 1 N; R = 5.5mm, L = 5N). The square in the top left corner represents the area analyzed by i-XPS.

5.4 Results

In Fig. 5.3, the coefficient of friction (COF) acquired during a test is displayed versus the number of turns for loads from 5 to 0.1 N. The COF is averaged over one full turn of the disk and displayed versus the turn number. The error bars represent the standard deviation during one full turn. It can be noticed that at 0.5 and 1N the COF decreases after the beginning of the test and shows a sudden increase after each fifth turn. This increase coincides with the 25μm steps that occur after every five turns. After the increase, the COF decreases again and at the end of a five-turn cycle the COF has reached a steady state. This steady state is assumed to indicate that a tribo film has been formed that is representative of the applied conditions. Thus the average of the COF during the last full turn of the test is given in Table 5.1 as a representation of the tribological properties of the tribo film. The COF shows a small decrease with increasing load.

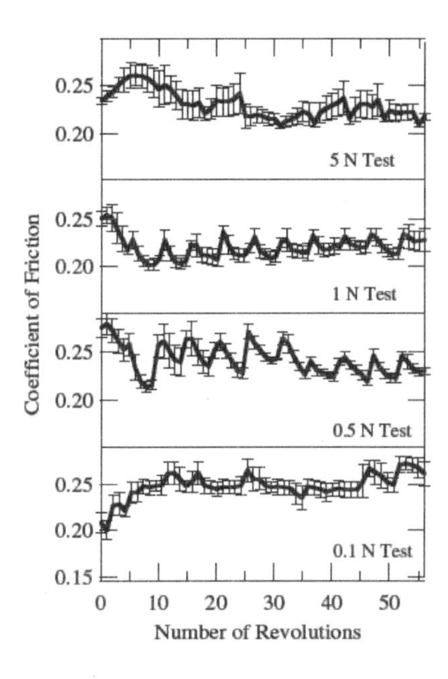

Fig. 5.3 Coefficient of friction (COF) of a tribotest for the loads from 5 to 0.1 N. The COF is averaged over 1 full turn of the disk; the error bars represent the standard deviation during one turn. The increase of the COF after the 25 μm steps after each fifth turn can be seen (see text).

The morphology of the wear tracks and the ball used for the tribotest were examined with an optical microscope. The ball shows a circular flat spot with a diameter of 150μm; which was worn off mainly during the running-in period of the tribotest. It defines the apparent contact area during subsequent tests. On the tribologically stressed disk, the areas tested with 5 and 1N are clearly visible by optical microscopy, but for lower loads no differences can be recognized between the contact and the non-contact area. Without additional information from imaging XPS it would be difficult to locate the areas tested at lower loads.

The total O(1s) intensity map of the tribologically stressed sample is presented in Fig. 5.4(a). Arcs of higher intensity can be seen running from the bottom left to top right, indicating the location of the tribologically stressed annuli. Each pixel of the O(1s) map contains the entire spectroscopic information for the O(1s) region. The spectra of the most intense and the least intense areas are extracted and displayed in Fig. 5.4(b). The dark areas (which correspond to a lower oxygen signal intensity) reveal a spectrum that is characteristic for the surface of an oxidized steel, showing a peak at 530 eV and a shoulder at a binding energy that is approximately 1.5 eV higher than the main peak. The bright areas show a different peak shape in the O(1s) region. The main peak is found at 531.7 eV plus a shoulder at 530 eV. The peak at 531.7 eV is characteristic for oxygen bound in a phosphate group, while the shoulder is due to contributing oxide. In the following, this shape of the O(1s) spectrum is referred to as being of the phosphate type.

The O(1s) map was further processed with the linear least-squares (LLS) algorithm of the PHI Multipak (V6.0) software. This algorithm fits the spectra of the selected areas (shown in Fig. 5.4(b)) with the spectra in each pixel of the map. The correlation is shown in the chemical-state maps (Fig. 5.4(c) and 5.4(d)). Bright areas in Fig. 4(c) show areas with high correlation with the oxide-type spectra, while bright areas in Fig. 5.4(d) show high correlation with the phosphate-type spectra. The phosphate-type spectrum is most prominently represented in the outermost test area (5N load) and decreases with decreasing load. The O(1s) peak shape characteristic of oxide (''oxide type'') is almost absent in the area tested with 5N but seems to be increasingly present at lower load and it is predominant in the non contact area.

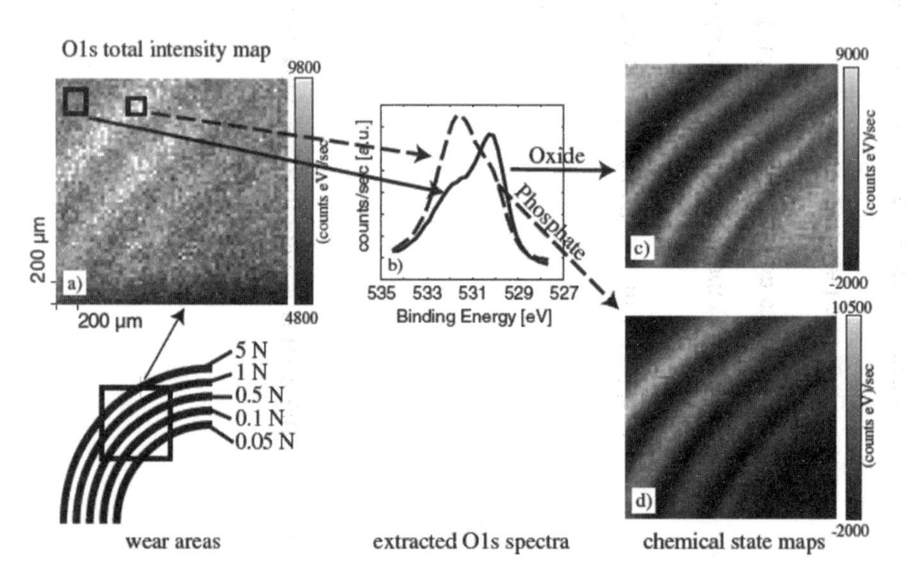

Fig. 5.4 In the total-intensity map (left), a variation in the total intensity of the O(1s) signal can be clearly observed to coincide with the tribotested regions. The O(1s) spectra shown in the graph in the middle represent the spectra extracted in the regions marked in the map. Two different signals can be observed, a "phosphate" signal and an "oxide" signal. The linear-least-squares routine is used to correlate these two signals with O(s) signal in each pixel of the map, producing chemical-state maps. Bright colors represent high correlation with the respective signal and thus indicate the distribution of the oxide or the phosphate.

From the chemical-state map it is possible to select ''areas of interest'' for more detailed (small-area) XPS analysis. The results are summarized in table 1 and a more detailed spectral analysis can be found elsewhere[5.13]. No differences in the binding energies of the spectra taken in the various tested areas could be found, although changes in intensity ratios were observed. With increasing applied load, the phosphorus-to-sulfur ratio in the tribofilm was observed to increase. The ratio of the oxygen bound in a phosphate group (component at 531.7 eV, see Fig. 5.4(c)) to phosphorus was found be close to 4: 1 for high loads, indicative of P_4^{3-}. The binding-energy value of the P(2p) peak (133.6 eV) is also in agreement with phosphorus being bound within a phosphate group[5.14].

5.5 Discussion

The combinatorial approach to tribological testing presented here involves the creation of a set of spatially separated areas on a single sample that have undergone tribological tests under a variety of loads. The frictional information gathered during tribological tests as a function of test conditions can then be mapped onto subsequent, spatially resolved surface-analytical data. This methodology can be considered as the generation of a parameter library on the sample, which then serves as a means to relate tribological conditions with both friction and subsequent tribo-chemical reactions.

As described above, the COF increases after every fifth turn of the disk due to the 25μm step that is performed to ensure a contact area wide enough for XPS analysis (see Fig. 5.2). At the end of a five-turn cycle, the COF reaches a steady-state value, indicating that a tribo-film characteristic of the applied load is formed and thus a representative surface analytical analysis can be carried out within these contact areas. This behavior of the COF is most evident in the 1N and 0.5N tests. It can be assumed that due to the high load, a surface film is formed more rapidly than at lower load. The resolution of the load cell measuring both frictional force and normal load is $\pm$5 mN: The friction force for the low-load experiments is on the order of 20 and 10 mN for the 100 and 50 mN tests, respectively, and thus the experimental error for the friction force is on the order of 25–50% for these loads. Despite the high experimental error at low loads, a slight reduction of the COF at higher loads compared to the lower loads can be seen for both experiments reported in Table 5.1.

Table 5.1 COF and elemental ratio for the various loads of two independent samples. The detailed analysis of the XP spectra is presented elsewhere[5.13].

Load	COF		P/S		Ophosphate/P	
0.05N	0.25	0.31	0.66	0.68	5.1	7.0
0.1N	0.26	0.31	0.71	1.13	5.0	4.7
0.5N	0.23	0.21	0.94	1.30	4.1	4.5
1N	0.22	0.20	0.95	1.42	4.0	4.5
5N	0.20	0.19	0.91	1.45	3.6	4.2

A possible shortcoming of the described combinatorial test method, which involves using the same pin and disk for both running-in and load tests, can arise from the fact that the tribological conditions at the contact between the pin and disk result from the nature of the surface films present on both the pin and the disk and not just from the disk. Indeed, any change in the surface film composition of the pin might also affect the subsequent results. This needs to be determined in a controlled experiment for each system investigated. To ascertain whether any such interference was significant in the present study, oscillating load tests have been performed, during which the load is cycled from the minimum value to the maximum value, in synchrony with the angular position on the disk. These tests have shown that the coefficient of friction changes as a function of the applied load in a similar way during each cycle, as well as showing symmetric behavior during increasing- and decreasing-load phases[5.15]. In the presence of ZnDTP, the COF showed higher values at the smaller loads and decreased with increasing load. Results obtained performing traditional single-wear track tests were in good agreement with those obtained during combinatorial tests for the same applied loads, confirming that the history of the pin did not influence subsequent tribological measurements in the system under investigation[5.15].

The decreasing oxygen-to-phosphorus ratio and the increasing phosphorus-to-sulfur ratio (see Table 5.1) indicate that with increasing load an increasing amount of phosphate is formed in the contact area. Detailed analysis of the S(2p) spectra indicate that some of the sulfur is present in the sulfide state and some is present as organo sulfur species[5.13]. Spectroscopic analysis of the Fe(2p) signals also indicates the presence of iron phosphate. Tribological films from ZnDTP have been reported to be polyphosphate films[5.4, 5.10]. They are formed under pure thermal or combined thermal and tribological stress[5.11]. At temperatures above 100 °C, tribo films with a thickness of a few tens of nanometers are found, while at lower temperatures thinner films are reported[5.16]. The absence of thermal activation in our experiments explains the lack of a thick polyphosphate film. Only the presence of a thin orthophosphate film is indicated.

5.6 Conclusions

Based on the above studies, the following conclusions are drawn[5.13]:

A combinatorial experiment has been successfully applied to the investigation of a tribological system. The advantage of the combinatorial approach is that multiple experiments can be efficiently combined on a single sample and readily compared. Tribological and spectroscopic results could be acquired in one experiment for a number of conditions and mapped onto the parameter library.

Useful information concerning the reactions occurring in the tribological contact could be derived from the experiment: the i-ZnDTP molecule reacts under tribological stress with the steel surface and forms an iron phosphate-containing film. The amount of phosphate film formed and the composition is dependent on the applied load. Sulfides and organo sulfur species are formed, but the amount of these species present is not as strongly load dependent as the amount of phosphate on the surface.

In future, the information density obtained in such experiments will need to be increased in order to realize the full potential of this type of experiment. The range of experimental parameters should also be enlarged in order to cover as many tribologically relevant regions of the parameter space as possible. Ideally, with one experiment the tribological conditions under which a given lubricant additive formulation shows desirable properties could be determined. This approach could therefore speed up the search for new lubricant-additive systems. The approach can also be extended to other (surface) analytical methods. Auger electron spectroscopy or time-of-flight secondary ion mass spectroscopy would have the advantage that the analysis could be performed with a higher lateral resolution and therefore allow a higher information density.

References

[5.1] X.D. Xiang, X.D. Sun, G. Briceno, Y.L. Lou, K.A. Wang, H.Y. Chang, W.G. Wallace-Freedman, S.W. Chen and P.G. Schultz, Science 268 (1995) 1738.

[5.2] R.B. van Dover, L.D. Schneemeyer and R.M. Fleming, Nature 392 (1998) 162.

[5.3] J.C. Zhao, J. Mater. Res. 16 (2001) 1565.

[5.4] A.J. Gellman and N.D. Spencer, J. Eng. Tribol. 216 (2002) 443.

[5.5] J.M. Martin, C. Grossiord, T. Le Mogne and J. Igarashi, Wear 245 (2000) 107.

[5.6] K. Matsumoto, A. Rossi and N.D. Spencer, *Proceedings of the International Tribology Conference Nagasaki, 2000 II* (2000) 1287.

[5.7] M. Eglin, A. Rossi and N.D. Spencer, in: *Proceedings of the 28th Leeds–Lyon Symposium on Tribology*, ed. D. Dowson et al. (Vienna, 2002) p. 49.

[5.8] J.-B.D. Green and G.U. Lee, Langmuir 16 (2000) 4009.

[5.9] S. Hogmark, S. Jacobson and O. Wä nstrand, in: *Proceedings of the 21st IRG-OECD Meeting,* ed. D.J. Schipper (Amsterdam, 1999).

[5.10] J.M. Martin, Tribol. Lett. 6 (1999) 1.

[5.11] Z. Yin, M. Kasrai, M. Fuller, G.M. Bancroft, K. Fyfe and K.H. Tan, Wear 202 (1997) 172.

[5.12] M.L.S. Fuller, M. Kasrai, G.M. Bancroft, K. Fyfe and K.H. Tan, Tribol. Int. 31 (1998) 627.

[5.13] M. Eglin, A. Rossi and N.D. Spencer, Tribol. Letter, Vol.15, No.3 (2003), pp 193-198.

[5.14] D. Schuetzle, R.O. Carter, J. Shyu, R.A. Dickie, J. Holubka and N.S. McIntyre, Appl. Spect. 40 (1986) 641.

[5.15] M. Eglin, PhD Thesis. Presented at the Department of Materials, ETH Zürich, 2003

[5.16] S.H. Choa, K.C. Ludema, G.E. Potter, B.M. Dekoven, T.A. Morgan and K.K. Kar, Wear 177 (1994) 33.

Section - III

Bearing under Electric Current

6

A State-of-the-Art Review of Bearings and Lubricants in Electrical Environment

6.1 Introduction

Recent investigations on response and performance of bearings and lubricants under the influence of electrical current have given insight into behaviour of bearings, deterioration of lubricants and the approach developed for the evaluation of various electrical parameters of different types of bearings. The multifold bearing analysis and its co-relation with the electrical analogy opens a hitherto unexplored era in Tribology, the tribology in electrical environment, which adds a different dimension to the tribological approach. This is because of the fact that genesis of intermolecular forces involves electrostatic attraction or repulsion during tribological interaction and thus it creates electrodynamic, magnetic and exchange forces between atoms. Thus, nearly, all tribological phenomenons, occurring in any interacting system, metal-lubricant-metal or metal-to-metal is electrical in nature. This leads to the dawn of the science of tribology in electrical environment. If the precise nature between microscopic contacts could be determined, then better understanding of friction could lead to industrial innovation for the development of improved lubricants, and wear resistant components. Furthermore, in any tribological interaction thermo-electrical component, capacitance component, acousto-electric component, galvanic component, inductive component of oil film thickness play a very significant role besides the conventional knowledge to understand friction and wear in presence of lubricant having different properties.

It is well known that the large number of components found in the degraded oils cannot be thermodynamically formed by normal chemical transformations, which are expected to be taking place at the tribological systems. Due to the tribological interaction between metal surface and oil, the metal surface acquires electrostatic charges at the random spots, which emits low charge particles. The frequency of emission depends on operating parameters, which governs the tribological behaviour of the interacting surfaces, and thus arises the need of fundamental micro and nano level investigations.

Before dealing with the behaviour and response of the bearings and lubricants under the influence of electrical environment, it must be understood why the bearings and lubricants are to work under such conditions. This happens because of the developed shaft voltage due to various causes in a machine. The causes and the phenomenon of the shaft voltages are summarized as: asymmetry of faults, i.e. winding faults, unbalanced supplies, electrostatic effects, air gap fields, magnetized shafts or other machine members, asymmetries of magnetized fields, etc[6.1]. The causes of shaft voltages can be grouped in four categories, namely:

- external causes,
- magnetic flux in the shaft,
- homopolar magnetic flux,
- Ring magnetic flux.

Friction between belt and pulley can set up an electrostatic voltage between shaft and bearings. Accidental grounding of a part of a rotor winding to the rotor core can lead to stray currents through the shaft and bearings and can result in the permanent magnetization of the shaft. Also, shaft voltage and current could be generated when the machine is started. Furthermore, homopolar flux from an air gap or rotor eccentricity can generate voltage[6.2]. Another cause of shaft voltage is the linkage of alternating flux with the shaft. The flux flows perpendicular to the axis of the shaft, and pulsates in the stator and rotor cores. It is caused by the asymmetries in the magnetic circuit of the machine such as[6.3]:

- uneven air gap and rotor eccentricity,
- spilt stator and rotor core,
- segmental punchings,
- axial holes through the cores for ventilation and clamping purposes,
- key ways for maintaining the core stackings,
- Segments of different permeability.

All the causes listed above develop a magnetic flux, which closes over a yoke and induces voltage on the shaft as the machine rotates. This results in a localized current at each bearing rather than a potential difference between the shaft ends. A current path, however, along shaft, bearings and frame results in a potential difference between the shaft ends[6.4]. This happens because of axial shaft flux caused by residual magnetization, rotor eccentricity, and asymmetrical rotor winding.

For the analysis and review of response and behaviour of bearings under electrical environment, papers are categorized into four main topics, that is: Rolling-Element Bearings, Lubricants, Hydodynamic Journal Bearings, and Thrust Bearings. Understanding all the papers listed in the references can not be summarized separately, furthermore, it would be tedious to read a neutral summaries of the published papers, so the collective summary of each topic is given in a broader prospective.

6.2 Rolling – Element Bearings

In a rolling-element bearing, at each revolution of the shaft, part of the circumference of the inner race passes through a zone of maximum radial force, and Hertzian pressure between the rolling-elements and raceways at the contact points leads to a maximum shear stress, and this occurs in the sub-surface at a depth approximately equal to half of the radius of contact surface[6.5]. It is generally at this point, that the failure of material, if occurring, will initiate. It has been highlighted that the process of deformation which leads to the formation of a corrugation pattern on the track surfaces is accelerated by: the passage of electrical current, corrosion and oxidation of surfaces, lubricant characteristics and quality of a bearing[6.6].

6.2.1 *Corrugation Pattern*

The mathematical formulations have been developed separately for roller bearing and ball bearing depending on the profile of contacts for evaluation of pitch of corrugations which depends on the bearing kinematics, frequency of rotation, position of plane of action of radial loading, bearing quality and lubricant characteristics. Also, formulations for the width of corrugations on the track surfaces have been worked out. The comparison of experimental and theoretical data for the pitch and width of corrugations have been reported. The mathematical formulations show that the width of corrugations on the track surfaces is not affected by the frequency of rotation and depends only on load conditions and bearing kinematics[6.6, 6.7]. It has been analyzed that the passage of current causes local surface heating which leads to low temperature tempering, and accelerates formation of corrugations with time for a bearing using low resistivity lubricant (10^5 ohm.m). After long operation, softer tempered surfaces of the races become harder, and thus harder and re-hardened particles due to localized high temperature and load eject from the craters which intensify the depth of corrugations.

6.2.2 *Flux Density Distribution*

Further, it is reported that under the influence of electrical current, a rolling – element bearing using low resistivity lubricant develops distribution of magnetic flux density on the track surfaces. This happens because of the passage of high intensity current through a bearing[6.8]. On the contrary, a bearing using high resistivity lubricant (10^9 ohm.m) does not develop significant flux density on its surface[6.9]. Theoretical and experimental investigations have been reported on the distribution of magnetic flux density on the track surfaces of races and rolling-elements of bearings[6.10]. It has been brought out that by using the developed model and by experimental determination of residual flux density on the track surfaces, the level of current flow through a bearing can be ascertained without the measurement of shaft voltage and bearing impedance[6.11]. A typical case study of failure of the bearings of the alternators has been analyzed by using the developed flux density distribution technique[6.12].

6.2.3 *Threshold Voltages, Impedance Response, Capacitance and Charge Accumulation*

Phenomenon of threshold voltages and investigations pertaining to first/second threshold voltages for the bearings operating under the influence of electrical currents has been reported. It has been shown that the threshold voltages depend on the lubricant resistivity, oil film thickness and operating parameters[6.13]. The detected threshold voltages are primarily responsible for the momentary flow of current and the further increase in current intensity with a slight change in potential drop across the bearings. The threshold voltage decreases as the load on a bearing increases at the constant speed. It is established that the increase of current intensity reduces the bearing impedance significantly. It has been emphasized that for the reliable operation of a bearing, the safe limit of the potential drop across the bearing elements should be less than the first threshold voltage[6.13].

Under the influence of potential drop across a bearing, the minimum film thickness between races and rolling-elements offers a maximum capacitance and minimum capacitive reactance depending on the permittivity of the lubricant. The electrical interaction between the races and the rolling-elements in the presence of oil film is like a resistor-capacitor (RC) circuit that offers impedance to the current flow. The capacitance and resistance for roller bearing and ball bearing have been separately determined analytically, and found to depend on film thickness, width of deformation and are governed by the permittivity and resistivity of the used lubricant[6.14, 6.15]. Also, charges stored on the bearing surfaces have been reported. It has been established that equivalent capacitance of bearing decreases with increasing speed at constant load. Furthermore, it has been analyzed that a bearing lubricated with high resistivity lubricant as opposed to low resistivity lubricant, with the same permittivity, behaves like a capacitor up to the first threshold voltage. Besides this, for a bearing to accumulate charges, the ratio of capacitive reactance to active resistance should be less than unity[6.14].

6.2.4 *Contact Temperature, Developed Stresses, Slip Band Initiation and Time/cycles for Appearance of Flutes and Craters on Track Surfaces*

Current passing through a bearing at the line/point contact between the tracks and rolling-elements, and the corresponding impedance at the junction generates heat and increases temperature instantaneously. This increases the contact stresses and enables to determine the number of cycles before the slip bands are initiated on the track surfaces of a bearing lubricated with low resistivity lubricant[6.16, 6.17]. The contact duration and temperature rise between track surface of races and a rolling – element have been theoretically determined depending on kinematics, number of rolling-elements in the loaded zone, material properties and depth of slip bands. Based on the developed stress levels on the track surfaces, duration before the slip bands are established has been determined[6.16]. Furthermore, a theoretical approach is developed by continuum theory of Griffith to assess the time span/cycles for the development of flutes/corrugations on the track surfaces after appearance of slip bands[6.18]. Similar approach using the effect of leakage of charge has been extended for the formation of craters on the track surfaces of a bearing lubricated with high resistivity lubricant[6.19]. Further, crater formation criterion and mechanism of electrical pitting on the lubricated surfaces has been analyzed[6.20, 6.21].

6.2.5 *Effect of Instantaneous Leakage of Stored Energy of Charge on Track Surfaces*

The effect of leakage of charge on the rise of contact temperature on the track surfaces of a bearing using high resistivity lubricant has been diagnosed. Based on these, mathematical formulations for the evaluation of contact stresses and minimum cycles for the appearance of craters has been developed[6.22]. The ratio of contact cycles required for charge accumulation and discharge of the accumulated charges on the bearing surfaces depending on the ratio of bearing to shaft voltage have been analyzed. The number of cycles and the number of starts and stops before initiation of craters on track surfaces as against the ratio of bearing to shaft voltage have been theoretically established to restrict the deterioration of bearings[6.23].

Besides the detailed investigations, as reported in different papers, the diagnosis and cause analysis of failure of rolling-element bearings used in electrical power equipments due to various unforeseen causes leading to current leakage have been discussed[6.24]. Studies have been conducted, to highlight the experimental techniques and diagnosis of the data to assess the causes of various failures of bearings[6.25].

6.2.6 *Localized Electrical Current*

Cause of generation of localized current in presence of shaft voltage has been established. A theoretical model has also been developed to determine the value of localized current density depending on dimensional parameters, shaft voltage, contact resistance, frequency of rotation of shaft and rolling-elements of a bearing. The time for appearance of flutes on the track surfaces can be estimated by bearing kinematics, existing potential difference between track surface of inner race and rolling-elements, value of localized current, properties of bearing material together with measured values of pitch of corrugations on the track surfaces[6.26].

6.2.7 *Bearing Stiffness*

An alternative approach and model has been developed to determine the effective stiffness of ball and roller bearings through evaluation of electrical parameters under different operating conditions. These parameters have been theoretically evaluated using width of deformation on the track surface of races, minimum film thickness, characteristics of the lubricants and bearing geometry. Besides this, with the application of another alternative approach, the stiffness of ball and roller bearings have been assessed approximately. This alternative approach is based on the inverse of the change of width of bearing deformation on the track surfaces of races and applied load.

The variation of stiffness has been determined with the applied loads under different speeds of operation using electrical analogy. The values of effective stiffness determined by these alternative approaches have been compared with the values determined by the existing conventional procedure, and found to have matching trend[6.27, 6.28].

6.3 Lubricants

The recent investigations indicate that the resistivity of a lubricant depends on antiscuff properties, viscosity, torque characteristics and consistency. Resistivity is a function of applied voltage and duration of exposure of voltage on the lubricants. The difference in resisivities among lubricants can be as high as 10^5 times. The change in resistivity depends on nature of impurities or by-products, and the type of additives present in the lubricant besides its density, compressibility and structure. Detailed investigations have been carried out on the behaviour of the resistivity of generally used industrial lubricants[6.29, 6.30]. It has been brought out experimentally that the low resistivity lubricant tends to "recoup" its resistivity when the applied voltage is switched–off. The percentage of recouping from the original value varies and depends on the stretching of the molecules[6.31]. It has also been determined that the low resistivity lithium base grease decomposes and lithium metal concentration in the aqueous solution increases relatively. The decomposition of carboxylic acid leads to corrosion of the track surfaces before pitting process is initiated. However, the applied voltage does not affect the oil content of the grease, but the carboxylate anion stretching and carboxylic group, present in the soap residue of the grease, undergo changes[6.32]. For further investigations using X-ray diffraction techniques indicated that the original structure of the soap residue of the fresh grease, i.e. lithium stearate changed to lithium palmitate, and lithium iron oxide and lithium zinc silicate were formed[6.33]. On the contrary such changes were not detected under rolling friction without the effect of electric current[6.34]. It has been further reported that under the influence of electric current, the formation of lithium hydroxide and lithium carbonate make the dielectric alkaline and corrode the bearing surfaces which finally lead to increased wear and failure of bearing[6.35]. Furthermore, authors have also reported the rheological physical studies of lubricating greases before and after use in the bearings[6.36].

6.4 Hydrodynamic Journal Bearings

In a hydrodynamic journal bearing, the zone of minimum film thickness i.e., load carrying oil film, varies along the circumference of a bearing through its length. It has been reported that this forms a capacitor of varying capacitance between the journal and the bearing depending on permittivity of the used lubricant, circumferential length of load carrying oil film, bearing length, eccentricity ratio, and the clearance ratio. Besides this, load carrying oil film offers resistance depending on operating parameters and resistivity of the lubricant. Thus, the load carrying oil film forms a resistor-capacitor (RC) circuit and offers an impedance to a current flow. By analyzing the RC circuit, behaviour of a journal bearing has been predicted. Theoretically the mathematical model has been developed to determine capacitance, active resistance, capacitive reactance and impedance under different parameters of operation to analyze bearing performance and safe load carrying capacity[6.37]. Also, reduction in bearing life under the influence of different levels of shaft voltages before the initiation of craters on the bearing surfaces has been developed, and has been compared to that of the bearings operating without the influence of shaft voltages. The mechanism of formation of craters on the bearing surfaces has been physically explained. Furthermore, it has been established that for safe reliable operation and adequate life of a bearing, a shaft voltage of 0.5 V must not be exceeded[6.38].

A study reported the capacitive effect and life estimation of a bearing on repeated starts and stops of a machine operating under the influence of shaft voltage gives in depth evaluation of bearing performance[6.39]. Increase in charge accumulation on the bearing liner with time is established theoretically when the machine is started, and the gradual leakage of the accumulated charges from the liner as the shaft voltage falls when the power supply to the machine is switched off. Under these conditions, the variation of shaft revolutions to accumulate charges and to discharge of the accumulated charges from the liner surface at various levels of bearing-to-shaft voltages has been analyzed and mathematically formulated. The variation of the safe limits of starts and stops as against the ratio of bearing-to- shaft voltage has also been reported. From the analysis, it is established that with the increase in ratio of bearing–to-shaft voltage, the ratio of shaft rotations to accumulate and discharge of the accumulated charges increases but the number of starts and stops to initiate craters on the liner surface decreases[6.39]. Besides this, a study showed the inductive effect of bearing under start/stop regime of a machine. The ratio pertaining to time required for the growth of bearing current to that for its decay under this regime at various ratios of transient to steady state value of bearing current has been analyzed, and the effect of self inductance on the bearing surfaces studied[6.40].

The development of theoretical model for assessing the minimum number of cycles for appearance of craters of various sizes on the liner surface under different levels of shaft voltages uses softening temperature of liner at the "high" points or prows of its surface[6.41]. Reported analysis has been used for the diagnosis of premature failure of the hydrodynamic journal bearings of a synchronous condenser failed within a few hours of operation. The number of cycles, thus established theoretically, matched with that of the operating cycles of the bearings failed prematurely[6.42, 6.43].

The similar approach identical to that of a cylindrical bearing for determination of electrical parameters has been extended for two-lobe, three–lobe and four–lobe bearings. Capacitance, resistance, capacitive reactance and impedance have been evaluated for each lobe separately under different operating parameters and equivalent values have been determined for two and three lobe bearings[6.44, 6.45]. Using the electrical parameters, an electrical analogy approach has been developed for evaluation of effective dynamic coefficients of two-lobe, three–lobe and four-lobe journal bearings[6.46, 6.47, 6.48]. Also, a developed approach has been reported for evaluation and comparison of effective dynamic coefficients of cylindrical and elliptical bearings[6.49]. Besides this, non-conventional electrical analogy approach has been further extended for determination of stiffness and damping coefficients in xx,yy,xy,yx directions and their comparison with the published conventional approach[6.50-6.53].

6.5 Hydrodynamic Thrust Bearings

Similar to journal bearings, in a thrust bearing, the variation in oil film between pads and the thrust collar forms a capacitor of varying capacitance from leading to trailing edge. This depends on permittivity of the lubricant, pad width, angle of tilt, and the ratio of oil film thickness at the leading to trailing edge. Besides this, variable oil film thickness offers variation in resistance along the pad profile depending on resistivity of the lubricant. A theoretical approach has been developed to determine capacitance, capacitive reactance, and other electrical parameters of a thrust bearing. Variation of capacitance and other parameters with angle of tilt have been mathematically formulated and analyzed[6.54]. The safe limit of shaft voltage has been assessed for a reliable operation of the thrust bearings operating under the influence of shaft voltages. For this analysis, charge leakage between high "points" of the thrust collar and a pad liner during momentary contact in the zone of load-carrying oil film is used to establish the heat generated and instantaneous temperature rise in each shaft rotation. The contact stresses by instantaneous temperature rise leading to craters formation have been analyzed, which consequently determined the reduction in bearing life. It has been established that at the shaft voltage of 2V, the percentage reduction in bearing life is 15.6 times as much as that at the shaft voltage of 0.25 V[6.55]. Minimum cycles before the formation of craters due to leakage of charge energy on the liner surface of tilting pads of a thrust bearing have also been analyzed[6.56, 6.57]. A developed theoretical model indicates the capacitive effect and life estimation of the pivoted pad thrust bearings on repeated starts and stops of a machine operating under the influence of shaft voltages. The analysis gives the time required for the charge accumulation and increase of charge with time on the surface of liners, and gradual leakage of the accumulated charges with time as the shaft voltage falls as soon as the power supply to the machine is switched off. This leads to determine the number of repeated starts and stops before initiation of craters on the liner surface of pads of a thrust bearing[6.58].

6.6 Conclusions

Based on the reviews given in this Chapter, the following conclusions are drawn[6.59]:

In short, the research in the area of Tribology in the Electrical Environment is in very initial stage of development concerning analysis of bearings, diagnosis and frictional processes. This work is a first step to compile the available literature on this field for the benefit mechanical engineers. As the need to conserve both energy and raw material are becoming very significant, the understanding of basic tribological electromagnetic phenomenon concerning friction/wear process and bearing behaviour/performance need to be further explored and accelerated in the engineering and scientific organizations.

References

[6.1] Anderson, S., "Passage of Electrical Current Through Rolling Bearings", The Ball Bearing Journal, No.153, pp 6-12 (1968).

[6.2] Kaufman, H. N., and Boyd, J., "The Conduction of Current in Bearings", ASLE Trans., 10, pp 226-234 (1959).

[6.3] Bradford, M., "Prediction of Bearing Wear due to Shaft Voltage in Electrical Machines", ERA Technology Limited (1984).

[6.4] Chu, P.S.Y. and Cameron, A., "Flow of Electrical Current Through Lubricated Contacts", ASLE Trans., 10, pp 226-234 (1967).

[6.5] Warnock, F.V. and Benham, P.P., "Mechanism of Solids and Strength of Materials", Sir Isaac Pitman and Sons, London (1996).

[6.6] Prashad, H., "Investigations of Corrugated Pattern on the Surfaces of Roller Bearings Operated Under the Influence of Electrical FIelds", Lubrication Engineering, Vol.44, Issue 8, pp 710-718, August 1988.

[6.7] Prashad, H., "Theoretical and Experimental Investigations on the Pitch and Width of Corrugations on the Surfaces of Ball Bearings", Journal of Wear, Vol.143, pp 1-14 (1991).

[6.8] Prashad, H., "The Effects of Current Leakage on Electro-adhesion Forces in Rolling Friction and Magnetic Flux Density Distribution on the Surfaces of Rolling Element Bearings", Transactions of the ASME, Journal of Tribology, Vol.110, pp 448-455, July 1988.

[6.9] Prashad, H., "Determination of Magnetic Flux Density on the Surfaces of Rolling-Element Bearings - An Investigation" BHEL journal, Vol.21, No.2, August 2000, pp49-67.

[6.10] Prashad, H., "Magnetic Flux Density Distribution on the Track Surface of Rolling-Element Bearings"—An Experimental and Theoretical Investigations", Tribology Transactions, Vol.39, Issue 2, pp 386-391, April 1996.

[6.11] Prashad, H., "Determination of Magnetic Flux Density on the Surfaces of Rolling-Element Bearings as an Indication of the Current that has Passed through Them", Tribology International 32, (1999),pp 455-467.

[6.12] Prashad, H., "Diagnosis of Failure of Rolling-Element Bearings of Alternators—A Study", Wear, Vol.198, pp 46-51, October 1996.

[6.13] Prashad, H., "Effect of Operating Parameters on the Threshold Voltages and Impedance Response of Non-Insulated Rolling Element Bearings under the Action of Electrical Currents", Wear, 117, pp 223-240 (1987).

[6.14] Prashad, H., "Theoretical Evaluation of Impedance, Capacitance and Charge Accumulation on Roller Bearings Operated under Electrical Fields", Wear, 125, pp 223-239 (1988).

[6.15] Prashad, H., "Theoretical Determination of Impedance, Resistance, Capacitive Reactance and Capacitance of Ball Bearings", BHEL Journal, Vol.14, Issue 2, pp 40-48 (1993).

[6.16] Prashad, H., "Analysis of the Effects of Electrical Current on Contact Temperature, Contact Stresses and Slip Bands Initiation on Roller Tracks of Roller Bearings", Wear, 131, pp 1 – 14 (1989).

[6.17] Prashad, H., "Analysis of the Effects of Electrical Current on Contact Temperature, Residual Stresses leading to slip Bands and initiation and formation of Corrugation Pattern on Ball Tracks of Ball Bearings", BHEL Journal Volume 11, No.1, pp 39-47, (1990).

[6.18] Prashad, H., "Determination of Time Span for Appearance of Flutes on Track Surface of Rolling-Element Bearings under the Influence of Electric Current", Presented in World Tribology Congress, London, 8-12 September, 1997 and published in Tribology Transactions, Vol. 41, issue 1, pp 103 – 109 (1998).

[6.19] Prashad, H., "Appearance of Craters on Track Surface of Rolling-Element Bearings by Spark Erosion", to appear in Tribology International, Vol.34, issue 1, pp 39-47 (2001).

[6.20] Chiou, Y.C., Lee, R.T., Lin, C.M.," Formation Criterion and Mechanism of Electrical Pitting on Lubricated Surface under A.C. Electrical Field", Wear, 235, pp 62-72 (1999).

[6.21] Prashad, H., Venugopal, K., "Formation of Craters on Track Surfaces of Rolling-Element Bearings due to Spark Erosion", BHEL Journal, Vol. 23, No. 1, pp 34-37, Feb 2002.

[6.22] Prashad, H., "Theoretical Analysis of the Effects of Instantaneous Charge Leakage on Roller Tracks of Roller Bearings Lubricated with High Resistivity Lubricants", ASME Transaction, Journal of Tribology, Vol.112, pp 37-43, January 1990.

[6.23] Prashad, H., "Theoretical Analysis of the Capacitive Effect of Roller Bearings on Repeated starts and stops of a machine operating under the Influence of Shaft voltages" Transactions of ASME, Journal of Tribology, Vol.114, pp 218-222, October (1992).

[6.24] Prashad, H., "Diagnosis and Cause Analysis of Rolling-Element Bearings Failure in Electrical Power Equipments Due to Current Leakage", Lubrication Engineering, STLE May (1999), pp 30-35 Also, presented in ASME/STLE Tribology Conference in Toronto, Ontario, Canada, October 26-28, 1998.

[6.25] Prashad, H.," Investigations and Diagnosis of Failure of Rolling-Element Bearings due to Unforeseen Causes – A Case Study", BHEL Journal, Vol. 20, No.1, March 1999,pp 59-67.

[6.26] Prashad, H., "Diagnosis of Rolling-Element Bearings by localized Electrical Current between Track Surfaces of Races and Rolling-Elements", ASME Journal of Tribology, Vol. 124, pp 468-473, July 2002.

[6.27] Prashad, H., "Alternative Approach to Determination of Stiffness of Ball Bearings", BHEL Journal, issue 24, No. 3, pp 17-22, December 2003.

[6.28] Prashad, H., "Determination of Stiffness of Roller Bearings – An Alternative Approach", IE (I) Journal- Mc, Vol. 84, pp 186-192, January 2004.

[6.29] Prashad, H., "Experimental Study on Influence of Electrical Fields on Behaviour of Greases in Statically Bounded Conditions and When used in Non-Insulated Bearings", BHEL Journal, Vol.7, No.3, pp 18-34 (1986)

[6.30] Prashad, H., Murthy, T.S.R., "Behaviour of Greases in Statically Bounded Conditions and When Used in Non-insulated Anti-friction Bearings under the Influence of Electrical Fields", Lubrication Engineering, Vol.44, No.3, pp 239-246 (1988).

[6.31] Prashad, H., "Variation and Recouping of Resistivity of Industrial Greases—An Experimental Investigation", Proceedings of International Symposium on Fuels and Lubricants, December 8-10, 1997, New Delhi.

[6.32] Prashad, H., "Investigations of Damaged Rolling-Element Bearings and Deterioration of Lubricants under the Influence of Electric Fields", Wear, 176, pp 151-161 (1994).

[6.33] Prashad, H., "Diagnosis of Deterioration of Lithium Greases used in Rolling-Element Bearings by X-Ray Diffractrometry", STLE Transactions, Volume 32, 2, pp 205-214, (1989).

[6.34] Prashad, H., Murthy, T.S.R., "The Deterioration of Lithium Greases Under the Influence of Electric Current - An Investigation", Presented at 10th Symposium on Lubricants, Additives, Waxes and Petroleum Speciality Products (LAWPSP), held on January 30,31 and February 1, 1997, at IIT, Mumbai. Published in Journal of Lubrication Science, France, 10-4, pp 323-342, August 1998.

[6.35] Prashad, H.,"Variation and Recovery of Resistivity of Industrial Greases - An Experimental Investigation ", Journal of Lubrication Science, France, 11- 1, pp 73-103, Nov. 1998.

[6.36] Mas Remy, Magnin Albert, "Rheological and Physical Studies of Lubricating Greases before and after Use in Bearings", Transactions of ASME, Journal of Tribology, Vol. 118, pp 681-686 (1996).

[6.37] Prashad, H., “Theoretical Evaluation of Capacitance, Capacitive Reactance and their Effects on Performance of Hydrodynamic Journal Bearings”, ASME Transaction, Journal of Tribology, Vol. 113,pp 762-767, October 1991.

[6.38] Prashad, H., “Theoretical Evaluation of Reduction in Life of Hydrodynamical Journal Bearings Operating Under the Influence of Different Levels of Shaft Voltages”, STLE Transactions, Vol. 34,4, pp 623-627 (1991).

[6.39] Prashad, H., Rao, K.N., “Analysis of Capacitive Effect and Life Estimation of Hydrodynamic Journal Bearings on Repeated Start and Stop of Machine Operating under the Influence of Shaft Voltages”, Tribology Transaction, Vol. 37, Issue 3, pp 641-645 (1994).

[6.40] Prashad, H., “Analysis of Inductive Effect of Bearings under the Influence of Shaft Voltages”, BHEL Journal, Vol.15, Issue I (1994).

[6.41] Prashad, H., “A Theoretical Model to Determine the Minimum Number of Shaft Revolutions/Cycles for Appearance of Craters of Various Sizes on the Liner Surface of a Hydrodynamic Journal Bearing Operating under the Influence of Shaft Voltages of Different Levels”, BHEL Journal, Vol.16, No.2, pp 27-33, Dec. 1995.

[6.42] Prashad, H., “Diagnosis of Bearing Problem of Synchronous Condenser—An Experimental and Theoretical Investigation”, Wear, 188, pp 97-101(1995).

[6.43] Prashad, H., “A Study of Electrical Pitting of Journal Bearing with Water Contaminated Lubricant”, Tribo-test Journal, Leaf Coppin (France), pp 115-124, 7-2 (December 2000).

[6.44] Prashad, H., “Theoretical Evaluation of Electrical Parameters of Two-Lobe Journal Bearings”, IE (I) Journal - Mc, Vol.77, pp 47-51, May 1996

[6.45] Prashad, H., “Assessment of Electrical Parameters of Three Lobe Journal Bearings— An Approach”, Mechanical Engineering Journal of Institution of Engineers (I), Volume 78, August 1997, pp 53-56.

[6.46] Prashad, H., “Evaluation of Dynamic Coefficients of a Two-Lobe Journal Bearing Using an Electrical Analogy Approach”, ASME Journal of Tribology, Vol.118, pp 657-662, January 1996.

[6.47] Prashad, H., “Assessment of Dynamic Coefficients of Three-Lobe Journal Bearings through Evaluation of Electrical Parameters—A New Approach”, BHEL Journal, Vol.17, pp 40-48 (1997).

[6.48] Prashad,H.,” Determination of Capacitance, Resistance and Dynamic Coefficients of Four-Lobe Journal Bearings through Electrical-Analogy”, IE(I) Journal, Vol. 81, May 2000, pp 30-36.

[6.49] Prashad, H., "Determination of Effective Stiffness of Cylindrical and Two-Lobe Journal Bearings by an Electrical Analogy and its Comparison with the Conventional Approach", IC-HBRSD 97, Second International Conference on Hydrodynamic Bearing Rotor System Dynamics, held on March 20-22, 1997 in Xi'an, P.R. China.

[6.50] Prashad, H., "Evaluation of Stiffness Coefficients of Cylindrical Journal Bearings by Electrical Analogy – A Non – Conventional Approach", IE(I) Journal, Vol. 81, Sept. 2000, pp 55-61.

[6.51] Prashad, H.," Evaluation of Damping Coefficients of Cylindrical Journal Bearings by Electrical Analogy – A Non – Conventional Approach", BHEL Journal, Vol. 22, issue 1, pp 51-60 (2001).

[6.52] Prashad, H., "Evaluation of Stiffness Coefficients of Cylindrical Journal Bearings by a Non-Conventional Approach", BHEL Journal, Vol. 24, No. 1, pp 34-45, January 2003.

[6.53] Prashad, H., "Evaluation of Damping Coefficients of Cylindrical Journal Bearings by Electrical Analogy - A Non-Conventional Approach", IE (I), Vol. 83, pp 72-77, July 2002.

[6.54] Prashad, H., "An Approach to Evaluate Capacitance, Capacitive Reactance and Resistance of Tilted Pads of a Thrust Bearing" STLE, Tribology Transactions, Vol.35, 3, pp 435-440 (1992).

[6.55] Prashad, H., "Analysis of the Effects of Shaft Voltages on Life Span of Pivoted Pad Thrust Bearings", BHEL Journal Vol.13, Issue 2, pp 1-12 (1992).

[6.56] Prashad, H., "Theoretical Model to Analyze the Minimum Cycles before Formation of Craters due to Leakage of Charge Energy on the liner surfaces of Tilting-Pads of a Thrust Bearing", Institution of Engineers (India), Vol.75, pp 82-86, August 1994.

[6.57] Prashad, H., "An Approach to Determine Minimum Cycles Before Formation of Craters Due to Leakage of Charge Energy on the Liner Surface of Tilting-Pads of a Thrust Bearing" NCIT-93, March 24 to 26, IIP, Dehradun.

[6.58] Prashad, H., "Analysis of Pivoted Pad Thrust Bearings on Repeated Start and Stop of a Machine Operating under the Influence of Shaft Voltages", Accepted for presentation in World Tribology Congress, London, 8-12 September, 1997, published in Journal of Institution of Mechanical Engineering IE (I), Vol. 79, pp42 - 47 (1998).

[6.59] Prashad, H., "Tribology in Electrical Environments", Lubrication Science, 13-4, 2001, pp 359-369.

7
Causes of Shaft Voltages in Rotating Machines

7.1 Introduction

Shaft voltages and their origin in rotating machines is a very complex phenomenon. It can occur due to variety of reasons like winding faults, inbalanced supplies, electro-static effects, air-gap fields, magnetized shaft or other machine components, dissymmetry of magnetic fields, etc. These reasons in electrical rotating machines result in a net flux or a current linking with the circuit consisting of shaft, bearing and frame. Shaft flux usually results in a localized current at each bearing rather than a potential difference between shaft ends. A current path along the shaft, bearings and frame results in a potential difference between shaft ends.

Certain important causes for the origin of shaft voltage in rotating machines are discussed in this chapter and highlighted as under:

7.2 Causes for Origin of Shaft Voltages

7.2.1 *Dis-Symmetry Effect*

Even though the potential is not directly applied to a shaft, bearing current can still arise through induced effects. This occurs due to dissymmetry of the magnetic circuits developed in the electrical machines, which closes over the yoke and induces the voltage on the shaft. The shaft voltage induced because of dissymmetry of the magnetic circuit is influenced by the rotor and stator arrangement of the machine. The principle, by which current is generated, is shown in Fig.7.1. The Figure shows a two-pole machine with north and south poles as indicated with stator comprising of two semi-circular sections. If the construction of the machine is perfect, the reluctance of the magnetic circuit will be uniform around the periphery of the stator and hence there will not be induced shaft voltage and the current through the bearings. If, however, there is some dis-symmetry in the magnetic circuit, then a shaft voltage may be induced. Various sources of dissymmetry may crop up in the construction of the machine, namely:

Difference in the spaces between the ends of laminations, non-symmetrical distribution of slots, etc as shown in Fig.7.1. Voltage is induced in the shaft because the value of each of the two fluxes Φ_1 and Φ_2 (Fig.7.1) varies during each revolution of the shaft. The magnitude of the flux Φ_1 or Φ_2 will decrease or increase depending on the position of the shaft, since, the flux flow is either through the small gap or the large gap depending on the location of the shaft. As the value of Φ_1 increases or decreases, it cuts the conducting material, causing an alternating emf to be set up between M and N (Fig.7.1). The emf generated by the change in the flux Φ_2 reinforces that generated by the change in the flux Φ_1. This generates an alternating potential between the shaft ends.

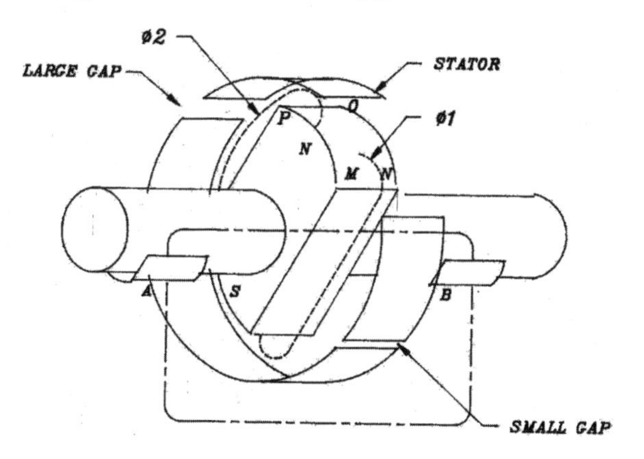

Fig. 7.1 Dissymmetry effect causing shaft voltage

7.2.2 *Shaft Magnetization Effect*

This is another source of bearing current in the electrical machines. This occurs due to any imbalanced ampere-turns, which surrounds the shaft. Such turns causes the shaft to become magnetized and set up a flux which flows from one bearing to another through the machine frame as shown in Fig.7.2. The portion of the shaft, which confines each bearing, cuts the flux as the shaft rotates and generates the potential difference on that portion of the shaft. This tends to produce bearing current, which flows from one bearing to another through shaft and the frame.

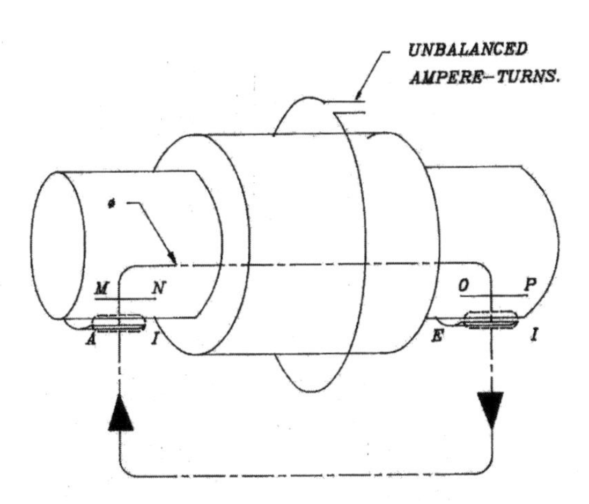

Fig. 7.2 Shaft magnetization effect

In addition, accidental grounding of a part of the rotor winding to the rotor core can lead to stray currents through the shaft and bearings. This may lead to the permanent magnetization of the shaft.

If the machine has a significant level of permanent magnetization, shaft voltage and current could be generated when the machine is rotated. The magnitude of these generated voltages and currents will depend on factors such as the strength of the residual magnetization, the relationship between various residual fields, and the particular component that has been magnetized. It also depends on the magnitude of the air gap between rotor and stator, the current path, and the variation in insulating properties of oil films in seals and bearings during operation.

On many occasions where bearing damage occurs, it is found that the shaft of the machine is magnetized. In such cases magnetic flux flows in the shaft from one bearing to another bearing as shown in the Fig.7.2. Hence, a magnetic circuit exists which the shaft, bearing and casing, forms. A local bearing current can flow due to this shaft voltage through the oil films. The voltage over the bearing may be very low but very high local bearing currents can flow if the resistance of oil film is very low.

Furthermore homo-polar flux can result from an air gap or rotor eccentricities as shown in Fig.7.3. The flux crosses the air gap and leads to local bearing currents (Fig.7.3). The homo-polar flux crossing the air gaps will generate an additional voltage causing the current to flow along the shaft, bearings, bedplate and the casing. At the inner edge of the bearings, the bearing currents due to magnetized shaft and the shaft current combine. Hence, this part of the bearing and the inner oil ring will be subjected to more damage than the outer section of the bearings.

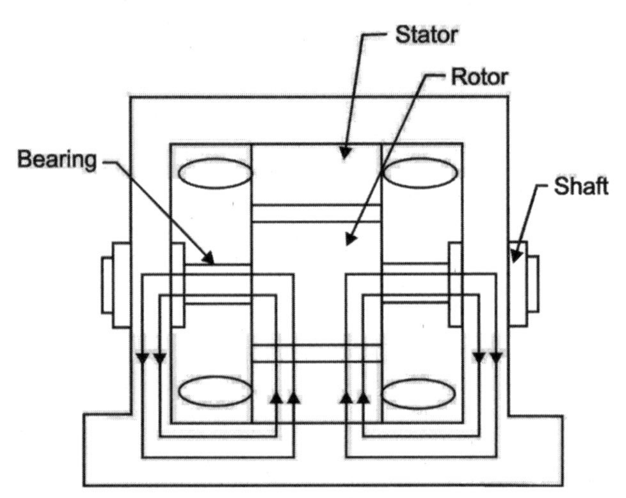

Fig. 7.3 Homo-polar fluxes between stator, rotor and shaft.

Another most important cause of the bearing currents is the linkage of an alternating flux with the shaft. The flux flows perpendicularly to the axis of the shaft and pulsates in the stator and rotor cores. It is caused by dis-symmetries in the magnetic circuits of the machines such as:

(i) Uneven air gaps and rotor eccentricity;
(ii) Split stator and rotor cores;
(iii) Segmental punchings;
(iv) Axial holes through the core for ventilation or clamping purposes;
(v) Keyways for mounting the core stakings;
(vi) Segments of different permeability.

The flux from each pole crosses the air gap and, if the magnetic path is symmetrical, will divide equally, half clockwise and half anti-clockwise as shown in Fig.7.4. However, if there is a difference in reluctance of the core in one direction compared with the other direction, there will not be an equal division of the flux, and there will exist a net flux linking with the shaft. This flux is alternating and will produce voltage in the circuit formed by the shaft, bearing and frame.

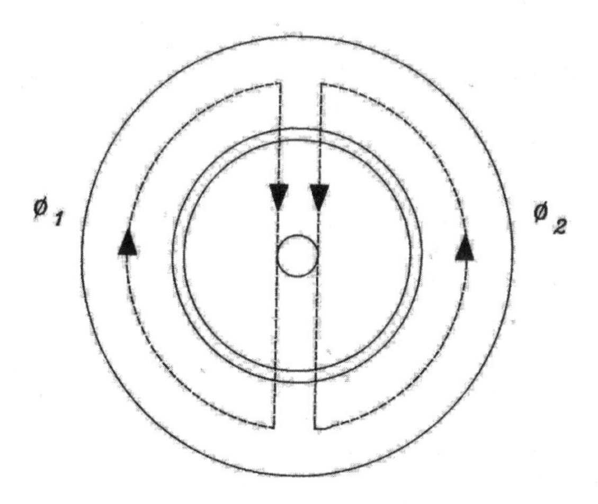

Fig. 7.4 Ring flux lining the shaft (if $\phi_1 \neq \phi_2$)

7.2.3 *Electrostatic Effect*

Due to electrostatic effect, charges build up on the shaft when the resistivity of the lubricant used in the bearings is such that no current flows due to low voltage. The minimum voltage, exceeding which the momentary flow of current is established, is called threshold voltage. When the potential difference across the film thickness exceeds the threshold voltage, the accumulated charge is dissipated through the bearings causing flow of current. Some of the important sources of such charges are discussed as below.

7.2.3.1 *Potential Developed by Impinging Particles*

Each particle of a finely divided material has equal number of positive and negative charges and this makes the material electrically neutral under equilibrium conditions. When such a stream of neutral particles strikes an object under certain conditions, the positive and negative charges of the particles, on impact make the object also charged in the same fashion.

7.2.3.2 *Potential Developed Due to Charged Particles*

Most of the lubricants are relatively poor conductors. The individual molecules of the lubricant act like the particles of a finely divided material. The lubricant while passing through the certain fine passages of the filter can charge the lubricant molecules. The charges on the molecules are likely to be retained because; in general the lubricants are non-conducting. The charged particles of the lubricant can transmit the charges to the surfaces of the journal, and also may charge the bearing surface, provided, the latter is insulated. As soon as this charge becomes high or exceeds the threshold voltage, the accumulated charges will be dissipated through the oil film.

7.2.3.3 *Potential Developed by Belts*

Belts are generally poor conductor of electricity. Charges originating at the point of contact of the belt and pulley are equal. However, if the condition of both the pulleys are different, then charges may build up on the shaft, provided, the oil film insulates the shaft and the charges may get dissipated through the bearings. This is shown diagrammatically in Fig.7.5.

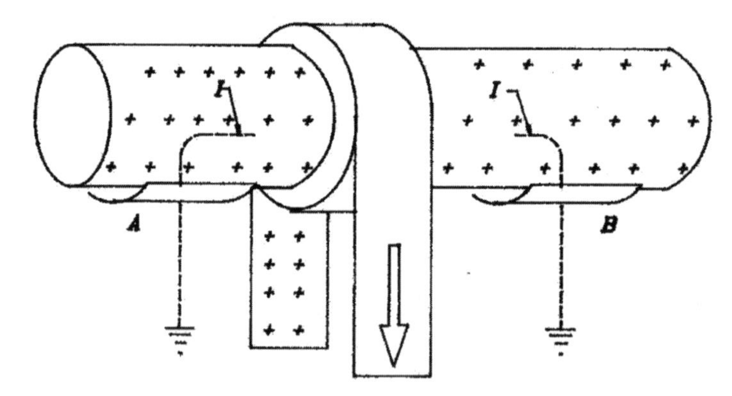

Fig. 7.5 Electrostatic effect-potential developed by charged belt.

7.2.4 *Shaft-Voltage Caused by Permanent Magnetization of Casing, Shaft or Pedestals*

In practice, the shaft, bedplate, pedestals, casings and other components are made of material, which possess good magnetic properties. In general, permanent magnetization is developed on shaft, casing, etc. due to heavy short circuit (like welding of components without proper ear thing) or any other abnormal operation of the generator. Under certain conditions, bedplate or other such components can produce an axial residual flux in shaft and this can produce self-excitation resulting in very large shaft and bearing currents. Such phenomenon leads to damage of shaft, bearings and other components. Failure of oil pumps and shafts of control-device in generating units have been reported due to self-excitation phenomenon.

Shaft voltage, as such, does not have much relevance with the electrical equipment. It is generated by residual magnetic fields in the stator and other components of a machine. The magnetic fields are the results of increased use of magnetic devices such as magnets used for lifting machines, magnetic chucks on machines, magnetic particle inspection, etc. and especially arc welding.

Every time electric arc welding is done around a machine, current travels through the casing and rotor creating residual magnetic field on these components. When a machine with significant residual magnetism rotates, electric voltage and current are generated in the same manner as in an electric generator or in an eddy current brake. The strength of these voltages and currents depends on the strength of the residual magnetic field. The path of the current flow and the voltage depends on the insulating properties of the oil films in bearings and seals, rotor speed and various other factors.

7.3 Factors Affecting Shaft Voltage

Important factors affecting the magnitude and frequency components of the shaft voltage generated due to magnetic dissymmetry depend on machine size, stator core construction, rotor alignment, number of poles, generator design and load conditions.

Larger units, in general, have greater magnitude of shaft voltage. This can be attributed to the greater magnitudes of air-gap flux, greater constructional dis-symmetries, and saturation of core. Besides this larger unit introduces non-uniformity in the air-gap length resulting in larger magnitudes of the shaft-voltage. The excessive eccentricity distorts the magnetic field considerably and enhances the effect of shaft voltage.

7.4 Reasons for Current Problems

The significant reasons for the epidemic shaft current problems are:

7.4.1 Magnetic tools and magnetic particle inspection are widely used with portable units. Rotor and stator components must be demagnetized after repair, manufacturing and inspection.

7.4.2 Electrical arc welding around the machine or use of "arc torch" to burn off foreign matter on a base plate. This leaves strong permanent residual magnetic fields.

7.4.3 A new unit, in most of the cases, has the less degree of residual magnetism. But it increases continually and steadily with time due to repairs, welding and use of magnetic tools. Finally, residual magnetism will reach a level where the fields are strong enough to reorient themselves in such a manner as to allow self-excitation of an area. However, there are no chances for the billions of minute magnets to align themselves. But due to shock, surge, oil whirl, vibrations, rubbing etc., these small magnets may align themselves in such a manner and increase field strength considerably.

This will generate high voltage as the rotating field reacts with the stationery fields. This may allow the current passage through bearings depending on the resistance of oil film thickness.

7.5 Passage of Current Through Bearings

The main effect of bearing current is the damage caused by arcing across the bearing surfaces. As electric current passes through the surface in contact, the flow of current concentrates through the asperity contact points, and the local current density increases to a very high value. The flow of current through the asperity contacts generates heat at the contact area. Based on the resistance of the oil film and the resistivity of the lubricant used, either slow passage of current or arcing takes place at the contact zone. The main consequence of current passage and the arcing leads to the wear of bearing surfaces due to the removal of fused metals in the arc, resulting in closely pitched marks and burned craters. Due to this roughening of the surface and mechanical wear is accelerated. The arcing, formation of craters, the mechanical roughing of the active surfaces of bearings, and the liberated metallic particles cause the oxidation of the lubricant and loss of purity of lubricant. This deterioration of the lubricant gradually increases the bearing destruction.

7.6 Bearing Electrical Parameters

Bearing current mainly depends on the magnitude of the shaft voltage and the bearing impedance. The magnitude of the shaft voltage is usually estimated as a function of the ring flux. The ring flux results from dissymmetry whose magnitude cannot be readily measured or estimated. Bearing impedance, resistance, and capacitance depends on many factors such as lubricant resistivity, lubricant permittivity/dielectric constant, bearing temperature, load and speed of operation. In general, bearing impedance reduces with increase in temperature and increases substantially with increase in speed. Different types of bearings are found to have different characteristics. The impedance of a ball bearing is different than a roller bearing. Hydrodynamic journal and thrust bearings used in large machines have electrical parameters and characteristics, which are different compared to rolling-element bearings. Knowledge and literature on the values of a bearing impedance, resistance, capacitance, capacitive reactance is very scarce and a much better understanding of the bearing electrical parameters is needed to fulfill the know how and know why gap.

In the absence of bearing insulation, a shaft voltage appears across the bearing oil film. The oil film thickness in a rolling-element bearing ranges from 0.001 mm to 0.02 mm; the area of contact depends on local surface roughness. The oil film will cease to be an insulator when the shaft voltage reaches the threshold value and flow of current through bearing occurs.

In addition presence of dirt, metallic particles and irregular oil film thickness makes the impedance of bearing circuit so low that small shaft voltages may cause substantial bearing currents. If these currents are suddenly reduced by the re-establishment of the non-conducting film in the bearing, the self-inductance of the single loop of the shaft, bearing and casing may then cause a relatively high-induced voltage across the oil film. This induced voltage may breakdown the oil film depending on lubricant characteristics and other operating parameters. The understanding of this phenomenon's in different bearings is very scare and need to be developed conceptually, theoretically and experimentally.

7.7 Conclusions

To bridge the above know-how gap in the literature detailed analysis has been carried out so that various bearing and their performance, behavior etc. could be understood more precisely under the influence of electrical current and shaft voltages[7.1]. The book on "Tribology in Electrical Environments" is the first of its kind to probe into this hitherto unprobed area in many aspects and dimensions.

References

[7.1] Prashad, H. "Tribology in Electrical Environments", Elsevier; Tribology and Interface Engineering Series, No.49, (2006).

8

Magnetic Flux Density Distribution on Track Surfaces of Rolling-Element Bearings

8.1 General

This Chapter deals with the investigations carried out on the various rolling element bearings after being operated under the influence of electric fields, and pure rolling friction on the roller bearing test machine. The significant magnetic flux density was detected on surfaces of the bearings lubricated with low-resistivity lubricant under the influence of electrical fields. No such phenomenon was observed either on bearings using high or low-resistivity lubricant under pure rolling friction or on bearings lubricated with high-resistivity lubricant under the influence of electrical current. New bearing surfaces do not show significant magnetic flux density but it have been detected after long operation on different motor bearings, lubricated with low-resistivity greases. The electro-adhesion forces in the bearings using low-resistivity lubricant increase under the influence of electrical fields in contrast to those with high-resistivity lubricants. Under the pure rolling friction resistivity of lubricants do not affect the electro-adhesion forces. The investigations reported in this Chapter along with the study of damaged/corrugated surfaces, deterioration of the used lubricants (Chapters 3 and 11), and flux density distribution in the bearing surfaces, the leakage of current leading to failure of the non-insulated motor bearings can be established.

The residual magnetic flux density on the track surfaces of a rolling-element bearing produces forces of attraction, hysteresis loss and eddy current, and can lead to premature bearing failure. This Chapter brings out a theoretical model to determine the magnetic flux density developed on the inner and outer surfaces of inner race and outer race, and on the surface of rolling-elements, of a rolling bearing operating under the influence of electric current. The flux density, analytically determined, is found to agree well with the measured flux density developed on the surfaces of races and rolling-elements of the bearings tested in a test rig under the influence of electric current. Also, the magnetic flux density on the surfaces of damaged bearings of motors and alternators has been measured, and the theoretical model has been used to determine the amount of current flow through the damaged bearings. The value of current flow through the bearings, thus established, has been found to be close to that evaluated by the measurement of shaft voltage and bearing resistance.

Furthermore, analysis given in this Chapter besides having a potential to ascertain the cause of failure by passage (leakage) of current, can establish the amount of the flow of the leakage current through the bearings by determining the magnetic flux density on the surfaces of rolling-element bearings. Also, the current flow, thus established, along with the measurement of the shaft voltage, lead to establish the bearing impedance.

8.2 Introduction

Normally rolling-element bearings are not expected to carry electric current, yet, there are instances in which they are to carry current. The current may flow in the bearings owing to several reasons:

(i) A bearing may carry current as a necessary part of an electrical circuit;
(ii) current may be self-induced as a result of the design characteristics of the machine; and
(iii) current may result from an electrostatic phenomenon.

When the rolling-element bearings of the motors are damaged in service, the question often arises as to whether the damage is due to leakage electrical current. Although, considerable investigations and analysis have been carried out dealing with bearing current still the diagnosis of the cause of failed bearings – using lubricants having different characteristics – due to electrical current, has not been well understood. Hence, a study was undertaken to understand the effect of electric fields on the roller bearings, lubricated with greases of different resistivities, by diagnosis of magnetic flux density on the bearing surfaces. This study also examined the change in the electro-adhesion forces in the bearings operated in pure rolling friction without electrical fields and under the influence of electrical fields by analysis of the used lubricants.

Shaft voltages exist in electrical machines as a result of asymmetry of faults, winding faults, unbalanced supplies, electrostatic effects, air-gap fields, magnetized shaft or other machine members, asymmetries of the magnetic fields, etc. The latter are caused by rotor eccentricity, poor alignment, manufacturing tolerances, uneven gaps, segmental lamination punching, variation in permeability, and various other unforeseen reasons. In general, magnetic flux develops in electric machines due to asymmetry of the magnetic circuit, which closes in the circumference over the yoke and induces voltage on the shaft. This results in localized current at each bearing rather than a potential difference between shaft ends. A current path, however, along shaft, bearings and frame results in a potential difference between shaft ends as discussed in[8.1-8.3].

This phenomenon can be elaborated as follows. When asymmetry of the magnetic circuits exists due to various reasons, as explained above, including that of rotor and stator sagging, it produces variable magnetic flux. Since shaft continuously rotates even under these conditions (under the influence of magnetic flux in the magnetic field), this leads to induce voltage on the shaft.

Since variable magnetic asymmetry is localized because of localized uneven gaps in segmental punching and tolerance, the voltage generated is localized and variable. The localized variable shaft voltage induces voltage on the rolling -elements and bearing outer race by mutual induction, and localized loop of current between shaft/inner race and rolling-elements, and rolling-elements and outer race appears depending on localized bearing impedance. This is a complex phenomenon. But, in principle, it happens as explained. The combination of various defects including axial shaft flux due to residual magnetization, rotor eccentricity and asymmetrical rotor winding increase shaft voltage, and a current path, along shaft bearings and frame, results in a potential difference between shaft ends.

At a certain threshold voltage, depending on the resistivity of the lubricant and operating conditions, electrical breakdown occurs and current flows through the bearing[8.4]. Thus, circular current in the inner race leaks through the rolling-elements to the outer race by following a path of least resistance, and establishes the field strength leading to development of residual magnetic flux on the track surface of races, rolling-elements and inner and outer surface of races in due course[8.5].

The investigations given in this Chapter were undertaken to determine and compare experimentally and theoretically the developed magnetic flux density distribution on the track surface, inner and outer surface of races, and rolling-elements under the influence of electric current. The flow of current through the damaged bearings of motors and alternators have been established by the developed theoretical model, by measurement of residual flux density on the track surfaces, inner/outer surfaces of races and rolling elements. The level of current, thus determined, has been compared with the value of current obtained by the measurement of shaft voltage and bearing resistance.

This Chapter does not deal with hydrodynamic bearings used in turbo-generators. It deals with rolling-element bearings used in motors/alternators. Frequent failure (30% Of total failure) has been reported because of this phenomenon.

8.3 Theoretical Model and Approach for Determination of Field Strength

In a rolling-element bearing, current enters the inner surface of the inner race, through its bore in a distributed form, and flows around and outward until it concentrates at one or several rolling contacts. As the rolling-elements orbit, the current carrying contact travels. Currents then flow through each of the conducting rolling-elements, between two diametrically opposite contact areas. As the rolling-elements rotate with respect to the contact areas, these currents gradually sweep out 360 degrees around a major circle. In the outer race, the same situation as in the inner race occurs, but from the contact surface outward rather than inward. If there were only one conducting rolling-element at a time, then two arc currents would travel in the races towards its contact areas -- one clockwise and one counter clockwise. Depending on the resistance, these two currents may or may not be equal in magnitude. If more than one rolling-element conducts, then of course, the situation is more complex. In the theoretical model, current flow in one direction, both in inner and outer race, is assumed.

8.4 Field Strength on Track Surface of Races and Rolling Elements

8.4.1 *Field Strength on the Track Surface of Inner Race due to the Flow of Circular Current in Outer Race and Rolling-Elements*

Under the effect of shaft voltage, circular current flows through the inner race before it leaks through rolling-elements to outer race and establishes the field strength in the bearing[8.4 to 8.6]. The field strength on the track surface of inner race depends on the flow of circular current in the arc of the outer race before it leaks to the ground through the path of least resistance, and on the flow of current through the rolling-elements before it passes to the outer race. The field strength follows Fleming's left hand rule. Field lines surrounding an arc of race due to flow of current are shown in Fig. 8.1.

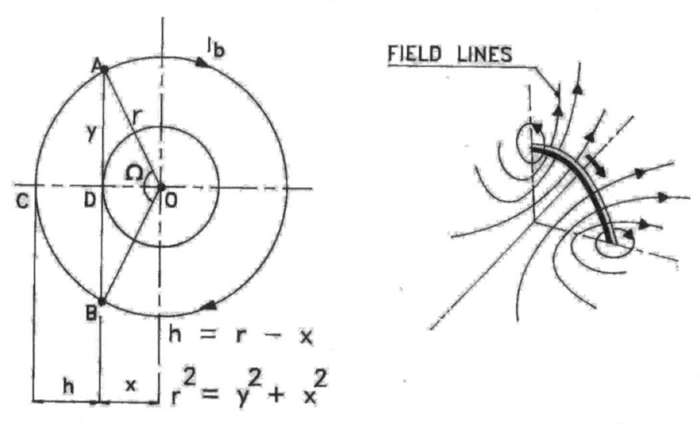

Fig. 8.1 Field strength at locations O and D, and field lines due to circular current I_b

8.4.1.1 *Due to Flow of Circular Current in Outer Race*

The developed field strength on the track surface of inner race, H_{iro}, due to the flow of circular current, I_b, in the outer race is determined by using R_{ir} for X and $(R_{or}^2 - R_{ir}^2)^{0.5}$ for Y in Equation (ix) of the Appendix, and as shown in Fig. 8.1, is given as:

$$H_{iro} = 2\,\pi I_b\,(R_{or}^2 - R_{ir}^2)/\,R_{or}^3 \qquad (8.1)$$

8.4.1.2 *Due to Flow of Circular Current in Rolling-Elements*

The circular current flows in a few rolling-elements before it leaks from the inner race to the outer race by following a path of least resistance. Since there are a number of rolling-elements in a bearing, the system of rolling-elements is treated as the ring of radius R equal to that of the bearing pitch radius. The field strength on the track surface of inner race due to the current flow in rolling-elements is determined similar to Equation (8.1) and is given as:

$$H_{irr} = 2\pi\, I_b\, (R^2 - R_{ir}^2)/R^3 \qquad (8.2)$$

8.4.1.3 *Equivalent Field Strength on Track Surface of Inner Race*

The equivalent field strength (H_{ir}) on the track surface of inner race due to the direction of the flow of circular current in the outer race and rolling-elements is the summation of H_{iro} and H_{irr}, since on rotation, rolling-elements change the polarity and outer race is stationery. So, the equivalent field strength will be either addition or difference of H_{iro} and H_{irr}, and is determined, using Equations (8.1) and (8.2), as :

$$H_{ir} = 2\pi I_b[R^2.\ R_{or}^2(R \pm R_{or}) - {}_{ir}^2(R^3 \pm R_{or}^3)]/R_{or}^3.R^3 \qquad (8.3)$$

8.4.2 *Field Strength on the Inner Surface of the Inner Race due to Flow of Circular Current in the Outer Race and Rolling-Elements*

8.4.2.1 *Due to Flow of Circular Current in Outer Race*

Similar to the Section 8.4.1, the developed field strength on the inner surface of inner race, H_{irio}, due to flow of circular current, I_b, in the outer race is determined by using R_{iri} for X and $(R_{or}^2 - R_{iri}^2)^{0.5}$ for Y in Equation (ix) of the Appendix, and as shown in Fig. 8.1, is given as :

$$H_{irio} = 2\pi\, I_b\, (R_{or}^2 - R_{iri}^2) / R_{or}^3 \qquad (8.4)$$

8.4.2.2 *Due to Flow of Circular Current in Rolling-Elements*

The field strength on the inner surface of inner race due to the current flow in rolling-elements, H_{irir}, is determined similar to Equation (8.2) by replacing R_{ir}^2 by R_{iri}^2, and is given as:

$$H_{irir} = 2\pi\ I_b\, (R^2 - R_{iri}^2) / R^3 \qquad (8.5)$$

8.4.2.3 *Equivalent Field Strength on Inner Surface of Inner Race*

The equivalent field strength, H_{iri}, on the inner surface of inner race is determined similar to Equation (8.3) by summation of H_{irio} and H_{irir} , and is given as:

$$H_{iri} = 2\pi\ I_b[R^2.\ R_{or}^2.(R \pm R_{or}) - R_{iri}^2(R^3 \pm R_{or}^3)]/R_{or}^3.R^3 \quad (8.6)$$

8.4.3 *Field Strength on Track Surface of Outer Race due to Flow of Circular Current in Inner Race and Rolling-Elements*

8.4.3.1 *Due to Flow of Circular Current in Inner Race*

The field strength on the track surface of the outer race H_{ori} due to the flow of current in the inner race is evaluated by using R_{or} for X and $R_{or}\ (R_{or}^2 - R_{ir}^2)^{0.5}/R_{ir}$ for Y (as determined in Fig.8.2) in Equation(ix) of the Appendix, and is determined as:

$$H_{ori} = 2\pi I_b\ R_{ir}\ (R_{or}^2 - R_{ir}^2)/\ R_{or}^4 \quad (8.7)$$

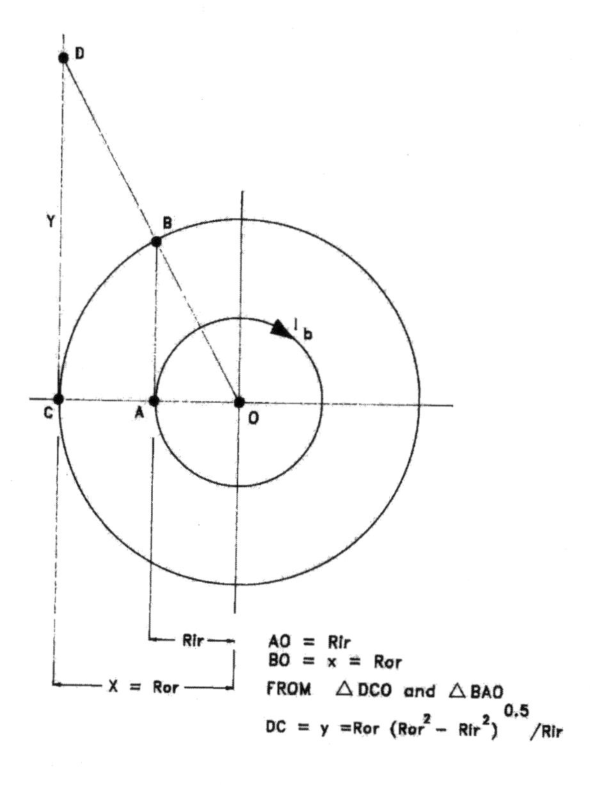

Fig. 8. 2 Field strength at locations C due to circular current I_b

8.4.3.2 *Due to Flow of Circular Current in Rolling-Elements*

The flow of current through the rolling elements creates the field strength both in the inner race and outer race. Similar to the Equation (8.2) for the field strength on the bearing inner race due to the flow of current in the rolling-elements, the field strength on the track surface of outer race is determined by changing R_{ir} by R in Equation (8.7), and is given as:

$$H_{orr} = 2\pi I_b R [R_{or}^2 - R^2]/R_{or}^4 \qquad (8.8)$$

8.4.3.3 *Equivalent Field Strength on the Track Surface of Outer Race*

The equivalent field strength on the track surface of outer race due to the flow of circular current in inner race and rolling-elements is determined as the summation of H_{ori} and H_{orr} by using Equations (8.7) and (8.8), since on rotation rolling-elements change the polarity with respect to rotating inner race. So, the equivalent field strength will be either difference or addition of H_{ori} and H_{orr}, and is given as:

$$H_{or} = 2\pi I_b [R_{or}^2 (R_{ir} \pm R) - (R_{ir}^3 \pm R^3)]/R_{or}^4 \qquad (8.9)$$

8.4.4 *Field Strength on the Outer Surface of the Outer Race due to Flow of Circular Current in Inner Race and Rolling-Elements*

8.4.4.1 *Due to Flow of Circular Current in Inner Race*

Similar to the Section 8.4.3, the developed field strength on the outer surface of outer race, H_{oroi},due to flow of circular current, I_b, in the inner race is determined by using R_{oro} for X and $R_{oro}(R_{oro}^2 - R_{ir}^2)^{0.5}/R_{ir}$ for Y (as determined by Fig.8.2) in Equation (ix) of the Appendix, and is determined as:

$$H_{oroi} = 2\,\pi\, I_b.R_{ir}\,(R_{oro}^2 - R_{ir}^2) / R_{oro}^4 \qquad (8.10)$$

8.4.4.2 *Due to Flow of Circular Current in Rolling-Elements*

The field strength on the outer surface of the outer race, H_{oror}, due to flow of current in rolling-elements is determined by changing R_{ir} by R in Equation (8.10), and is given as:

$$H_{oror} = 2\pi\, I_b.R.(R_{oro}^2 - R^2) / R_{oro}^4 \qquad (8.11)$$

8.4.4.3 *Equivalent Field Strength on the outer Surface of outer Race*

The equivalent field strength, H_{oro}, on the outer surface of outer race is the summation of H_{oroi} and H_{oror}, and is determined similar to Equation (8.9), and is given as:

$$H_{oro} = 2\pi\, I_b[R_{oro}^2(R_{ir} \pm R) - (R_{ir}^3 \pm R^3)] / R_{oro}^4 \qquad (8.12)$$

8.4.5 *Field Strength on the Rolling-Elements due to flow of Circular Current in Inner and Outer Races of Bearing*

8.4.5.1 *Due to Flow of Circular Current in Outer Race*

The field strength on the rolling-elements due to flow of current in outer race is determined using R in place of R_{ir} in Equation (8.1) and is given as:

$$H_{ror} = 2\pi\ I_b (R_{or}^2 - R^2)/R_{or}^3 \quad (8.13)$$

8.4.5.2 *Due to Flow of Circular Current in Inner Race*

The field strength on the surface of rolling-elements due to flow of current in inner race is determined by using R in place of R_{or} in Equation (8.7) and is given as :

$$H_{rir} = 2\pi I_b\ R_{ir}\ (R^2 - R_{ir}^2)/R^4 \quad (8.14)$$

8.4.5.3 *Equivalent Field Strength on Rolling-Elements*

On rotation, rolling-elements change the polarity, and outer race being stationary, the equivalent field strength on the surface of rolling-elements depends on the direction of current flow in the races. The equivalent field strength will be either difference or the addition of H_{rir} and H_{ror} . So by using Equations (8.13) and (8.14), it is given as:

$$H_r = 2\pi I_b[R^2(R_{ir}.R_{or}^3 \pm R^4) - R_{or}^2(R_{or}.R_{ir}^3 \pm R^4)]/R_{or}^3.R^4 \quad (8.15)$$

8.5 Magnetic Flux Density on Bearing Surfaces

8.5.1 *On Track Surface of Inner Race*

The magnetic flux density on track surface of bearing inner race, B_{ir}, due to field strength, H_{ir}, in an oil medium of relative permeability U_r with respect to free space, is given by:

$$B_{ir} = U.U_r\ H_{ir} = 4\ \pi\ x10^{-7}\ U_r.\ H_{ir}\ \text{(in tesla)} \quad (8.16)$$

On using Equation (8.3) for equivalent field strength on track surface of inner race (H_{ir}), the flux density on track surface of inner race (in gauss),B_{ir}, is determined as:

$$B_{ir} = 78.96x10^{-3}\ U_r.I_b\ [R^2\ R_{or}^2\ (R \pm R_{or}) - R_{ir}^2\ (R^3 \pm R_{or}^3)]/\ R^3.R_{or}^3 \quad (8.17)$$

8.5.2 *On Inner Surface of Inner Race*

Using Equation (8.6), flux density, B_{iri}, on inner surface of inner race is determined by the relation given in Equation (8.16), which yields:

$$B_{iri} = 78.96x10^{-3}U_rI_b[R^2.R_{or}^2(R \pm R_{or}) - R_{iri}^2(R^3 \pm R_{or}^3)]/R_{or}^3.R^3 \quad (8.18)$$

8.5.3 On Track Surface of Outer Race

Using Equations (8.9) and (8.16), the flux density on track surface bearing outer race (in gauss) is determined as:

$$B_{or} = 78.96 \times 10^{-3}\, U_r.I_b[R_{or}^2\,(R_{ir} \pm R)-(R_{ir}^3 \pm R^3)]/R_{or}^4 \qquad (8.19)$$

8.5.4 On Outer Surface of Outer Race

Similarly, using Equations (8.12) and (8.16), flux density on outer surface of outer race is determined as:

$$B_{oro} = 78.96 \times 10^{-3} U_r.I_b[R_{oro}^2(R_{ir} \pm R)-(R_{ir}^3 \pm R^3)]/R_{oro}^4 \qquad (8.20)$$

8.5.5 On Surface of Rolling-Elements

Using Equations (8.15) and (8.16), the minimum/maximum residual magnetic flux density on a few rolling-elements (in gauss) is determined as:

$$B_r = 78.96 \times 10^{-3}\, U_r I_b\, [R^2(R_{ir}\, R_{or}^3 \pm R^4) - R_{or}^2(R_{or} R_{ir}^3 \pm R^4]/R_{or}^3 R^4 \qquad (8.21)$$

It may be noted that the theoretical and average measured flux densities match closely. According to Table 8.1, the theoretical and the average measured flux densities differ by less than a factor of 1.3 in all cases.

Table 8.1 Comparison of theoretical and measured values of magnetic flux density on the surfaces of races and rolling-elements of various bearings

BEARING TYPE	BEARING PARAMETERS AND OPERATING CONDITIONS	LOCATION	FLUX DENSITY (GAUSS)		FLOW OF CURRENT THROUGH BEARING
			THEORETICAL	MEASURED (Average)	
NU 326	R_{ir} = 83.5 mm	B_{ir}	30.04-4.06	34.50	
	R_{or} = 121.5 mm				
	R = 102.5 mm	B_{iri}	45.65-0.18		50 A (measured)
	R_{oro} = 140.0 mm				
	R_{iri} = 65.0 mm				
	After testing NU 326	B_{or}	19.725-3.93	15 - 25	
	Bearing for 250 hr	B_{oro}	21.10 -1.10		
	on exposure of 50 A (a.c.) at 1.12-2.30 V	B_r	19.94 -1.99	14 - 18	
	on Bearing test machine at 1100 rev min^{-1} under 5000 N horizontal, 10000 N radial loads			(on a few rolling elements)	

Table 8.1 *Contd...*

BEARING TYPE	BEARING PARAMETERS AND OPERATING CONDITIONS	LOCATION	FLUX DENSITY (GAUSS)		FLOW OF CURRENT THROUGH BEARING
			THEORETICAL	MEASURED (Average)	
NU 228	R_{ir} = 79 mm				
	R_{or} = 93 mm				
	R = 86 mm	B_{ir}	9.51-2.5	12.2	
	R_{iri} = 70 mm	B_{iri}	16.95-1.45		
	R_{oro} = 125 mm				
	After about 6000 hr of motor operating at 2880 rpm, bearing damage detected. Shaft voltage and resistance measured approx. 2.5 V and 0.1 ohm, respectively	B_{or} B_{oro} B_r	7.90-2.18 11.72-0.27 6.35-0.22	9.0 5-7 (on a few rolling elements)	25 A (measured) 32.11 (calculated)
NU 311	R_{ir} = 35 mm	B_{ir}	15.64-2.29	14	12 A (measured)
	R_{or} = 52 mm	B_{iri}	26.20-0.05		
	R = 43.5 mm				
	R_{iri} = 27.5 mm				
	R_{oro} = 60 mm				
	A number of bearings of alternator operating at 750 rpm damaged at different intervals varying between 500 hr and 2500 hr of operation. Shaft voltage and resistance measured approximately 6V and 0.5 ohm, respectively.	B_{or} B_{oro} B_r	11.27-2.13 11.51-0.65 11.64-0.70	10 10 -12 (on a few rolling elements)	 10.65 A (calculated)
NU 230	R_{ir} = 91 mm	B_{ir}	15-2.93	15	32.6A (calculated)
	R_{iri} = 75 mm	B_{iri}	25.11-0.98	$\simeq$20	
	R = 105 mm				
	R_{or} = 119 mm				
	R_{oro} = 135 mm				

Table 8.1 *Contd...*

A number of bearings of 1700 kW motors operating at 2880 rpm damaged at different intervals of operation under	B_{or} B_{oro}	10.44-2.61 12.71-1.30
the influence of different levels of shaft voltages.	B_r	9.78-0.65

8.6 Electro-Adhesion Forces

8.6.1 *Role and Assessment of Electro-adhesion Forces in Rolling Friction*

The genesis of intermolecular forces during rolling fiction involves electrostatic attraction or repulsion between electrons and nuclei, and interaction of electro-dynamic, magnetic and exchange forces between the atoms[8.7]. The higher temperature rise of bearings operated under electrical field (45° C temperature rise of the outer race of NU 326 bearing as against 25° C under pure rolling friction under identical operating conditions) suggests the reduction in the fatigue life, and increase in resistance to the rolling due to the forces of electro-adhesion. However, the function of a lubricant is to reduce the adhesion forces across an interface[8.8].

The mechanism of adhesion, friction, and wear on the bearing surfaces in presence of lubricating film is quite complex. The process of adhesion involves the formation of a junction between the asperities contact, which may finally lead to elastic and plastic deformation under load[8.2,8.9]. The energies of atomic nature are exchanged at the asperities, which may be affected by cage and roller slip due to close interaction of rolling elements with the races[8.10].

It is rather difficult to estimate the electro-adhesion forces in the rolling friction. But, these can be assessed with a reasonable amount accuracy by SRV analysis; the change in coefficient of friction, profile depth and ball scar diameter of the used greases recovered from the active zone of the bearings. This is because the activity of zinc additive i.e., zinc dithiosphosphate or zinc dialkyldithiophosphate (ZPTPs) used as multifunctional additives in the grease, under pure rolling friction protects the rubbing metal surfaces and contributes to friction and wear reduction, and depends, partly, on the amount of additive on these surfaces. Physisorption and chemisorption processes precede the chemical reactions with metal; therefore, it is probable that load-carrying capacity is related to these processes. Correlation between ZDTPs adsorption data and wear is shown and also discussed that ZDTPs are reversibly physisorbed on iron at 25° C, but at 50° C under go chemisorption reactions[8.11]. On the other hand decomposition of ZDTPs in the lithium base greases under the influence of electrical fields leads to the formation of lithium zinc silicate ($Li_{3.6}\,Zn_{0.2}\,SiO_2$) in presence of high relative percentage of free lithium and silica impurity in the grease under high temperature in the asperity contacts along with the formation of gama lithium iron oxide (γ-Li Fe O_2). Besides this the original structure of lithium stearate changes to lithium palmitate as discussed in Chapter 3. On the contrary, these changes are not detected under pure rolling friction[8.12].

The above change in the lubricating medium of the bearings operated under electrical fields are reflected in SRV analysis (Table 8.2), and are related to electro-adhesion forces in the rolling friction. This change is contributed by medium-metal interaction, rate of chemical reaction, affinity of lubricating medium components for the metal, availability of free metal and temperature rise. If the grease, recovered from such bearings is put in a new bearing, the bearing may have premature failure due to higher temperature rise even under pure rolling friction.

8.6.2 *Role of Lubricant in Rolling Friction*

The bearing using high-resistivity (10^{11} ohm-cm) and low-conductivity lubricant accumulate electrical charges on the surfaces up to the threshold limit at which discharge leaks at the asperity contacts. On the contrary, a bearing using low-resistivity (10^{7} ohm-cm) and high conductivity lubricant does not accumulate electrical charges, but the 'silent' discharges occurs between the interacting surfaces through the lubricant, which gradually gets decomposed[8.12]. The type of lubricant affects a bearing in a characteristic manner and develops distinguished failure pattern on the surfaces (Chapter 11).

A bearing is influenced by adsorption potential and chemical activity of the rubbing surfaces, boundary layer properties of the lubricant, the temperature gradient, and the electrical field strength in the crevice-shaped space between rollers and races. The phenomenon in the rolling friction is similar to the adhesion on separation like gas discharge luminescence, emission of high energetic electrons and X-ray radiation[8.7].

8.7 Experimental Facilities and Investigations

8.7.1 *Bearing Test Machine*

The bearing test machine developed in-house[8.2],[8.3], and[8.4], was used to evaluate the performance of various sizes of rolling-element bearings at different operating parameters, and to programmed investigations in tribological areas.

For evaluating the bearing performance under shaft current, a silver-lined slip-ring assembly was used. The current is allowed through the shaft, a test bearing, and the housing to complete the electrical circuit; the support bearings were fully insulated to avoid the passage of current. For applying different levels of potential across the test bearing, a variable transformer and dimmerstat were used.

8.7.2 *Schmierstoff - / Lubricant - / Material (SRV) Test System*

The SRV test system has been used to evaluate the characteristics of the fresh and used greases. After sampling the small quantities of the grease from bearing's active zone (less than 0.1 gms), the lubricity has been examined by determining friction coefficient and wear, under100 N load and 1000 μm amplitude and 50 Hz frequency by running ball on flat surface for 2 hours test duration at 50° C on SRV. Table 8.2 shows analysis of the fresh and deteriorated grease samples.

Table 8.2 SRV analysis of greases

		Fresh grease before filling in the bearings		Grease from active zone of bearings after 250 h of operation under current		Grease from active zone of bearings after 250 h of operation under pure rolling friction
S. No.	Test data	Low resistivity	High resistivity	Low resistivity	High resistivity	Low resistivity
1.	Coefficient of friction μ(min) μ(max)	0.105 0.1125	0.125 0.132	0.132 0.40	0.122 0.145	0.10 0.120
2.	Profile depth point (μm)	0.5	0.90	1.6	0.90	0.50
3.	Ball scar diameter (mm)	0.52	0.48	0.66	0.48	0.48
4.	Max. ratio of coefficient of friction as compared to fresh grease	-	-	3.81	1.16	1.14
5.	Ratio of profile depth as compare to fresh grease	-	-	3.2	1.0	1.0
6.	Ratio of ball scar diameter as compared to fresh grease	-	-	1.27	1.0	1.0

8.8 Test Conditions and Procedure

8.8.1 *Non-insulated Bearings under Shaft Current*

The bearing NU 326, lubricated with low resistivity grease (10^5 ohm. m) were tested at 1100 rpm under 10^4 N and 5 x 10^3 N radial loads (90 deg to each other) for a duration of 250 hours (at 1.12 to 2.3 V) by passing 50 A (alternating current). Current of 50 A was controlled throughout the test. Corrugations and corrosions were detected on the track surface of races and rolling-elements of the bearing[8.2, 8.3]. The tests were repeated on new bearings without the passage of electric current.

8.8.2 *Study of Magnetic Flux Density on Bearing Surfaces*

The surface of bearings, tested under the influence of electric current and without the passage of current, and used damaged bearings of motors and alternators have been examined. The magnetic flux density distribution at various angular locations on the lower and upper width sides on inner and outer surfaces of the races have been studied with M/s Bell Inc. – make Hall Probe and a Gauss Meter. The same was studied on track surfaces, inner surface of inner race and outer surface of outer race of the various bearings. The technical particulars of the Hall Probe and Gauss Meter are given in Table 8.3.

Figure 8.3 shows the probe location for the measurement of flux density emanating from the track surface, at various angular locations. Figure 8.4 shows the residual flux density distribution on the upper and lower width sides of the inner race of NU 326 bearing after passing 50A for 250 hours. Figure 8.5 shows the same on inner race of a damaged bearing, NU 228, used in a motor for about 6000 hours. Figures 8.6 to 8.9 show the residual flux density distribution on different surfaces of damaged NU 230 bearings of a motor. Residual flux density distribution on damaged NU 311 bearings of alternator was also studied.

Table 8.3 Technical particulars of hall probe and Gauss meter

1. Field Range :	0 - 10 gauss, 0 - 100 gauss different ranges with 10 X probe
2. Linearity % of Reading :	0.1% to 10 KG
3. Instrument Accuracy :	± 0.05% of reading plus ; ± 0.02% of FS plus probe error
4. Stability	–20° to 70° C, time ± 0.1% FS maximum

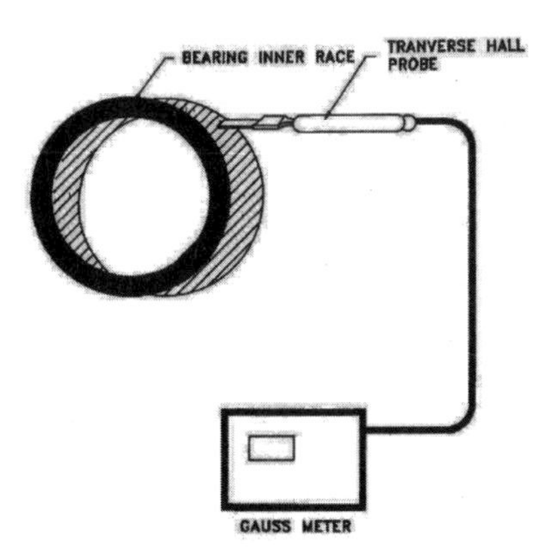

Fig. 8.3 Probe location on bearing track surface for measurement of flux density

8.9 Theoretical and Experimental Data on Flux Density

8.9.1 *Experimental*

8.9.1.1 *Bearing Lubricated With Low-Resistivity Lubricant*

Significant magnetic flux density distribution as shown in Fig. 8.4, along with corrosion and corrugations, were detected on the track surfaces of races and rolling-elements of the NU 326 roller bearing after operation under electrical current[8.3, 8.4]. On the contrary, these changes were not detected on bearings after operation without the influence of electric current. A maximum residual flux density of 2.5 gauss on NU 326 bearing surfaces was found without the influence of electric current, which was more or less the same as on the new bearings type NU 326, NU 230,NU 311, and NU 228.

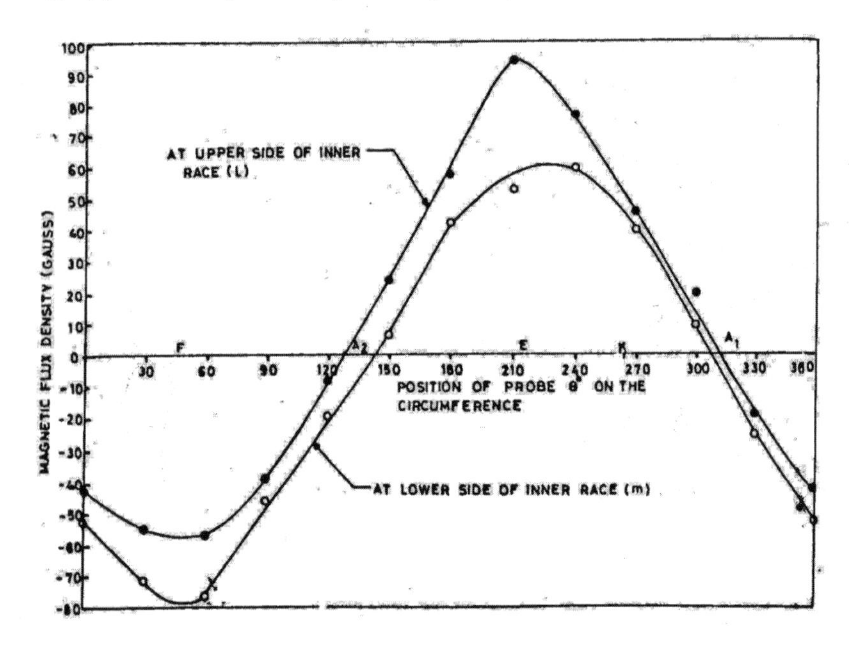

Fig. 8.4 Magnetic flux density distribution around the track surface of inner race of NU 326 bearing after passing 50 A (A.C.) at 1.2 to 2.3 volts for 250 hours

Significant changes were found in SRV analysis of the friction and wear characteristics of the used greases (Table 8.2). On the contrary, these changes were not detected after operation with pure rolling friction. Under the influence of electrical fields corrosion and corrugations were detected on 6326 ball bearing surfaces, but the significant flux density distribution was not detected.

8.9.1.2 *Bearings Lubricated With High - Resistivity Lubricant*

Under the influence of electrical fields and pure rolling friction, the significant flux density (2.2 gauss) on the surface of roller bearing NU 330 and NU 2215 were not detected, and also significant changes in SRV analysis of the used grease (Table 8.2). The bearing NU 2215 was damaged under electrical fields, but no corrugation pattern and corrosion were observed (Fig. 11.13 of Chapter 11). No damage were found showing corrugations on the NU 330 motor bearing surfaces.

8.9.2 *Theoretical*

The theoretical values of magnetic flux density, for a certain measured value of bearing current, were determined on the inner and outer surface of inner and outer races, and rolling-elements of different bearings (NU 326, NU 228, NU 230 and NU 311) by using Equations (8.17) to (8.21), and are shown in Table 8.1. The residual flux density, thus determined, on the NU 326 bearing was compared with the measured values of the tested bearing under the influence of 50 A current (Fig.8.4). Conversely, the flow of current through the damaged bearings type NU 228 and NU 230 of the motors was analytically established by measuring the flux density on the bearing surfaces using Fig.8.5 and Figs.8.6 to 8.9, respectively. The flow of current through NU 311 alternator bearing was also established. The same was compared with the measured values of current obtained from measured values of shaft voltage and bearing resistance.

8.10 Discussion on Investigations

8.10.1 *Magnetic Flux Density Distribution on Races and Rolling-Elements of Test Bearing*

The magnetic flux density on the inner race of the tested NU 326 roller bearing under the influence of electric current varies between 95 and -80 gauss (Fig. 8.4). The two positions (`$A_2$' and `A_1') of zero flux density at 180 deg. apart (at 130 deg and 310 deg angular locations) indicate that the inner race has become a two-pole magnet with `E' and `F' (90° away from `$A_1$' and `A_2') as the points of maximum flux density (Fig. 8.4).

The current from the inner to the outer race tends to flow through the asperity contacts through a path of least resistance towards `$A_2$' (Fig. 8.4). The current entering the outer race at the farthest point from `A_2' (at 180 degrees opposite) tends to flow in the clockwise and anti-clockwise directions to reach `$A_2$'. The ring current thus flows in the outer race and creates the residual alternating magnetic flux density distribution on the inner race and rolling-elements. The difference in the direction of current flow, at points 90 degrees away from `A_2', creates equal and opposite residual flux densities on the inner race (the maximum being 95 gauss at `E' and -80 gauss at `F', both 180 deg. apart). The maximum difference of 35 gauss at `E' and -25 gauss at F between the curves L and m, across the width of inner race, is attributed to the difference in the film thickness that could have occurred due to misalignment, and corrosion on the surfaces due to electrochemical decomposition of the lubricant[8.7].

The average residual flux density has been measured on the track surface of inner race of NU 326 bearing as 34.5 gauss (Fig. 8.4). The flux density on the track surface of outer race varies between 15 and 25 gauss and that on a few rolling-elements vary between 14-18, and most of the rolling-elements have flux density between 2 and 3 gauss. The average to minimum flux density determined by the developed relations vary between 30.04-4.06, 19.725-3.93 and 19.94-1.99 gauss on the track surfaces of inner race, outer race and on rolling-elements respectively, under the influence of 50 A electric current , as shown in Table 8.1. Theoretical values of the similar flux density on the inner surface of inner race, B_{iri}, and outer surface of outer race, B_{oro}, are found to vary as 45.65-0.18, 21.10-1.10, gauss, respectively, and are observed to match the measured values (Table - 8.1).

8.10.2 *Residual Magnetic Flux Density Distribution on Races and Rolling-Elements of Damaged Bearings of Motor and Alternator*

8.10.2.1 *Damaged NU 228 Bearing of a Motor*

The residual flux density distribution on the track surface of inner race of the damaged NU 228 bearing varies between + 20 and - 42 gauss as shown in Fig. 8.5. The average residual flux density on the track surface of inner and outer races were detected as 12.2 and 9 gauss, respectively, and on a few rolling-elements as 5-7 gauss (maximum). However, most of the rolling-elements generally have residual flux density of 1-2 gauss. From the measured residual flux density data and using Equations (8.17) to (8.21), the current flow through the motor bearing was estimated as 32.11 A. This matches approximately with a current flow of 25 A determined by the measurement of resistance and the shaft voltage (Table 8.1). The analytically determined flux density on the track surface of inner race, outer race and on a few rolling-elements varies between 9.51-2.5, 7.90-2.18, and 6.35-0.22 gauss, respectively, using 25 A flow of current through the bearing. The residual flux density on the inner surface of inner race and outer surface of outer race has been determined to vary between 16.96-1.45, 11.72-0.27, respectively. These values match closely with that of the average measured values (Table 8.1).

The two positions C and D (260 degree apart) as the points of maximum residual flux density, with the positions E and F (180 ° apart) as the locations of zero flux density indicate that the inner ring has become a two- pole magnet (Fig. 8.5). The circular current flows in the outer race in such a way before it is grounded that it creates the maximum residual flux at locations C and D with a zero flux at locations E and F in the inner race. The maximum difference in flux density of 16 gauss, between curves J and K across the width of inner race, is attributed to misalignment and difference in film thickness (Fig. 8.5).

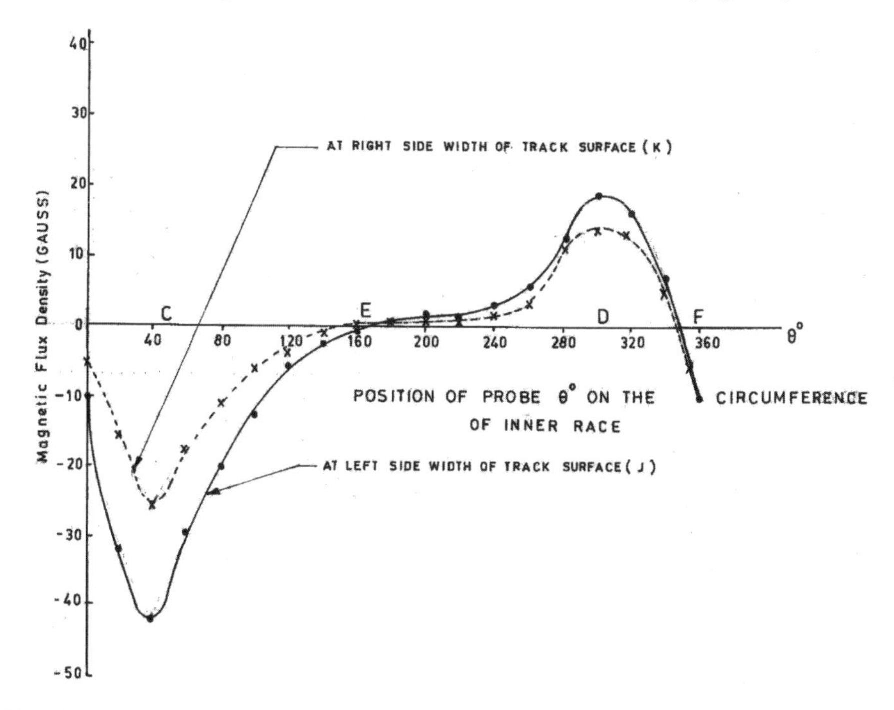

Fig. 8.5 Magnetic flux density distribution on width of inner race track surface of NU 228 damaged motor bearing

8.10.2.2 Damaged NU 230 Bearing of a Motor

The average residual flux density on the track surface of inner race, B_{ir}, and inner surface of inner race, B_{iri}, of the damaged NU 230 bearing is measured as 15 and 20 gauss, respectively, as shown in Figs. 8.6 and 8.7. This matches closely with the calculated average to minimum values of 15-2.93, and 25.11-0.98 gauss, for B_{ir} and B_{iri}, respectively, with a flow of current as 32.6 A (Table 8.1). Residual flux density variation on the track surface and outer surface of outer race (B_{or} and B_{oro}), and on rolling-elements, B_r, is analytically determined as 10.44-2.61, 12.71-1.30, and 9.78-0.65 gauss, respectively, which have been found to match closely with the measured values.

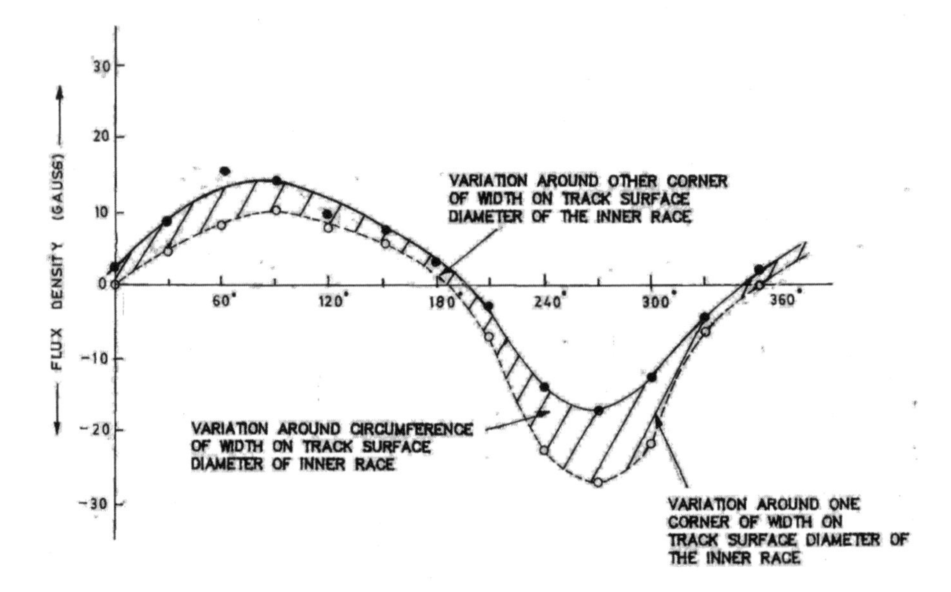

Fig. 8.6 Variation of flux density along the circumference of the width of track, surface diameter of the inner race of NU 230 bearing

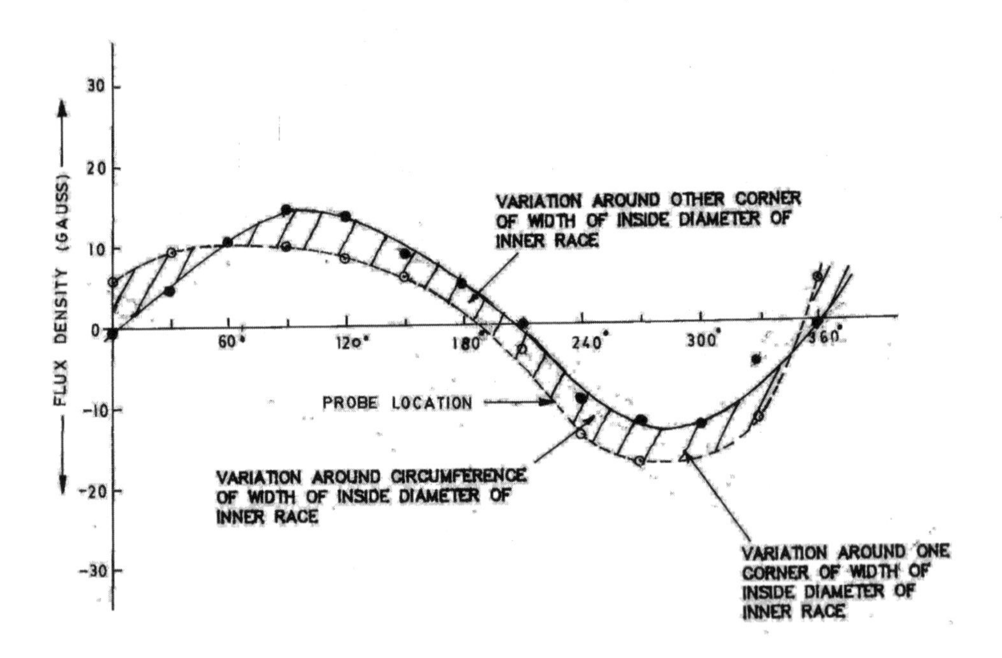

Fig. 8.7 Variation of flux density along the circumference of the width of inside diameter of the inner race of NU 230 bearing

Figs. 8.8 and 8.9, showing the variation of flux density along the circumference of the thickness of the inner race on both sides indicate a pattern similar to that of the variation on the track surface and inner surface of the inner race (Figs. 8.6 and 8.7).

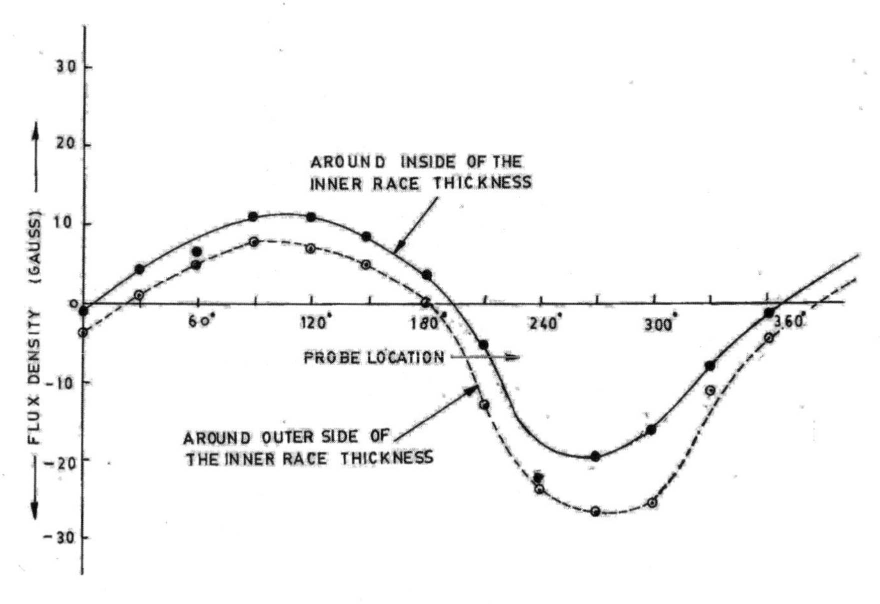

Fig. 8.8 Variation of flux density along the circumference of the thickness of the inner race of NU 230 bearing (side A)

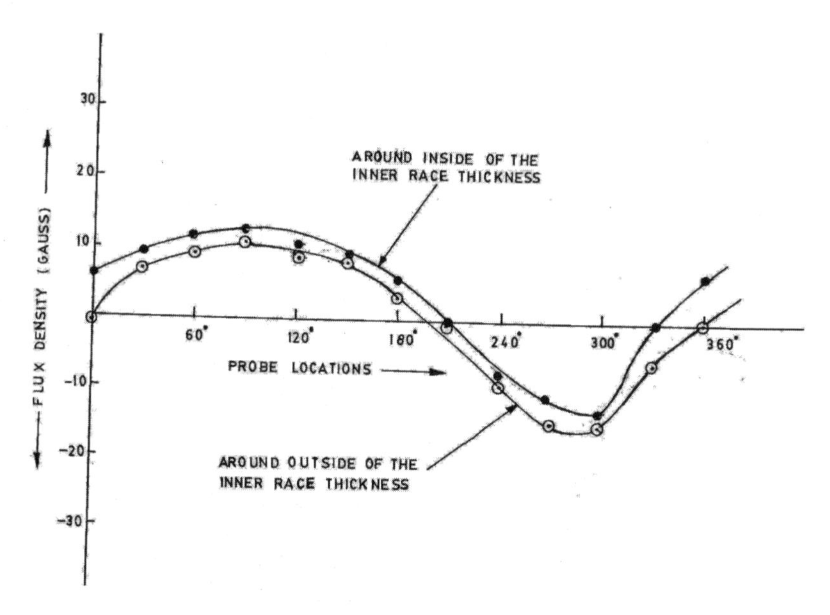

Fig. 8.9 Variation of flux density along the circumference of the thickness of the inner race of NU 230 bearing (side B)

8.10.2.3 *Damaged NU 311 Bearings of Alternator*

The residual magnetic flux density has been measured on various damaged bearings of type NU 311. The average residual flux density on the track surface of inner race, outer race and rolling-elements was measured as 14, 10, 10-12 gauss, respectively, which matches the theoretically determined average to minimum flux density variation of 15.64-2.29, 11.27-2.13 and 11.64-0.70 gauss respectively, obtained by taking the flow of current as 12 A determined through the measured shaft voltage of 6V under the bearing resistance of 0.05 ohm, as shown in Table 8.1. Also, residual flux density variation of average to minimum on inner surface of inner race, B_{iri}, and outer surface of outer race, B_{oro}, has been analytically determined as 26.20-0.05, and 11.51-0.65, gauss, respectively.

In general, it may be noted that the residual flux density on inner surface of inner race and outer surface of outer race is more than that on the respective track surfaces of races of the bearings (Table 8.1).This is because of the effect of flow of current and the positional location of the surface with respect to the bearing center.

8.10.2.4 *Electro-adhesion Forces in Bearings*

Under the influence of electrical fields, the increase in ratio of maximum coefficient of friction, profile depth and ball scar diameter of the used and fresh greases in SRV analysis are 3.81, 3.2, and 1.27 in the bearings using low-resistivity greases as against 1.16, 1.0 and 1.0, respectively with greases of high resistivity (Table 8.2). This may be correlated with decomposition of the grease and presence of lithium iron oxide and lithium zinc silicate in the medium[8.12]. This suggests that the electro-adhesion forces increase in the bearings using low-resistivity grease as compared to greases of high-resistivity under pure rolling friction (Table 8.2). The deterioration of the low resistivity grease gradually corrodes bearing surfaces, which finally leads to increased wear and failure of a bearing[8.2 & 8.12].

8.10.2.5 *Effect of lubricants on Bearings and Flux Density*

Under the influence of the electrical current, bearing lubricated with low-resistivity lubricant allows the passage of current, which corrodes the bearing surfaces and flux density is developed on the track surfaces. On the contrary under the influence of electrical fields, a bearing lubricated with high-resistivity grease accumulates electrical charges on the surfaces, and does not allow passage of current, thereby significant magnetic flux density is not developed. However, when the accumulated charges exceed the threshold value, the sudden discharge takes place at the asperity contacts, which damages the bearing surfaces[8.4].

The current start flowing when high points on the asperities of interacting surfaces come close or when conducting particles bridge the oil film. These conducting paths are broken as the asperities become separated either by higher film thickness, or by vibration effects, or a combination of both, the arcing results, which damages the surfaces by the arc welding and high temperature effect (Chapter 11).

8.10.3 *Magnetic Flux Density – An Overall View*

The developed magnetic flux density on the races and rolling-elements is governed by the field intensity due to the flow of electric current and the permeability of the medium. The flow of current in a bearing depends on the shaft voltage, lubricant characteristics and operating parameters[8.2, 8.3]. These are the variable parameters during the course of operation, and accordingly the field intensity becomes variable. Furthermore, the leakage/flow of current from inner race to rolling-elements and to the outer race follows the path of least resistance, which itself is a variable function and is governed by different parameters including that of asperity contacts and temperature-rise besides rolling-element frequency, cage and roller slip and the machine characteristics[8.4]. To develop a theoretical model for determination of field intensity due to the variable circular current is a complex investigation. However, the developed model is an initial approach work in this direction, which facilitates determining field intensity depending on radius of track surfaces, including that of the pitch radius, and the flow of current through the bearing. The flux density varies on the races and the rolling-elements. This is because of relative dimensions of races and rolling-elements and unsteady flow of current under the dynamic conditions of operation of a rolling-element bearing.

8.10.4 *Effects of Residual Magnetic Flux on Bearings*

The residual magnetic flux density on the track surfaces of a rolling-element bearing can lead to premature failure through the mechanical forces or heat generated. The flux passing through a bearing produces forces of attraction, hysteresis loss and eddy current[8.13]. The parasitic energy losses are supplied from the mechanical shaft power, and, therefore, the torque required to turn the shaft increases.

The bearings made of hard steel have relatively large magnetic hysteresis loss. The loss can be due to variation in magnitude of flux. During rotation of a bearing, load is distributed between a limited number of rolling-elements. This may further lead to instantaneous change in flux and, hence, cyclic change in attractive forces on bearing elements. This may cause premature failure of bearings due to non-uniform wear of bearing surfaces[8.13].

8.11 Conclusions

Based on the above study and investigations, following conclusions and recommendations are drawn[8.5, 8.14, 8.15, 8.16]:

1. Under the effect of current, significant flux density is not developed on the surfaces of ball bearing using grease of low-resistivity.
2. Under the influence of electric current, a bearing using high-resistivity grease does not develop significant magnetic flux density distribution on its surfaces.
3. Under the influence of electrical fields, there occurs changes in electro-adhesion forces when bearings are lubricated with low-resistivity grease than with grease of high-resistivity. Whereas, pure rolling friction does not affect electro-adhesion forces considerably.
4. Detection of corrugations on bearing surfaces does not necessarily indicate damage by electrical current.
5. By the study of magnetic flux density distribution along with the study of damaged and corrugated bearings surfaces, and also by the analysis of deterioration of greases used in the bearings, a diagnosis of leakage current through a non-insulated bearing of an electrical motor can be established.
6. Under the influence of electrical current, a rolling-element bearing using low-resistivity lubricant develops field intensity which leads to magnetic flux density distribution on the surfaces of races and rolling-elements.
7. The flow of current in inner race and rolling-elements affects the flux density distribution on outer race track and outer surfaces, and flow of current in outer race and rolling-elements affects the flux density distribution on inner race track and inner surfaces, and current flow in the races affects the flux density on rolling-elements.
8. The developed flux density on the different surface of races and rolling-elements also depends on track radii, respective race radii, pitch radius and relative permeability of the lubricant.
9. In general, the developed flux density is higher on track surface of inner race as compared to that of outer race and rolling-elements. Also, residual flux density on the inner surface of inner race and outer surface of outer race is more than that on the respective track surfaces of the races.
10. By using the developed analytical model, and by experimental determination of the flux density on the surfaces of the damaged bearing, the level of flow of electric current through the bearing can be ascertained without the measurement of shaft voltage and bearing impedance.
11. The magnetic flux density on races and rolling-elements, analytically determined by the developed model, match closely with the experimentally evaluated average residual flux density on the bearing elements.

Study of the magnetic flux density under the influence of variable current density and mutual interaction of flux densities on the races and rolling-elements of a bearing under different modes of operation, is a complex analytical model, and needs further investigations.

References

[8.1] Boyd, J. and Kaufman, H.N., "The Causes and the Control of Electrical Currents in Bearings", Lubr. Eng., Vol.15, 1, 1959, pp 28-35.

[8.2] Prashad, H., "Investigations of Corrugated Pattern on the Surface of Roller Bearings Operated Under the Influence of Electrical Fields", Lubr. Engr., Vol.44, 8, 1988, pp 710-718.

[8.3] Prashad, H., "Investigations of Damaged Rolling-Element Bearings and Deterioration of Lubricants under the Influence of Electric Fields", Wear, Vol.176, 1994, pp 151-161.

[8.4] Prashad, H., "Effects of Operating Parameters on the Threshold Voltages and Impedance Response of Non-Insulated Rolling-Element Bearings under the action of Electrical Current", Wear, Vol 117, 2, 1987, pp 223-240.

[8.5] Prashad, H., "The Effects of Current Leakage on Electroadhesion Forces in Rolling Friction and Magnetic Flux Density Distribution on the Surface of Rolling-Element Bearings", ASME, Journal of Tribology, Vol.110, 1988, pp 448-455.

[8.6] Starling G. Sydney, “Electricity and Magnetism", published by Longmans, Green and Co. London, 1960.

[8.7] Postnikov, S.N., “Electrophysical and Electrochemical phenomena in Friction, Curing and Lubrication”, VNR, 1978.

[8.8] Buckley, H., and Donald, “Surface Effect in Adhesion, Friction, Wear and Lubrication”, Tribology Series 5, Elsevier Scientific Publishing Company, Oxford, 1981.

[8.9] Brainin, E. I., Zbarskii, L.I., Ostankovich, E. V., Pyosetskoya, L.I., and Spotkai, G. V., “Investigations of Damage to Roller Bearings of Electrical Machines under the Action of Current”, Electrotacknika, Vol. 54, No. 1, 1983, pp 43-47.

[8.10] Prashad, H., “The Effect of Cage and Roller Slip on Defect Frequency Response of Rolling Element Bearings”, Presented in ASME/ASLE, Tribology Conference in Pittsburgh, Pennsylvania, Oct, 20-22, 1986. Also published in ASLE Transactions, Vol.30, 3, 1987, pp360-368

[8.11] Plaza, S., “The Adsorption of Zinc Dibutyldithiophosphates on Iron and Iron Oxide Powders”, ASLE Transactions, Vol. 30, No. 2, 1987, pp 233-240.

[8.12] Prashad, H., Murthy, T.S.R., "Behaviour of Greases in Statically Bounded Conditions and when used in non-insulated Anti-friction Bearings under the Influence of Electrical Fields", Lubr. Engg. 44(3), 1988, pp 239-246.

[8.13] Tasker, J.L. and Graham, R.S., “Effects of Magnetic Flux on Rolling-Element Bearings", IEE Conference, Sept. 17-19, 1985, Publication 254, pp 152-156.

[8.14] Prashad, H., “Magnetic Flux Density Distribution on the Track Surface of Rolling -Element Bearings –An Experimental and Theoretical Investigation” Tribology Transactions, Vol. 39, 2, 1996, pp. 386-391.

[8.15] Prashad. H., “Determination of Magnetic Flux Density on the Surfaces of Rolling-Element Bearings as an Indication of the Current that has Passed Through Them – An Investigation”, Tribology International, Vol. 32, 1999, pp 455-467.

[8.16] Prashad, H., “Determination of Magnetic Flux Density on the Surfaces of Rolling-Element Bearings – An Investigation”, BHEL Journal, Vol. 21, No.2, August 2000, pp 49-66.

Appendix

Calculation of Field Strength at the Centre Due to Circular Current

If a current of strength I_b flows in the race of a rolling-element bearing (in a circular path), it can be treated as a circular magnetic shell of strength ρ. And strength ρ can be taken equivalent to I_b[8.6].

To find a magnetic potential at a point O on the axis at the bearing centre due to flow of current in the race, the solid angle, Ω , subtended by the circle is determined (Fig.8.1). To do this,a sphere with centre O is drawn such that circle of radius r lies on the sphere. The solid angle (Ω) subtended by the slice ACB as shown in Fig.8.1, is given as:

$$\Omega = \frac{\text{Area of slice ACB}}{r^2} \quad \text{(i)}$$

The area (A) of slice of a sphere lying between two parallel planes is equal to the area of circumscribing cylinder between the plane, whose axis is perpendicular to these planes. So

$$A = 2\,\pi\,r.h \quad \text{(ii)}$$

where

$$h = r - X \quad \text{(iii)}$$

and

$$\Omega = \frac{2\,\pi\, r.\, h}{r^2} = 2\,\pi\,(1 - X/r) \qquad \text{(iv)}$$

But, magnetic potential at the bearing centre is given as:

$$V = \rho\,\Omega = I_b\Omega. = 2\pi I_b\,(1 - X/r) \qquad \text{(v)}$$

Since from Fig. 8.1

$$r^2 = X^2 + Y^2, \qquad \text{(vi)}$$

Magnetic potential at the bearing centre is given as (by Equations v and vi)

$$V = 2\,\pi\, I_b\,[1 - X\,(X^2 + Y^2)^{-1/2}] \qquad \text{(vii)}$$

By symmetry, it is evident that the magnetic field strength due to the circular current in the inner as well as outer race is directed along the axis, and its value is therefore given as[8.6]:

$$H = - dV/dX \qquad \text{(viii)}$$

Using Equation (vii), the field strength is expressed as

$$H = 2\,\pi Y^2\, I_b\,(X^2 + Y^2)^{-3/2} \qquad \text{(ix)}$$

For a point at the centre of bearing, when X = O, the field strength is given as:

$$H_o = \frac{2\,\pi\, I_b}{Y} \qquad \text{(x)}$$

Nomenclature

A - area of element
B_{ir} - magnetic flux density on track surface of inner race
B_{iri} - magnetic flux density on inner surface of inner race
B_{or} - magnetic flux density on track surface of outer race
B_{oro} - magnetic flux density on outer surface of outer race
B_r - magnetic flux density on rolling-elements
h - height of element
H - field strength
H_o - field strength at the centre
H_{iro} - field strength on track surface of inner race due to the flow of circular current in outer race
H_{irio} - field strength on the inner surface of inner race due to flow of circular current in outer race
H_{irr} - field strength on track surface of inner race due to flow of circular current in rolling-elements
H_{irir} - field strength on the inner surface of inner race due to flow of circular current in rolling-elements
H_{ir} - equivalent field strength on track surface of inner race
H_{iri} - equivalent field strength on inner surface of inner race
H_{ror} - field strength on rolling-elements due to flow of circular current in outer race
H_{rir} - field strength on rolling elements due to flow of circular current in inner race
H_r - equivalent field strength on rolling-elements
H_{ori} - field strength on track surface of outer race due to flow of circular current in inner race
H_{oroi} - field strength on outer surface of outer race due to flow of current in inner race
H_{orr} - field strength on track surface of outer race due to flow of circular current in rolling-elements
H_{or} - equivalent field strength on track surface of outer race
H_{oror} - field strength on outer surface of outer race due to flow of circular current in rolling-elements
H_{oro} - equivalent field strength on outer surface of outer race
I_b - bearing current

R_{ir}	-	track radius of inner race
R_{iri}	-	inside radius of inner race
R_{or}	-	track radius of outer race
R_{oro}	-	outside radius of outer race
R	-	pitch radius of bearing
r	-	radius of circle
v	-	magnetic potential
X	-	distance from bearing centre
Y	-	length of the element
Ω	-	solid angle
ρ	-	magnetic strength
U_r	-	relative permeability oil with respect to free space (=1)
U_o	-	permeability of free space (4×10^{-7} henry/m)

9

Unusual Causes of Rolling-Element Bearings Failure

9.1 General

This chapter highlights the investigations pertaining to the diagnosis of rolling-element bearings of motors and alternators failure due to the causes generally unforeseen during design and operation. However, in general the diagnosis of the failure of the bearings has been well established in literature. The unforeseen causes of failure have been established. These generally unforeseen causes were found to be due to the passage of electric current through bearings of the motors and alternators and deterioration to the same in due course.

The vibration and shaft voltage data, characteristics of the grease used in various rolling-element bearings were analyzed and unforeseen causes of failure of bearings diagnosed.

9.2 Introduction

9.2.1 *Causes of Shaft Voltages and Flow of Current through Bearings*

Phenomenon of Shaft Voltages exists in electrical machines, which causes flow of current through the bearings depending on the resistance of the bearing circuits. This has been discussed in the Chapter 7 in details[9.1-9.15].

All the causes of bearing current develop a magnetic flux, which closes in the circumference over the yoke and induces the voltage on the shaft as the machine rotates. This results in a localized current at each bearing rather than a potential difference between the shaft ends. A current path, however, along shaft, bearings and frame results in a potential difference between the shaft ends[9.3].

At a certain threshold voltage, depending on the resistivity of the lubricant and operating conditions, current flows through the bearing[9.4]. Thus, the flow of circular current in the inner race leaks through the rolling-elements to the outer race by following a path of least resistance and establishes a field strength leading to the development of magnetic flux on the track surface of races and rolling-elements[9.5] as discussed in Chapter 8.

Studies have been carried out by various authors on the causes and control of electrical currents in bearings[9.6], flow of current through lubricated contacts[9.7], the effect of electrical current on bearing life[9.8], the effect of operating parameters on an impedance response[9.4], and the deterioration of lubricants used in non-insulated bearings[9.10, 9.11, 9.12]. These have been discussed in Chapter 11.

The surveys of the failure of rolling-element bearings indicate various causes including that of failure due to corrugations[9.16, 9.13, 9.14]. The mechanism of formation of pattern of corrugations on the track surfaces and related investigations on roller bearing surfaces have been explained in Chapters 10 and 11.

This Chapter deals the investigations pertaining to generally unforeseen and unusual causes, which lead to the premature failure of rolling-element bearings in motors and alternators.

9.3 Bearing Arrangement and Nature of Bearing Failure

The motors with rated capacity 2100 KW operating at 1494 rpm have three bearing arrangement. The motors are used to drive primary air fans of 500-MW generators.

On the non-drive end (NDE) of the motor, the bearing type NU 228 is used and is insulated. On the drive end (DE) of the motor, roller bearing type NU 232 and ball bearing type NU 6326 are used for taking both radial and axial loads simultaneously. All the bearings are grease lubricated and the motors are designed for continuous operation with the periodic re-lubrication. Bearings are also instrumented for continuous measurement of the temperature during operation. Fig. 10.1 of the Chapter 10 shows the schematic diagram of NDE bearing arrangement in the motor.

The alternators were designed for the output voltage of 24V with 4.5 KW rating. The alternators use bearing type NU 311 on the drive end and 6309 on the non-drive end. The alternators are used for charging batteries and operation of lighting.

After commissioning, a few motors and alternators, bearings failed prematurely. The nature of the failure was the formation of corrugations and flutings on the roller tracks of the races besides corrosion on the raceways and rolling-elements, irrespective of the make of the bearings used. However, depth of corrugations and degree of development of corrugations were different. The color of the grease was also found to have got blackened.

9.4 Investigations and Data Collection

The investigations pertaining to shaft voltage, vibration and shock pulse levels were carried out on various motors both on non-drive end and drive end bearings.

9.4.1 *Measurement of Shaft Voltages*

The shaft voltages were measured between non-drive end and drive end shaft, drive end shaft to ground, and non-drive end shaft to ground. Table 9.1 shows shaft voltage data after removing the grounding brush. Table 9.2 indicates frequency analysis of different shaft voltages after spectrum analysis of recorded shaft voltages on tape recorder using shaft probes. Shaft voltages were also measured on retaining the grounding brush and keeping the brush intact (Fig. 10.1 of Chapter 10).

9.4.2 *Measurement of Vibration levels and spectrum analysis*

Vibration levels were measured in mm/s on both non-drive end and drive end bearings using B and K vibration measuring instruments. Table 9.3 shows the vibration levels in horizontal, vertical and axial directions.

9.4.3 *Inspection*

The dimensional accuracy of new bearings were checked and found to be in line with the specifications. Also, the dimensional accuracy and metallurgical examinations of the components of the failed bearings indicate the results are in line with the specified norms.

9.4.4 *Failed bearings and lubricants*

Various failed bearings were examined. Sample of fresh grease and used grease from the rolling-elements of the failed bearings were collected and analyzed.

9.4.5 *Study of Bearing Location in Alternators*

In certain original design of alternators, it was found that the rolling-element bearing type NU 311, on the drive end was mounted in the housing as shown in the Fig. 9.1. In the modified design of the alternators for the compact design, bearings were provided in the projection of the end cover (Fig. 9.2). But the projection was towards the inside of the alternator, so the bearing was located almost under the influence of the field coil, as shown in Fig. 9.2. The bearing elements were found to magnetize. The magnetic flux density on the track surface of some bearings was measured as high as 40G by gauss meter. In general, flux density vary between 10 and 40G. Also, the stray voltage of the 2-6V and 6-20V was measured under 'no load' and 'load' conditions, respectively.

9.4.6 *Checking of Grease Pipe Contacting the Base Frame*

In a few motors it was found the outlet grease pipe is either in direct contact with the base plate or through the accumulation of grease outlet from the bearing (Fig. 10.1 of Chapter 10).

9.4.7 *Checking of Grease Leakage through Seals*

Excess of grease leaking through seals gets collected just beside the bearing housing and thus the contact between the bearing housing and the base plate is made through the contaminated grease even if the bearing pedestal is insulated from the base plate as shown in Fig. 10.1 of Chapter 10. The collection of dirt/sludge etc between the corners of the base at bearing pedestal and the base plate were also detected particularly in the thermal power houses.

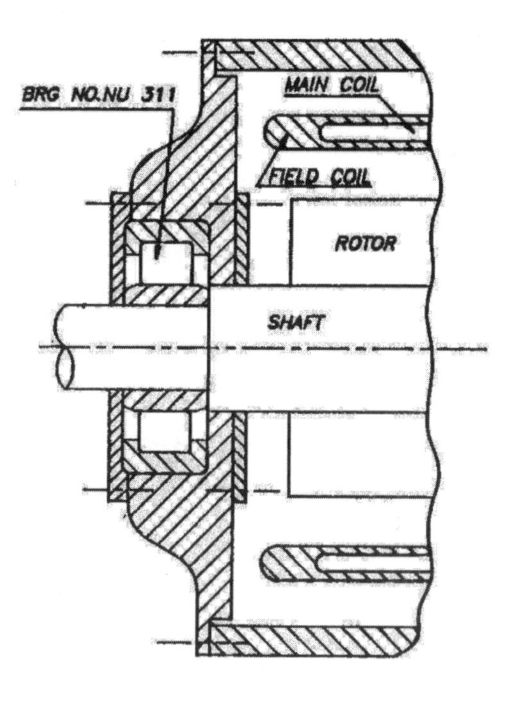

Fig. 9.1 Design, configuration showing location of bearing, field and main coils of original alternator

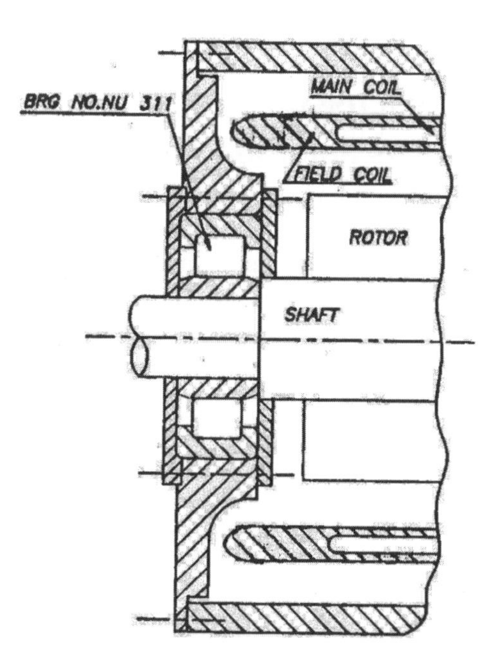

Fig. 9.2 Design, configuration showing location of bearing, field and main coils of modified alternator

9.4.8 *Checking of Un-shielded Instrumentation Cables*

Instrumentation cable used for measurement of bearing temperature some time gets unshielded due to various unforeseen reasons and cable comes partially in contact with the base plate/bearing housing. A few of such case have been found and investigated. Unshielded instrumentation cables were established as the cause of bearing failure after investigations.

9.4.9 *Checking of Improper Contact of Grounding Brush with Shaft*

Due to prolonged operation, dust collection, improper gap between brush and shaft, and improper maintenance of the grounding brush looses the grip with the rotating shaft. This was found in various locations.

9.4.10 *Checking of Damage of Bearing Insulation*

The prolong operation of bearings under vibration, lead to the ageing of insulation, depending on quality of insulating material. This results in the breakage/crack of the bearing insulation. Consequently it generates the resistance free path of bearing current and lead to catastrophic failure of the bearings (Fig.10.1 of Chapter 10)

9.4.11 Checking of Passage of Current through Connecting Bolts, Nuts etc

If the connecting bolts, nuts, joints of bearing pedestal and base plate are not properly shielded, the current is found to pass through the bearing even if the bearing insulation pads are properly maintained (Fig. 10.1 of Chapter 10).

9.5 Result and Discussions

9.5.1 *Shaft Voltages and their Frequencies*

The measurement of shaft voltages with and without the grounding brush indicated that the gap between the shaft and grounding brush was not set precisely. Also, the brush was not maintained and cleaned properly. That is why the brush was not able to ground the shaft voltage properly. Table 9.1 indicates that the different levels of shaft voltages exist in all the motors because of the various causes as brought out and explained in the Chapter 7. The minimum voltage of motor `B' between NDE to ground was measured as 0.030V as against 0.64V between NDE and DE shaft, and 0.65V between DE to ground (Table 9.1). This indicates that the insulation at NDE bearing of motor `B' is bridged on the path. This has been confirmed on dismantling the bearings of motor `B', and breakage of the insulation was detected. The maximum voltage of motor `C' between NDE and ground was measured as 1.15V. The drive end shaft to ground voltage of 0.4V of motor `F' indicates the passage of feeble current through the bearings as compared to motors `A', `C', `D' and `E'. The investigations indicated the partial unshielded instrumentation cable contacting the bearing housing was the source of the passage of current through the bearing of the motor `F'.

Table 9.1 Overall values of shaft voltages of different motors without grounding brush (in V)

Motor	Non-Drive End to Drive End Shaft	Non-Drive End to Ground	Drive-end Shaft to Ground
A	1.004	1.004	1.0
B	0.64	0.030	0.65
C	1.14	1.15	0.47
D	0.85	0.85	0.58
E	0.65	0.80	0.82
F	0.74	0.91	0.40

The frequency analysis of the shaft voltage signals shown in Table 9.2 indicates that the major component of shaft voltage consists of the magnitude of the slot passing frequency at 195 Hz, 1150 Hz and 1445 Hz and line passing frequency component of 50 Hz. The voltage of slot passing frequency component at 1445 Hz between NDE to DE vary between 0.17 and 0.37 V, and between NDE to ground vary between 0.014V and 0.65 V. The voltage at line frequency component of 50 Hz, between NDE to DE, and NDE to ground vary between 0.04V and 0.87 V, 0.014V and 0.55 V, respectively.

Table 9.2 Frequency analysis of shaft voltages (V) of different motors

Frequency		Non-drive end to Drive end				Non-drive end to Ground				
(Hz)	50	195	1150	1445	Over all	50	195	1150	1445	Over all
Motor										
A	0.04		0.10	0.33		0.13	0.03	0.28	0.55	0.74
B	0.67		0.07	0.17	0.81	0.008	0.002	0.009	0.014	0.025
C	0.96	0.15	0.19	0.46	1.44	0.062	0.18	0.31	0.65	0.91
D						0.77	0.1	0.21	0.65	1.22
E	0.18	0.029	0.11	0.37	0.525	0.06	0.026	0.123	0.47	0.53
F	0.87	0.0212	0.053	0.19	0.978	0.55		0.076	0.22	0.763

The presence of different magnitudes at the above frequencies of the shaft voltage is generated due to the magnetic asymmetry at unequal air gap created by static and dynamic eccentricity. This may occur due to the exceeding allowable limit of quality control norms and may be controlled to some extent by high precision erection of the machines[9.5].

9.5.2 *Vibration Analysis*

The vibration levels and noise emission indicate the bearing condition. For an unhealthy rolling-element bearing levels of high frequency components and overall vibration levels increase considerably. Initially incipient damage like micros palls on the track surface of rolling-element bearings generates high frequency vibrations. However, when the defects in the roller track increase and grow in size, then the magnitude of low frequency components increases[9.16].

From the magnitude of vibration levels, it is evident that the overall vibration levels in axial and radial directions are within acceptable limits as per ISO 10816, Part 1-5 (Table 9.3). The presence of the principal slot harmonic frequency components and their side bands between 1300 Hz and 1600 Hz exist in all the motor bearings with varying higher levels in axial direction. This indicates the existence of varying degree of magnetic asymmetry of air gaps due to static and dynamic eccentricity in all the motors. This originates different levels of shaft voltages[9.1, 9.2, 9.3] as shown in the Tables 9.1 and 9.2, and lead to bearing failure due to the passage of electric current through bearings due to the various unforeseen causes as brought out in section 9.3 of this Chapter [Chapter 3].

Overall vibration levels in all the motors were in normal limits except in motors `B' and `F', where overall vibration levels were exceeded up to 2mm/s as shown in Table 9.3. Moreover, the major constituents of the spectrum of motor `B' and `F' found to exist in low frequency range, which indicates that the incipient damage has already set in these motor bearings. Shock Pulse Levels of motors `B' & `F' also indicate initial incipient stage of damage of bearings, where as condition of all other motor bearings is normal but needs immediate re-lubrication.

Table 9.3 Vibration levels of different motors in mm/S (rms)

Motor	Non-drive end			Drive end			Remarks
	H	V	A	H	V	A	
A	0.7	1.2	0.7	1.5	1.5	0.8	H,V,A are Horizontal Vertical and Axial, levels of vibration.
B	0.7	0.65	0.7	2.0	1.6	1.1	As per ISO 10816, PART 1-5 bearings are in good condition. Up to vibration levels of 1.8mm/s and above, these are in satisfactory condition.
C	0.8	0.8	0.65	1.2	1.1	0.8	
D	0.6	0.6	0.55	1.0	0.8	0.7	
E	0.65	0.35	0.35	0.8	0.9	0.5	
F	0.6	1.0	1.0	2.0	1.5	1.0	

9.5.3 *Passage of Current through Bearings*

Whenever a grease outlet pipe either in direct contact with the base plate or through the accumulation of grease outlet from the pipe and the base plate are in contact, this makes the closed circular path of current flow through the rotating shaft and the bearings. This happens even if the bearing insulation and/or insulation pads are intact. Similarly, accumulation of grease leakage from bearing seals on the base plate and pedestal makes the passage for feeble current flow. Furthermore, accumulation of excessive dirt/sludge between the corners base of bearing pedestal and base plate creates the path for a current flow even in presence of the insulation pads. This happens when the bearing is not insulated in the housings as shown in Fig. 10.1 of Chapter 10.

Under the condition of improper contact of grounding brush with shaft, brush is not able to make the path of least resistance for grounding the shaft current. The current then tend to pass through the bearing or tend to create localized loop in the bearing depending on the bearing impedance. This led to deteriorate the bearing condition and causes the bearing failure in due course. This happens when the bearing housing and bearing pedestal are not properly insulated.

Some times unshielded instrumentation cable contacting the base plate also create the path of least resistance for the flow of electric current through the bearings apart from the puncturing of the bearing insulation. This was established as the cause of failure of motor bearings `F'.

9.5.4 *Magnetic Flux Density*

The presence of magnetic flux density along with the corrugation pattern and corrosion on the bearing surfaces indicate the damage due to electric current as explained in Chapter 8[9.14],[9.15]. The presence of corrugation pattern without significant flux density distribution indicates plastic deformation of the bearing surfaces by mechanical loading, accompanied by the flexibility of the supporting structure, and is influenced by the frequency of rotation of rolling-elements in the inner race.

From the Fig. 9.2 it is evident that the non-insulated bearing NU 311 is located under the field winding in the alternator design. This makes the NU 311 bearing of alternator permanently magnetized and magnetic flux density develops on the bearing elements[9.5]. If the bearing is located away from the influence of the field coils, as shown in Fig. 9.1, magnetic flux density is not developed on the bearing elements[9.14].

When the outer race of the bearing is magnetized and the inner race and rolling-elements rotating inside the outer race, voltage is generated by electromagnetic principles and the flow of current starts through the inner race and rolling-elements depending on the impedance of the oil film thickness and threshold voltage phenomenon[9.4]. It is confirmed by the stray voltages of 2-6V and 6-20V of the alternator measured under `no load' and `load' conditions, respectively. This shows dissymmetry of the magnetized field, rotor eccentricity apart from the other causes given in references[9.3, 9.5]. However, in addition to other manufacturing errors, the major contribution to the higher voltage between the shaft and the body is attributed to the bearing location under the influence at the field coil, which is able to magnetize the bearing and damages in due course as explained in Chapter 8[9.5, 9.14, 9.15].

9.5.5 *Analysis of Failed Bearings*

The corrugation and ridges found on the track surface of the races of all the failed bearings besides the corrosion on the track surfaces. The track surface of the bearings are corroded because of the decomposition of the grease, and formation of corrugations on the track surfaces by low temperature tempering and Hertz ion pressure on the race ways[9.1,9.3], which finally results in the reduction of bearing fatigue life and failure in due course as dealt in references[9.8,9.9].

9.5.6 *Effects of Bearing Current on Lubricant*

The zinc additive, i.e., Zinc dithiosphosphate or zinc dialkyldithiophosphate (ZDTP), used as multifunctional additive in the grease, under rolling friction protects the rubbing metal surfaces and contributes to friction and wear reduction, which depends partly on the amount of additive on these surfaces. Decomposition of ZDTP in the lithium base greases under the influence of electric fields leads to the formation of lithium zinc silicate ($Li_{3.6}\,Zn_{0.2}\,SiO_2$) in the presence of a high relative percentage of free lithium and silica impurity in the grease under high temperature in the asperity contacts along with the formation of gamma lithium iron oxide ($r - LiFeO_2$). During the process lithium hydroxide is also formed which corrodes the bearing surfaces. And the original structure of lithium stearate changes to lithium palmitate. On the contrary, these changes are not detected under rolling friction. The used grease taken from the motor bearing `B' showed these changes, similar to that described in references[9.10, 9.11] and Chapter 3.

9.5.7 *Process of Bearing Failure under the Influence of Leakage Current*

When the current leaks through the roller bearing in which low resistivity (10^5 Ohm-m) grease has been used, a "silent" discharge passes through the bearing elements. This creates a magnetic flux density distribution on the bearing surfaces. Also, when a bearing is located and operates under the influence of magnetic field, voltage is generated and current flows through the bearing depending on the bearing impedance and the threshold voltage phenomenon. Under these conditions, this leads in the initial stages to the electrochemical decomposition of the grease, which corrodes the bearing surfaces[9.10] as explained in Chapter 11. And then gradual formation of flutings and corrugations on the surfaces[9.3, 9.7, 9.13] occurs as found in the bearings of motors `B' and `F'. Subsequently, wear increases and bearing fails.

9.6 Conclusions

From the above investigations and analysis, the following conclusions are drawn[9.17, 9.18]:

(i) Current passes through the bearing because of puncturing of bearing housing insulation, grease outlet pipe touching motor base frame and improper contact of grounding brush with the shaft.

(ii) Current can also pass through a bearing on such occasions when unshielded instrumentation cable touches the bearing, even if the bearing insulation is healthy.

(iii) If the bearing is not insulated in the housing, accumulation of dirt/sludge between the pedestal and base plate creates the path for leakage current through the bearing even if pedestal insulation pads are healthy.

(iv) Under the influence of field coils, a roller bearing develops magnetic flux. During operation of a bearing under the influence of magnetic field, voltage is generated. Also, stray voltage on the bearing is developed.

(v) Current passes through a bearing depending on the bearing impedance and threshold voltage phenomenon.

(vi) Levels of measured voltage between bearing and ground, between bearing and shaft indicate the condition of the bearing insulation. Vibration levels indicate the bearing condition.

(vii) A varying magnetic flux density of 10 to 40G is developed on the track surface of races of bearings depending on the flow of current.

(viii) In the case of bearings using grease of low resistivity, failure occurs under the `silent' electric discharge due to chemical decomposition, formation of flutings and corrugations on the bearing surfaces.

(ix) To avoid the bearing failure under the influence of electric current, bearings should not be located under the influence of field coils in the alternators.

(x) Bearings should be properly insulated and shielded for any possible means of leakage of electric current.

By the detection of the flux density on the bearing surfaces along with the corrugated pattern on the track surfaces, by analysis of deterioration of greases used in the bearings, and by measurement of stray and shaft voltages, the failure of bearings by the passage of electric current can be established. And the extend of deterioration is ascertained by the vibration analysis. Then, the bearings can be shielded for any possible means of leakage of current as reported for a trouble free operational life.

References

[9.1] Prashad H., "Investigations of Damaged Rolling-Element Bearings and Deterioration of Lubricants under the Influence of Electric Current," Wear 176 (1994), pp 151-161.

[9.2] Bradford, M., "Prediction of Bearing Wear due to Shaft Voltage in Electrical Machines " ERA Technology Limited, 1984.

[9.3] Prashad H., "Investigations on Corrugated pattern on the surface of Roller Bearings Operated under the Influence of Electrical Fields" Lubr. Engg, Vol. 44, 8, 1988, pp 710-718.

[9.4] Prashad H., Effects of operating parameters on the Threshold Voltages and Impedance Response of Non-insulated Rolling - Element Bearings under the Action of Electric Current", Wear 117, (1987) pp 223-240.

[9.5] Prashad H., "Magnetic Flux Density Distribution on the Track Surface of Rolling-Element Bearings - An Experimental and Theoretical Investigation", Tribology Transaction, Volume 39, 2, (1996) pp 386-391.

[9.6] Morgan, A. W. and Whillie, D., "A Survey of Rolling Bearing Failures", Proc. I. Mech. E.F, 184 (1969-70), pp 48-56.

[9.7] Anderson, S., "Passage of Electric Current through Rolling Bearings", The Ball Bearing Journal No. 153 (1968) pp 6-12.

[9.8] Simpson F.E., Crump J.J., "Effects of Electric Currents on the Life of Rolling contact Bearings," Proc. Lubrication and wear cony. Bournmouth, 1963, Institution of Mechanical Engineers, London, 1963, Paper 27.

[9.9] Prashad H., "Analysis of the Effects of on Electric Current on Contact Temperature Contact Stresses and Slip Band Initiation on the Roller Tracks of Roller Bearings ". Wear 131, (1989) pp 1-14.

[9.10] Prashad H., "Diagnosis of Deterioration of Lithium Greases used in Rolling - Element Bearings by X-ray Diffractometry", Tribol. Trans, Vol. 32, 2, 1989, pp 205-214.

[9.11] Komatsuzaki, S., "Bearing Damage by Electrical Wear and its Effects on Deterioration of Lubricating Grease". Lubrication Engineering, Vol 43, (1987), pp 25-30.

[9.12] Mas Remy, Albert Magnin, "Rheological and Physical Studies of lubricating greases before and after use in bearings" ASME, journal of Tribology, July 1996, Vol 118, pp 681-686.

[9.13] Winder L.R., Wolfe, O.J., "Valuable Results from Bearing Damage Analysis", Met.Proc. April 1968, pp 52-59.

[9.14] Prashad H., "Diagnosis of Failure of Rolling-Element Bearings of Alternators - A study ", Wear, 1997. Vol. 198, Issue 1-2, October 1996 pp 49-51.

[9.15] Prashad H., "The Effect of current leakage on Electro-adhension Forces in Rolling - Friction and Magnetic Flux Density Distribution on the Surface of Rolling-Element Bearing", ASME, Journal of Tribology, 110 (1988) pp 448-455.

[9.16] Prashad H., et al, "Diagnostic Monitoring of Rolling-Element Bearing by High-Frequency Resource Technique" ASLE Transactions Vol.28, 4 (1985) pp 39-448.

[9.17] Prashad,H., "Investigations and Diagnosis of Failure of Rolling-Element Bearings Due to Unforeseen Causes—A case Study", BHEL Journal, Vol.20,No.1,1999,pp 59-67.

[9.18] Prashad,H., Diagnosis and Cause Analysis of Rolling-Element Bearings Failure in Electric Power Equipments Due to Current Passage", Journal of Lubrication Engineering, STLE,5, May 1999, p30-35.

10 Localized Electrical Current in Rolling-Element Bearings

10.1 A General Review

The diagnosis and cause analysis of rolling-element bearing failure have been well studied and established in the literature. Failure of bearings due to unforeseen causes have been reported as: puncturing of bearing insulation; grease deterioration; grease pipe contacting the motor base frame; unshielded instrumentation cable; the bearing operating under the influence of magnetic flux, etc. These causes lead to the passage of electric current through the bearings of motors and alternators, which causes them to deteriorate in due course. Bearing failure due to localized electrical current between track surfaces of races and rolling-elements has not been hitherto diagnosed and analysed.

This study explains the cause of generation of localized current in the presence of shaft voltage. It also brings out the developed theoretical model to determine the value of localized current density depending on dimensional parameters, shaft voltage, contact resistance, frequency of rotation of shaft and rolling-elements of a bearing. Furthermore, failure caused by flow of localized current has been experimentally investigated.

10.2 Introduction

10.2.1 *Causes of shaft voltages and flow of current through bearings*

Shaft voltages exist in electrical machines as a result of asymmetries of various faults, e.g. winding faults, unbalanced supplies, electrostatic effects, air-gap fields, magnetized shaft and asymmetries of magnetic fields. The causes of shaft voltage can be grouped into four categories:

External causes;
magnetic flux in the shaft;
homo-polar magnetic flux;
ring magnetic flux.

Furthermore, friction between the belt and pulley can set up an electrostatic voltage between the shaft and bearings. Accidental grounding of a part of the rotor winding to the rotor core can lead to stray currents through the shaft and bearings, and can result in the permanent magnetization of the shaft. Also, a shaft voltage and current could be generated when the machine is rotated.

Homo-polar flux can result from an air gap or rotor eccentricity, and this can generate voltage[10.1,10.2].

The most important cause of bearing current is the linkage of alternating flux with the shaft. The flux flows perpendicularly to the axis of the shaft and pulsates in the stator and rotor cores. It is caused by asymmetries in the magnetic circuit of the machine, such as:

Uneven air gap and rotor eccentricity;
split stator and rotor core;
segmented punching;
axial holes through the cores for ventilation or clamping purposes;
key ways for maintaining the core stackings; and
segments of differing permeability.

All the causes listed above result in developing a magnetic flux, which closes in the circumference over the yoke and induces voltage on the shaft as the machine rotates. This results in a localized current at each bearing rather than a potential difference between the shaft ends. A current path, however, along shaft, bearings and frame results in a potential difference between the shaft ends[10.3].

It has been reported that at a certain threshold voltage, depending on the resistivity of the lubricant and operating conditions, current flows through the bearing[10.4]. Also, it was established by Busse *et al.*[10.5] that the dielectric strength of the lubricant is the ability to withstand voltage without breakdown. Prashad[10.6] found that the flow of circular current in the inner race leaks through the rolling-elements to the outer race by following a path of least resistance and establishes a field strength, leading to the development of magnetic flux on the track surface of races and rolling-elements. This has been experimentally investigated.

Studies were reported on the causes and control of electrical currents in bearings by Morgan and Wyllie[10.7]. Andreason[10.8] determined the flow of current through lubricated contacts. Simpson and Crump[10.9] studied the effect of electrical current on bearing life. The effect of operating parameters on an impedance response of a rolling element bearing has been analysed by Prashad[10.4,10.10]. The deterioration of lubricants used in non-insulated bearings investigated by Komatsuzaki[10.11], Prashad[10.12] and Remy and Magnin[10.13]. Investigations were also carried out to evaluate the pitting mechanism on the lubricated surfaces under an AC electric current by Lin *et al.*[10.14] and Chiou *et al.*[10.15]

The surveys of the failure of rolling-element bearings indicate various causes, including that of failure due to corrugations by Winder and Wolfe[16] and Prashad[10.17,10.18]. The mechanism of corrugations formed on the track surfaces and related investigations have been reported by the author[10.3,10.19]. In addition to the well-established causes, the bearing may also fail due to causes that are generally unforeseen during design and operation. These causes have been discussed by Prashad[10.20].

This chapter presents, with the help of a case study, the failure of bearings of motors by the effect of a localized electrical current between the track surface of races and rolling-elements under the conditions when the other causes of bearing failure have been ruled out by all possible investigations. The principle for the generation of a localized current is established and analysed along with the development of theoretical model to determine its value.

10.3 Bearing arrangement and the nature of bearing failure

Failure of bearings was detected in a few motors. These motors, rated 2100 kW and operating at 1494 rpm, have a three-bearing arrangement. The motors are used to drive primary air fans.

At the non-drive end (NDE) of the motor, bearing type NU 228 is used and is insulated. Fig. 10.1 shows the arrangement of the bearing in the motor.

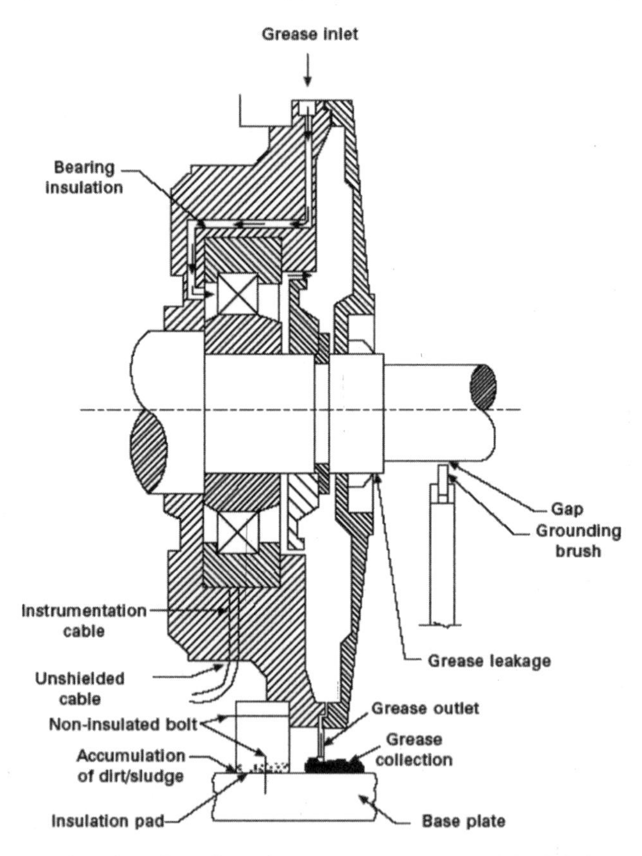

Fig. 10.1 Schematic showing bearing arrangement with lubricant flow path, unshielded instrumentation cable and other unforeseen causes leading to the passage of electric current.

At the drive-end (DE) of the motor, roller bearing type NU 232 and ball bearing type 6332 are used for taking both radial and axial loads simultaneously. All the bearings are grease-lubricated, and the motors are designed for continuous operation with periodic re-lubrication. Thermocouples were provided for continuous measurement of the temperature during operation.

After commissioning, a few motor bearings on both NDE and DE failed prematurely. The nature of the failure was due to the formation of corrugations and flutings on the roller tracks of races besides corrosion on the raceways and rolling-elements, irrespective of the make of the bearings used. A typical corrugation pattern on the track surface of a failed bearing is shown in Fig. 10.2. However, the depth of corrugations and degree of development of corrugations and flux density were different on various failed bearings. The colour of the grease between the rolling-elements was also found to change, by turning black.

Fig. 10.2 Typical corrugation pattern on the track surface of a failed bearing under the influence of electrical current.

10.4 Investigations, observations and data collection

Various investigations pertaining to shaft voltage and vibrations were carried out on bearings of different motors, both on NDE and DE, and various other related aspects were inspected. The shaft voltages were measured between NDE and DE shaft, DE shaft-to-ground, and NDE shaft-to-ground. Vibration levels were measured in micrometres and mm s^{-1} at both NDE and DE bearings. The dimensional accuracy of new bearings were checked and found to be in line with the specifications. Inspection of dimensional accuracy and metallurgical examinations of the components of the failed bearings also revealed consistency with the specifications. Various failed bearings were examined. Samples of fresh grease and used grease from the rolling-elements of the failed bearings were collected and analysed.

Potential causes of bearing failure, which might have led to passage of current though bearings, were ruled out as follows (Fig. 10.1):

- Grease pipe was not in contact with the base frame in any of the motors.
- Current path through bearings via collection of leaked grease and base frame was not detected.

- All the instrumentation cables were completely shielded.
- Insulation of the bearings was fully intact in all the failed motor bearings. This was established by value of high resistance measured across the bearing insulation.
- Passage of current through connecting bolts, nuts, etc. could not be established in any of the failed motor bearings.

The shaft voltage at the site on these failed motor bearings was measured as 1600mV before and after the replacement of failed bearings. The shaft voltage at the works during the test run was as high as 500mV for these motors. The difference of shaft voltage measured at the sites and the works may be attributed to the analogue/digital devices used as well as the coupled mode of operation and other unforeseen causes. For other normal running motors, shaft voltage as measured was less than 200mV. The failure of NU 228 bearings was reported at the site on these motors after a run of approximately 300 h. A grounding brush was not used in all such motors. Furthermore, there was not adequate space available on the shaft to install the grounding brush.

10.5 Theoretical model and approach to determine the flow of localized current in a bearing

In a rolling-element bearing using low-resistivity lubricant ($10^5 \Omega m$) in the presence of shaft voltage, current can flow through the bearings. This occurs in all such cases when current finds a low-resistance path to flow as discussed in the above Section. However, when the current does not find a way to pass through any of the paths as listed in the above Section , and the shaft voltage is more than the threshold voltage/safe voltage (>200mV), a substantial level of localized current between the rolling-elements and the track surface of the inner race may appear which may cause the bearing to deteriorate and the lubricant to disintegrate with the passage of time, causing flux density distribution, corrosion and blackening of the track surface as reported by Prashad[10.20]. This reduces the fatigue life of bearings and leads to premature failure.

The existing shaft voltage on the rotating shaft induces a voltage on the track surface of rolling-elements. Since the rolling-element rotates at a much higher frequency than the shaft rotating speed[19,21] this generates a higher voltage on the surface of rolling-elements as per the principles of electric fields and electromagnetism[10.22]. Thus, a potential difference between the inner race and rolling-elements develops. Revolution of rolling-elements along the track surface further affects the induced voltage on the track surface of the rolling-elements as well as their rotation around their own axis. This makes the analytical model quite complex. However, the potential difference between rotating shaft/inner race and rolling-elements leads to the passage of localized current at the contact surface, depending on the contact resistance between the contact zone of track surface of the inner race and the rolling-elements. The induced voltage phenomenon also occurs between the track surface of the outer race and the rolling-elements. Furthermore, the flow of localized current between the inner race and the rolling-elements, depending on the different points of contacts during rotation, turns into a circular current in the inner race and rolling-elements quite frequently in the course of operation. The proximity contact of rolling-elements with the outer race may lead to a flow of current through the outer race for a short duration. This flow of partial/full circular current establishes the field strength, leading to the development of residual magnetic flux on the track surface of races and rolling-elements in due course[10.6].

It may be imagined that current enters the track surface of the inner race because of the developed potential drop between rolling-elements and track surface of the inner race in the loaded zone at the surfaces of contact points between them, in a distributed form. The current flows around and outward until it concentrates at one or several rolling-contacts. As the rolling-elements orbit, the current carrying elements travel. Current then flows through each of the conducting rolling-elements, between two diametrically opposite contact areas. As the rolling-elements rotate with respect to the contact areas, these currents gradually might sweep out 360° around a major circle. If there was only one conducting rolling-element at a time, then two arc currents would travel in the race towards its contact areas, one clockwise and one counterclockwise. Depending on the resistance, these two currents may or may not be equal in magnitude. If more than one rolling-element contacts, then, of course, the situation is more complex. In the theoretical model, current flow in one direction in the inner race is assumed for analysis. The flow and break of flow of localized electrical current continue to depend on condition and regime of operation.

10.5.1 *Determination of speed of rolling-elements*

Under pure rolling, the absolute velocity at a point located on the circumference of a rolling-element is equal to the circumferential speed of the inner race. Simultaneously, an opposite point (at 180°) of the same rolling-element contacts the stationary outer race. The absolute speed of each point on the rolling-element is a combination of the speed of the rolling-element around its axis and the speed of the set of rolling-elements around the bearing axis. The absolute velocities of the two points 180° apart on the diameter of the rolling-element are parallel and vary linearly between the velocities at the points of contact on the inner and outer races.

With the above analysis, the rotating speed frequency of the rolling-elements has been determined as:[10.19]

$$f_b = 0.5\, f_s(D/d)\,[1 - (d/D)^2 \cos^2 \alpha] \tag{10.1}$$

The ratio of the speed of rotation of rolling-elements and shaft speed is given as:

$$V_{bs} = f_b / f_s = 0.5\, D/d[1 - (d/D)^2 \cos^2 \alpha] \tag{10.2}$$

For a radial cylindrical roller bearing, $\alpha = 0$. So:

$$V_{bs} = 0.5[(D^2 - d^2)/dD] \qquad (10.3)$$

10.5.2 *Induced voltage on rolling-elements*

As the shaft voltage E develops and changes with time, the voltage on the track surface of the inner race E_{ir} changes accordingly. This is because the inner race is press fitted on the shaft and is considered integral with the shaft. The voltage E_{ir} induces the voltage on the rolling-elements rotating around their own axis and moving along the track surface of the inner and outer races. This makes the phenomenon of induced voltage very complex.

The rotating rolling-elements loop may be considered as a type of AC generator or, in other words, rolling-elements develop voltage depending on the speed of rotation. The voltage in absolute terms is more on those surfaces of the rolling-elements, which are at right angles to the existing field, because of the shaft voltage on the track surface of the inner race and is zero on the parallel surfaces to the field. The developed voltage on rolling-elements broadly depends on the area of such a loop formed by rotating rolling-elements and not on their shapes. The voltages thus developed on the surfaces of rolling-elements E_{ir} depend on the rotating speed of the rolling-elements as well as other electromagnetic factors. Considering the ratio of the speed of rolling-elements with respect to the shaft speed V_{bs} the voltage on the rolling-elements E_r is taken as the product of V_{bs} and the shaft voltage E using principles of electrical current theory[10.22]. This leads us to determine the potential difference between rolling-elements and track surface of inner race as:

$$E_{rir} = E_r - E_{ir} \qquad (10.4)$$

On using Equation (10.3), the potential difference E_{rir} is determined as

$$E_{rir} = E\, V_{bs} - E \qquad (10.5)$$

where $E_{ir} = E$ and $E_r = E\, V_{bs}$

or,

$$E_{rir} = E\left[\frac{(D^2 - d^2 - 2Dd)}{2Dd}\right] \qquad (10.6)$$

Furthermore, development of the same induced voltage as on rolling-elements occurs on the track surface of the outer race under the influence of voltage existing on the surfaces of rolling-elements. Thus, the potential difference between the rolling-elements and the outer race will be insignificant as the outer race of the bearing is stationary[10.22].

10.5.3 *Bearing resistance*

The resistance between the track surface of the inner race and the rolling-elements depends on bearing kinematics, lubricant film thickness and width of contact between track surface and rolling-elements and the number of rolling-elements in loaded zone. For the bearing using low-resistivity lubricant ($10^5 \Omega$m), the resistance (R_c) of the bearing as determined is found to vary between 0.1 and 0.5Ω.[4]

10.5.4 *Localized electrical current in bearings*

Localized current between the track surface of the inner race and the rolling - elements is determined using Equation (10.6) and the contact resistance (R_c), and is given as:

$$I_b = \frac{E_{rir}}{R_c} = \frac{E}{R_c}\left[\frac{D^2 - d^2 - 2Dd}{2Dd}\right] \tag{10.7}$$

10. 6 Field strength on track surfaces of races and rolling-elements

Because of the presence of localized electrical current in the partial/complete arc of the inner race and rolling-elements, the field strength develops on the track surfaces of rolling-elements and the inner race in due course. The field strength on these elements can be determined separately as reported by Prashad[10.6].

10.6.1 *Field strength on the track surface of inner race*

The field strength H_{irr} on the track surface of the inner race due to flow of current in rolling-elements has been analysed, and is determined as[10.6]:

$$H_{irr} = 2\pi I_b[R^2 - R_{ir}^2]/R^3 \tag{10.8}$$

10.6.2 *Field strength on rolling-elements*

On rotation, rolling-elements change polarity and, the outer race being stationary, the field strength on the surface of rolling-elements depends on the flow of circular current in the inner race and is given as[10.6]:

The chances of development of significant field strength on the track surface of the outer race due to the flow of intermittent local current in the track arcs of inner race and rolling-elements are quite remote, and hence it is not considered in the analysis. However, sometimes development of traces of flux density on the outer race track surface cannot be ruled out.

$$H_{irr} = 2\pi I_b[R^2 - R_{ir}^2]/R^4 \tag{10.9}$$

10.7 Magnetic Flux Density

The field strength on the track surface of the inner race and rolling-elements of the bearing gives rise to the development of magnetic flux density on these surfaces. These can be determined analytically as reported by Prashad[10.6].

10.7.1 *Magnetic flux density on the track surface of the inner race*

The magnetic flux density on the track surface of the inner race of the bearing B_{ir} due to field strength H_{irr} in an oil medium of relative permeability U_r, with respect to free space is given (in tesla) by Prashad[10.6]:

$$B_{ir} = U_o U_r H_{ir} = 4\pi \times 10^{-7}\, U_r H_{irr} \qquad (10.10)$$

which is determined (in gauss) using Equation (10.8) as:

$$B_{ir} = 78.96 \times 10^{-3}\, U_r I_b (R^2 - R_{ir}^2)/R^3 \qquad (10.11)$$

10.7.2 *Magnetic flux density on the track surface of rolling-elements*

On using Equations (10.9) and (10.10), the residual flux density on a few rolling-elements (in gauss) is determined as:

$$B_r = 78.96 \times 10^{-3}\, U_r I_b R_{ir} (R^2 - R_{ir}^2)/R^4 \qquad (10.12)$$

In may be noted that the residual magnetization of steel, which remains after an alternating magnetizing current is switched off, bears no simple theoretical relationship to the magnetization during passage of a direct current of the same effective value.

10.8 Determination of time span for the appearance of flutes on the track surfaces

Instant thermal stresses due to thermal transients on the roller track of races caused by roller contact under the influence of electrical current depend on an instant rise in temperature. As the temperature rise stabilizes, the contact thermal stresses increase and affect the fatigue life. The duration the bearing would have taken after the formation of slip bands and before the appearance of flutes/corrugations on the track surface of the inner race is determined as[10.23]:

$$t = \frac{(\pi\partial_{corr})^2\,(BD_{ir} + BD_{or} + LdN)\,(BD_{ir}\Delta_{ir}^{-1} + BD_{or}\Delta_{or}^{-1} + LdN\,\Delta_r^{-1})}{2E_{rir} I_b Y}$$

10.9 Data Deduction

All the failed bearings both on NDE and DE ends of the three motors were examined. In these bearings, the damage as reported in Section 10.3 was studied. Investigations were carried out on NU 228 bearings and flux density distribution on their track surfaces was detected using a Hall probe, similar to the procedure reported by the author[10.6]. The pitch of corrugations on the track surfaces of the NU 228 bearing was measured. The theoretical time span for the formation of corrugations after the slip band formation was determined using Equation (10.13). The dimensional and operating parameters of bearing type NU 228 and various values of measured/analytical parameters are given in Tables 10.1 and 10.2, respectively.

Table 10.1 Dimensional and operating parameters of NU 228 bearing

B	=	42mm
R_{ir}	=	79mm
R_{oro}	=	125mm
R_{or}	=	116mm
R	=	97.5mm
d	=	37mm
D	=	195mm
D_{ir}	=	158mm
D_{or}	=	232mm
L/B	=	1
N	=	14
n	=	1500 rev min^{-1}

Table 10.2 Experimental and theoretical data

E	=	1.6V
R_c	=	0.20Ω
E_{rir}	=	2.46V
E_{rir} / E	=	1.54
I_b	=	12.30A
Y	=	210 × 10^{9}Nm^{-2}
B_{ir}	=	3.6 G (theoretical)
B_{ir}	=	5 G (maximum on measurement)
B_r	=	3 G (experimental on a few rolling-elements)
B_r	=	2.7 G (theoretical)
Δ_{ir}	=	0.25mm
Δ_{or}	=	not detected
Δ_r	=	not detected
∂_{corr}	=	700 × 106 Nm^{-2}
t	=	107.45h
t_1	=	300h

10.10 Results and Discussion

10.10.1 *Potential difference between rolling-elements and track surface of the inner race and flow of localized current between them*

The potential difference between the rolling-elements and the track surface of the inner race of an insulated bearing working under the influence of shaft voltage depends on shaft voltage, pitch diameter and diameter of rolling-elements as per the derived relations in Equation (10.6). Bearing type NU 228, working under the influence of shaft voltage of 1.6V, develops a potential difference of 2.46V between the rolling-elements and the track surface of the inner race. The ratio of this potential difference to the shaft voltage E_{rir}/E_r has been worked out as 1.54 (Table 10.2). This ratio depends on the speed of rotation of rolling-elements to the shaft speed and is a function of dimensional parameters of a roller bearing, Equation (10.3). Furthermore, localized electric current between the track surface of the inner race and rolling-elements depends on the potential difference E_{rir} and contact resistance R_c. It is a function of shaft voltage E contact resistance R_c and dimensional parameters of the bearing Equation (10.7).

The intensity of local current has been determined theoretically as 12.30 A for the existing shaft voltage of 1.6V and resistance R_c of 0.2Ω. (Table 10.2). The localized electrical current damages the track surfaces of the bearing in due course and leads to form corrugations. In contrast, for shaft voltages of 500 and 200mV, the value of local current has been determined as 3.84 and 1.54 A, respectively. These low values of current do not affect the bearing during operation. In addition, the outer race being stationary, no significant potential difference is generated between the outer race and the rolling-elements, and thus current does not pass between them. However, a momentary circular current, through an arc of the outer race, can flow as and when proximity contact of the outer race and rolling-elements takes place in a sector owing to instabilities/vibrations during operation.

10.10.2 Residual flux density distribution on the track surface of inner race and rolling-elements

A flow of current between the track surface of inner race and rolling-elements due to the existing potential drop leads to the generation of a residual flux density distribution on the track surface of inner race and rolling-elements in due course. The flux density on the track surface of the inner race has been determined theoretically as 3.6G, using Equation (10.8). The maximum value of flux density was determined as 5G by measurement (Table 10.1). On a few rolling-elements flux density was determined and found to be a maximum of 4G. Theoretically, flux density was determined as 2.7G on rolling-elements using Equation (10.9). The presence of flux density by measurement and theoretical investigations indicates that the localized current passed through the bearing arc partially[10.6]. This created corrugations/flutes on inner race track surface of the bearings and deteriorated the lubricant, which has corroded the surfaces of bearings and led to failure by the reduction of fatigue life[10.10, 10.12, 1014].

10.10.3 Assessment of time span before appearance of flutes on the track surfaces

As a result of the shearing of atomic planes within the crystals, some crystals on the roller track of the inner or outer race develop slip bands. The slip bands are formed in the subsurface prior to the appearance of flutes/corrugations on the track surfaces. The formation of slip bands is initiated by shear stress caused by operating parameters, and is accelerated by the passage of electric current, corrosion and oxidation of track surfaces. Furthermore, formation of slip bands depends on lubricant characteristics and quality of a bearing. The following may take place before the initiation of corrugations:

- Generation of persistent slip bands (PSB).
- Crack form along PSB and initiated from the tip of slip bands.
- Formation and propagation of flutes/corrugations on the track surface.

After the cracks/slip bands under the track surfaces are initiated, the process that governs the propagation leading to the formation of corrugations/flutes is considered by the continuum theory of Griffith[10.10, 10. 23]. The time a bearing takes before formation of flutes on the track surfaces after the slip bands formation depends on the dimensions of the track surface, pitch of corrugations, number of rolling-elements, potential difference between the track surface and rolling-elements, intensity of local current and properties of the bearing material. The pitch of corrugations on rolling-elements and outer race was not found and hence the values of Δ_r^{-1}, Δ_{or}^{-1} are negligible, Equation (10.13). This time span t for the initiation of corrugations is determined as 107.45 h, using Equation (10.13), without considering the time for formation of slip bands (Table 10.2). The net time for the development of slip bands, including that of flute formation on track surface of the inner race after commissioning, is found to be approximately 300 h according to site data as given in Table 10.2. This might match practically with the actual time of initiation of flutes after formation of slip bands, as determined analytically.

10.10.4 *Bearing failure under localized current*

Electrical current damage of a bearing is of two types. In the first type, when the low-resistivity lubricant (< 10^5 Ωm) is used in the roller bearings, a silent discharge occurs through the bearing elements under the influence of electrical current. This breaks down the used lubricant and corrodes the surfaces of the bearings, which lowers the fatigue life of the bearings as discussed by Prashad[10.3,10.12]. Furthermore, the passage of current through a bearing increases the bearing's operating temperature. Subsequently, thermal stresses are increased and surface heating takes place, which leads to low-temperature tempering of the track surfaces. This accelerates the formation of slip bands and corrugations on the track surfaces in due course[10.10, 10.18]. Magnetic flux density is also developed on the track surfaces, and the original structure of the lubricant undergoes changes[10.6, 10.12].

In the second type of bearing failure, when the high-resistivity lubricant (> 10^9 Ωm) is used in the roller bearings, an accumulation of charges occurs on the track surfaces until it reaches a threshold critical value, when breakdown takes place. This leads to damage of the track surface caused by arcing. This is accompanied by mass transfer and elevated local temperature on the asperity of the contact surfaces[10.4, 10.25].

10.11 Conclusions

Based on the above analysis, the following conclusions are drawn:[10.26]

- Under the influence of higher shaft voltage, a rolling-element bearing using low-resistivity lubricant may deteriorate because of the development of a potential difference between the track surface of the inner race and rolling-elements. This leads to the passage of localized current, depending on the contact resistance between them.
- The potential difference between the track surface of the inner race and rolling-elements develops because of the higher frequency of rotation of rolling-elements compared with the shaft speed. The potential difference, thus developed, depends on shaft voltage and bearing kinematics.
- The ratio of potential difference between track surfaces of inner race and rolling-elements to that of shaft voltage is a function of pitch and rolling-elements diameter of a bearing. For the NU 228 bearing, it is determined as 1.54.
- The development of magnetic flux density on the track surface of the inner race and rolling-elements indicates flow of locally generated current between them.
- The time of appearance of flutes on the track surface can be estimated by bearing kinematics, existing potential difference between track surface of inner race and rolling-elements, value of localized current, properties of bearing material, together with measured values of pitch of the corrugations on the track surfaces.
- The failure of a bearing by the localized current can be avoided by limiting the shaft voltage to a maximum of 200mV. If a higher value of shaft voltage persists, it is preferable to ground a brush. However, altering the path of flow of circular current, through dismantling and reassembling of the bearing as a means of modulating the track surfaces under interaction, the effect of localized current can sometimes be minimized.
- Localized current in rolling-element bearings is a complex phenomenon and needs the solution of complex analytical models and investigations.

References

[10.1] Prashad, H., 'Investigations of Damaged Rolling-element Bearings and Deterioration of Lubricants under the Influence of Electric Current,' *Wear*, **176**, 151–161, 1994.

[10.2] Bradford, M., 'Prediction of Bearing Wear due to shaft Voltage in Electrical Machines', *Technical Report No. 84–007*, ERA Technology Limited, England, pp. 49–53, 1984.

[10.3] Prashad, H., 'Investigations on Corrugated Pattern on the Surface of Roller Bearings Operated Under the Influence of Electrical Fields', *Lubr. Eng.*, **44**(8), 710–718, 1988.

[10.4] Prashad, H., 'Effects of Operating Parameters on the Threshold Voltages and Impedance Response of Non-insulated Rolling-element Bearings under the Action of Electric Current', *Wear*, **117**, 223–240, 1987.

[10.5] Busse, D., Erdman, J., Kerkman, R.J., Schlegel, D. and Skibinski, G., 'System Electrical Parameters and their Effects on Bearing Currents', *IEEE Trans. Ind. Appl.* **33**(2) 577–584, 1997.

[10.6] Prashad, H., 'Determination of Magnetic Flux Density on the Surfaces of Rollingelement Bearings as an Indication of the Current that has Passed Through Them –An Investigation', *Tribol. Int.*, **32**, 455–467, 1999.

[10.7] Morgan, A.W. and Wyllie, D., 'A Survey of Rolling Bearing Failures', *Proc. Inst. Mech. Eng., Part F* **184**, 48–56, 1969–70.

[10.8] Andreason, S., 'Passage of Electric Current through Rolling Bearings', *Ball Bearing J.* **153**, 6–12, 1968.

[10.9] Simpson, F.E. and Crump, W.J.J., 'Effects of Electric Currents on the Life of Rolling Contact Bearings', *Proc. Lubrication and Wear Convention*, Bournemouth, Inst. Mech. Eng., London, Paper 27, pp. 296–304, 1963.

[10.10] Prashad, H., 'Analysis of the Effects of Electric Current on Contact Temperature Contact Stresses and Slip Band Initiation on the Roller Tracks of Roller Bearings', *Wear*, **131**, 1–14, 1989.

[10.11] Komatsuzaki, S., 'Bearing Damage by Electrical Wear and its Effects on Deterioration of Lubricating Grease', *Lubr. Eng.*, **43**(1), 25–30, 1987.

[10.12] Prashad, H., 'Diagnosis of Deterioration of Lithium Greases used in Rollingelement Bearings by X-ray Diffractometry', *Tribol. Trans.*, **32** (2), 205–214, 1989.

[10.13] Remy, M. and Magnin, A., 'Rheological and Physical Studies of Lubricating Greases Before and After Use in Bearings', *Trans. ASME, J. Tribol.*, **118**, 681–686, 1996.

[10.14] Lin, C.M., Chiou, Y.C. and Lee, R.T., 'Pitting Mechanism on Lubricated Surface of Babbit Alloy/Bearing Steel Pair under AC Electric Field', *Wear*, **249**, 133–142, 2001.

[10.15] Chiou, Y.C., Lee, R.T. and Lin, C.M., 'Formation Criterion and Mechanism of Electrical Pitting on the Lubricated Surface under AC Electrical Field', *Wear*, **236**, 62–72, 1999.

[10.16] Winder, L.R. and Wolfe, O.J., 'Valuable Results from Bearing Damage Analysis', *Metal Progress*, **4**, 52–59, 1968.

[10.17] Prashad, H., 'Diagnosis of Failure of Rolling-element Bearings of Alternators – A Study', *Wear*, **198**, 46–51, 1996.

[10.18] Prashad, H., 'The Effect of Current Leakage on Electro-adhesion Forces in Rolling – Friction and Magnetic Flux Density Distribution on the Surface of Rollingelement Bearing', *Trans. ASME, J. Tribol.*, **110**, 448–455, 1998.

[10.19] Prashad, H., Ghosh, M. and Biswas, S., 'Diagnostic Monitoring of Rolling-element Bearing by High-frequency Resource Technique', *ASLE Trans.*, **28**(4), 439–448, 1985.

[10.20] Prashad, H., 'Diagnosis and Cause Analysis of Rolling-element Bearing Failure in Electrical Power Equipment Due to Current Leakage', *Lubr. Eng.*, **55**(5), 30–35, 1999.

[10.21] Prashad, H., 'The Effect of Cage and Roller Slip on the Measured Defect Frequency Reference of Rolling-element Bearings', *ASLE Trans*, **30**(3), 360–66, 1987.

[10.22] Starling, S.G., *Electricity and Magnetism*, 7th ed., Longmans, Green and Co., New York, 1960.

[10.23] Prashad, H., 'Determination of Time Span for Appearance of Flutes on Track Surface of Rolling-element Bearings Under the Influence of Electrical Current', *Tribol. Trans*, **41**(1), 103–109, 1998.

[10.24] Prashad, H. and Murthy, T.S.R., 'Behavior of Greases in Statically Bounded Conditions and When Used in Non-insulated Anti-friction Bearings Under the Influence of Electrical Fields', *Lubr. Eng.*, **44**(3), pp. 239–246, 1988.

[10.25] Prashad, H., 'Theoretical Analysis of the Effects of Instantaneous Charge Leakage on Roller Track of Roller Bearings Lubricated with High Resistivity Lubricants', *Trans. ASME J. Tribol.*, **112**, 37–43, 1990.

[10.26] Prashad, H., 'Diagnosis of Rolling-element Bearings Failure by Localized Current Between Track Surfaces of Races and Rolling-Elements', *Trans. ASME J. Tribol.*, **124**, 468–473, 2002.

Nomenclature

B	width of track surface
B_{ir}	magnetic flux density on track surface of inner race
B_{or}	magnetic flux density on track surface of outer race
B_r	magnetic flux density on rolling-elements
d	diameter of rolling-element
D	pitch diameter
D_{ir}	outside diameter of inner race
D_{or}	inner diameter of outer race
E	shaft voltage
E_r	voltage on rolling-elements
E_{ir}	voltage on track surface of rotating inner race
E_{rir}	potential difference between rolling-elements and track surface of inner race
f_s	shaft rotational frequency
f_b	rolling-element frequency
H_{irr}	field strength on track surface of inner race due to flow of current in rolling-elements
H_{rir}	field strength on surface of rolling-elements due to flow of current in inner race
I_b	bearing current
L	length of rolling-element
N	number of rolling-elements in a bearing
n	rpm
R_c	bearing resistance
R_{ir}	track radius of inner race
R_{or}	track radius of outer race
R_{oro}	outside radius of outer race
R	pitch radius of bearing
t	time required for appearance of flutes after slip band formation on track surfaces
t_1	net time of bearing operation after commissioning for inspection
U_r	relative permeability oil with respect to free space (=1)

U_o	permeability of free space (4×10^{-7} Hm^{-1})
V_{bs}	ratio of speed of rolling-elements and shaft speed (f_b/f_s)
Y	Young's modulus of elasticity
$\propto$	contact angle
∂_{corr}	stress for flute appearance from opening of the tip of flute from an existing slip band on the track surface
Δ_{ir}, Δ_{or}, Δ_r	pitch of corrugation on inner race, outer race and rolling-element, respectively

11 Rolling-Element Bearings Under the Influence of an Electric Current

11.1 A General Review

In this chapter the response and performance of roller bearings operating under the influence of an electrical current are analysed and the effect of electrical current on the lubricating greases, analysis of pitch and width of corrugations formed on the roller track of races, threshold voltage phenomenon, impedance response and electro-adhesion forces in rolling friction are discussed. The roles of bearing kinematics and operating conditions on the capacitance, impedance and charge accumulation are discussed. The methodology to determine contact stresses, rise in contact temperature and the number of cycles before the slip band/crater initiation on the track surfaces is also assessed. The effects of the capacitive response of bearings on repeated starts and stops of a machine and instantaneous temperature rise owing to discharge of the accumulated charges are highlighted. This chapter gives an analysis of various aspects of roller bearings working under the influence of different levels of shaft voltages and a discussion of the mechanism of bearing failure and makes it possible to predict the life of a roller bearing lubricated with lubricants of different resistivities. It also investigates bearing response and provides the potential to analyse the performance of a roller bearing operating under the influence of electric currents.

11.2 Introduction

Various surveys have indicated that about 30% of all motor failures are due to bearing damage accounted for by the bearing current. Insulating the bearings from the shaft and machine frame can prevent this. Such insulation is an additional complication in machine design and construction, and is difficult to incorporate in flameproof enclosures. Intermittent shorting of this insulation might occur inadvertently in service, leading to sparking, which is dangerous in a hazardous environment. This also re-establishes the bearing current and tends to make the bearings deteriorate. The effect of electrical currents on bearings in different modes of operation needs to be investigated thoroughly.

Magnetic flux develops in electrical machines due to dissymmetry of the magnetic circuits, which close in the circumference over the yoke, and induces a voltage on the shaft. Shaft voltages can be established in rotating machines of any type, but more particularly in electrical machines. The causes of such voltages can be categorized as external causes, e.g. electrostatic effects, operational faults, magnetic flux in the shaft, homo-polar magnetic flux and ring magnetic flux. In general, shaft voltages exist in electrical machines as a result of asymmetry of faults, i.e. winding faults, unbalanced supplies, air-gap fields, magnetized shaft or other machine members, asymmetries of magnetic fields, etc. Asymmetries are caused by rotor eccentricity, poor alignment, manufacturing tolerances, uneven air gaps, segmental lamination punching, variation in permeability and various other unforeseen reasons. These asymmetries in electrical machines result in a net flux or a current linking with the circuit consisting of shaft, bearing and frame. Shaft flux results in localized currents at each bearing rather than a potential difference between shaft ends. Furthermore, under the influence of a potential drop across a roller bearing, the varying film thickness between races and the rollers form a capacitor of varying capacitance, depending on the permittivity of the lubricant. The minimum film thickness between the races and the rollers offers maximum capacitance and minimum capacitive reactance. The active resistance offered by a bearing is minimum at the minimum film thickness; however, it is primarily governed by the resistivity of the lubricant. Under the influence of the shaft voltage, the electrical interaction between the races and the rollers in the presence of the oil film is like a resistor–capacitor (RC) circuit and offers impedance to the current flow.

Fatigue and surface distress usually describe the limits for reliable operation of a rolling-element bearing. Fatigue initiating in the subsurface is better understood than the fatigue initiated on the surface itself, i.e. surface distress. In general, analysis of bearing failures seldom indicates subsurface-initiated fatigue as the reason for failure. The most common form of bearing failure is caused by surface distress. This includes different types of damage including micro-pitting, smearing, indentations and plastic deformation as well as surface corrosion. Obviously, as per the application demands, bearings work under higher temperatures and heavily stressed conditions. Such conditions are unavoidable when bearings are to operate under the influence of shaft voltage and other rigorous operating conditions. This leads to technical innovation covering materials, modified bearing surface coatings and lubricants. Furthermore, this opens a new phase for the evaluation of bearings. In addition, under the effect of a shaft voltage, dirt, metallic particles and irregular lubricant film permit the lubricant oil film to be pierced by the electric current. Under these conditions, the impedance of the bearing circuit becomes so low that small shaft voltages may cause substantial bearing currents. If currents are reduced by using high-resistivity lubricant or by establishing a non-conducting oil film in the bearing, the self-inductance of the single loop of the shaft, bearing and casing may then cause a relatively higher induced voltage across the oil film. If this induced voltage is exceeded, the threshold voltage may again break down the lubricating film. Until then, when the voltage is below the threshold voltage, the bearing gives the capacitive response and stores the charges between the rolling elements and the roller track of races.

Much of the work referred to above has been carried out and the purpose of this chapter is to highlight the recent investigations on response and performance of bearings under the influence of shaft voltages. The literature on this subject is very scarce and efforts are being made to understand the phenomenon. It is often asked whether the bearing damage is due to leakage of electrical current. In general, when the bearings carry current as a necessary part of an electrical circuit, the current is self-induced as a result of the design characteristics of the machine, or the current is due to an electrostatic phenomenon, the bearing current affects the lubricating media (grease, oil), bearing surfaces and response characteristics of the bearings. The investigations carried out on lubricants and bearing surfaces indicate in-depth behaviour of bearings under the influence of an electrical current, which has been highlighted in the following sections.

11.3 Behaviour of Grease in Non-insulated Bearings

Recent investigations indicate that the resistivity of the grease depends on its EP properties, viscosity, torque characteristics and consistency. The resistivity of grease also varies with respect to voltage and time. The difference in resistivities among greases can be as high as 10^5 times. The Pl.check subscript on 10 change in resistivity depends on the nature of the impurities or by-products and the types of additive in the greases, besides their density, compressibility and structure. Low-resistivity grease (10^5 Ωm) tends to pl.check subscript every where as comparison to original 'recoup' its resistivity when the applied field is switched off. The percentage of 'recouping' varies from 18 to 82 and depends on the stretching of the molecules. Under the influence of an electric field, the carboxylic group at 2660 cm^{-1} of the low-resistivity lithium base grease decomposes and the lithium metal concentration in the aqueous solution increases relatively. The decomposition of carboxylic acid leads to corrosion of the bearing track surfaces before the pitting process is initiated. However, the oil content of the grease is not effected by the electrical field, but the carboxylate anion stretching and carboxylic group present in the soap residue undergo changes. Figs. 11.1 and 11.2 indicate variation of resistivity of lithium base grease 'A' with time at different potential drops across electrodes, and variation as well as 'recouping' of resistivity with time under 50V potential drop across electrodes, respectively. Figs 11.3 and 11.4 show IR spectra of the soap residue of fresh and used greases[11.1-11.5].

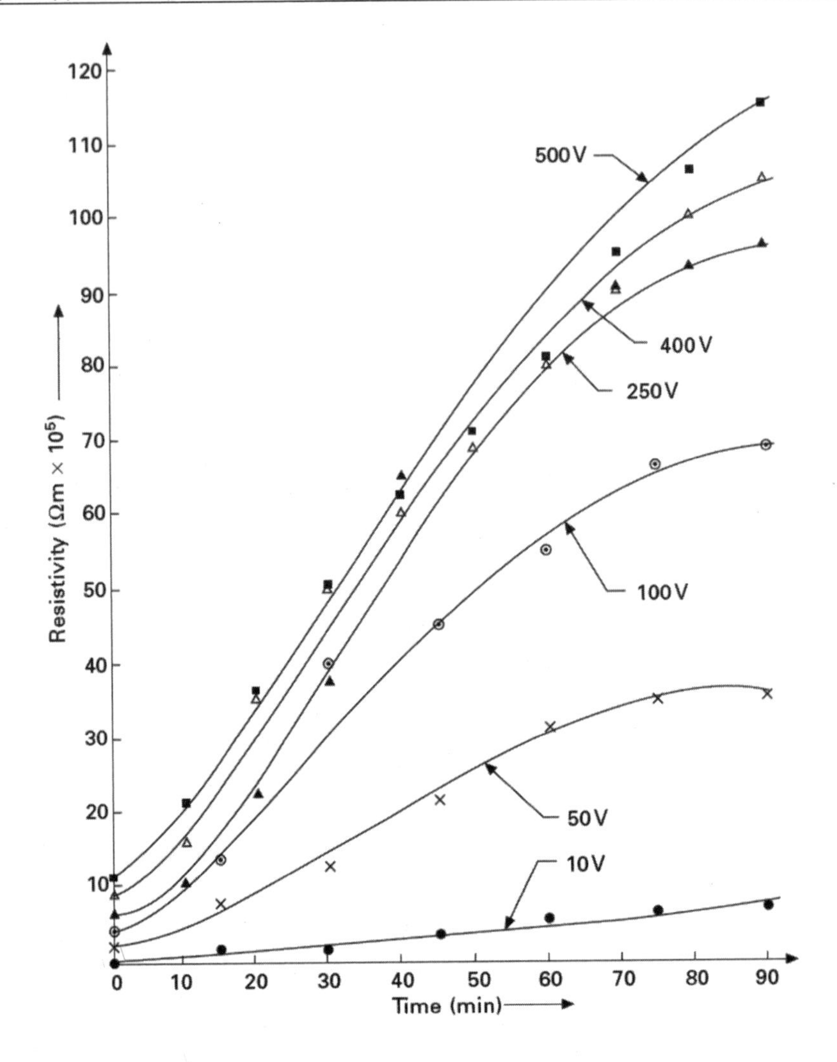

Fig. 11.1 Variation of resistivity of grease 'A' with time at different potential drops across electrodes.

The original structure of the grease and structure of the grease after operation of a rolling-element bearing under electrical field have also been studied by X-ray diffraction techniques[11.6, 11.7]. This indicates that the original structure of the soap residue of the fresh lithium grease is lithium stearate ($C_{18}H_{35}LiO_2$), which changes to lithium palmitate ($C_{16}H_{31}LiO_2$), a lower fraction of hydrocarbon, after operation of a bearing under electrical fields; the original structure of the grease is not changed under pure rolling friction. Also, gamma lithium iron oxide and lithium zinc silicate ($Li_{3.6}Zn_{0.2}SiO_2$) are formed in the presence of Zn, Fe and SiO_2 (Fig. 11.5 and 11.6). In contrast, under pure rolling friction only lithium iron oxide is detected. The crystalline structure of a fresh lithium grease in a bearing changes to an amorphous structure. Under the influence of an electrical current, the formation of lithium hydroxide and lithium carbonate makes the dielectric alkaline and corrodes the bearing surfaces, which lead to increased wear and failure of a bearing.

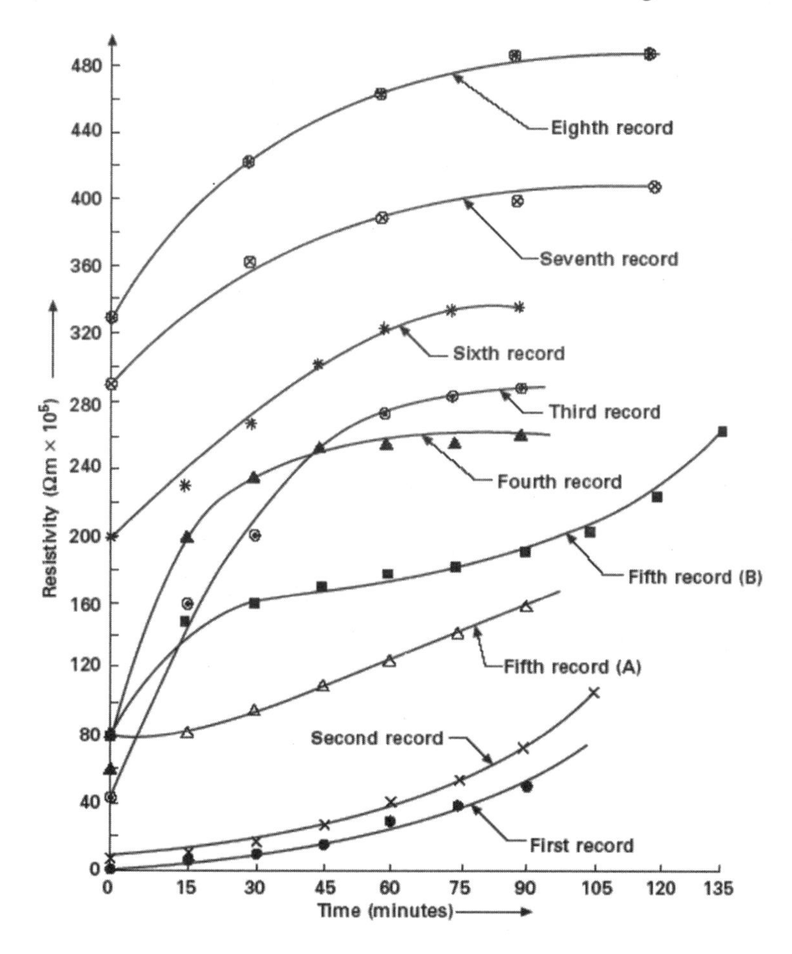

Fig. 11.2 Variation and recovery of resistivity of grease 'A' with time under 50V potential drop across electrodes.

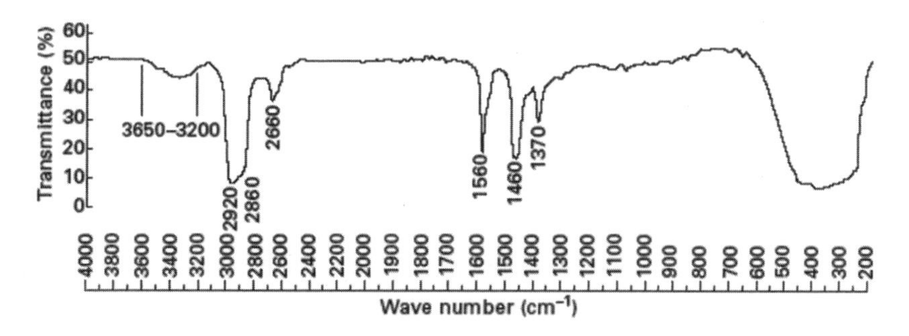

Fig. 11.3 IR spectrum of soap residue extracted from fresh grease 'A'.

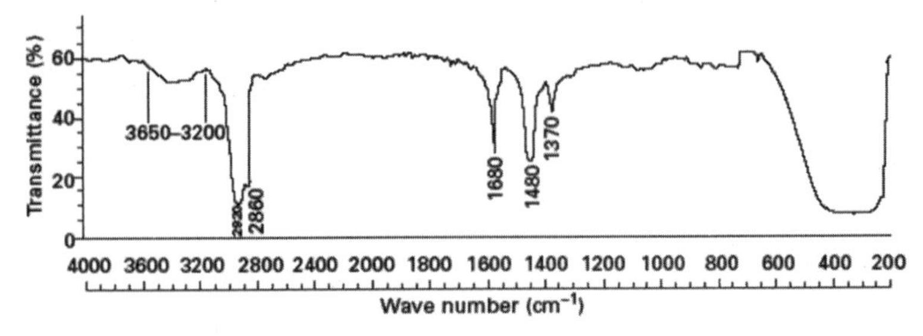

Fig. 11.4 IR spectrum of grease 'A' taken from NU 326 bearing after exposure to 50A current (AC) for 41 hours.

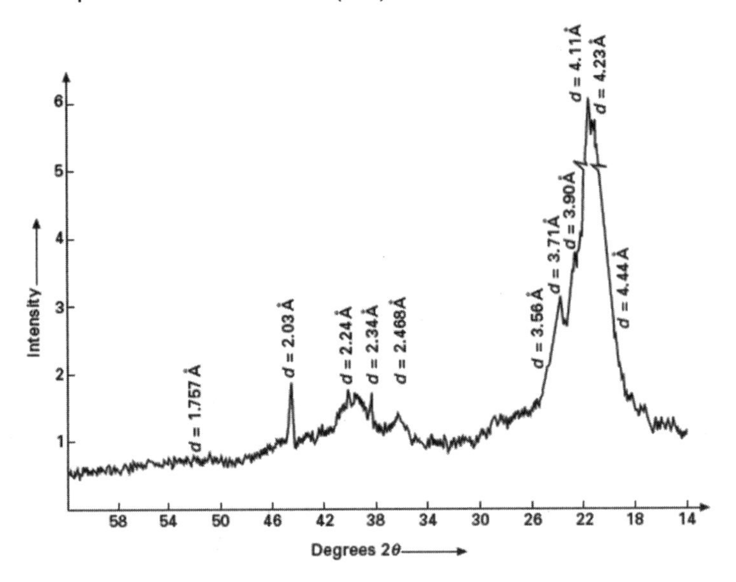

Fig. 11.5 X-ray diffraction pattern of soap residue for fresh grease 'A'.

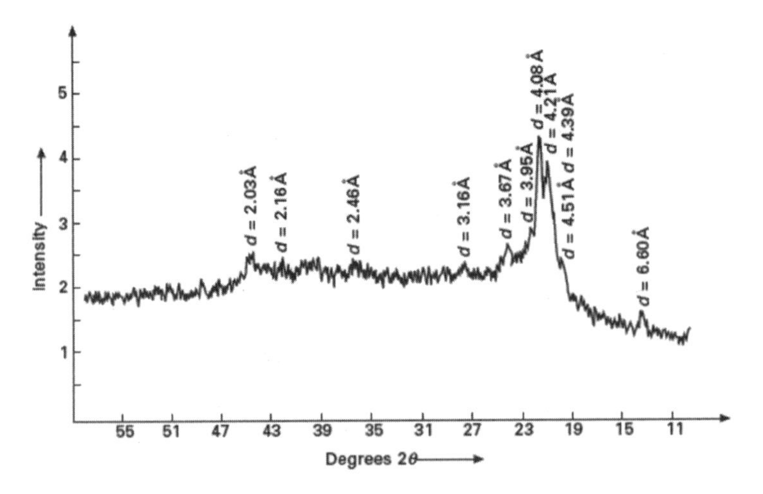

Fig. 11.6 X-ray diffraction pattern of soap residue of grease from NU 326 bearing after operation under electrical fields for 250 h.

11.4 Effect of current on formation of corrugated patterns on the roller track of races of roller bearings

The passage of current causes local surface heating, which leads to low temperature tempering, and accelerates formation of corrugations with time (Fig. 11.7). As rolling continues, corrosion increases due to the decomposition of grease 'A' and small particles of material are pulled out from track surfaces at the points of asperity contacts between the races and rollers, which develop scores on the surfaces, besides forming corrugations/flutings on the surfaces. After long operation, the softer tempered surfaces of the races become harder, and thus harder/re-hardened particles due to localized high temperature and load erupt from the craters and intensify the depth of corrugations.

In the presence of low-resistivity lubricant ($10^5\Omega$m), in the close asperity contacts between rolling elements and races, current intensity is increased by short-circuiting at the line of contacts that leads to the formation of corrugations in due course. After the formation of corrugations in one half of the races – the resistance becomes higher due to a decrease in contact area – the current leaks through the other half, and thus fully fledged corrugations are formed on the surfaces.

At each revolution of the shaft, part of the circumference of the inner race passes through a zone of maximum radial force, and Hertizian pressure between the rolling elements and raceways (at the line contacts) leads to a maximum shear stress. The maximum shear stress is taken as the criterion for yielding and this occurs in the subsurface at a depth approximately equal to half the radius of the contact surface. It is generally at this point that failure of material, if occurring, will initiate. As soon as the fatigue spall appears on the surface, the actual area of the asperity contact between the rolling-element and the race is reduced, which reduces electrical contact resistance and increases flow of current. Furthermore, there is a gradual increase in the width of corrugation by deformation at the asperity contact due to an increase in contact pressure per unit area, which leads to further reduction in electrical contact resistance and increase in flow of high-intensity current. These intensify the corrugation pattern in due course.

When a bearing is significantly loaded, the deformation is made by rolling elements (K) on the races in the loaded zone. The process of deformation, which leads to the formation of a corrugation pattern on the surfaces at the line contacts due to a decrease in contact resistance, is accelerated by the passage of high-intensity current, corrosion and oxidation of surfaces, lubricant characteristics and quality of a bearing.

Fig. 11.7 Corrugation pattern on the inner race of NU 326 motor bearing after about 6000 hr of operation.

The pitch of corrugations on the roller tracks depends on the bearing kinematics, the frequency of rotation, the position of plane of action of radial loading, the bearing quality and the lubricant characteristics, and is given as[11.8, 11.9]:

$$\Delta_{ir} = \pi Dd / SF_s Kp(D + d) \quad (11.1)$$

$$\Delta_{or} = \pi Dd / SF_s Kp(D - d) \quad (11.2)$$

$$\Delta_{re} = \pi d / SF_s p \quad (11.3)$$

The width of corrugations on the surface is not affected by frequency of rotation and depends on the load conditions and bearing kinematics. The width of corrugations on the inner and outer races is given as[10.8, 10.9]:

$$W_{ir} = 2.15 \, [Pd \, (D - d)/pKELD]^{1/2} \quad (11.4)$$

$$W_{or} = 2.15 \, [Pd(D + d)/pKELD]^{1/2} \quad (11.5)$$

$$W_{re} = W_{or} + \theta W_{ir} \quad (11.6)$$

where θ, the overlapping coefficient, varies from 0 to 1.

The pitch and width of corrugations are smaller on the inner race than on the outer race. Also pitch and width of corrugations on rollers are affected by corrugation pattern already formed on the races.

11.5 Effect of current leakage on electro-adhesion forces in rolling friction and magnetic flux density distribution on bearing surfaces

The mechanism of adhesion, friction and wear on bearing surfaces in the presence of lubricating film is quite complex. The process of adhesion involves the formation of a junction between the asperities contact, which may finally lead to elastic and plastic deformation under load. Energies of atomic nature are exchanged at the asperities, which may be affected by cage and roller slip due to close interaction of rolling elements with the races[11.10].

It is rather difficult to estimate the electro-adhesion forces in the rolling friction. But these can be assessed with a reasonable accuracy by SRV (Schmierstoff –lubricant–material) analysis; the change in coefficient of friction, profile depth and ball scar diameter of the used greases recovered from the active zone of the bearings. This is because of the activity of the zinc additive, i.e. zinc dithiosphosphate or zinc dialkyldithiophosphate (ZDTP) used as a multifunction additive in the grease. Under pure rolling friction it protects the rubbing metal surfaces and contributes to friction and wear reduction, and depends, partly, on the amount of additive absorbed on these surfaces. Physisorption and chemisorption processes precede the chemical reactions with the metal; therefore, it is probable that load-carrying capacity is related to these processes. Correlation between ZDTP adsorption data and wear shows that ZDTP is reversibly physiosorbed on iron at 25 °C, but at 50 °C undergoes chemisorption reactions. On the other hand, decomposition of ZDTP in the lithium base greases under the influence of electrical fields leads to the formation of lithium zinc silicate ($Li_{3.6}Zn_{0.2}SiO_2$) in the presence of high relative percentage of free lithium and silica impurity in the grease under a high temperature in the asperity contacts along with the formation of gamma lithium iron oxide. Besides this, the original structure of lithium stearate changes to lithium palmitate. These changes are not detected under pure rolling friction.

The above changes in the lubricating medium of the bearings operated under electrical fields are reflected in SRV analysis, and are related to electro-adhesion forces in the rolling friction. Also, these are contributed by the medium–metal interaction, rate of chemical reaction, affinity of lubricating medium components for the metal, availability of free metal and temperature rise. If the grease recovered from such bearings is put in a new bearing, the bearing may fail prematurely owing to higher temperature rise even under pure rolling friction[11.11].

It is established that under the influence of an electrical field, a roller bearing using low-resistivity grease develops alternating magnetic flux density distribution on the inner race and rolling elements (Fig. 11.8).[11.11] The maximum magnetic flux density on the inner race has been detected up to 95G, and on rolling elements up to 18G. However, significant flux density is not developed on the surfaces of ball bearings. Also, a bearing using high-resistivity grease does not develop significant flux density distribution on its surfaces[11.12-11.15].

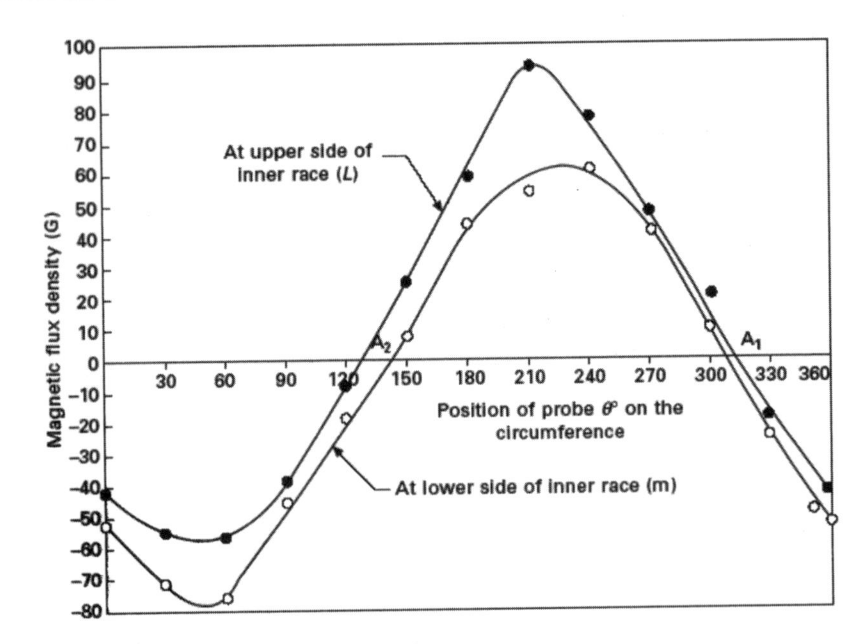

Fig. 11.8 Magnetic flux density distribution around the track surface of inner race of NU 326 bearing after passing 50A (AC) at 1.2 to 2.3 V for 250 h (bearing using grease 'A').

Besides this, there occur changes in electro-adhesion forces when bearings are lubricated with low-resistivity grease (10^5 Ωm) rather than grease of high resistivity (10^9 Ωm). Pure rolling friction does not greatly affect the electro-adhesion forces. In general, by the study of magnetic flux density distribution along with the study of damaged and corrugated bearing surfaces, and also by the analysis of deterioration of greases used in bearings[11.1, 11.5], a diagnosis of leakage current through a non-insulated bearing of an electrical motor can be established[11.11-11.15].

11.6 Effect of operating parameters on the threshold voltages and impedance response of noninsulated rolling-element bearings

Investigations reveal that the first and second threshold voltages (V_{t1} and V_{t2}) appear under the influence of electrical currents in the bearings, depending on the lubricant resistivity, oil film thickness, bearing conditions and the operating parameters. The detected threshold voltages are primarily responsible for momentary flow of current and the further increase in current intensity with a slight change in potential drop across the bearings (Fig. 11.9). The impedance of the bearings becomes negligible as the current intensity across the bearing increases (Fig. 11.10). However, the impedance is more affected by the speed and film thickness than by the load on the bearings.

The threshold voltages (V_{t1} and V_{t2}) decrease as the load on the bearing is gradually increased at constant speed. It is found that the threshold coefficients (V_{t1k} and V_{t2k}) are almost constant at a particular operating speed and the change in load does not affect the threshold coefficients. The threshold coefficients are given by[11.16]:

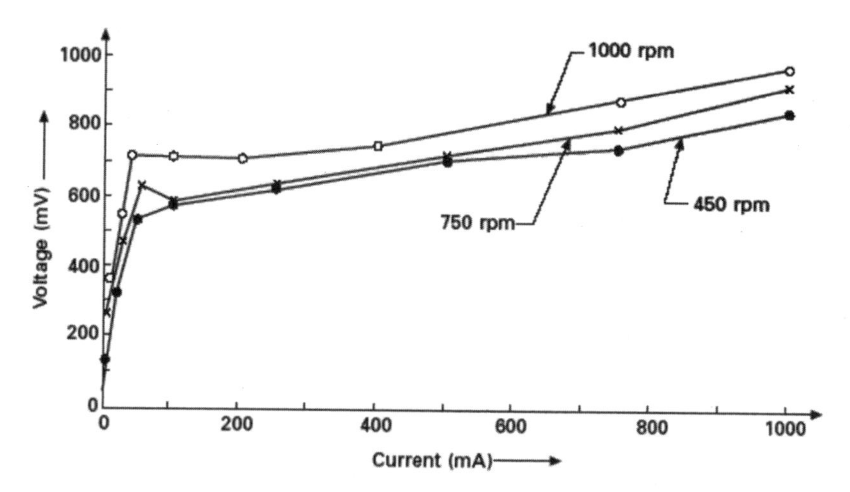

Fig. 11.9 Variation of bearing current with voltage of NU 330 bearing using lubricant 'B' at different speeds at 750kgf load.

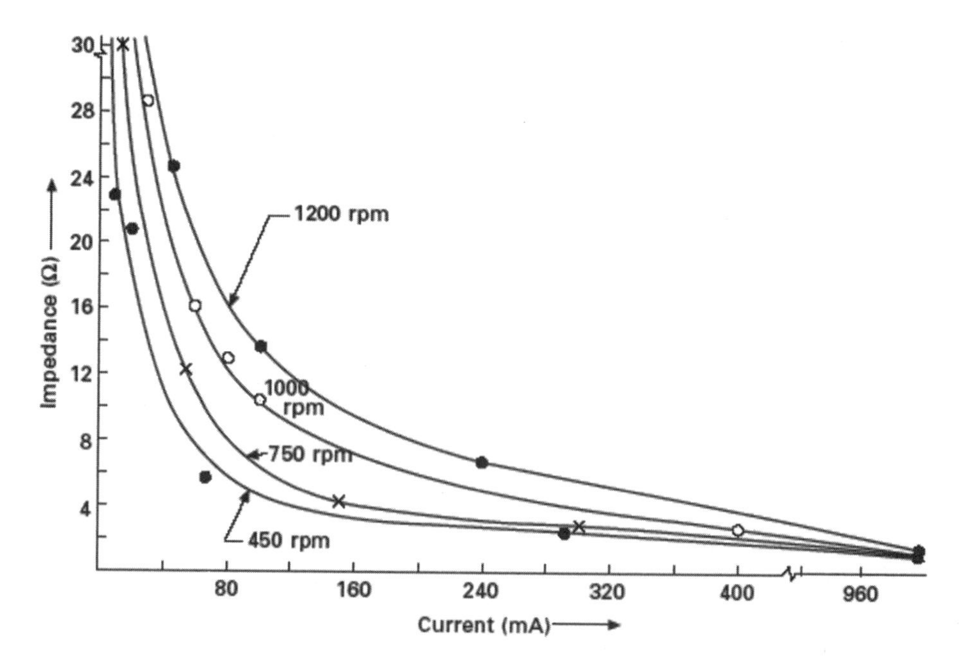

Fig. 11.10 Variation of bearing impedance with current at different speeds and 450kgf load – for NU 330 bearing (using lubricant 'B').

$$V_{\mathrm{t1k}} = V_{\mathrm{t1}} P^{0.3} \quad (11.7)$$

$$V_{\mathrm{t2k}} = V_{\mathrm{t2}} P^{0.3} \quad (11.8)$$

In general, the impedance of bearings varies with the parameters of operation and resistivity of the lubricant. The variation in load at constant speed has very little effect on the bearing impedance at the fixed levels of current intensity compared with the change in speed at fixed load. Also, an increase in current intensity reduces the bearing impedance very significantly irrespective of the operating parameters. For reliable operation, the safe limit of the potential drop across the bearing elements should be less than the first threshold voltage.

11.7 Impedance, capacitance and charge accumulation on roller bearings

Under the influence of potential drop across a roller bearing, the minimum film thickness between the races and the rollers offers maximum capacitance and minimum capacitive reactance depending on the permittivity of the lubricant. The electrical interaction between the races and the rollers in the presence of the oil film is like a resistance capacitor (RC) circuit and offers an impedance to the current flow[11.17].

In short, under different operating conditions, the capacitance between the outer race and a roller is higher than that between the inner race and a roller. The capacitance and resistance of a bearing depend on the film thickness and width of deformation, and are governed by the permittivity and resistivity of the lubricant. The capacitance and resistance of races can be determined by the following formulae[11.17]:

$$C_{\mathrm{ir}} = 2\xi L(\beta h_0)^{-1/2} \tan^{-1}[0.5\, W_{\mathrm{ir}} (\beta/h_0)^{1/2}] \quad (11.9)$$

$$C_{\mathrm{or}} = 2\xi L(\delta h_0)^{-1/2} \tan^{-1}[0.5 W_{\mathrm{or}} (\delta/h_0)^{1/2}] \quad (11.10)$$

$$R_{\mathrm{ir}} = \frac{\rho(\beta h_0)^{1/2}}{2L \tan^{-1}[0.5W_{\mathrm{ir}}(\beta/h_0)^{1/2}]} \quad (11.11)$$

$$R_{\mathrm{or}} = \frac{\rho(\delta h_0)^{1/2}}{2L \tan^{-1}[0.5W_{\mathrm{or}}(\beta/h_0)^{1/2}]} \quad (11.12)$$

The equivalent bearing capacitance is given as:

$$C_{\mathrm{b}} = K/W(X_{\mathrm{cir}} + X_{\mathrm{cor}}) \quad (11.13)$$

where

$$X_{\mathrm{cir}} = \frac{1}{WC_{\mathrm{ir}}} \text{ and } X_{\mathrm{cor}} = \frac{1}{WC_{\mathrm{or}}} \quad (11.14)$$

Stored charges in the bearing depend on bearing capacitance and charge increases with applied voltage provided the dielectric is not dissociated. The stored charge in a bearing is given as:

$$Q = VC_{\mathrm{b}} \quad (11.15)$$

It is established that the equivalent capacitance of a bearing decreases with increasing speed at constant load but increases with load at constant speed (Fig. 11.11). Also, a bearing lubricated with high-resistivity lubricant as opposed to low-resistivity lubricant, with the same permittivity, behaves like a capacitor up to the first threshold voltage. In addition to this, for a bearing to accumulate charges, the ratio of capacitive reactance to active resistance should be less than unity.

11.8 Contact temperature, contact stresses and slip bands initiation on roller track of races

Current passing through a bearing at the line contacts between the roller tracks and rollers, and the corresponding impedance generates heat and increases the temperature instantaneously. This increases the contact stresses and enables the determination of the number of cycles before the slip bands are initiated on the track surfaces of a bearing lubricated with low-resistivity lubricant ($10^7 \Omega$cm)[11.18].

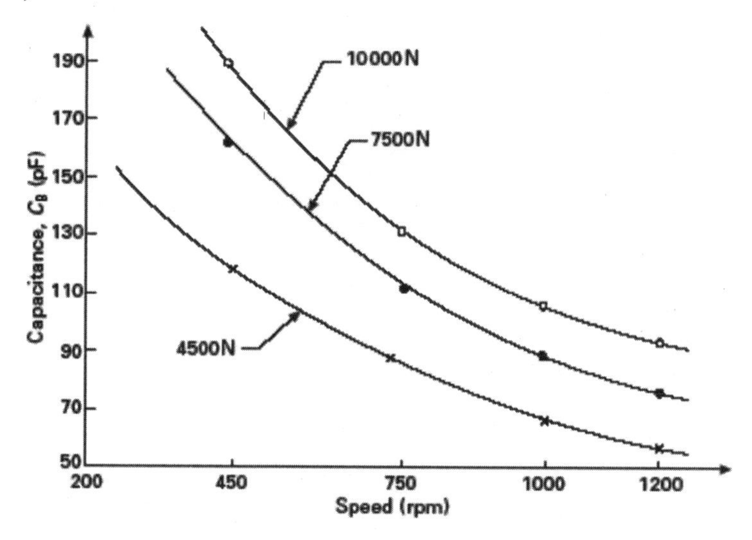

Fig 11.11 Variation of capacitance with speed at different radial loads in NU 330 type bearing using a high-resistivity lubricant 'B'.

The duration of contact between roller tracks of races and a roller has been theoretically determined. This depends on the pitch diameter and the roller diameter of a bearing. It is higher for the roller track of the outer race and a roller than for the roller track of the inner race and a roller. It decreases with rotating speed frequency but increases with width of contact. The values can be determined as:

$$t_{ir} = 2DW_{ir}/\pi F_s(3D + d)(D - d) \qquad (11.16)$$

and

$$t_{or} = 2DW_{or}/\pi F_s(D^2 - d^2) \qquad (11.17)$$

The temperature rise at each line contact between the roller track of races and a roller in each shaft rotation depends on the bearing kinematics, the depth of slip bands, the number for rollers in the loaded zone and the properties of the bearing material. It decreases with an increase in frequency of rotation but increases with the resistance of the bearing and with the current. The temperature rise of the inner and outer race tracks is given as[11.18]:

$$T_{\text{irn}} = T_{\text{orn}} = I^2 R_b K/\pi F_s dLH\, \varepsilon C \qquad (11.18)$$

Stresses on the roller tracks are determined as:

$$\alpha_{\text{irn}} = \frac{\alpha E(T_{\text{irn}} - T_a)}{1-\mu} \qquad (11.19)$$

It is established that the slip bands (Fig. 11.12) on the roller track of races of the NU 326 bearing at a current of 50 A are initiated after fewer than 41h of operation, 15.15 h or 10^6 cycles after the bearing temperature is stabilized. At a lower current intensity, more time elapses before slip bands appear on the roller tracks.

Fig. 11.12 Enlarged view of slip bands on inner race of NU 326 roller bearing after exposure to a current of 50A (AC) for 41 h (lubricant A).

11.9 Effects of instantaneous charge leakage on roller tracks of roller bearings lubricated with high-resistivity lubricants

The effects of instantaneous charge leakage on the rise of contact temperature between rollers and roller track of races of roller bearings lubricated with high-resistivity lubricant have been analysed. Estimates of the leakage of charge between roller tracks and rollers during momentary asperity contacts along with an expression for the instantaneous contact resistance between roller and race have been used to establish heat generated and instantaneous temperature rise in the contact zone on the roller tracks in each shaft rotation. Using this temperature rise, the contact stresses are determined and the minimum number of cycles calculated before craters appear on the roller track of the races (Fig. 11.13).

Fig. 11.13 Damaged inner race of NU 2215 bearing (using lubricant 'B') under electrical fields.

The instantaneous temperature rise due to sudden charge leakage during each contact of a roller with the roller track of inner race and outer race is determined as[11.19]:

$$T_{\mathrm{ir}} = \frac{4Q^2 D}{\pi F_{\mathrm{s}} C_{\mathrm{b}}^2 L_{\mathrm{a}} R_{\mathrm{s}} H \varepsilon C (3D + d)(D - d)} \tag{11.20}$$

$$T_{\mathrm{or}} = \frac{4Q^2 D}{\pi F_{\mathrm{s}} C_{\mathrm{b}}^2 L_{\mathrm{a}} R_{\mathrm{s}} H \varepsilon C (D^2 - d^2)} \tag{11.21}$$

A bearing with a higher capacitance and less charge accumulation takes a longer time before the craters are initiated on the roller tracks.

11.10 Capacitive effects of roller bearings on repeated starts and stops of a machine

Recent investigations have defined the time required for the charge accumulation and increase of charge with time on the bearing surfaces based on the bearing capacitance, the resistance of film thickness and the shaft voltage. Also, the investigation reveal the effect of gradual leakage of the accumulated charges with time as the shaft voltage falls as soon as the power supply to the machine is put off.

The ratio of contact cycles required for the charge accumulation and gradual discharge of the accumulated charges on the bearing surfaces depending on the bearing to shaft voltage has been analysed. The number of cycles and number of repeated starts and stops before initiation of craters on roller track of races, as against the bearing to shaft voltage, have been theoretically established to restrict the deterioration and damage of the bearings.

The time required in developing charges (Q_{ir} and Q_{or}) on a roller and roller track of inner and outer races is given as[11.20]:

$$T_{cir} = -C_{ir} R_{ir} \log_e (1 - a) \tag{11.22}$$

$$T_{cor} = -C_{or} R_{or} \log_e (1 - a) \tag{11.23}$$

Similarly, the time required to discharge the accumulated charges from roller track of inner as well as outer races and a roller/rollers is given as[11.20]:

$$T_{dir} = -C_{ir} R_{ir} \log_e a \tag{11.24}$$

$$T_{dor} = -C_{or} R_{or} \log_e a \tag{11.25}$$

where $a = V/V_o$ (ratio of bearing to shaft voltages).

The number of starts and stops before initiation of craters on the roller track of inner and outer races (N_{ssi} and N_{sso}) can be determined as[11.20]:

$$N_{ssi} = -\frac{C_{si}}{F_s C_{ir} R_{ir} \log_e a(1-a) 20 F_s} \tag{11.26}$$

and

$$N_{sso} = -\frac{C_{so}}{F_s C_{or} R_{or} \log_e a(1-a) 20 F_s} \tag{11.27}$$

It is shown that with an increase of bearing to shaft voltage, the number of starts and stops to initiate craters on the roller track of races decreases. For the NU 330 bearing, the number of starts and stops decreases from 803.63 to 463.50 as the ratio of bearing to shaft voltage increases from 0.5 to 0.9 (Fig. 11.14 and 11.15).

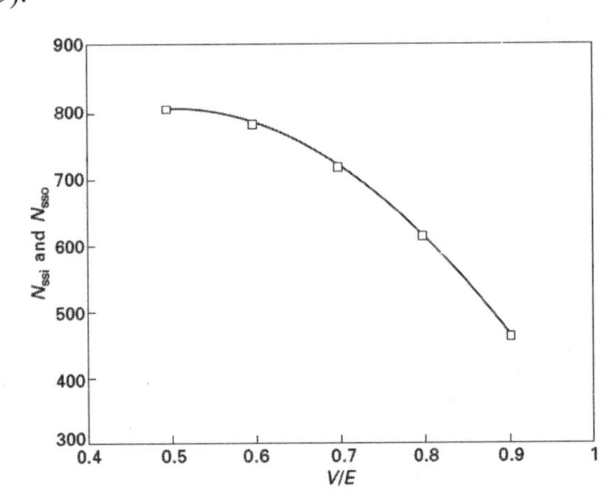

Fig. 11.14 Variation of number of starts and stops of a motor before formation of craters on roller track of inner and outer races (N_{ssi} and N_{sso}) at various levels of bearing the shaft voltage V/E of the NU 330 bearing operating under the influence of electrical currents.

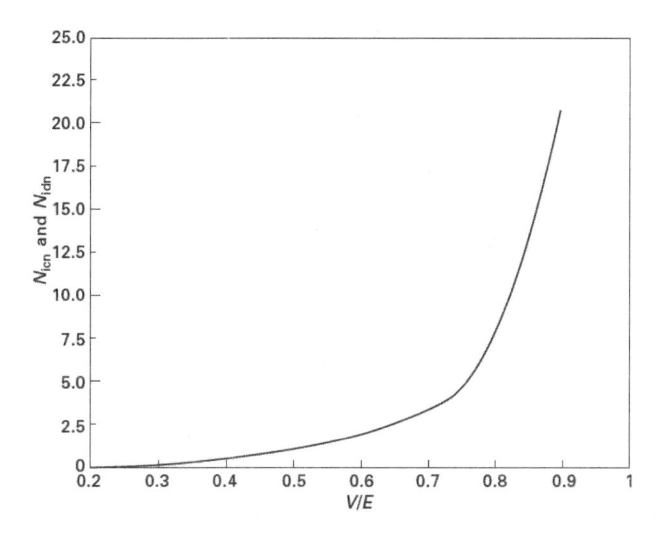

Fig. 11.15 Variation of the ratio of shaft revolutions to accumulate and discharge of accumulated charges (Nicn/ Nidn) at various levels of bearing to shaft voltages (V/ E) on roller tracks of inner and outer races of a roller bearing operating under the influence of electrical current.

11.11 Mechanism of bearing failures

When current leaks through the roller bearing in which low-resistivity lubricant has been used, a silent discharge passes through the bearing elements. This creates magnetic flux density distribution on the surfaces. Initially, this is accompanied by electrochemical decomposition of the grease and corrosion on the bearing surfaces and then gradual formation of slip bands at the line contacts, which lead to the formation of flutings and corrugations, and subsequently wear increases and the bearing fails[11.18]. In contrast, when high resistivity lubricant is used, the charges accumulate on the bearing surfaces due to polarization till they reach the threshold critical value at which the feeble current is conducted through a bearing[11.19], which is not able to result in significant flux density distribution on the surfaces. However, mass transfer at the elevated local temperature accompanies this in a few cycles of operation on asperities of the interacting surfaces by instantaneous discharge. This reduces the fatigue life and initiates crater formation on the surface by the welding effect and causes failure of a bearing.

11.12 Conclusions

Recent investigations on responses and performance of bearings under the influence of shaft voltage have given insight into bearing behaviour, deterioration of lubricant and lubricant decomposition. Besides this, failure of bearing can be established under bearing current. The effect of bearing capacitance on charge accumulation and of electro-adhesion forces on the magnetic flux density distribution is understood. The contact temperature and number of cycles before the slip bands and craters initiation on roller tracks, leading to the reduction in the life of bearing operating under the influence of shaft voltages, have been theoretically established. Analysis of the safe limit of a number of starts and stops of a machine influencing the bearing deterioration and crater formation acts as a potential tool to diagnose the bearing performance under different levels of bearing to shaft voltages. But the problem remains as to how to increase bearing life under the effect of shaft voltage. This may be achieved by the improved lubricant characteristics, bearing design or by alteration of bearing material properties. These aspects require further investigation to complete the 'know-how' gap. However, bearing insulation is the only solution presently available to tackle this problem.

References

[11.1] Prashad, H. and Murthy, T.S.R., 'Behaviour of Greases in Statically Bounded Conditions and When Used in Non-insulated Anti-friction Bearings Under the Influence of Electrical Fields', *Lubr. Eng.*, **44**(3), 239–246, 1988.

[11.2] Prashad, H., 'Experimental Study on Influence of Electrical Fields on Behaviour of Grease in Statically Bounded Conditions and When Used in Non-insulated Bearings', *BHEL J.*, **7**(3), 18–34, 1996.

[11.3] Polacios, J.M., 'Elasto-hydrodynamic Films of Lithium Greases', *Macanique, Materiaux, Electricite (GAMI)*, 176, 365–366, 1980 (also published in *NLGI Spokesman*, March 1981).

[11.4] Pastnikov, S.N., 'Electrophysical and Electrochemical Phenomena in Friction,Cutting and Lubrication', VNR, 1978.

[11.5] Prashad, H., 'Variation and Recovery of Resistivity of Greases – An Experimental Investigation', *J. Lubr. Sci. (France)*, 11–1, 73–103, November 1998.

[11.6] Prashad, H., 'Diagnosis of Lithium Greases Used in Rolling-element Bearings by X-Ray Diffractometry', *STLE Trans*, **32**(2), 205–214, 1989.

[11.7] Prashad, H. and Murthy, T.S.R., 'Deterioration of Lithium Greases Under the Influence of Electrical Current – An Investigation', *J. Lubr. Sci. (France)*, 10–4, 323–342, August 1998.

[11.8] Prashad, H., 'Investigations of Damaged Rolling-element Bearings and Deterioration Response and performance of a rolling-element bearing 103 of Lubricants Under the Influence of Electrical Fields', *Wear*, 176, 151–161, 1994.

[11.9] Prashad, H., 'Investigations of Corrugated Pattern on the Surfaces of Roller Bearings Operated Under the Influence of Electrical Fields', *ASME/ASLE Tribology Conference*, San Antonio Marriott (5–8, Oct. 1987), also published in *Lubr. Eng.*, **44**, 710–718. 1988.

[11.10] Prashad, H., 'The Effect of Cage and Roller Slip on the Measured Defect Frequency Response of Rolling-element Bearings', *ASLE Trans*, **30**(3), 360–367, July 1987.

[11.11] Prashad, H., 'The Effects of Current Leakage on Electroadhesion Forces in Rolling Friction and Magnetic Flux Density Distribution on the Surface of Rolling-element Bearings', *Trans. ASME, J. Tribol.*, **110**, 448–455, 1988.

[11.12] Tasker, J.L. and Graham, R.S., 'Effects of Magnetic Flux on Rolling-element/Bearings', *IEE Conference*, 17–19, Sept. 1985, Publication 254, pp. 152–156.

[11.13] Prashad, H., 'Magnetic Flux Density Distribution on the Track Surface of Rollingelement Bearings – An Experimental and Theoretical Investigation', *Tribol. Trans*, **39**(2), 386–391, 1996.

[11.14] Prashad, H., 'Determination of Magnetic Flux Density on the Surfaces of Rollingelement Bearings as an Indication of the Current That Has Passed Through Them – An Investigation', *Tribol. Int.*, **32**, 455–467, 1999.

[11.15] Prashad, H., 'Determination of Magnetic Flux Density on the Surfaces of Rollingelement Bearings – An Investigation', *BHEL J.*, **21**(2), 49–66, August 2000.

[11.16] Prashad, H., 'Effects of Operating Parameters on the Threshold Voltages and Impedance Response on Non-insulated Rolling-Element Bearings under the Influence of Electrical Currents', *Wear*, **117**, 223–240, 1987.

[11.17] Prashad, H., 'Theoretical Analysis of Impedance, Capacitance and Charge Accumulation of Roller Bearings Operated Under Electrical Fields', *Wear*, **125**, 223–239, 1988. Also in *BHEL J.*, **9**(2), 21–30, 1988.

[11.18] Prashad, H., 'Analysis of the Effects of Electrical Currents on Contact Temperature, Residual Stresses, and Slip Bands Initiation on Roller Tracks of Roller Bearings', *Wear*, **131**, 1–14, 1989.

[11.19] Prashad, H., 'Theoretical Analysis of the Effects of Instantaneous Charge Leakage on Roller Bearings Lubricated with High Resistivity Lubricants under the Influence of Electric Current', *Trans. ASME, J. Tribol.*, **112**, 37–43, 1990.

[11.20] Prashad, H., 'Theoretical Analysis of Capacitive Effect of Roller Bearings on Repeated Starts and Stops of a Machine Operating under the Influence of Shaft Voltages', *Trans. ASME, J. Tribol.*, **114**, 818–822, October 1992.

Nomenclature

a	ratio of potential difference across bearing to shaft voltage
C	specific heat of bearing material
C_b	equivalent bearing capacitance
C_{ir}, C_{or}	capacitance between inner race and a roller, and outer race and a roller, respectively
C_{si}, C_{so}	number of cycles before initiation of craters on roller track of inner race, and outer race, respectively
d	diameter of rolling element
D	pitch diameter
E	Young's modulus of elasticity
F_s	shaft rotational frequency
h_o	minimum oil film thickness
H	depth of crater/slip bands on roller track of races
I	bearing current
K	number of rolling elements in the loaded zone
L	length of rolling element
L_a	summation of the length of asperity contact on circumference of roller track during contact with a roller
N_{icn}	number of shaft revolutions to accumulate charge
N_{idn}	number of shaft revolutions to discharge accumulate charge
N_{ssi}, N_{sso}	number of starts and stops before the formation of craters on roller track of inner race and outer race, respectively
p	position of plane of action of radial loading ($p = 1, 2, 3, ...$)
P	resultant load on bearing
Q	stored electrical charge
Q_{ir}	charge on roller track of inner race
Q_{or}	charge on roller track of outer race
r	radius of rolling element
R	outer radius of inner race
R_i	inside radius of outer race
R_b, R_s	equivalent bearing resistance under operating and static conditions, respectively
R_{ir}, R_{or}	resistance between roller track of inner race and a roller, and outer race and a roller, respectively
S	bearing coefficient(s)

t_{ir}, t_{or}	duration of each line contact between roller track of inner race and a roller, and outer race and a roller, respectively
Ta	ambient temperature
T_{dir}, T_{dor}	time required to discharge the accumulated charges from roller track of inner race and a roller, and outer race and a roller, respectively
T_{ir}, T_{or}	instantaneous temperature rise of roller track of inner race and outer race, respectively, due to charge leakage during each contact with a roller
T_{irn}, T_{orn}	instant temperature rise of roller track of inner race and outer race, respectively, in each shaft rotation
V	applied voltage/voltage across bearing
V_o	shaft voltage
V_{t1}, V_{t2}	first and second threshold voltages, respectively
V_{t1k}, V_{t2k}	first and second threshold voltage coefficients, respectively
W_{ir}, W_{or}	width of corrugations on roller track of inner race and outer race, respectively
W_{re}	width of corrugations on rolling elements
W	$2\pi f$
X_{cir}, X_{cor}	capacitive reactance between inner race and a roller, and outer race and a roller, respectively
ρ	resistivity of lubricant
α	coefficient of thermal expansion
α_{irn}, α_{orn}	tangential stress on roller track of inner race and outer race, respectively
μ	Poisson ratio
ε	density of bearing material
ξ	permittivity (dielectric constant) of lubricant
Δ_{ir}, Δ_{or}	pitch on corrugations of roller track of inner race and outer race, respectively
Δre	pitch of corrugations of the surface of rolling elements
θ	overlapping coefficient
β	$0.5\ [1/r + 1/R]$ inner race constant
δ	$0.5\ [1/r - 1/R_i]$ outer race constant

Section - IV

Bio-tribology

12

Tribometrology of Skin

12.1 A General Review

The quantitative assessment of both skin health and skin care products is suggested based on skin tribological properties. Simultaneous multi-sensor measurements of both coefficient of friction and contact electrical impedance allow for fast and quantitative evaluation of such skin conditions as dryness and moisturization, and early diagnosis of skin diseases or of the deterioration in skin functions at a stage that may not be easily discernable visibly. It may be instrumental in developing and testing skin cosmetics and medicine.

12.2 Introduction

Skin health and beauty is a concern for people of all ages. Physiologically, skin is the first line of defense against any environment, and it is repeatedly subjected to physical and chemical damage. Over the course of time a person's skin undergoes changes, and to maintain skin health, it is important to quantitatively follow these changes. Sometimes these skin changes are visible, such as wrinkles, blemishes or rashes. In other cases, the changes may not be easily discernable without a quantitative assessment of skin properties.

Most people invest in skin care products, and it is important to provide quantitative comparison between them. Some products inhibit water evaporation from the skin, some compounds absorb and directly release hydrating elements into the skin, other compounds provide a greasy texture, while others elicit a more sticky texture. Quantification of these properties is crucial in the development of new skin care products.

The bulk of dermatological observations are based on qualitative visual or papillatory data. The experience with skin bioengineering proves the sharpness of the biometric focus provided by appropriate instrumentation and execution. Skin tribology has not been widely utilized as appropriate, robust, and facile instrumentation (and validation was not available). The unique multi-sensing technology described in this chapter should fill this void in the fields of dermato-physiology, dermato-pharmocology, dermato-toxicology and cosmetology.

Currently, there is no widespread diagnostic technique or method available to quantitatively relate skin properties to skin health. Since skin is a surface, it can be conveniently analyzed and described in tribological terms. Friction and electrical measurements have already been utilized as analytical techniques for skin health research. Utilization of our multi-sensors technology with real-time high frequency data acquisition allows for dramatic improvement in data quality of the tribological *in vivo* and *in-vitro* assessment of human skin.

Knowledge of skin's tribological properties, namely, surface coefficient of friction and surface contact electrical impedance, provides a quantitative assessment of skin health. This work allows for a fast and quantitative assessment of such skin conditions as dryness and moisturization, early diagnosis of skin diseases or of the deterioration in skin functions at a stage that may not be easily discernable visibly. It may be instrumental in developing and testing skin cosmetics and medicine.

12.3 Review of Friction Measurements

When a person feels her or his skin with her or his finger, the resultant perception of the skin property is nothing but friction between the finger and the skin. Thus, friction measurements represent the most straightforward way to exactly mimic the person's feeling of her or his skin conditions. Friction studies can be conducted non-invasively and can give an invaluable measure of the skin's health. For example, Naylor[12.22] showed that moistened skin has elevated friction, El-Shimi[12.11] demonstrated that drier skin has lower friction. According to Wolfram[12.28], Appeldoorn and Barnett[12.1], friction provides a good quantitative measure of skin assessment.

To measure friction, a probe is brought into contact with and is moved relative to the skin. A friction force, which opposes relative movement between the probe and skin, is monitored and then used to calculate the friction coefficient (see Table 12.1). A perpendicular normal load varies from author to author and is poorly controlled with either static weights or spring.

Table 12.1 Findings of In-Vivo Skin Friction Studies

Authors	Probe Size, Shape	Probe Material	Probe Motion	Setup of normal load	Friction Coefficient
Comaish and Bottoms[12.5]	15 mm Ring	Teflon, Nylon, Polyethylene, Wool	Linear	Static Weights	0.2 (Teflon), 0.45 (Nylon), 0.3 (Polyethylene), 0.4 (Wool)
Naylor[12.22]	8 mm Sphere	Polyethylene	Linear, Reciprocaing	Static Weights	0.5-0.6
Prall[12.25]	Disc	Glass	Rotational	Spring Load	0.4
El-Shimi[12.11]	12 mm Hemi-sphere	Stainless Steel (rough), Stainless Steel (smooth)	Rotational	Static Weights	0.2-0.4 (Rough) 0.3-0.6 (Smooth)
Highley et al.[12.15]	Disc	Nylon	Rotational	Spring Load	0.2-0.3

Table 12.1 C

Authors	Probe Size, Shape	Probe Material	Probe Motion	Setup of normal load	Friction Coefficient
Cua et al.[12.8]	15 mm Disc	Teflon	Rotational	Spring Load	0.34 (Forehead) 0.26 (Volar Forearm) 0.21 (Palm) 0.12 (Abdomen) 0.25 (Upper Back)
Asserin et al.[12.2]	3 mm Sphere	Ruby	Linear	Balloon; Static Weights	0.7
Johnson et al.[12.17]	Lens	Glass	Linear, Reciprocating	Static Weights	0.3-0.4 (Dry Skin)
Elsner et al[12.12]	15 mm Disc	Teflon	Rotational	Spring Load	0.48 (Forearm) 0.66 (Vulva)

There are two types of designs for the test apparatus that have dominated the earlier studies of skin friction, namely a probe moving across the skin either linearly or rotary. For example, Comaish and Bottoms[12.5] used one of the simplest linear designs: they moved the probe across the skin by attaching it to a pan of weights via pulley. Weights were placed in the pan such that the probe slides over the skin at a constant velocity. The dynamic friction coefficient was calculated by dividing the total weight in the pan by the normal load on the probe. More sophisticated linear designs provided motorized movement of the probe, either unidirectional or reciprocating; the motorization afforded greater control of the constant velocity of the probe. Also, in modern designs, strain gauges measure the friction force as the probe moves along the skin.

The rotating-probe designs use a rotating wheel or disc pressed onto the surface of the skin with a known normal load. Highley et al.[12.15] measured the frictional resistance by determining the angular recoil of the instrument as the wheel contacted the skin, by monitoring the light via a photocell. Comaish et al.[12.5] developed a portable device with a torsion spring to measure skin friction.

The major problems of these studies have been data repeatability from test to test and reproducibility from person to person, day to day and apparatus to apparatus. Indeed, one can hardly expect the same results from probes of very different materials, shapes and dimensions, performing different motions with different speeds and accelerations.

For example, smoother probes showed higher friction for both stainless steel (El-Shimi[12.11]) and nylon (Comaish and Bottoms[12.5]) probes, which is related to the larger contact area and adhesion. Even for the same probes and apparatus, a large portion of these problems remains due to the absence of effective real-time closed-loop load control, which is critical on the non-flat and not-so-smooth skin. Monitoring the normal force on the probe is absolutely essential for getting accurate and repeatable friction coefficient measurements. The two methods used to set up the normal load, static weights and springs, had no real-time normal load monitoring, but just incorrectly assumed that the normal load stays constant during the test. Normal load maintenance is a source of data variation on the non-flat skin, as the probe encounters dips (valleys) and raises (peaks), and so the normal load fluctuates during the test.

Previous studies have focused on correlating the friction with age, gender, anatomical site, and hydration.

Friction varies with anatomical site. Cua et al.[12.8, 12.9] found friction coefficients to vary from 0.12 on the abdomen to 0.34 on the forehead. Elsner et al.[12.12] measured the vulvar friction coefficient at 0.66 and the forearm friction coefficient at 0.48. Manuskiatti et al.[12.20] observed significant differences in skin roughness at various anatomical sites. Differences in environmental influences (i.e. sun exposure) and hydration may contribute to this. Elsner et al.[12.12] showed that a more-hydrated vulvar skin had a 35% higher friction than a forearm, which is in agreement with hydration studies that contend that skin has an increased friction under increased hydration.

With respect to age, friction measurement results are contradictory. Cua et al.[12.8] showed no differences in friction with respect to age on the ankle. Elsner et al.[12.12] and Asserin et al.[12.2] observed no age-related differences in vulvar friction, but higher forearm friction in younger subjects; they postulated that the skin exposed to sunlight undergoes photo aging and thus, forearm skin shows aging, while the light-protected vulvar skin does not.

Cua et al.[12.9] found no significant skin friction differences between the genders. There are no studies addressing race as pertains to friction, but Manuskiatti et al.[12.20] found no differences in skin roughness between black and white skin.

Much of the reviewed research has been devoted to ascertaining how the application of certain ingredients influences the skin surface, of interest to the cosmetic/moisturizer and lubricant industries. El-Shimi[12.11] and Comaish and Bottoms[12.5] showed that friction decreased with the application of talc powder, by 50% for dynamic friction and by 30% for static friction with a polyethylene probe; however, they also found that wetting the talc powder caused an increase in friction. Friction drops after the application of oils and oil-based lubricants, but then eventually increases (Comaish and Bottoms[12.5]). Prall[12.25] and Nacht et al.[12.21] found that friction rises with the addition of emollients and creams in a similar fashion to water; however, the cream effects lasted for hours while the water effects lasted for minutes. Hills et al.[12.16] observe that at elevated temperature of 45 °C most emollients lowered friction to a greater degree than at a room temperature of 18 °C.

12.4 Review of Electrical Measurements

Another measurement technique used to assess the skin is based on its electrical properties: capacitance, conductance, and impedance. The dry stratum corneum is a dielectric medium. Addition of water makes the stratum corneum responsive to an electrical field (Leveque and De Rigal[12.18]). The electrical methods are mostly used for assessment of skin hydration.

Capacitance measurements involve two oppositely charged plates held in close proximity. An electric field is formed between them, and the maximum charge on each plate is known as the capacitance. When dielectric materials are introduced into the gap between the two plates, they increase capacitance; for example, water increases capacitance by a factor of 81 as compared to vacuum. Thus, capacitance measurements are convenient for monitoring a hydration level. A number of capacitance studies of the effects of age, gender, anatomical site, clinical skin dryness, and moisturizer applications are outlined in Table 12.2.

Table 12.2 Findings in Skin Capacitance Studies

Comparative Study	Authors	Capacitance Findings
Age	1. Cua et al.[12.8] 2. Elsner et al.[12.12] 3. Frödin et al.[12.13]	1. Difference on the palm 2. No difference 3. No difference
Gender	1. Cua et al.[12.9]	1. No difference
Anatomical Site	1. Cua et al.[12.9] 2. Elsner et al.[12.12] 3. Blichmann et al.[12.4] 4. Tagami et al.[12.26]	1. Differences among various anatomical sites 2. Differences between vulva and forearm 3. Differences between palm and forearm 4. Differences among various anatomical sites
Clinical Skin Dryness	1. Lodén et al.[12.19] 2. Tagami et al.[12.26] 3. Hashimoto-Kymasaka et al.[12.14]	1. Decrease with increased skin dryness 2. Decreased with increased skin dryness 3. Decreased in psoriatic lesions, same in suction blisters
Moisturizer Application	1. Blichmann et al.[12.4] 2. Frödin et al.[12.13]	1. Significant increase in skin treated with cream 2. Significant increase 2 hours after cream application; significant decrease on the day after stopping treatment
Water Hydration	1. Blichmann et al.[12.4] 2. Tagami et al.[12.26]	1. Increase, then return to pre-hydration values after 7 min 2. Increase, then returned to pre-hydration values after 3 min

Electrical resistance of skin to an electrical current is measured at low currents (a few microamperes), so that the skin is not harmed in the measurement process. Studies in skin resistance and conductance are summarized in Table 12.3.

Most of the electrical assessment has involved measurements of either capacitance or resistance (conductance), but a few studies have looked at a parameter that combines them both: electrical impedance. Nicander et al.[12.23] investigated impedance differences of skin across anatomical locations, age, and gender, and found the impedance to vary by anatomical site, but not by gender. In a subsequent study, Nicander and Ollmar[12.24] observed seasonal variations in the electrical impedance on all of the tested anatomical sites except for the neck.

Table 12.3 Findings in Skin Resistance Studies

Comparative Study	Authors	Conductance Findings
Gender	1. Cua et al.[12.9]	1. No significant differences
Anatomical Site	1. Blichmann et al.[12.4] 2. Tagami et al.[12.27]	1. Differences between palm and forearm 2. Differences among various anatomical sites
Clinical Skin Dryness	1. Tagami et al.[12.27] 2. Hashimoto-Kymasaka et al.[12.14]	1. Decrease with increased dryness 2. Decreased in psoriatic lesions and increased in suction blisters
Moisturizer Application	1. Tagami et al.[12.27] 2. Blichmann et al.[12.4]	1. Increase in moisturized skin 2. 13-fold increase, dropped to a 2-fold increase after 15 min
Water Hydration	1. Blichmann et al.[12.4] 2. Tagami et al.[12.27]	1. Increase, then return to pre-hydration levels after 3 min 2. Increase, then return to pre-hydration levels after 4 min

For reasonable repeatability of electrical data, ambient humidity has to be controlled to control the amount of water in the stratum corneum; relative humidity above 60% should be avoided. The apparatus must be placed on hair-free skin; too much coarse hair or an inclined position reduces the accuracy of results. Skin should be exposed to the ambient conditions for 5 min before taking measurements. There must be at least 10 sec waiting time between measurements, because repeated measurements on the same skin location may change skin conditions.

12.5 State-of-the Art Tribo-Metrology

The multi-sensing technology has been implemented on a commercially available, portable Skin Micro-Tribometer model UMT of CETR, a photo of which is shown in Fig. 12.1. It provides comprehensive tribological measurements for different types of samples in a variety of biomedical applications. For example, it is successfully utilized for testing bathroom tissues on skin, soap on skin, surgical staples and sutures, medical needles, shaving blades, after-shave lotions, tooth pastes and tooth brushes.

The Skin Micro-Tribometer has significant technological and design advantages over previous devices used to measure skin friction or electrical properties. In performing its test functions, the UMT is capable of providing precision linear, rotational and reciprocating motions with programmable speeds in the range from 0.1 μm/s to 10 m/s. A normal load is tightly controlled with a closed-loop servomechanism, and can be programmed to be either constant or changing gradually or by steps, with user-defined tolerances, in the total range from 0.1 mN to 1 kN. A number of tribological parameters, namely friction force and coefficient, normal load, electrical contact resistance, capacitance or impedance, skin deformation (elastic, plastic, creep) or wear depth, temperature, and contact acoustic emission can all be measured and recorded simultaneously, with a sampling rate of 20 kHz. Digital video with magnifying optics is also readily available.

The UMT is utilized for both *in-vitro* testing on the artificial or cut-off skin (laid on the sample table seen in Fig. 12.1) and *in-vivo* testing on people's arms (see photo in Fig. 12.2), fingers (see photo in Fig. 12.3) and other body parts (not shown). Comfortable hand and arm supports (see Figs. 12.2 and 12.3) allow for *in vivo* skin tests of these limb sites on the bench-top tester configuration. Special hand-held adapters are used for measurements on the face, back, legs and other anatomical sites.

Fig. 12.1 Skin Micro-Tribometer Model UMT

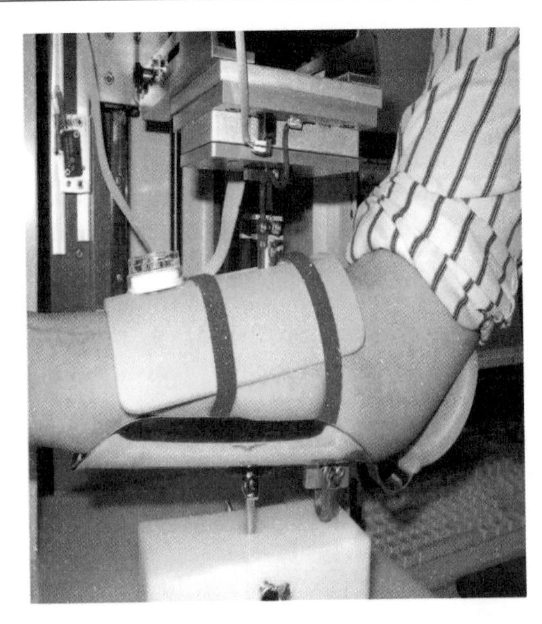

Fig. 12.2 Skin Micro-Tribometer Measures Forearm In-Vivo

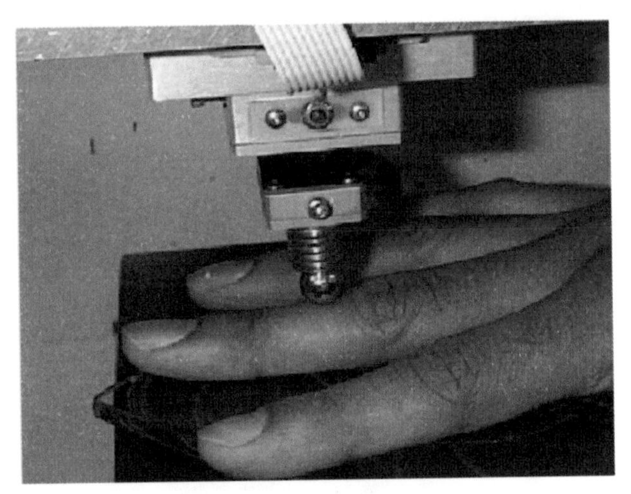

Fig. 12.3 Skin Micro-Tribometer Measures Finger Skin In-Vivo

The measurement probe chosen for most of the measurements in this work consisted of a 12 mm copper cylindrical friction/electrical probe, attached to a proprietary suspension system.

In some measurements, nylon, Teflon and stainless steel balls were used. In each test, the probe was pressed onto the skin with a constant load of 0.2 N, so that the contact pressure was enough to maintain a constant contact with the skin, but not too high to avoid skin macro-deformations and focus the measurements on the skin surface only. For studies involving under-skin layers and tissues, higher loads were used.

The probe was moved across the skin in a straight line at the speed of 1 mm/s for 10 mm. The slow motion typically produces most repeatable results, and so seems to be most useful for dermatological applications. Some tests were performed at higher speeds of 1 cm/s and even 5 cm/s on longer traces, to better simulate the motion of a person's finger when checking her or his skin, which may be important for characterization and marketing of cosmetic products.

During probe sliding, the dynamic friction coefficient was measured with a proprietary strain-gauge sensor, monitoring simultaneously and independently both normal load and friction force, with the resolution of 0.2 mN. In-situ electrical measurements were performed by applying an alternating current of a small constant amplitude 10 μamps, with frequency of 10 kHz, measuring the voltage across the probe, and calculating the electrical impedance. Next, the probe was lifted off the skin and the measurement repeated twice, for the total of three data points that were then averaged for each test.

An additional parameter obtained from the friction curves was the friction variation coefficient, calculated as its amplitude-to-mean ratio (Fig. 12.4). As the friction variation coefficient increases, the skin is expected to be stickier or rougher.

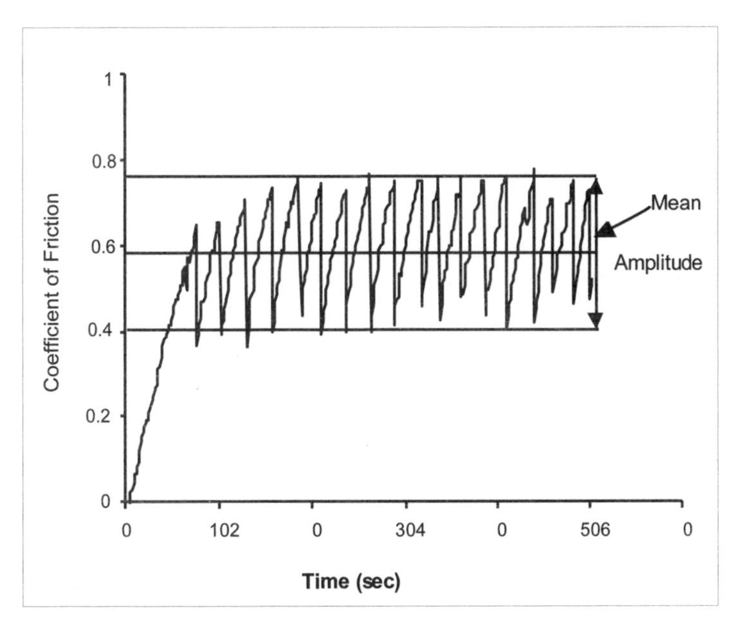

Fig. 12.4 Calculation of the Amplitude/Mean Ratio

12.6 In-Vivo Measurements of Skin Interventions

Sixty healthy (by history) adult volunteers were tested. A volunteer was self-classified to a certain race when all four grandparents were similarly identified. Age groups were split into the following: young (18-40 years old), middle

(41-59 years old), and old (over 60 years of age). The tests were carried out in a controlled room with constant temperature and humidity. Volunteers were asked to refrain from wearing creams prior to coming to the test and asked to rest for 30 minutes upon arriving at the clinic. The test sites on their arms were cleaned with isopropyl alcohol prior to testing.

The tests were conducted on the right and left volar forearms. Any visible hair was gently clipped to prevent its influence on the measurements. The forearm was chosen for the tests due to low amount of hair there, relative ease in using the measurement apparatus, and consistent test results. Four sites along the right and left forearm were measured as outlined in Fig. 12.5. Different treatments were administered at each site of the forearm, including no treatment, occlusion by wrapping the arm in polyvinylidene chloride or PVDC (saran wrap) for 30 min to prevent water loss, glycerin applied at 3 mg/cm^2, and petrolatum applied at 0.5 mg/cm^2 (interchanged between the sites). Measurements of the petrolatum and glycerin treated sites were taken after leaving the treatments on for 1 min and dabbing the skin with a paper towel to remove excess. Results are shown in Figures 6 and 7 as percent variation from the values recorded for untreated skin.

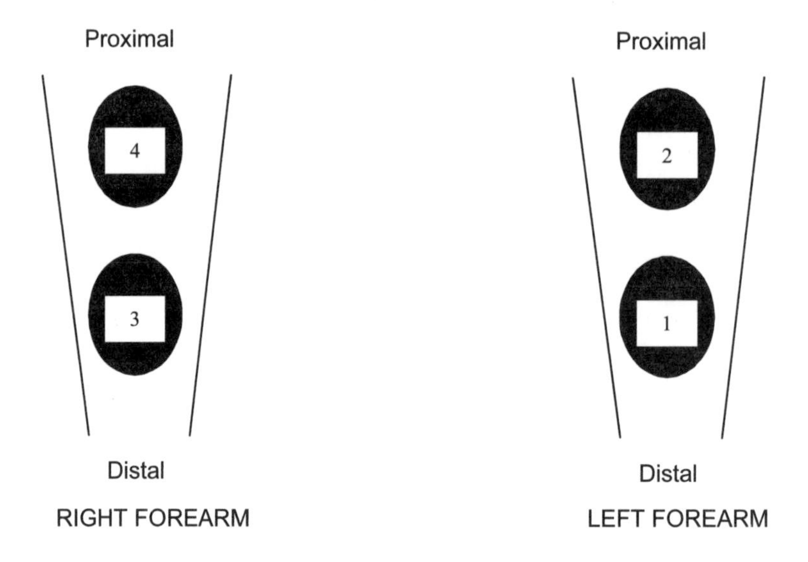

Fig. 12.5 Layout of Test Sites on Right and Left Volar Forearms

No significant differences were found in either friction or electrical parameters of volar forearms between age, gender, or ethnicity. The untreated skin on the volar forearm produces similar frictional and electrical characteristics across gender, age, and ethnicity, and so can be used as a site for comparative testing of skin care products. Since there is some variation with anatomical site on the volar forearm, comparisons among chemicals on the same volunteer can be done by comparing similar anatomical sites on the right and left volar forearm. The similarity between young and old volar forearm skin may be related to the suggestion that the volar forearm is partly protected from the sun. In fact, it corresponds to the previous studies that skin hidden from sun exposure, like that on the volar forearm, showed no significant differences between young and old people (Cua et al.[12.9]; Elsner et al.[12.12]; Nicander et al.[12.23]. Thus, the volar forearm may be a good anatomical site for testing skin care products.

There were substantial differences between the distal (sites 1 and 3 in Fig. 12.5) and proximal (sites 2 and 4) volar forearm sites, with the friction coefficient higher by about 30% and the electrical impedance lower by about 15% for the proximal sites as compared to the distal sites. This difference may be due to both hydration and smoothness of the skin along the volar forearm. Thus, researchers have to be very careful in choosing the forearm site for skin testing.

All of the skin treatments showed no significant differences between gender, age and ethnicity, but quite dramatic and well-expected differences as a result of the different interventions.

Occluded skin showed an increase in friction and a decrease in impedance (Figs 12.6 and 12.7). As the PVDC covers the skin, it prevents the water evaporation, so the trapped water decreases the electrical impedance. The increased hydration makes the skin stickier due to water-mediated adhesion between the probe and the skin, which results in increased both friction and friction variation coefficient for the occluded skin (Fig. 12.8).

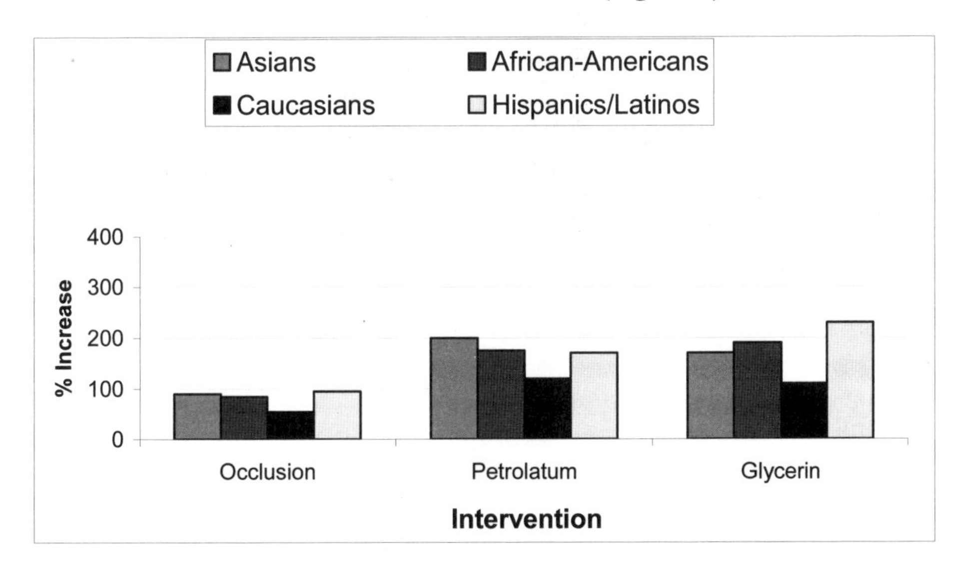

Fig. 12.6 Percent Increase in Coefficient of Friction for Three Skin Treatments, Comparing to Untreated Skin

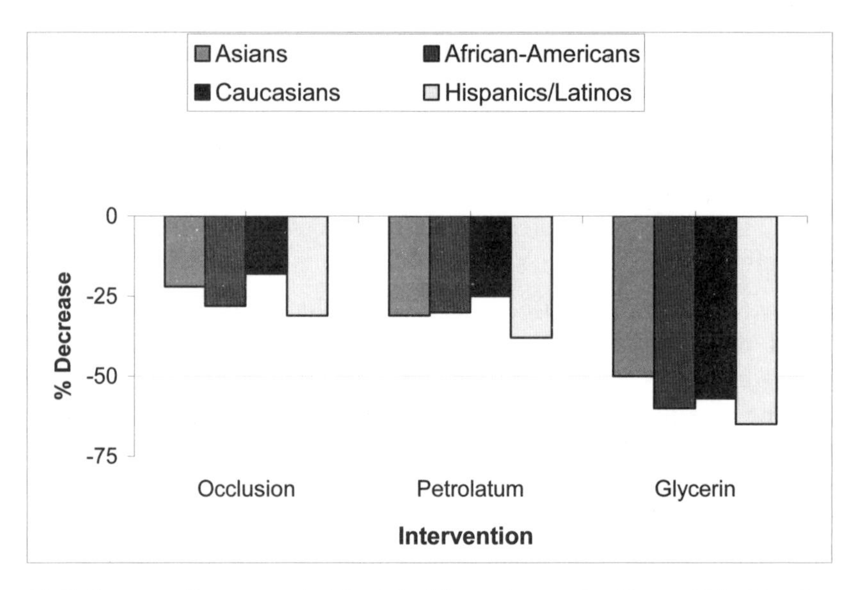

Fig. 12.7 Percent Decrease in Electrical Impedance for Three Skin Treatments, Comparing to Untreated Skin

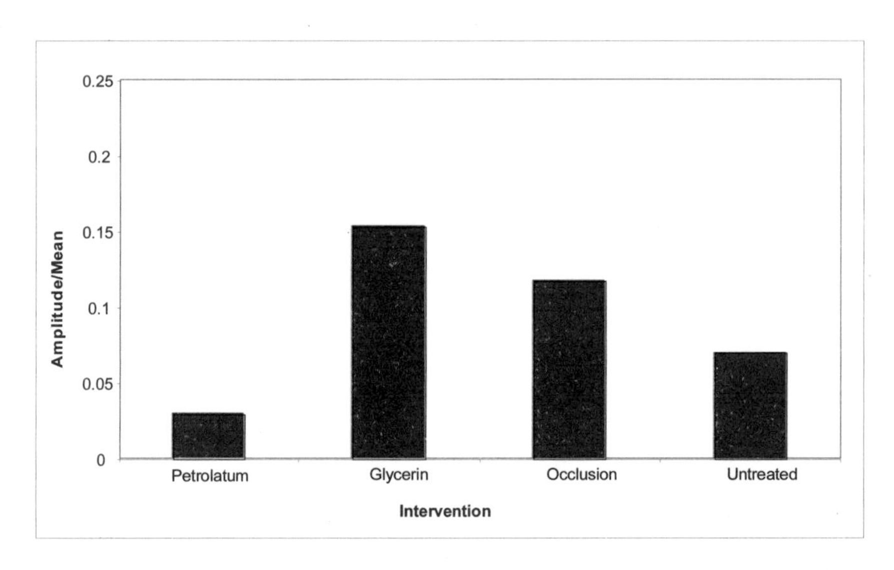

Fig. 12.8 Amplitude-to-Mean Ratio for Three Skin Treatments

Petrolatum coated the skin and served as a barrier for the water loss. Similarly to the PVDC, it is reported to increase skin hydration primarily through this occlusive effect. Indeed, petrolatum application lowered the skin impedance by an amount similar to that for the occlusion (Fig. 12.7). Friction for the petrolatum-treated skin, however, was higher than after the PVDC occlusion (Fig. 12.6), indicating that the petrolatum may have been absorbed into the skin, unlike the PVDC. The friction variation coefficient indicated that the petrolatum lowered the skin stickiness (Fig. 12.8). These results give a solid quantitative support for the qualitative perception that petrolatum is greasy and tends to make skin more slippery.

Glycerin increased friction similarly to that of petrolatum. However, it dropped impedance to a much greater degree than either petrolatum or PVDC occlusion did. This may be reflective of the higher rate and amount of glycerin absorbed directly into the skin, which may allow for higher skin hydration, as measured by the lower skin impedance. The highest friction variation coefficient showed that glycerin increased the skin stickiness the most (Fig. 12.8).

So, three surface interventions were quantitatively differentiated with the simultaneous use of the electrical and friction measurements. The electrical impedance provided a measure of the water levels under the skin surface and revealed the treatment ability to absorb into the skin, whereas the friction measurement revealed the treatment ability to affect the exposed surface of the stratum corneum. For example, the PVDC wrap caused the least decrease in skin impedance (Fig. 12.7) and the smallest increase in friction (Fig. 12.6) comparing to the other interventions, which suggests that the occlusion was not effectively absorbed into the skin (impedance) and so affects the surface moderately (friction). Both glycerin and petrolatum raised friction by a similar amount (Fig. 12.6), but the petrolatum did not absorb as readily into the skin as glycerin, evidenced by its lesser effect on the electrical impedance (Fig. 12.7); also, its substantially lower friction variation coefficient shows that petrolatum makes the skin greasier than with glycerin (Fig. 12.8).

12.7 In-Vivo Measurements of Skin Moisturizers

Hydration is a complex phenomena influenced by intrinsic (i.e. age, anatomical site) and extrinsic (i.e. ambient humidity, chemical exposure) factors. Earlier studies have revealed that drier skin had lowered friction, whereas hydrated skin had increased friction (Table 12.1). Indeed, water increases adhesive forces between skin and a probe, as well as softens the skin, which in turn increases contact area and friction between a probe and the skin, too. Thus, there is an increased frictional resistance between the hydrated skin and a probe. Since water evaporates within minutes, the skin returns to its pre-hydration state in a few minutes; a dried skin becomes less supple and allows the probe to glide more easily over it, which results in a lower friction (Blichmann and Serup[12.3]; Blichmann et al.[12.4]; Comaish and Bottoms[12.5]; Courage[12.7]; Denda[12.10]; Loden et al.[12.19]; Tagami et al.[12.26].

The above studies were focused on an intermediate level of hydration, when the skin is moistened without an appreciable "slippery" layer of water on its surface. In general, a skin response to water is much more complex, as the very wet skin also has low friction due to the hydrodynamic effects, while the very dry (clinically dry) skin becomes rough and increases a mechanical component of friction.

We performed an experimental comparison of three different moisturizing creams: a common older-formulation one, an advanced daytime cream, and the most advanced highest-performance might time cream. Ten healthy adult volunteers were tested on the same locations of both their right and left volar forearms. Each test included three sequential unidirectional runs, done before cream application (to establish the reference levels of the test parameters), then every 5 minutes for 1 hour, then after 5 hours, finally next day after 24 hours.

The frictional and electrical data were very consistent between all the volunteers, and summarized in Fig. 12.9. One can see that all three moisturizers had the same qualitative effect of increasing friction and decreasing electrical impedance, but quantitatively differed substantially. The common low-performance cream lasted for less than an hour; the advanced cream lasted for several hours, and the effect of the high-performance cream was still measurable a day later. The repeatability and reproducibility of these results was within 10%, which is sufficient for dermatological studies. These tests provide a scientific justification to the pricing of these creams, as they quantify their functional quality.

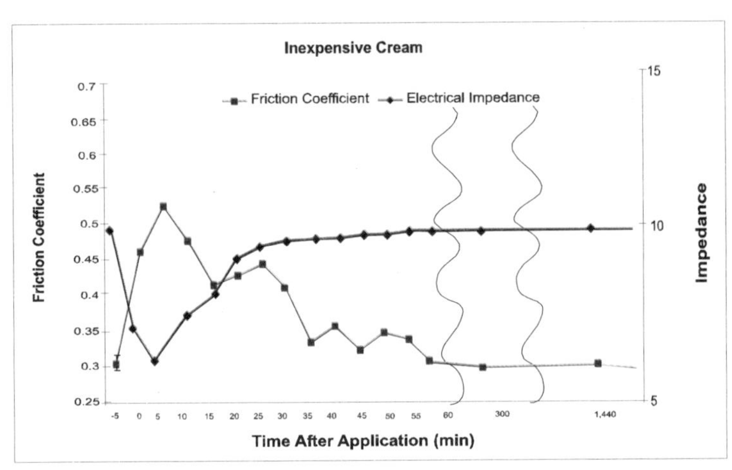

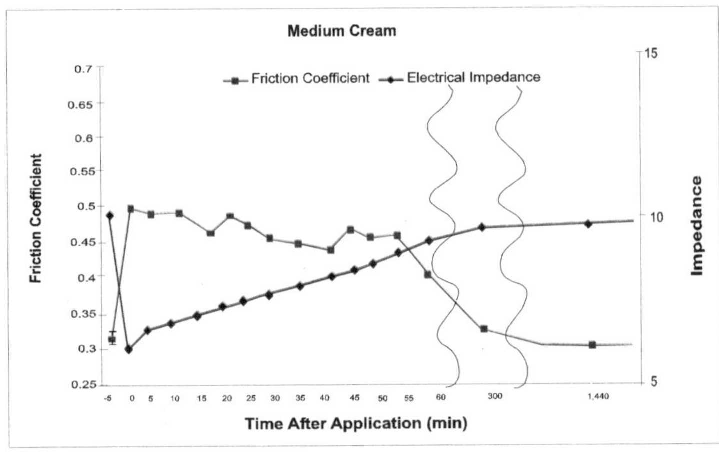

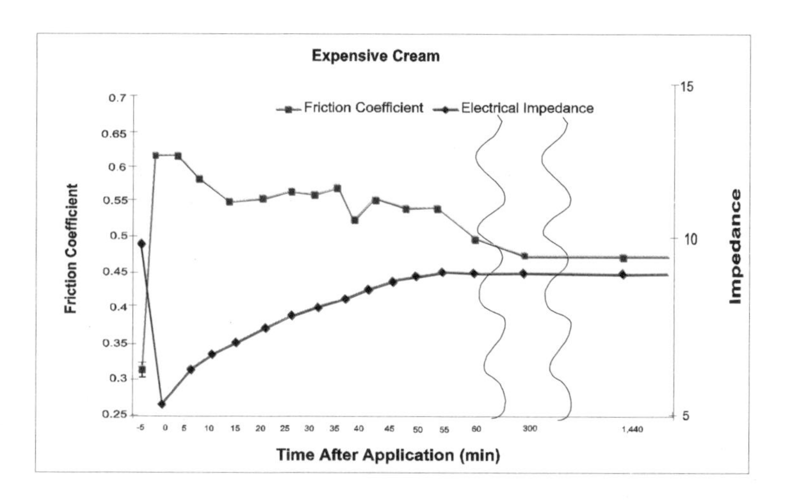

Fig. 12.9 Effects on Skin of Three Moisturizers

12.8 Conclusions

Investigations reported in this chapter confirm earlier studies that differences in skin due to aging, health, hydration and other factors are reflected in skin frictional and electrical characteristics. The novel technology and instrumentation for simultaneous measurements of friction coefficient and electrical impedance is an effective tool of skin assessment. It can serve as a quantitative method of evaluating skin health in dermatology on various body parts of different people.

It is also useful for quantitative functional evaluation of skin care products like creams, lotions, soaps, etc.

The experimental data suggest that there is little variation in volar forearm skin parameters across gender, age, and ethnicity, which makes this anatomical site promising for testing cosmetics.

References

[12.1] Appeldoorn, J., Barnett, G., 1963, Frictional Aspects of Emollience, *Proc. Sci. Sect. Toilet Goods Association*, 40, pp 28-35.

[12.2] Asserin, J., Zahouani, H., Humbert, Ph., Couturaud, V., Mougin, D., 2000, Measurement of the friction coefficient of the human skin *in vivo*. Quantification of the cutaneous smoothness, *Colloids and Surfaces B: Biointerfaces*, 19, pp. 1-12.

[12.3] Blichmann, C.W., Serup, J., 1988, Assessment of Skin Moisture, *Acta Derm Venereol* (Stockh), 68, pp. 284-290.

[12.4] Blichmann, C.W., Serup, J., Winther, A., 1989, Effects of Single Application of a Moisturizer: Evaporation of Emulsion Water, Skin Surface Temperature, Electrical Conductance, Electrical Capacitance, and Skin Surface (Emulsion) Lipids, *Acta Derm Venereol* (Stockh), 69, pp. 327-330.

[12.5] Comaish, S., Bottoms, E., 1971, The skin and friction: deviations from Amonton's laws, and the effects if hydration and lubrication, *British Journal Dermatology*, 84, pp. 37-43.

[12.6] Comaish, J.S., Harborow, P.R.H., Hofman, D.A., 1973, A hand-held friction meter, *British Journal Dermatology*, 89, pp. 33-35.

[12.7] Courage, W., 1994, *Bioengineering of the Skin: Water and the Stratum Corneum*, Eds. Elsner, P., Berardesca, E., Maibach, H.I., CRC Press, Boca Raton, pp. 171-175.

[12.8] Cua, A.B., Wilheim, K.P., Maibach, H.I., 1990, Friction properties of human skin: relation to age, sex and anatomical region, stratum corneum hydration and transepidermal water loss, *British Journal of Dermatology*, 123, pp. 473-479.

[12.9] Cua, A.B., Wilhelm, K.P, Maibach, H.I., 1995, Skin Surface Lipid and Skin Friction: Relation to Age, Sex, and Anatomical Region, *Skin Pharmacy*, 8, pp. 246-251.

[12.10] Denda, M., 2000, *Dry Skin and Moisturizers: Chemistry and Function*, Eds. Lodén, M., Maibach, H., CRC Press, Boca Raton, pp. 147-153.

[12.11] El-Shimi, A.F., 1977, In vivo skin friction measurements, *Journal Society Cosmetic Chemists*, 28, pp. 37-51.

[12.12] Elsner, P., Wilhelm, D., Maibach, H.I., 1990, Friction Properties of Human Forearm and Vulvar Skin: Influence of Age and Correlation with Transepidermal Water Loss and Capacitance, *Dermatologica,* 181, pp. 88-91.

[12.13] Frödin, T., Helander, P., Molin, L., Skogh, M., 1988, Hydration of Human Stratum Corneum Studied In vivo by Optothermal Infrared Spectroscopy, Electrical Capacitance Measurement, and Evaporimetry, *Acta Derm Venereol* (Stockh), 68, pp. 461-467.

[12.14] Hashimoto-Kumasaka, K., Takahashi, K., Tagami, H., 1993, Electrical Measurement of the Water Content of the Stratum Corneum In vivo and In vitro under Various Conditions: Comparison between Skin Surface Hygrometer and Corneometer in Evaluation of the Skin Surface Hydration State, *Acta Derm Venereol* (Stockh), 73, pp. 335-336.

[12.15] Highley, D.R., Coomey, M., DenBeste, M., Wolfram, L.J., 1977, Frictional Properties of Skin, *Journal Investigations Dermatology*, 1977; 69, pp. 303-305.

[12.16] Hills, R.J., Unsworth, A., Ive, F.A., 1994, A comparative study of the frictional properties of emollient bath additives using porcine skin, *British Journal Dermatology*, 130, p. 37-41.

[12.17] Johnson, S.A., Gorman, D.M., Adams, M.J., Briscoe, B.J., 1993, The friction and lubrication of human stratum corneum, *Thin Films in Tribology*, D.Dowson et al. Eds., Proceedings of the 19th Leeds-Lyon Symposium on Tribology, Elsevier Science Publishers, B.V., pp. 663-672.

[12.18] Leveque, J. L, De Rigal, J., 1983, Impedance methods for Studying Skin Moisturization, *Journal Society of Cosmetic Chemists*, 34, pp. 419-428.

[12.19] Lodén, M., Olsson, H., Axéll, T., Linde, Y.W., 1992, Friction, capacitance and transepidermal water loss in dry atopic and normal skin, *British Journal Dermatology*, 126, pp. 137-141.

[12.20] Manuskiatti, W., Schwindt, D.A., Maibach, H.I., 1998, Influence of Age, Anatomic Site and Race on Skin Roughness and Scaliness, *Dermatology*, 196, pp. 401-407.

[12.21] Nacht, S., Close, J., Yeung, D., Gans, E.H., 1981, Skin friction coefficient: changes induced by skin hydration and emollient application and correlation with perceived skin feel, *Journal Society Cosmetic Chemists*, 32, pp. 55-65.

[12.22] Naylor, P.F.D., 1955, The skin surface and friction, *British Journal Dermatology*, 67, pp. 239-248.

[12.23] Nicander, I., Nyrén, M., Emtestam, L., Ollmar, S., 1997, Baseline electrical impedance measurements at various skin sites – related to age and gender. *Skin Research and Technology*, 3, pp. 252-258.

[12.24] Nicander, I., Ollmar, S., 2000, Electrical Impedance measurements at different skin sites related to seasonal variations, *Skin Research and Technology*, 6, pp. 81-86.

[12.25] Prall, J.K., 1973, Instrumental evaluation of the effects of cosmetic products on skin surfaces with particular reference to smoothness, *Journal Society Cosmetic Chemists*, 24, pp. 693-707.

[12.26] Tagami, T., 1994, *Bioengineering of the Skin: Water and the Stratum Corneum*, Eds. Elsner, P., Berardesca, E., Maibach, H.I., CRC Press, Boca Raton, pp. 171-175.

[12.27] Tagami, H., Ohi, M., Iwatsuki, K., Kanamaru, Y., Yamada, M., Ichijo, B., 1980, Evaluation of the Skin Surface Hydration in Vivo by Electrical Measurements, *Journal of Investigation Dermatology*, 75, pp. 500-507.

[12.28] Wolfram, L.J., 1983, Friction of skin, *Journal Society Cosmetic Chemists*, 34, pp. 465-476.

13 TRIBOLOGICAL TESTING OF SKIN PRODUCTS: GENDER, AGE, AND ETHNICITY ON THE VOLAR FOREARM

13.1 Background

A few studies have focused on the simultaneous measurement of the friction and electrical properties of skin. This work investigates the feasibility of using these measurements to differentiate between the effects of chemicals commonly applied to the skin. In addition, this study also compares the condition of the skin and its response to application of chemicals across gender, ethnicity, and age at the volar forearm. Friction and electrical tests were performed on 59 healthy volunteers with the UMT Series Micro-Tribometer (UMT) of CETR. A 13 mm diameter copper cylindrical friction/electrical probe was pressed onto the skin with a weight of 20 g and moved across the skin at a constant velocity of 0.4 mm/s. Each volunteer served as his or her own control. The friction and electrical impedance measurements were performed for polyvinylidene chloride (PVDC) occlusion and for the application of glycerin and petrolatum.

No differences were found across age, gender, or ethnicity at the volar forearm. Polyvinylidene chloride (PVDC) occlusion showed a small increase in the friction and a small decrease in the electrical impedance; petrolatum increased the friction by a greater amount but its effect on the impedance was comparable to PVDC occlusion; glycerin increased the friction by an amount comparable to petrolatum, but it decreased the impedance to a much greater degree than petrolatum or the PVDC occlusion. An amplitude/mean measurement of the friction curves of glycerin and petrolatum showed that glycerin has a significantly higher amplitude/mean than petrolatum.

The properties of the volar forearm appear to be independent of age, gender, and ethnicity. Also, the simultaneous measurement of friction and electrical impedance was useful in differentiating between compounds administered to the skin.

13.2 Introduction

SKIN HEALTH is a major concern for people of all ages, gender, and ethnicity. As a result, most people invest in skin products and it is important to provide quantitative comparisons between them. Some compounds like petrolatum (Vaseline) increase skin hydration by inhibiting water evaporation from the skin surface; other compounds will absorb into the skin to carry and directly release hydrative elements into the skin. Also, some compounds provide a greasy texture while others elicit a more sticky texture. Quantification of these properties will be important in the development of new skin care.

The most convenient testing apparatus will be one that is noninvasive for simplicity of measurement. Previous noninvasive tests have focused on friction, Sivamani RK et al.[13.1] and electrical measurements such as capacitance[13.2-13.8], conductance[13.2, 13.4, 13.5, 13.7, 13.8], and impedance[13.9, 13.10]. Several tests have focused on using an individual test to measure the influence of different skin care products[13.5-13.7, 13.11-13.19]. However, differentiation between products is much harder when only using one assay.

This study ascertains if using several measurements simultaneously could be of advantage in tracking changes in the skin after product application. The assays include evaluation of friction coefficient, electrical impedance, and an amplitude/mean calculation, explained later, of the friction curves[13.14].

13.3 Methods

13.3.1 *Volunteers*

Fifty-nine healthy, by history, adult volunteers were tested. Table 13.1 shows the demographics. volunteer was self-classified to a certain race when all four grandparents were similarly identified. Age groups were split into the following categories: young (18–40 years old), middle (41–59 years old), and old (greater than 60 years of age). The tests were performed in a controlled room with constant temperature (21–26 °C) and relative humidity (50–70%). Volunteers were asked to refrain from using creams prior to coming for the test and were asked to rest for 30 min after arriving at the clinic. The test sites on their arms were cleaned with isopropyl alcohol prior to testing.

Table 13.1 Volunteer profile

Volunteer	Caucasian	African -American	Hispanic	Asian
Young (18–40 years old)				
Male	6	3	4	3
Female	3	3	4	3
Middle (41–59 years old)				
Male	2	2*	0	1
Female	3	1	1	1
Old (> 60 years old)				
Male	3	3	3*	1
Female	5	3	3	0

* Two volunteers were not used due to pre-existing skin conditions.

13.3.2 *Anatomical site and applied interventions*

Tests were conducted on the right and left volar forearms. Any visible hairs were gently clipped to prevent their influence on the friction and electrical measurements. The forearm was chosen for measurement due to a decreased amount of hair found there and the relative ease in using the measurement apparatus. Four sites along the right and left forearm were measured as outlined in Fig. 12.5 of the Chapter 12. This indicates Outline of tests performed on the right and left volar forearm. Site 1 tested untreated skin, site 2 treated occluded skin, and sites 3 and 4 were interchanged between glycerin and petrolatum treatments.

Different treatments were administered at each site of the forearm and these included occlusion, glycerin, and petrolatum. Occlusion was achieved by wrapping the arm in polyvinylidene chloride (PVDC) (saran wrap) for 30 min to prevent water loss. The USP grade glycerin solution was mixed as suggested by the manufacturer of the glycerin from the local pharmacy (one part glycerin to two parts water). The glycerin solution was applied at 3 mg/cm^2. Petrolatum was also acquired from the local pharmacy, and was applied at 0.5 mg/cm^2. Measurements of the petrolatum and glycerin-treated sites were taken after leaving the treatments on for 1 min and dabbing the skin with a paper towel to remove the excess. Glycerin was applied to site 3 in 50% of the volunteers and to site 4 in 50% of the volunteers (the same was performed for petrolatum) so that the effects of glycerin and petrolatum would not be affected by the choice of anatomical site along the volar forearm.

13.4 Results

13.4.1 *Untreated skin*

No significant differences were found for the friction coefficient or the electrical impedance of volar forearm between age, gender, or ethnicity (data shown for age in Fig. 12.1). However, there was a significant difference between the distal volar forearm (sites 1 and 3 in Fig. 12.5 of Chapter 12) and the proximal volar forearm (sites 2 and 4 in Fig. 12.5 of Chapter 12). The friction coefficient was elevated and the electrical impedance was decreased for the proximal volar forearm (Fig. 12.2 and 12.3 of Chapter 12). This result suggests that there is variation in hydration and perceived smoothness of the skin along the volar forearm. Amplitude/mean measurements showed that there were no significant variations among the anatomical sites (Fig. 12.4 of Chapter 12).

13.4.2 *Interventions*

All interventions showed no significant differences between gender or age. Differences in measurements arose as a result of the different interventions applied.

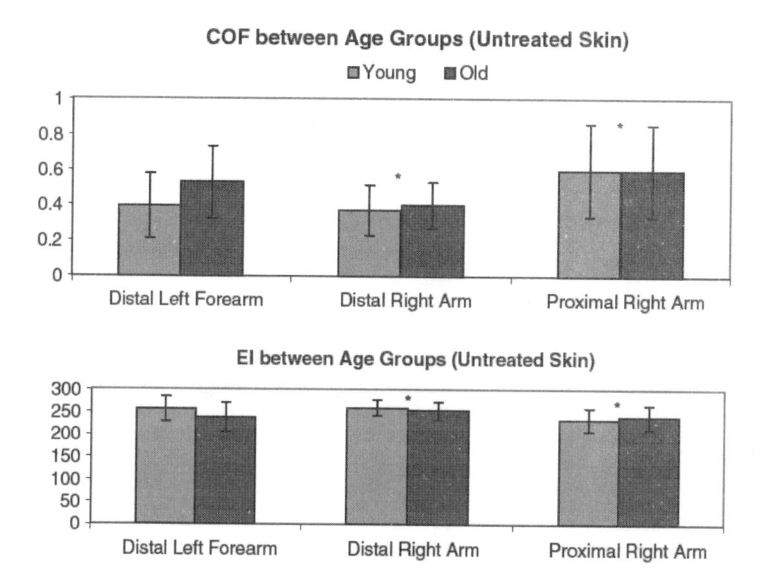

Fig. 13.1 Age-related comparisons of friction and electrical impedance. No significant differences were apparent between old and young skin on the volar forearm. Within each category, the proximal right arm friction and electrical impedance measurements were different from the distal right arm (P<0.001).

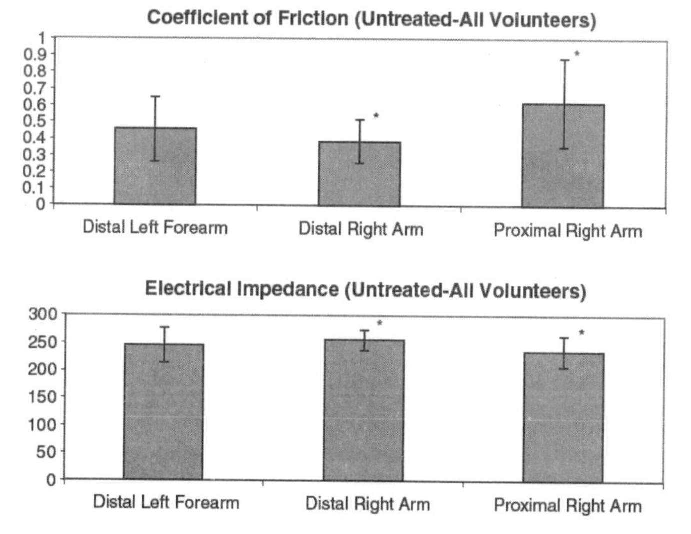

Fig. 13.2 Friction coefficient and electrical impedance. There were no significant differences between the distal left volar forearm and the distal right volar forearm. The proximal right volar forearm had a significantly higher friction coefficient and a significantly lower electrical impedance when compared to the distal right volar forearm and the proximal right arm friction and electrical impedance measurements were different from the distal right arm (P<0.001).

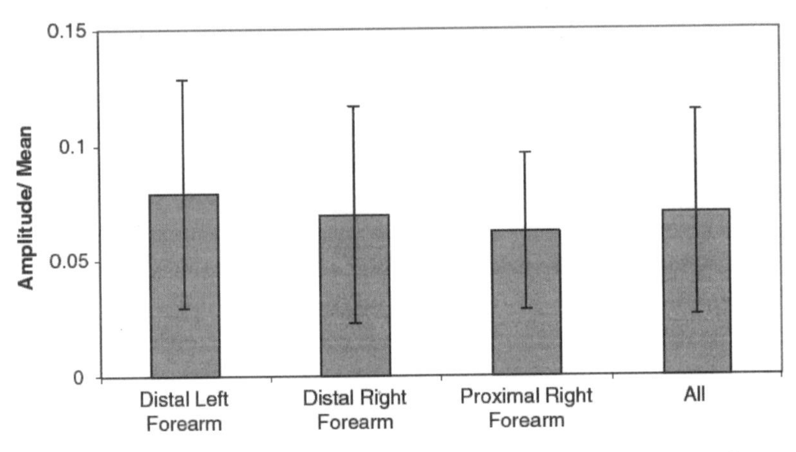

Fig. 13.3 Amplitude/mean on the untreated volar forearm. No significant differences were found for different anatomical sites between the left and right volar forearm or between distal and proximal sites on the same volar forearm.

13.5 Apparatus and measurements

The friction and electrical measurements utilized the UMT Series Micro-Tribometer (UMT) of CETR[13.11], a computer-controlled bench-top instrument adapted to measure tribological parameters on the skin (Fig. 12.1 of Chapter 12). The UMT provides precision translational, rotational, and linear motions to a variety of friction pair specimens, with speeds ranging from 0.1 to 10 m/s. A normal load is applied by a closed-loop servomechanism in the instrument's upper stage, and can be kept constant or linearly increasing. The range capacity for the applied normal force is from 0.5 mN to 200 N. The UMT measures and records the friction force and normal load at a total sampling rate of 20 kHz.

The measurement probe chosen consisted of a 13 mm diameter copper cylindrical friction/electrical probe (Fig. 12.2 and 12.3 of Chapter 12) attached to a suspension system. These figures show UMT test setup for volar forearm. The forearm is strapped into place with a holder to immobilize the forearm. The copper probe is then brought down to carry out the friction and electrical measurements.

The suspension system is in turn attached to a strain gauge-based force sensor. For each measurement, the probe was pressed onto the skin of the volar forearm with 20 g, maintained constant by the servo feedback loop. Then, the probe was moved proximally on the forearm across the skin linearly for 10 mm at a constant speed of 0.4 mm/s. Electrical measurements were performed by applying an alternating current of 10 μA, at a frequency of 10 kHz. Next, the probe was lifted off the skin and the measurement was repeated thrice. Parameters calculated were the dynamic coefficient of friction and the electrical impedance.

An additional parameter calculated from the friction coefficient measurement curves was the amplitude/mean for untreated skin and the various interventions (Fig. 12.4 of Chapter 12). This Fig. 12.4 shows calculation of the amplitude/mean measurement. The mean refers to the mean value of the measured friction coefficient as indicated on the graph. The amplitude refers to the deviation seen during the friction coefficient measurement as indicated on the graph. Then, the amplitude is divided by the mean to calculate the amplitude/mean. It has been suggested that this value represents the smoothness of the skin surface[13.11, 13.14].

This measurement is representative of the stickiness of the skin surface. As the amplitude/mean number increases, the skin is thought to be stickier and have more surface roughness[13.11, 13.14].

13.5.1 *Polyvinylidene chloride occlusion (applied to site 2 on all volunteers)*

Occluded skin showed an increase in friction coefficient and a decrease in electrical impedance (Figs. 13.1 and 13.2).

As the PVDC covers the skin, it prevents the evaporation of water from the skin surface and the water remains trapped. As a result, the skin becomes more hydrated and the presence of water decreases the electrical impedance as observed. Also, with increased hydration, the skin surface is more 'sticky' due to water adhesion between the probe and the skin, and the contact area will increase. Both these effects result in an increased friction coefficient over untreated skin measurements. The amplitude/mean measurement for occluded skin was higher than for untreated skin (Fig. 13.6).

13.5.2 *Petrolatum (applied to site 3 in half the volunteers and site 4 for the other half)*

When applied, the petrolatum will coat the skin surface and inhibit loss of water through this barrier. Similar to the PVDC, it is reported to increase skin hydration primarily through this occlusive effect[13.14]. The electrical impedance measurement showed that petrolatum application lowered the skin impedance by a similar amount as occlusion (Fig. 13.5).

This similarity may be due to the fact that petrolatum is thought to hydrate through occlusion like the PVDC. However, the friction coefficient for the petrolatum treated skin was much higher than for PVDC occlusion (Fig. 13.4), indicating that the petrolatum may have been absorbed into the skin, unlike the PVDC. The amplitude/mean measurement showed that the petrolatum acted to lower the stickiness of the surface (Fig. 13.3). This result is quantitative support for the qualitative perception that petrolatum is greasy and tends to make the surface more slippery.

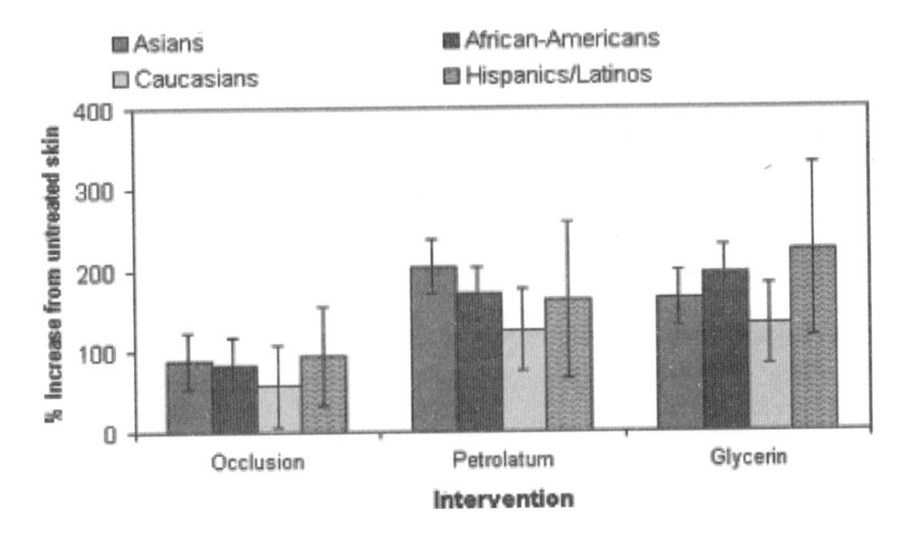

Fig. 13.4 Coefficient of friction across ethnicity. Data represent increases in friction when compared to untreated skin of the volar forearm. No significant differences were found between the different ethnic groups. Petrolatum and glycerin increased the friction coefficient significantly more than PVDC occlusion ($P<0.01$). The increase in the friction coefficient due to petrolatum was not significantly different from the effect of glycerin.

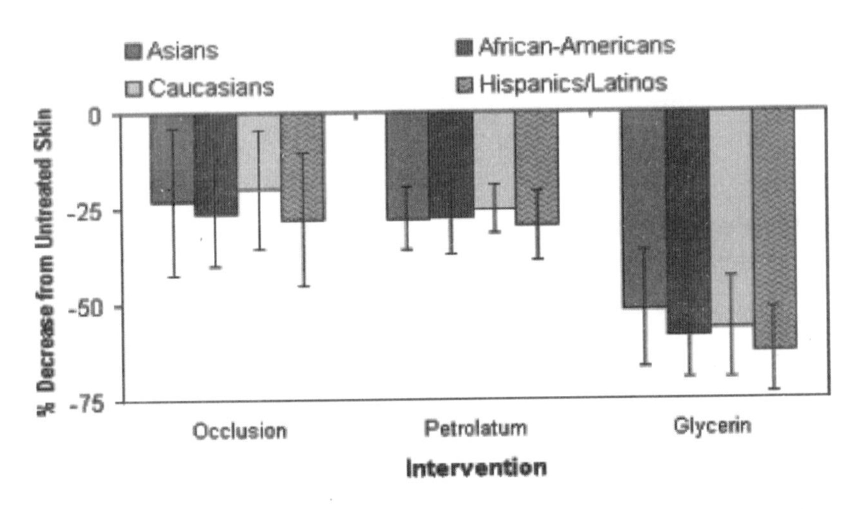

Fig. 13.5 Change in electrical impedance across ethnicity. Data represent decreases in electrical impedance when compared to untreated skin of the volar forearm. No significant differences were found between the different ethnic groups. Glycerin lowered the electrical impedance significantly more than PVDC occlusion or petrolatum ($P<0.01$). The decrease in the electrical impedance due to PVDC occlusion was not significantly different from the effect of petrolatum.

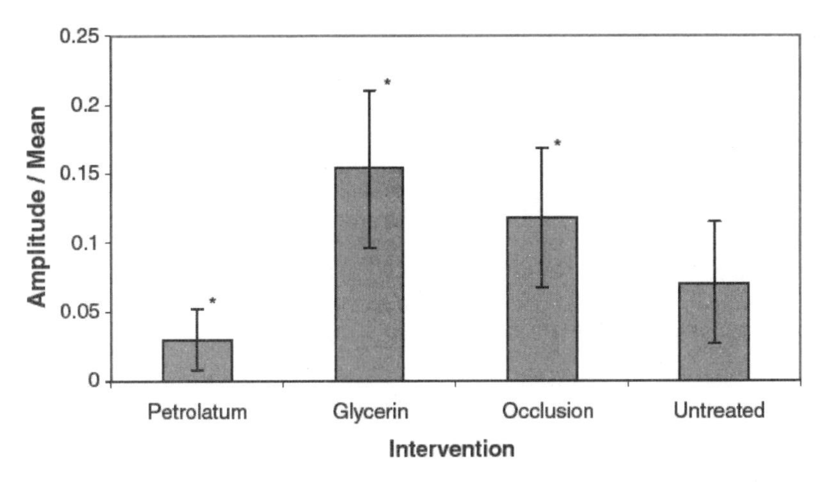

Fig. 13.6 Amplitude/mean measurements for interventions. The application of glycerin and the PVDC occlusion increased the amplitude/mean of the volar forearm. Also, the addition of glycerin raised the amplitude/mean significantly more than PVDC occlusion. Petrolatum significantly decreased the amplitude/mean and this is quantitative evidence of petrolatum's greasiness ($P<0.001$).

13.5.3 *Glycerin (applied to site 3 in half the volunteers and site 4 for the other half)*

The friction coefficient decreased by about 70% after glycerin application, and this decrease was similar to the decrease in friction coefficient seen with petrolatum application. However, the drop in electrical impedance was much greater for the glycerin application as opposed to the petrolatum or the PVDC occlusion. This may be a reflection of the speed and the amount of glycerin that is absorbed directly into the skin. This may allow for the skin to hydrate to a greater degree as measured by the lower skin impedance. The amplitude/mean measurement showed that glycerin application increased the skin stickiness above both untreated skin and occluded skin (Fig. 13.6).

13.6 Discussion

Noninvasive methods to measure the skin function can greatly ease the process of testing skin and products. Friction and electrical measurements have been explored as avenues to measure skin health, but few studies have explored differentiation between different skin effects with different chemicals through the simultaneous use of friction and electrical measurements. This study suggests that the untreated measurements of the skin on the volar forearm are consistent across gender, age, and ethnicity and can be used as a site for testing skin chemicals against one another. Since there is some variation with anatomical site on the volar forearm[13.20], comparisons among chemicals on the same volunteer can be made by comparing similar anatomical sites on the right and left volar forearm.

The properties of the skin on the volar forearm do not seem to change across gender, age, or ethnicity (data shown for age in Fig. 13.1). The lack of differences between young and old volar forearm skin was surprising since collagen breakdown and cross linking would lead to changes in the properties of the skin. The results from this study suggest that the volar forearm is somehow protected, perhaps from the sun. In fact, it has been shown in previous studies that skin that is hidden from sun exposure shows no significant differences between young and old people[13.3]. Another study of friction properties also found that there were no age-related differences on the volar forearm[13.21]. Egawa et al.[13.21] and this study suggest that the volar forearm may be an ideal anatomical site for studying the effects of skin products, since other factors will be minimized at this location. Three interventions were quantitatively differentiated using the electrical and friction measurements. The electrical impedance provided a measure of the water levels under the skin surface and revealed the intervention's ability to 'absorb' into the skin. The friction measurement revealed the intervention's ability to affect the exposed surface of the stratum corneum.

In the case of the PVDC occlusion, skin impedance did not greatly decrease (Fig. 13.5), the friction coefficient increased slightly compared to the other interventions (Fig. 13.4), and the amplitude/mean measurement showed that the change in skin stickiness was not as drastic as the other interventions (Fig. 13.6). This suggests that occlusion is not effectively 'absorbed' into the skin (impedance), it affects the surface by a moderate amount (friction coefficient), and it tends to make the surface stickier with the water build-up (amplitude/mean). Similar analysis can be extended to glycerin and petrolatum to differentiate between the two. Friction measurements show that both chemicals have increased the friction coefficient and it would seem that each chemical's effect is similar. However, comparisons along the change in skin impedance and the amplitude/mean measurement show that one can differentiate between the two compounds' skin effects. Petrolatum does not absorb as readily into the skin as glycerin and coats the exposed surface of the skin, evidenced by its lesser effect on the impedance (Fig. 13.5). The lowered amplitude/mean measurement for petrolatum shows that it makes the skin greasier than glycerin (Fig. 13.6).

13.7 Conclusions

From the analysis of the test data discussed in this chapter, the following conclusions are drawn[13.22]:

These data suggest that there is little variation in volar forearm skin across gender, age, and ethnicity and this site would be effective for the testing of skin and cosmetic products. Tribological measurements, namely friction and electrical measurements can be used in conjunction to gather differentiable data for various skin compounds. The data showed that PVDC occlusion, glycerin solution, and petrolatum were different when comparing across friction coefficient, electrical impedance, and amplitude/mean measurements. We do not wish to over generalize on the basis of sample size (59) examined; larger cohorts may show subtle differences not ascertained here.

References

[13.1] Sivamani RK, Goodman J, Gitis NV, Maibach HI. Coefficient of friction: tribological studies in man – an overview. Skin Res Technol 2003;9:227-234.

[13.2] Cua A, Wilheim KP, Maibach HI. Friction properties of human skin: relation to age, sex and anatomical region, stratum corneum hydration and transepidermal water loss. Br J Dermatol 1990; 123: 473–479.

[13.3] Elsner P, Wilhelm D, Maibach HI. Friction properties of human forearm and vulvar skin: influence of age and correlation with transepidermal water loss and capacitance. Dermatologica 1990; 181: 88–91.

[13.4] Blichmann CW, Serup J. Assessment of skin moisture. Acta Derm Venereol (Stockh) 1988; 68: 284–290.

[13.5] Blichmann CW, Serup J, Winther A. Effects of single application of a moisturizer: evaporation of emulsion water, skin surface temperature, electrical conductance, electrical capacitance, and skin surface (emulsion) lipids. Acta Derm Venereol (Stockh) 1989; 69 : 327–330.

[13.6] Fro¨din T, Helander P, Molin L, Skogh M. Hydration of human stratum corneum studied *in vivo* by optothermal infrared spectroscopy, electrical capacitance measurement, and evaporimetry. Acta Derm Venereol (Stockh) 1988; 68: 461–467.

[13.7] Tagami H, Ohi M, Iwatsuki K, Kanamaru Y, Yamada M, Ichijo B. Evaluation of the skin surface hydration *in vivo* by electrical measurements. J Invest Dermatol 1980; 75 : 500–507.

[13.8] Hashimoto-Kumasaka K, Takahashi K, Tagami H. Electrical measurement of the water content of the stratum corneum *in vivo* and in vitro under various conditions: comparison between skin surface hygrometer and corneometer in evaluation of the skin surface hydration state. Acta Derm Venereol (Stockh) 1993; 73: 335–336.

[13.9] Nicander I, Nyre´n M, Emtestam L, Ollmar S. Baseline electrical impedance measurements at various skin sites – related to age and gender. Skin Res Technol 1997; 3: 252–258.

[13.10] Nicander I, Ollmar S. Electrical impedance measurements at different skin sites related to seasonal variations. Skin Res Techno 2000; 6: 81–86.

[13.11] Sivamani RK, Goodman J, Gitis NV, Maibach HI. Friction coefficient of skin in real-time. Skin Res Technol (in press).

[13.12] El-Shimi AF. *In vivo* skin friction measurements. J Soc Cosmet Chem 1977; 28: 37–51.

[13.13] Comaish S, Bottoms E. The skin and friction: deviations from Amonton's laws, and the effects of hydration and lubrication. Br J Dermatol 1971; 84: 37–43.

[13.14] Koudine AA, Barquins M, Anthoine Ph, Auberst L, Leveque J-L. Frictional properties of skin: proposal of a new approach. Int. J Cosmet Sci 2000; 22: 11–20.

[13.15] Highley DR, Coomey M, DenBeste M, Wolfram LJ. Frictional Properties of Skin. J Invest Dermatol 1977; 69: 303–305.

[13.16] Prall JK. Instrumental evaluation of the effects of cosmetic products on skin surfaces with particular reference to smoothness. J Soc Cosmet Chem 1973; 24: 693–707.

[13.17] Johnson SA, Gorman DM, Adams MJ, Briscoe BJ. The friction and lubrication of human stratum corneum, thin films in tribology. In: Dowson D, et al: eds. Proceedings of the 19th Leeds–Lyon Symposium on Tribology. Amsterdam: Elsevier Science Publishers, B.V., 1993: 663–672.

[13.18] Nacht S, Close J, Yeung D, Gans EH. Skin friction coefficient: changes induced by skin hydration and emollient application and correlation with perceived skin feel. J Soc Cosmet Chem 1981; 32: 55–65.

[13.19] Hills RJ, Unsworth A, Ive FA. A comparative study of the frictional properties of emollient bath additives using porcine skin. Br J Dermatol 1994; 130: 37–41.

[13.20] Ale SI, Laugier JK, Maibach HI. Spacial variability of basal skin chromametry on the ventral forearm of healthy volunteers. Arch Dermatol Res 1996; 288: 774–777.

[13.21] Egawa M, Oguri M, Hirao T, Takahashi M, Miyakawa M. The evaluation of skin friction using a frictional feel analyzer. Skin Res Technol 2002; 8: 41–51.

[13.22] Shivamani K.Raja,Wu Gabriel, Gitis V.Norm, Maibach I. Haward, Tribological Testing of Skin Products: Gender, Age, and Ethnicity of the Volar Forearm, Skin Res. Technol. 2003; 9: 1-7

14 TRIBOMETROLOGICAL STUDIES IN BIOENGINEERING

14.1 A General Review

A novel experimental technology for studies of biological and bioengineered materials and structures has been developed. It is based on precision servo-control of forces or displacements and simultaneous real-time multi-sensor monitoring of deformations, forces and torques in all directions, contact acoustic emission, contact or surface electrical resistance or impedance, temperature, and digital microscopy, optionally accompanied with periodic atomic force microscopy of the test surfaces. Examples of the implementation of this technology in the Biomechanical Micro-tester mod. UMT are described for such diverse biotech applications as studies of prosthetic devices, skin, hydrogels, beauty care products, shaving blades and creams, toothpastes and dental materials, surgical sutures and needles, balloons and stent grafts, bio-fluids and bio-tissues.

14.2 Introduction

Biological and bioengineered materials and coatings present a wide spectrum of durability challenges. Over 200 bones and 600 muscles in the human body can fail due to numerous reasons. The biological lubricants produced by the body, including saliva, tears and synovial fluid, can dry up or deteriorate because of different effects. Artificial hips, knees, elbows and fingers cannot yet fully mimic the functional and tribological performance of the biological joints. Artificial heart valves, blood pumps and kidneys require much testing and improvement. Ocular tribology deals with friction in contact lenses and eyelids, lubricated with tears and complicated by eye blinking. Durability of artificial teeth from porcelain, gold and acrylic has to be increased to the level of tooth enamel. Surgical needles and shaving blades have to maintain their sharpness so their cutting force stays low. Surgical sutures and cosmetic products should maintain low friction, balloons and stent grafts require good durability and flexibility.

All these diverse tasks require much testing and understanding. The traditional engineering problems of correlating bench-top material test results with the real-life performance of components from the same materials are even more important in bioengineering, where full and comprehensive simulation of the *in-vivo* conditions is often impossible. The scientists need better test equipment, instrumentation and procedures to mimic the human body more accurately. To meet this demand, this chapter describes a multi-sensing technology, realized in a tester of modular design, which seems to provide the most useful test platform for the successful solution of the above challenges.

14.3 State-of-the Art Tribo-Metrology

The multi-sensing technology, which includes simultaneous real-time monitoring of deformations, forces and torques in all directions, contact acoustic emission, contact or surface electrical resistance or impedance, temperature, and digital microscopy, optionally accompanied with periodic atomic force microscopy of the test surfaces, allows for the most comprehensive evaluation of materials and components. The more signals are monitored, the better off we are in our process and materials control and optimization. These multi-sensor measurements have been implemented on a commercially available Bio-Mechanical Tester mod. UMT of CETR, a photo of which is shown in Fig. 12.1 of the Chapter 12. It provides comprehensive tribological measurements for various types of samples in a variety of biomedical applications. For example, it is successfully utilized for testing bathroom tissues on skin, soap on skin, surgical staples and sutures, medical needles, shaving blades, after-shave lotions, tooth pastes and tooth brushes, stent grafts and balloons.

The UMT is capable of providing precision linear and rotational, including reciprocation, motions with programmable speeds and accelerations in the range from 0.1 μm/s to 10 m/s (8 orders of magnitude) and programmable positions (displacements) in the range from 0.5 μm to 150 mm (over 5 orders of magnitude). A normal load is controlled via a closed-loop servomechanism with user-defined tolerances, and can be programmed to be either constant or changing gradually or by steps in the range from 10 μN to 1 kN (8 orders of magnitude). A number of parameters, including forces and displacements in all X, Y and Z directions, electrical contact or surface resistance, capacitance or impedance, deformation (elastic, plastic, creep) or wear depth, temperature, and contact acoustic emission can all be measured and recorded simultaneously, with a sampling rate of 20 kHz. Also, digital video with optical microscopy and scanning force microscopy are readily available.

14.4 Dermatological Studies

Skin health and beauty is a concern for people of all ages. Physiologically, skin is the first line of defense against any environment, and it is repeatedly subjected to physical and chemical damage. Over the course of time a person's skin undergoes changes, and to maintain skin health, it is important to quantitatively follow these changes. Sometimes these skin changes are visible, such as wrinkles, blemishes or rashes. In other cases, the changes may not be easily discernable without a quantitative assessment of skin properties. Some skin care products inhibit water evaporation from the skin, some compounds absorb and release hydrating elements into the skin, some compounds provide a greasy texture, while others elicit a more sticky texture. Quantification of these properties is crucial in the development of new skin care products.

The bulk of dermatological observations are based on qualitative visual or papillatory data. The unique multi-sensing technology described in this Chapter fills the void in the fields of dermato-physiology, dermato-pharmocology, dermato-toxicology and cosmetology. Utilization of the multi-sensing technology with real-time high frequency data acquisition provides a fast and quantitative assessment of skin dryness and moisturization, early diagnosis of skin diseases or of the deterioration in skin functions at a stage that may not be easily discernable visibly, development of skin cosmetics and medicine.

The UMT has been utilized for both *in-vitro* testing on the artificial or cut-off skin, *in-vivo* testing on people's arms, and fingers etc have been discussed in Chapter 12. The measurement probe, a 12 mm copper/Teflon cylinder attached to a proprietary suspension system (though in some tests Teflon and stainless steel balls were used), was pressed onto the skin with a load of 0.1 N, sufficient to maintain a contact with the skin, but not too high to avoid skin macro-deformations and focus the measurements on the skin surface only. For studies involving under-skin layers and tissues, higher loads up to 5 N were used. The probe was moved across the skin at the speed of 1 mm/s for 10 mm (though some tests were performed at higher speeds of 5 cm/s on longer traces, to better simulate the motion of a person's finger when checking her or his skin, which may be important for characterization and marketing of cosmetic products). During probe sliding, the dynamic friction coefficient was measured with a proprietary strain-gauge sensor, monitoring simultaneously and independently both normal load and friction force, with the resolution of 0.1 mN; also, a friction variation coefficient was calculated as its amplitude-to-mean ratio : as the friction variation coefficient increases, the skin is expected to be stickier or rougher. In-situ electrical measurements were performed by applying an alternating current of a small amplitude 10 μAmps, with frequency of 10 kHz, measuring the voltage and calculating the electrical impedance (Chapter 12).

14.5 Hip Replacement Studies

Artificial hips have been one of the most successful orthopedic devices, and are implanted in quantities of hundreds of thousands annually. Still, they last less than 15 years and have to be replaced via another major surgery. The cobalt-chromium or alumina head wears a socket made from either ultra-high molecular weight cross-linked polyethylene or ceramics, even in the presence of a synovial fluid

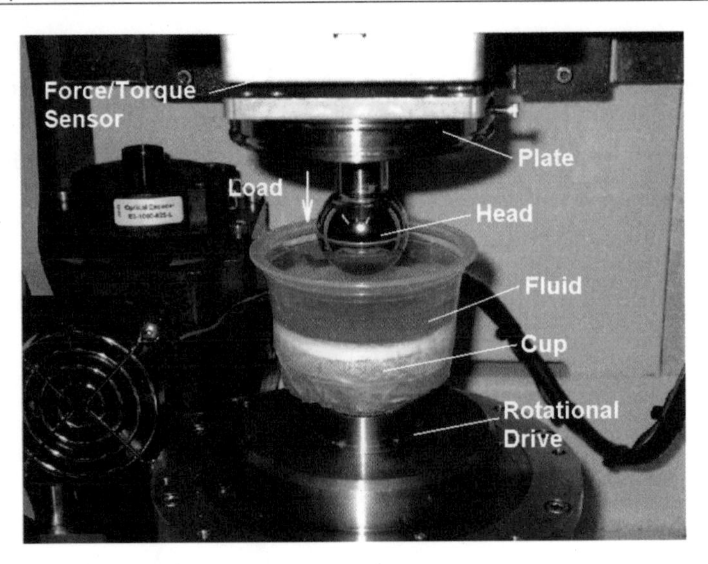

Fig. 14.1 Hip Joint Test Setup

We performed hip-simulating tests with in-situ monitoring tribological parameters of the metal head - polymer cup hip joint with bovine serum. The UMT can easily perform a combination of several simultaneous motions of both upper and lower specimens; the setup shown in Figure 14.1. The metal head was mounted on a tapered shaft with a flat plate, attached to a force/torque sensor (for simultaneous normal load and friction torque measurements) on a tilting and reciprocating drive, the polymer cup was mounted in a fluid container on a rotational drive. A small proprietary acoustic emission sensor was attached to the flat plate on the side opposite to the shaft with the head. The samples were fully immersed in a bovine serum. An applied vertical load was cycled from 0.3 to 1 kN, using a closed-loop feedback servo-control. To accelerate wear processes, the test was performed in a start-stop mode, with the drives cyclically accelerating, spinning for 30 sec, and decelerating to a full stop, while the bovine serum was not replenished and allowed to dry out, thus accelerating particle formation and accumulation.

During the tests, a normal force (F_z), friction torque (Tz), and AE signal were continuously monitored (since the polymer cup is non-conductive, contact electrical resistance was not monitored). During each start, the friction torque increased almost instantly and reached its maximum value corresponding to the full static friction (stiction). Then during the spin, the friction torque reduced, and when the drives stopped, friction torque dropped down almost to zero. In the beginning of the test, when the wear was insignificant, the AE signal remained at its near-zero level. The average values of Tz and AE from each cycle are plotted versus test time in Figure 14.2. The friction torque Tz grew within the first few hours due to run-in processes, then more or less stabilized for several hours while AE signal remained low. After some time, friction started to increase, while acoustics remained low, indicating a low level of mechanical interactions. Soon thereafter,, the friction torque increased sharply and reached its maximum value, the AE signal showed occasional peaks and then increased its level dramatically (30 times higher than the initial level), reflecting a failure. The post-test sample observation showed that the bovine serum dried out and left a lot of residual particles, which caused the increases in both friction and acoustics. After fresh fluid was added into the cup, and the test continued, both friction and acoustics reduced comparing to their levels at the end of the first test, however, staying higher and less-stable than their pre-failure levels in the first test.

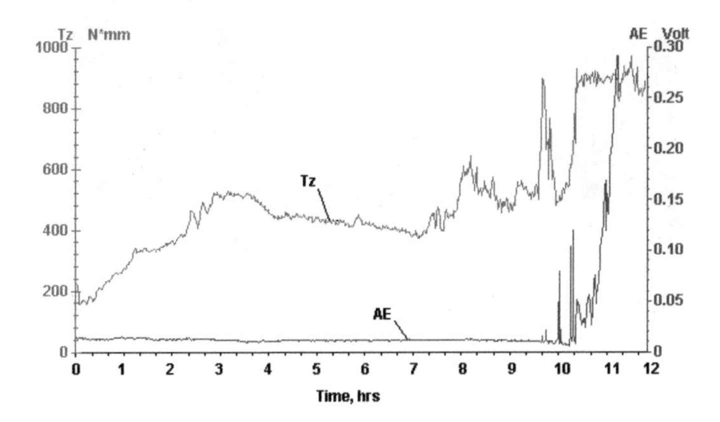

Fig. 14.2 Friction and AE Changes During Hip Replacement Accelerated Test

The above confirms that both tribological parameters of friction and acoustic emission are sensitive to the surface wear processes and can be used for in-situ wear monitoring during orthopedic life simulating tests. The friction torque seemed to show slightly higher sensitivity to wear during the break-in and pre-failure modes, the acoustic signal seemed to show slightly higher sensitivity to the presence of particles.

14.6 Testing of Tribological Properties of Contact Lenses

Hydrogel contact lenses have to have low static friction to ensure easy replacement, but sufficient friction to avoid slipping off. We tested several conventional and experimental lenses with curved surfaces of 14 mm diameter, with water content of 10-50%. The counter-samples were polished stainless steel and glass discs. The friction tests were performed on the setup shown in Figure 14.3. A contact lens was mounted in a holder with a stainless steel ball for a sample support, and attached to a dual-force sensor (for simultaneous normal load and friction force measurements) as the stationary upper specimen. The counter-disc was mounted on a rotational drive as the rotating lower specimen. A vertical load was maintained constant during the test, using a closed-loop feedback servo-control. Each lens was immersed in its packing solution, applied onto the disc surface; after each test the disc was cleaned and wiped. The tests were done at the same speed of 15 cm/s and three levels of load, 5 cN, 10 cN and 20 cN, for 30 s at each level.

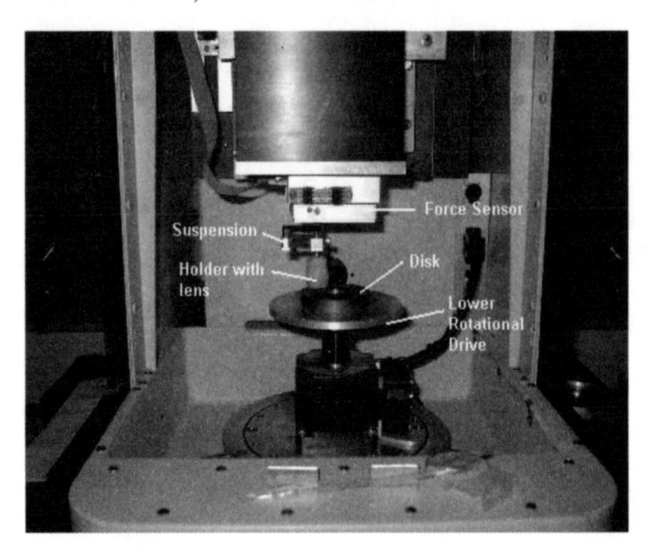

Fig. 14.3 Friction Test Setup for Contact Lenses

During the tests, a coefficient of friction (COF) was continuously monitored. Its typical real-time graph is shown in Figure 14.4. In the beginning of the test, when the disc just starts to move, friction increases instantly and reaches the maximum value (COF Peak) called stiction, or full static friction. During the test, with the disc spinning at constant speed, friction drops and remains at a certain sliding level (COF Average). The friction standard deviation from the average level (COF Chatter) characterizes smoothness of the motion and presence or absence of stick-slip.

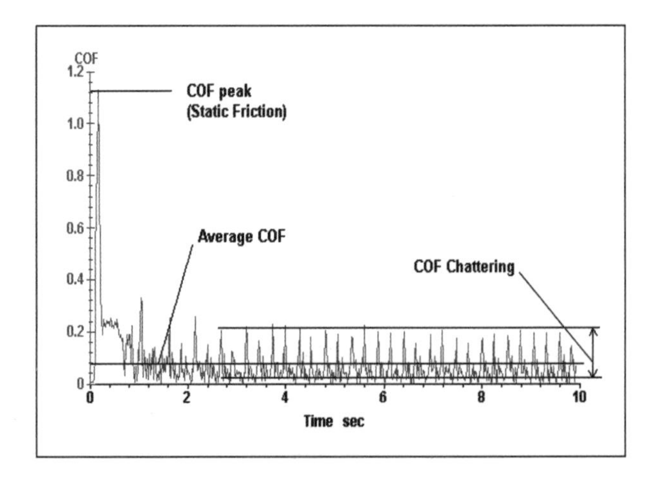

Fig. 14.4 Friction Changes During Tests of Contact Lenses

The graphs in Figures 14.5a to 14.5c show friction of seven types of contact lenses at three loads, on the glass disc. Focus contact lenses had overall the highest level of both sliding friction and friction chatter, but moderate static friction. Pure-vision lenses had higher average friction and friction chatter on steel, while lower friction and chatter on glass. AcuVue lenses had low friction at low load, but higher friction at higher loads. Biomedics lenses had low levels of sliding friction and friction chatter, but high static friction on steel. Pro-clear lenses also had low level of sliding friction and friction chatter, but high static friction.

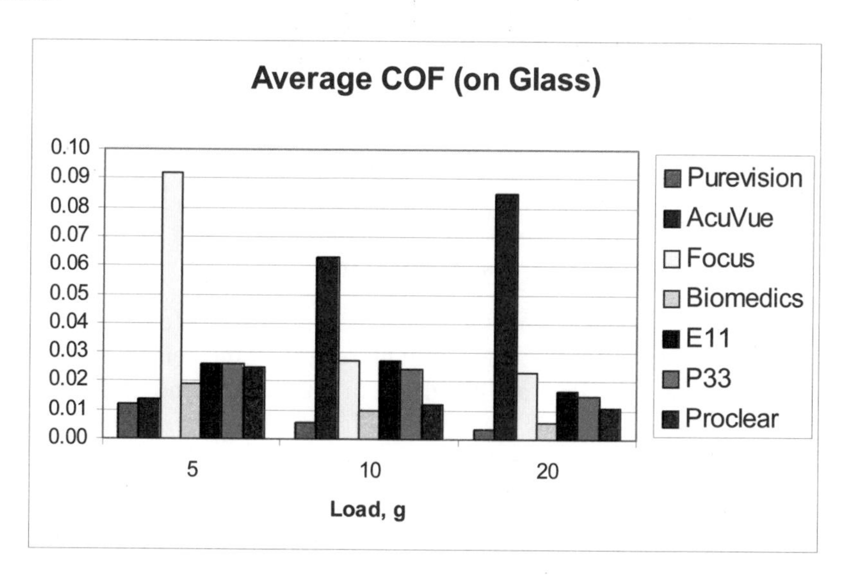

Fig. 14.5 a Sliding Average Friction of Lenses on Glass

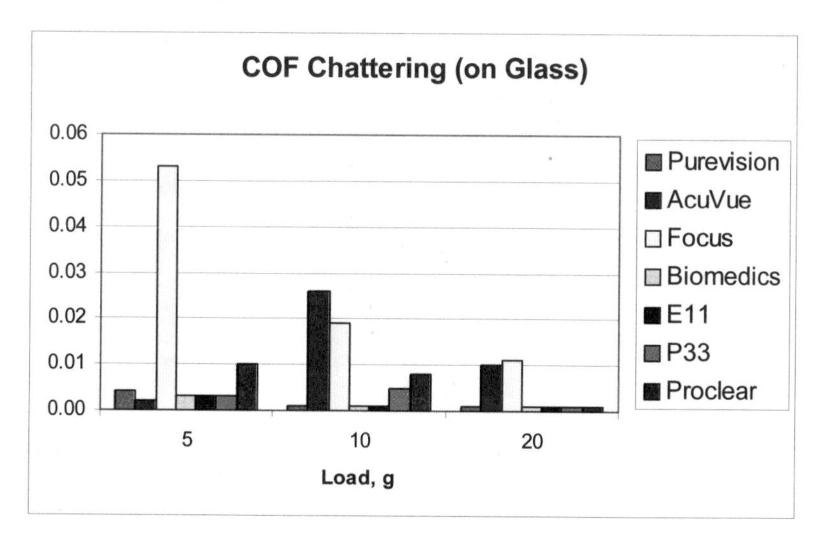

Fig. 14.5 b Friction Chatter of Lenses on Glass

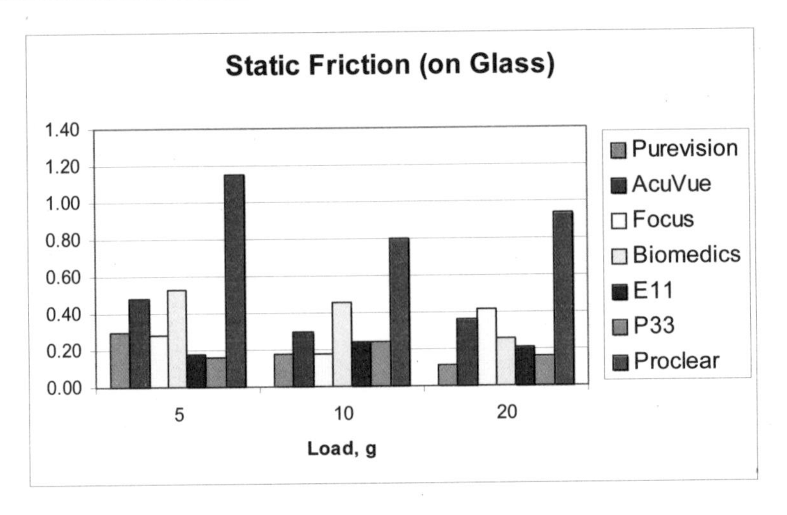

Fig. 14.5 c Static Friction of Lenses on Glass

14.7 Testing of Surgical Needles

Two important functional parameters were measured for medical needles, the puncturing force into the skin, which may be related to the pain experienced by a patient and so has to be low, and the depth of needle penetration into the skin. The schematic of needle testing is shown in Figure 14.6. A needle was mounted on the upper holder and brought down, at the constant speed, to punch into the artificial or cut-off skin, fixed on the lower table. Either the penetration depth is measured under a constant load, or the load is measured at a predetermined penetration depth. A close-loop feedback servo-control ensures either constant load or displacement. The needle is then pulled out for next penetration.

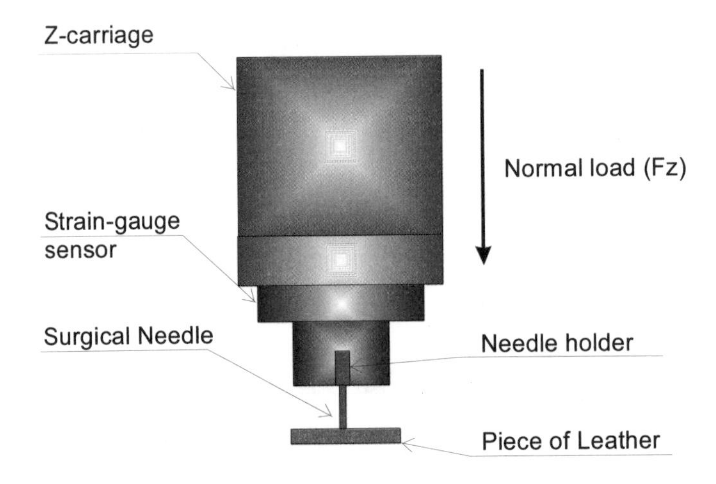

Fig. 14.6 Setup for Needle Testing

At the constant load of 20 cN, we measured the change in penetration depth over 20 consecutive punctures. The plots of depth versus a number of punctures are shown in Figure 14.7. The depth decreased with a number of penetration times, indicating that the needles became dull. The Needle 1 had better sharpness durability than Needle 2 over time.

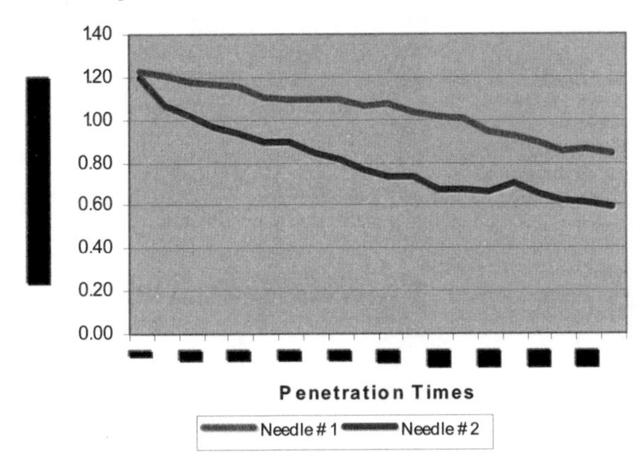

Fig. 14.7 Needle Penetration Depth (mm) vs. Puncture Times

14.8 Tests of Surgical Sutures

Medical surgical sutures are tested for durability against either each other or a stainless steel rod. The schematic of suture testing is shown in Figure 14.8 where either the rod (left) or suture (right) mounted on the upper sample holder moves on the lower suture back and forth for a reciprocating length of 5 mm at a frequency of 1 Hz under a constant normal load of 5 cN. When the suture is worn to break, the normal force drops to zero, since there is no support for the abrasion. The time to abrade the suture to break is measured, based on monitoring friction force, normal force, and wear depth.

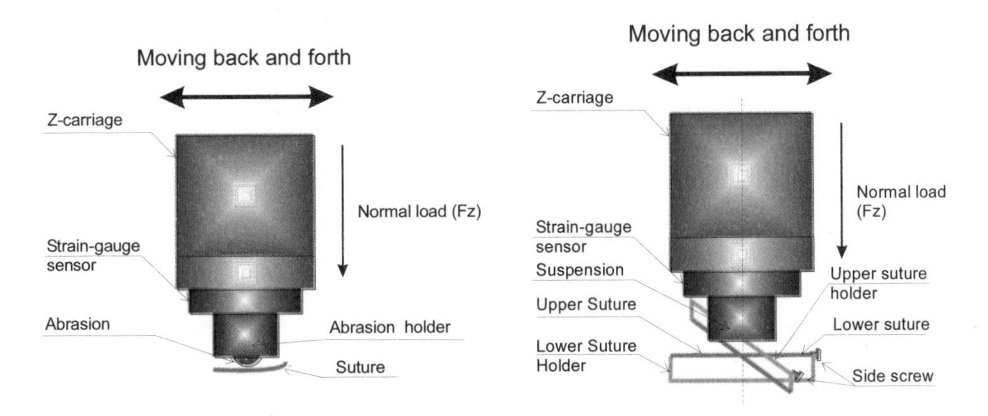

Fig. 14.8 Setups for Rod-on-Suture (left) and Suture-on-Suture (right) Tests

As shown in Figure 14.9, suture 1 exhibited better performance in both longer abrasion to failure and lower friction coefficient than suture 2.

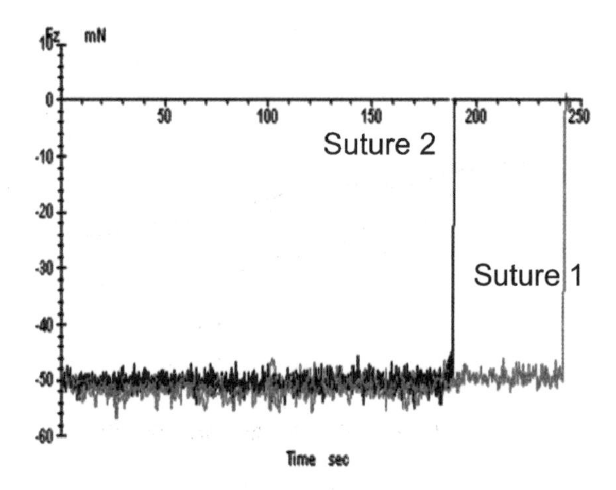

Fig. 14.9 Normal load versus Abrasion Time

14.9 Balloon Flexibility

Medical balloons are tested for their elasticity. The schematic of balloon flexibility testing is shown in Figure 14.10 A balloon was mounted on the lower sample holder with three different methods: three point, cantilever, and flexible. A flat head screw was mounted on the upper holder and moved down to the surface of the balloon; this position was set as the zero position. Then, the displacement of screw was set to 0.5 mm and the screw was moved down and up twice several times at a speed of 0.05 mm/sec. Balloon A is more flexible than Balloon B, since for the same force, Balloon A exhibited a larger deflection.

14.10 Stent graft tests

Stent graft material (polyester woven cloth) was tested for durability. The schematic for graft material test is shown in Figure 14.11 A triangular piece of stent sheath was mounted on the upper sample holder and reciprocated over the graft material for a length of 5 mm at a frequency of 1.5 Hz under a constant normal load of 0.3 N. When the graft sample was worn through, the normal force and upper carriage position changed suddenly, since the support for the stent triangle also decreased suddenly. During a test, carriage position, friction force, normal force, time and coefficient of friction were recorded. The elapsed time for the graft material to wear through indicated abrasion resistance.

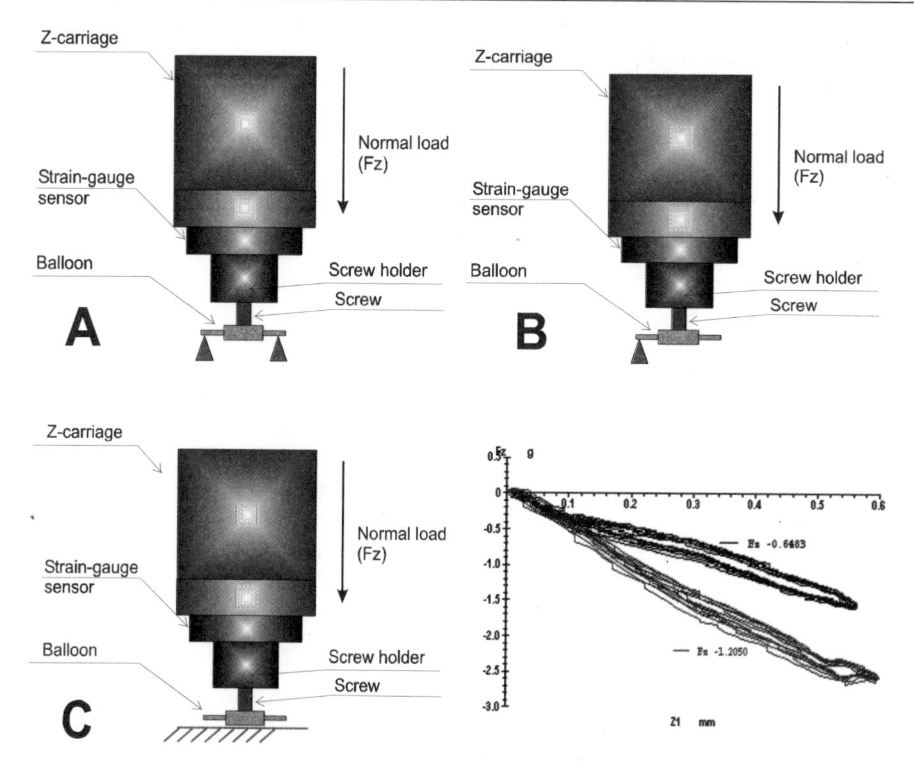

Fig. 14.10 Test Setup for balloon flexibility: A-three point; B-cantilever; C-flexible basis

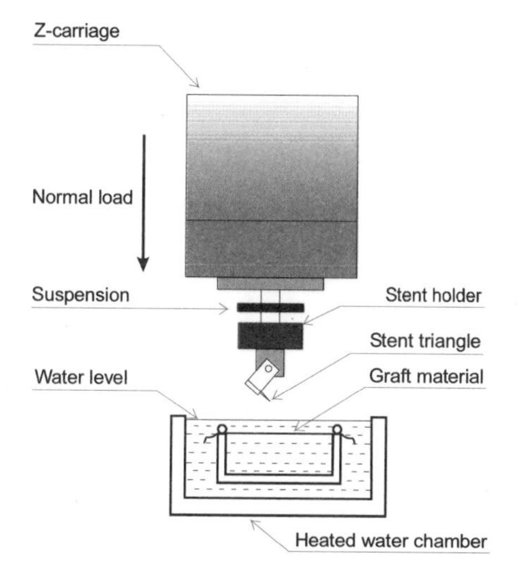

Fig. 14.11 Setup for Stent Graft Testing

14.11 Conclusions

The described tests show how bio-materials can be evaluated for their tribological and mechanical properties, including abrasion and puncture resistance, elasticity, wear and friction, all with a single instrument. Other material properties, like fatigue, hardness and scratch resistance, can also be measured.

15

AN OVERVIEW OF TRIBOLOGICAL STUDIES ON COEFFICIENT OF FRICTION OF HUMAN SKIN

15.1 Background

Compared to other studies of skin, relatively few studies have focused on the friction of skin. This chapter reviews existing skin friction, emphasizing test apparatuses and parameters that have added to information regarding the friction coefficient. This review also outlines factors that are important for future friction studies. However past studies have utilized numerous designs for a test apparatus, including probe geometry and material, as well as various probe motions (rotational vs. linear). Most tests were performed *in vivo*; a few were performed *in vitro* and on porcine skin.

Differences in probe material, geometry and smoothness affect friction coefficient measurements. An increase in skin hydration, either through water or through moisturizer application, increases its friction coefficient; a decrease in skin hydration, either through clinical dermatitis or through alcohol addition, decreases the coefficient. Differences are present between anatomical sites. Conflicting results are found regarding age and no differences are apparent as a result of gender or race.

In short skin friction appears to depend on several factors – such as age, anatomical site and skin hydration. The choice of the probe and the test apparatus also influence the measurement.

15.2 Introduction

Physiologically, the skin is the first line of defense against the environment and it is repeatedly subjected to physical and chemical damage. The skin's mechanical properties – such as its friction characteristics – can alter under this repeated damage. Mechanically, friction allows us to keep from slipping as we step out of the shower, hold the Styrofoam cups of coffee, or turn the steering wheel in our cars. Because the skin is a surface itself, it is convenient to analyze and describe it in terms of a surface phenomenon – such as friction; friction studies on skin provide valuable insight into how the skin interacts with other surfaces. Friction also provides information about the skin under various conditions – for example, age and gender – and under various chemical treatments – for example, lotions and moisturizers. Studying the friction of skin supplements other mechanical tests. An advantage of friction studies is that they can be performed with non-invasive methods and give a measure of the skin's health – for example, skin hydration. Naylor[15.1] showed that moistened skin has an elevated friction response and El-Shimi[15.2] demonstrated that drier skin has a lowered friction response. Friction provides a quantitative measurement to assess skin condition.

The friction parameter generally measured is the coefficient of friction. In order to measure the friction coefficient, one surface is brought into contact with another and moved relative to it. When the two surfaces contact, the perpendicular force is defined as the normal force N. The tangential friction force F is that force which opposes relative movement between the two surfaces. From Amonton's law, the coefficient of friction μ is defined as the ratio of the friction force to the normal force:

$$\mu = F / N$$

The friction coefficient can be measured in two ways: (i) the static friction coefficient μ_s and (ii) the dynamic or kinetic friction coefficient μ_k.

The static friction coefficient is defined as the ratio of the force required to *initiate* relative movement to the normal force between the surfaces; the dynamic or kinetic friction coefficient is defined as the ratio of the friction force to the normal force when the two surfaces are moving relative to each other. Much of the research has been focused on the dynamic friction coefficient wherein the two surfaces move at a relative *constant velocity*. Most of the friction studies on skin have dealt with the dynamic friction coefficient and the subscript k is usually dropped. This overview references the dynamic coefficient of friction unless otherwise noted.

According to Amonton's law, the dynamic friction coefficient remains unchanged regardless of the probe velocity or applied normal load in making the measurement. Amonton's laws hold true in the case of solids with limited elastic properties. Although Naylor[15.1] concluded Amonton's law to be true, later studies by El-Shimi[15.2], Comaish and Bottoms[15.3] and Koudine et al.[15.4] found that skin deviates from Amonton's law, because their studies found the friction coefficient to be inversely proportional to load. El-Shimi[15.2] and Comaish and Bottoms[15. 3] reasoned that the rise in friction coefficient with decreasing load resulted from the viscoelastic nature of the skin allowing for a non-linear deformation of the skin.

15.3 Materials and Methods

15.3.1 *Experimental design*

Various experimental designs have been devised in order to measure the friction on skin. They focus on measuring friction by pressing a probe onto the skin with a known normal force, and then detecting the skin's frictional resistance to movement of the probe. The designs fall into two categories:

- A probe moved across the skin in a linear fashion.
- A rotating probe in contact with the skin surface.

In the linear designs, the probe movement is accomplished in several ways. Comaish and Bottoms[15.3] utilized one of the simplest linear designs; they moved the probe across the skin by attaching it to a pan of weights by means of a pulley. Their design is illustrated schematically in Fig. 15.1. Weights are placed in the pan such that the probe slides over the skin at a constant velocity. This allows for the calculation of the dynamic friction coefficient by dividing the total weight in the pan by the normal load on the probe. However, there are many inaccuracies involved in this method as there is no monitor or control of probe speed or normal force.

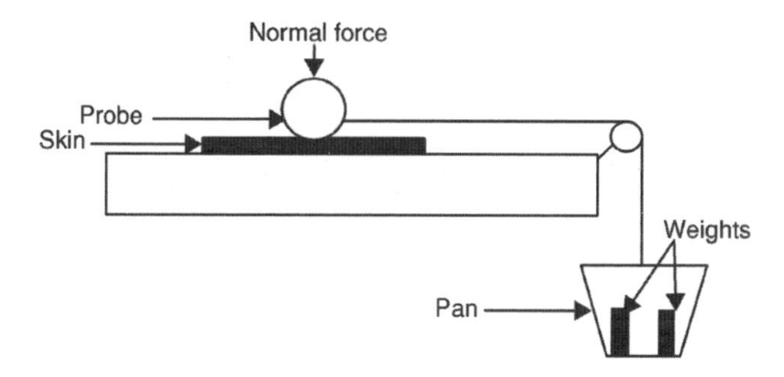

Fig. 15.1 Schematic apparatus used by Comaish and Bottoms (3)

More sophisticated linear designs followed the design used by Comaish and Bottoms[15.3], but provided motorized unidirectional movement of the probe or the use of a reciprocating motor in order to move the probe back and forth. In both designs the motorization afforded greater control in maintaining the velocity of the probe. Strain gauges were used in order to measure the friction force as the probe moved along the skin surface.

The second design category measures friction with a rotating wheel pressed onto the surface of the skin with a known normal force. Highley et al.[15.5] measured the frictional resistance by determining the angular recoil of the instrument as the wheel contacted the skin. They measured this angular recoil by recording the proportion of light that hit a dual element photocell. An electrical signal was then generated in proportion to the frictional resistance. Comaish et al.[15.6] developed a portable, hand-held device (Newcastle Friction Meter) that relied on a torsion spring in order to measure the skin's frictional resistance. The devices are surveyed in Table 15.1.

An important part of designing a friction measurement apparatus is the choice of probe size, shape and material. Since friction is an interaction between two surfaces, the probe geometry and material can affect the values calculated for the friction coefficient of the other surface. Several shapes and materials have been used as outlined in Table 15.1. The results will be more accurate when the probe's normal force is maintained at a constant value or continuously monitored; previous methods used to maintain the normal force include spring mechanisms or static weights to weigh down the probe (Table 15.1). These parameters are later revised critically.

Table 15.1 Probe and apparatus used in order to measure the dynamic friction coefficient μ of untreated 'normal' skin *in vivo*

Author	Probe size and shape	Probe material	Motion of test apparatus	Maintenance of normal load
Naylor[15.1]	8 mm diameter, sphere	Polyethylene	Linear, reciprocating	Static weights
El-Shimi[15.2]	12 mm diameter, hemisphere	Stainless steel (rough) Stainless steel (smooth)	Rotational	Static weights
Comaish and Bottoms[15.3]	15 mm diameter, annular ring	Teflon, nylon, polyethylene, wool	Linear	Static weights
Koudine et al.[15.4]	Hemisphere, lens	Glass	Linear	Static weights; balance beam
Highley et al.[15.5]	Disc	Nylon	Rotational	Spring load
Prall[15.7]	Disc	Glass	Rotational	Spring load
Cua et al.[15.8]	15 mm diameter, disc	Teflon	Rotational	Spring load
Johnson et al.[15.9]	8 mm (radius of curvature), lens	Glass	Linear, reciprocating	Static weights
Asserin et al.[15.10]	3 mm diameter, sphere	Ruby	Linear	Balloon; static weights
Eisner et al.[15.11]	15 mm diameter, disc	Teflon	Rotational	Spring load
Sivamani et al. [15.17]	10 mm diameter, sphere	Stainless steel	Linear	Computer-controlled servo-feedback

Much effort has been made in understanding how skin friction changes with differing biological conditions and upon the application of various products to the skin surface. These studies are of interest to various companies that manufacture products meant as skin topical agents, because friction measurements can provide clues regarding the effectiveness of their products. Previous studies are outlined in Table 15.2.

Table 15.2 Comparative studies on the human skin friction coefficient

Author	Comparative studies
Naylor[15.1]	Hydration
El-Shimi[15.2]	Hydration Lubricants/emollients/moisturizers Probes
Comaish and Bottoms[15.3]	Hydration Lubricants/emollients/moisturizers Probes
Koudine et al.[15.4]	Hydration Lubricants/emollients/moisturizers Anatomical site
Highley et al.[15.5]	Hydration Lubricants/emollients/moisturizers
Prall[15.7]	Hydration Lubricants/emollients/moisturizers
Cua et al.[15.8]	Hydration Age Gender Anatomical site
Johnson et al.[15.9]	Hydration pH Lubricants/emollients/moisturizers
Asserin et al.[15.10]	Age
Elsner et al.[15.11]	Age Anatomical site Race
Loden et al.[15.12]	Hydration
Sulzberger et al.[15.13]	Hydration
Nacht et al.[15.14]	Hydration Lubricants/emollients/moisturizers
Hills et al.[15.15]	Hydration Lubricants/emollients/moisturizers
Sivamani et al.[15.17]	Hydration Lubricants/emollients/moisturizers

All studies were performed in vivo except Comaish and Bottoms[15.3] who performed some *in vitro* tests on human skin and Hills et al.[15.15] utilized porcine skin in their *in vitro* tests.

15.3.2 *Hydration*

Hydration is a complex phenomenon influenced by intrinsic – that is, age, anatomical site – and extrinsic – that is, ambient humidity, chemical exposure – factors. These factors can affect the mechanical properties of skin and research carried out in order to correlate hydration levels with the skin's friction coefficient. Hydration studies have investigated how increases and decreases in skin hydration correlated with the friction coefficient. In past studies, researchers generally induced increases in skin hydration through water exposure. However, decreases in skin hydration were not experimentally induced and dehydration studies were performed between subjects with 'normal' skin and subjects that had clinically 'dry' skin[15.2, 15.12].

15.3.3 *Lubricants/emollients/moisturizers*

Much of the reviewed research has been devoted to ascertaining how the application of certain ingredients influences the skin surface, which is of interest to the cosmetic/moisturizer and lubricant industry. The studies focused on the effects of talcum powder[15.2, 15.3], oils[15.2, 15.3, 15.5, 15.14] and skin creams/moisturizers[15.7, 15.14]. Hills et al.[15.15] analyzed how changes in the friction coefficient, following emollient application, differed with temperature.

15.3.4 *Probes*

As mentioned earlier, the probe geometry and material influence the measured value of the friction coefficient because friction is a probe-skin interaction phenomenon. Few studies have examined probe effects; El-Shimi[15.2] studied probe roughness and Comaish and Bottoms[15.3] probe roughness and material.

15.4 Discussion

Friction is an important characteristic of skin, because it allows us to execute many of our daily activities. In addition, friction studies offer insight into how the skin surface changes across age, gender, race, anatomical site and chemical applications. These studies can provide better information about expected skin variations in the population and why certain topical applications are more effective than others. Comparative studies are particularly useful in following how the skin's mechanical properties change under various conditions.

Previous studies report a range of values for the skin's friction coefficient. Dynamic friction coefficient measurements (Table 15.3) fall in the range 0.12-0.7; however, most fall in a narrower range of 0.2-0.5 (Fig. 15.2). Besides natural variations in skin, the wide range in results may be as a result of differences in probe movement, geometry and material, and controlled monitoring of the normal force. In the reviewed friction measurement apparatuses, the two types of probe movements utilized were rotational and linear (Table 15.1). The linear probe constantly moves over 'untested' skin and the rotational probe spins over 'tested' skin. The different movements can lead to discrepancies in reported values for the skin friction coefficient. Another important source of variation may be in the ability to control the normal force while the probe is moving over the skin surface. The skin friction instruments are designed in order to measure the frictional resistance of the skin and it is assumed that the normal force is constant. During a test the normal force may not remain constant as a result of many factors – for example, uneven skin surface, inaccurate spring and/or a non-uniform distribution of static weights placed above the probe. Therefore, the assumption of a constant normal force may be incorrect and can lead to inaccuracy and variation in the calculated friction coefficient. A third source for variation is the choice of the probe material. Because friction is a surface phenomenon between two materials, the choice of the probe will influence the numerical value obtained for the friction coefficient.

Table 15.3 Reported values of the dynamic friction coefficient μ for untreated normal skin *in vivo*

Author	μ
Nayior[15.1]	0.5-0.6
Ei-Shirni[15.2]	0.2-0.4 (stainiess steei rough)
	0.3-0.6 (stainiess steei smooth)
Comaish and Bottoms[15.3]	0.2 (tefion)
	0.45 (nyion)
	0.3 (poiyethyiene)
	0.4 (wooi)
Koudine et al.[15.4]	0.24 (dorsai forearm)
	0.64 (voiar forearm)
Highley et al.[15.5]	0.2—0.3
Prall[15.7]	0.4
Cua et al.[15.8]	0.34 (forehead)
	0.26 (voiar forearm)
	0.21 (pairn)
	0.12 (abdomen)
	0.25 (upper back)
Johnson et al.[15.9]	0.3—0.4
Asserin et al.[15.10]	0.7
Elsner et al.[15.11]	0.48 (forearm)
	0.66 (vuiva)
Sivamani et al.[15.17]	0.33—0.55

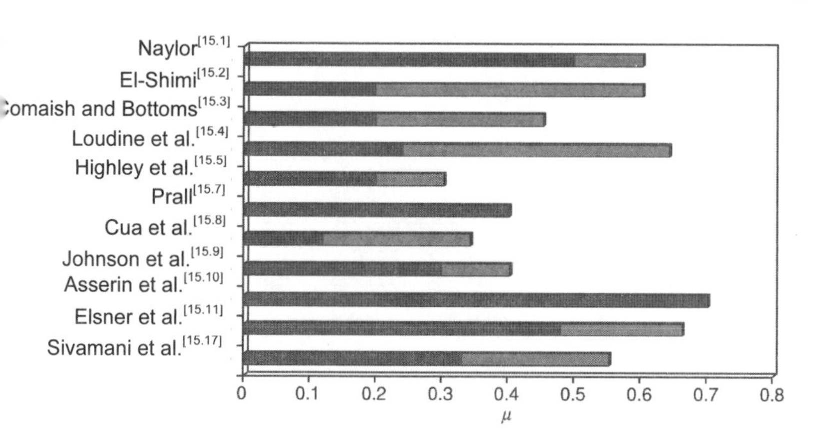

Fig. 15.2 Outline of the ranges in the dynamic coefficient of friction. These ranges reflect measurement of untreated 'normal' skin friction in vivo.

15.4.1 *Hydration Studies*

Hydration studies reveal that drier skin has lowered friction while hydrated skin has an in- creased amount of friction (Table 15.4). However, the skin response is more complex, because very wet skin also has a lowered friction coefficient much like the characteristics of dry skin[15.16]. Most studies focus on an intermediate zone of hydration where the skin has been moistened without an appreciable 'slippery' layer of water on the skin. Results in Table 15.4 show that the increases in friction are varied and this possibly results from the various probes used. Although the addition of water increases the friction coefficient, this effect lasts for a period of minutes before the skin returns to its 'normal' state[15.2, 15.5, 15.14, 15.17]. The water has an effect of softening the skin and this in turn allows for greater contact area between the probe and the skin. Water results in adhesive forces between the water and the probe. Thus, there is more frictional resistance between the skin and the probe and results in a higher friction coefficient[15.18]. Because the water evaporates over an order of minutes, the skin returns to its 'normal' state in the same time frame. For dry skin, the skin becomes less supple and the probe does not achieve as much contact area and this allows the probe to glide more easily over the skin surface. This results in a lowered friction coefficient as seen in the isopropyl study[15.17] and in prior studies involving subjects with clinically dry skin[15.2, 15.12]. The agreement between the experimentally induced dry skin and clinical dry skin is expected[15. 19].

Table 15.4 Comparative studies of the changes in dynamic friction coefficient μ with increasing hydration (hydration) and decreasing hydration (dryness)

Author	Probe material	% increase because of hydration $(\mu_{moist}/\mu_{normal})/\mu_{normal}\} \times 100$	% decrease because of dryness $\{(\mu_{normal}/\mu_{dry})/\mu_{normal}\} \times 100$
Naylor[15.1]	Polyethylene	80	–
El-Shimi[15.2]	Stainless steel (rough); Stainless steel (smooth)	100-200 (stainless steel rough)	28 (stainless steel rough); 41 (stainless steel smooth)
Comaish and Bottoms[15.3]	Wool; Teflon	40* (Wool); 400* (Teflon)	–
Highley et al.[15.5]	Nylon	500	–
Prall[15.7]	Glass	200	–
Johnson et al.[15.9]	Glass	100-223	–
Loden et al. [15.12]	Stainless steel	–	33 (hand); 41 (hack); 14 (hrm)
Nacht et al.[15.14]	Teflon	45	–
Sivamani et al.[15.17]	Stainless steel	55 (*in vitro*)	10 (*in vivo*)

*Comaish and Bottoms[15.3] studied the change in the static friction coefficient in their hydration study.

15.4.2 *Lubricants/emollients/moisturizers*

The studies on lubricants, emollients and moisturizers are important for cosmetics and products developed in order to make the skin look and feel healthier. The literature reports that the important qualitative characteristics in skin topical agents are skin smoothness, greasiness and moisturization[15.18, 15.20]. Previous research has tried to describe these subjective, qualitative descriptions in a quantitative fashion by correlating them against the friction coefficient. Prall[15.7] was unable to make a direct correlation of skin smoothness with friction coefficient until he added skin topography and hardness to the analysis. Nacht et al.[15.14] found a linear correlation between perceived greasiness and the friction coefficient (Fig. 15.4).

15.4.3 *Talcum powder*

El-Shimi[15.2] and Comaish and Bottoms[15.3] showed that the friction coefficient decreased after the application of powder. El-Shimi[15.2] found that the friction coefficient decreased by 50% after application; Comaish and Bottoms[15.3], in analyzing the static friction coefficient, observed an insignificant change for a wool probe and a 30% decrease in friction with a polyethylene probe. However, they also found that wetting the talcum powder caused an increase in the measured friction.

15.4.4 *Lubricant oils*

A lowering in the friction coefficient is the initial effect after the application of oils and oil-based lubricants[15.2, 15.5, 15.14]. Nacht et al.[15.14] and Highley et al.[15.5] also showed that after the initial decrease in friction, the oils eventually raised the skin's friction coefficient. The results of the lubricant cosmetic studies by Nacht et al.[15.14] are shown in Fig. 15.3.

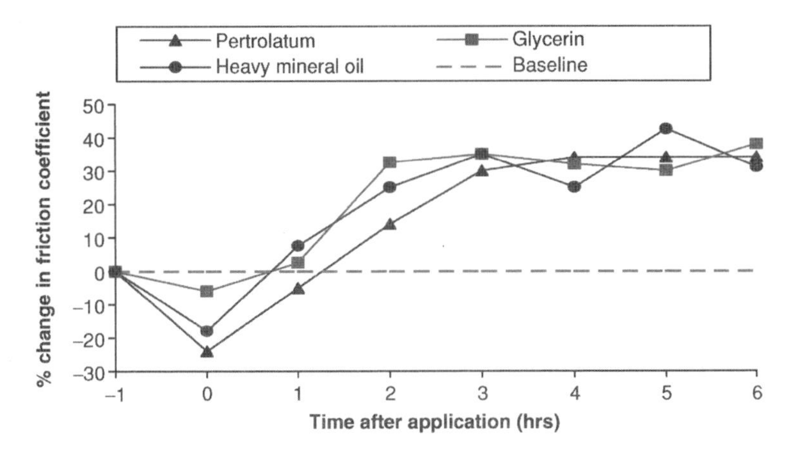

Fig. 15.3 Effect of lubricant cosmetic ingredient on skin friction coefficient. Amount applied of each material: approx. 2 mg cm^{-2}. Reproduced from Nacht et al.[15.14] (mean of five subjects but P-value was not published). Time = 1 is immediately prior to application; Time = 0 is immediately after application.

15.4.5 *Emollients and moisturizers*

Prall[15.7] and Nacht et al.[15.14] found that the friction coefficient rises with the addition of emollients and creams in a fashion similar to water. However, the effects of the creams lasted for hours, whereas the water effects lasted for about 5-20 mm[15.7, 15.17]. Hills et al.[15.15] also studied emollients, but they examined how various emollients compared against one another and how changes in temperature changed the friction coefficient. At a higher temperature (45 °C), most emollients lowered the friction coefficient to a greater degree than at a lower temperature (18 °C).

When lubricant/moisturizers are applied to the skin, the skin friction is affected in three general ways[15.14, 15.18].

- A large, immediate increase in the friction coefficient, similar to water application, that follows with a slow decrease in the friction coefficient. These agents can be interpreted to act by immediate hydration of the skin through some aqueous means in order to give the immediate increase in friction. In Fig. 15.5, cream B falls into this category and in Fig. 15.4, creams A, B and C represent this type of lubricant/moisturizer.
- An initial decrease in the friction coefficient that is followed by an overall increase in the friction coefficient over time. These agents are fairly greasy products (Fig. 15.3) and this greasiness causes the immediate decrease in the friction coefficient. The eventual rise in the friction coefficient is probably because of the increase in skin hydration through the occlusive effects of these agents[15.21]. Representations of a few ingredients that elicit this response are in Fig. 15.3 and represented as cream F in Fig. 15.4.

Fig. 15.4 Correlation between changes in the friction coefficient and the sensory perception of greasiness. A, B, C, D, E and F represent different creams that were applied to the skin. The reported percent change in the friction coefficient is immediately after application and the greasiness scores were subjective evaluations (From Nacht et at.[15.14]).

- A small, immediate increase in the friction coefficient increases slowly with time. These agents are interpreted to act as a combination of effects seen in the previous two cases. These lubricants/moisturizers have ingredients and agents that serve to both hydrate the skin through some aqueous method and prevent water loss through some occlusive mechanism. Because of the presence of these occlusive agents, which tend to be more slippery, the immediate rise in the friction coefficient is lower than in products that fall into the first category listed above. In Fig. 15.5, this is seen in cream A and in Fig. 15.4, this is seen in creams D and E.

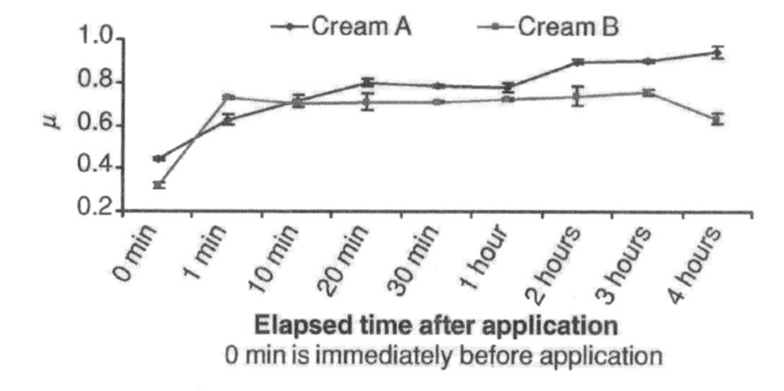

Fig. 15.5 Effects on the dynamic friction coefficient after applying moisturizing creams. The cream was applied to the back of the finger and then monitored for 4 h as shown above. Cream A was Loréal® Plentitude Hydra-Renewal Cream, a slow-acting, long-duration moisturizer. Cream B was Loréal® Plentitude Excell-A^3 Alpha Hydroxy Cream, a fast-acting, short-duration moisturizer; 0 min is immediately before application. Each data point represents the average of four measurements; ($n = 2$; $P < 0.05$ for 2, 3, and 4 h marks).

15.4.6 *Probes*

El-Shimi[15.2] and Comaish and Bottoms[15.3] compared probes (Tables 15.3 and 15.4) and found that smoother probes gave higher friction coefficient measurements. El-Shimi[15.2] noted that higher friction coefficient measurements were made with a smoother stainless steel probe as opposed to a roughened stainless steel probe. Comaish and Bottoms[15.3] found a similar result with two types of nylon probes: a sheet probe and a knitted probe. The sheet probe (the smoother of the two) gave a higher friction coefficient measurement. El-Shimi[15.2] postulates that the smoother probe forms more contact points with the skin and has a greater skin contact area than the rougher probe, resulting in more resistance from the skin and a larger measurement for the friction coefficient.

15.4.7 *Anatomic region, age, gender and race*

Few studies address the effects of anatomic region, age, gender, or race as they pertain to the friction coefficient. To date, no significant differences have been found with regard to gender[15.8, 15.22] or race[15.23]. Age-related studies have been contradictory where some authors found no difference[15.8, 15.22] and others found differences[15.10, 15.11].

The friction coefficient varies with anatomical site. Cua et al.[15.8, 15.22] found that friction coefficients varied from 0.12 on the abdomen to 0.34 on the forehead. Eisner et al.[15.11] measured the vulvar friction coefficient at 0.66, whereas the forearm friction coefficient was 0.48. Manuskiatti et ai.[15.23] studied skin roughness and found significant differences in skin roughness at various anatomical sites. Differences in environmental influences – that is, sun exposure – and hydration may account for this. Elsner et al.[15.11] showed that the more-hydrated vulvar skin had a 35% higher friction coefficient than the forearm, in agreement with hydration studies that contend the skin has an increased friction coefficient under increased hydration.

With respect to age, friction measurement results are contradictory. Cua et al.[15.22] showed no differences in friction with respect to age except for friction measurements on the ankle. Eisner et al.[15.11] also performed age-related tests and found no differences in the vulvar friction coefficient, but observed a higher forearm friction coefficient in younger subjects. They postulate that the skin on parts of the body that become exposed to sunlight can undergo photoaging and thus, forearm skin shows evidence of age-related differences while the light-protected vulvar skin does not[15.11]. Asserin et al.[15.10] concluded that younger subjects had a higher forearm friction coefficient than older subjects.

There are few gender-related friction studies. Cua et al.[15.8, 15.22] found no significant friction differences between the genders. There are no studies addressing race as it pertains to friction, but Manuskiatti et al.[15.23] looked for racial (black and white skin) differences in skin roughness and scaliness. Their studies indicated no significant differences.

15.5 Conclusions

Based on the above studies, the following conclusions are drawn[15.24]:

Although there have been limited studies dealing with the measurement of the skin friction coefficient, past studies and present study[15.17] show that differences in skin, because of various factors – such as age and hydration – can be correlated with the friction coefficient. Friction coefficient studies can serve as a quantitative method to investigate how skin differs on various parts of the body and how it differs between different people. It is also a useful method for tracking the changes resulting from the environmental and chemical treatments – such as sunlight – and when various chemicals are applied to the skin – such as soaps, lubricants and skin creams. The reviewed studies show that friction is an important parameter for understanding the skin's mechanical state. The reviewed studies also indicate that the design of the test apparatus is an extremely important factor, because test design parameters can also have an influence on friction measurements. A better appreciation of the importance of the friction coefficient will become clearer as measurement methods improve and allow for greater accuracy.

References

[15.1] Naylor PFD. The skin surface and friction. Br J Dermatol 1955; 67: 239-248.

[15.2] El-Shimi AF. *In vivo* skin friction measurements. J Soc Cosmet Chem 1977; 28: 37-51.

[15.3] Comaish 5, Bottoms E. The skin and friction: deviations from Amonton's laws, and the effects of hydration and lubrication. Br J Dermatol 1971; 84: 37-43.

[15.4] Koudine AA, Barquins M, Anthoine PH, Auberst L, Leveque J-L. Frictional properties of skin: proposal of a new approach. Tnt J Cosmet Sci 2000; 22: 11-20.

[15.5] Highley DR, Coomey M, DenBeste M, Wolfram U. Frictional properties of skin. J Invest Dermatol 1977; 69: 303-305.

[15.6] Comaish JS, Harborow PRH, Hofman DA. A hand-held friction meter. Br J Dermatol 1973; 89: 33-35.

[15.7] Prall JK. Instrumental evaluation of the effects of cosmetic products on skin surfaces with particular reference to smoothness. J Soc Cosmet Chem 1973; 24: 693-707.

[15.8] Cua A, Wilheim KP, Maibach HI. Frictional properties of human skin: relation to age, sex and anatomical region, stratum corneum hydration and transepidermal water loss. Br J Dermatol 1990; 123: 473-479.

[15.9] Johnson SA, Gorman DM, Adams MJ, Briscoe BJ. The friction and lubrication of human stratum corneum. In: Dowson D, et al., eds. Thin Films in Tribology. Proceedings of the 19th Leeds-Lyon Symposium on Tribology. Elsevier; 1993. p. 663-672.

[15.10] Asserin J, Zahouani H, Humbert PH, Couturaud V. Mougin D. Measurement of the friction coefficient of the human skin in vivo. Quantification of the cutaneous smoothness. Colloids Surf B: Biointerfaces 2000; 19: 1-12.

[15.11] Elsner P. Wilhelm D, Maibach HI. Frictional properties of human forearm and vulvar skin: influence of age and correlation with transepidermal water loss and capacitance. Dermatologica 1990; 181: 88-91.

[15.12] Loden M, Olsson H, Axéll T, Linde YW. Friction, capacitance and transepidermal water loss (TEWL) in dry atopic and normal skin. Br J Dermatol 1992; 126: 137-141.

[15.13] Sulzberger MB, Cortese TA, Fishman L Jr, Wiley H. Studies on blisters produced by friction. J Invest Dermatol 1966; 47: 456-465.

[15.14] Nacht 5, Close J, Yeung D, Gans EH. Skin friction coefficient: changes induced by skin hydration and emollient application and correlation with perceived skin feel. J Soc Cosmet Chem 1981; 32: 55-65.

[15.15] Hills RJ, Unsworth A, lye FA. A comparative study of the frictional properties of emollient bath additives using porcine skin. Br J Dermatol 1994; 130: 37-41.

[15.16] Dawson, D. In: Wilhelm K-P, Elsner P. Berardesca E, Maibach H, eds. Bioengineering of the skin: skin surface imaging and analysis. Boca Raton: CRC Press; 1997. p. 159-179.

[15.17] Sivamani RK, Goodman J, Gitis NV, Maibach HI. Friction coefficient of skin in real-time. In press.

[15.18] Wolfram U. Friction of skin. J Soc Cosmet Chem 1983; 34: 465-476.

[15.19] Denda M. In: Lodén M, Maibach H, eds. Dry skin and moisturizers: chemistry and function. Boca Raton: CRC Press; 2000. p. 147-153.

[15.20] Wolfram U. In: Leveque J-L, ed. Cutaneous investigation in health and disease: noninvasive methods and instrumentation, Chapter 3, New York, NY: Marcel Dekker 1989.

[15.21] Zhai H, Maibach HI. Effects of skin occlusion on percutaneous absorption: an overview. Skin Pharmacol Appl Skin Physiol 2001; 14: 1-10.

[15.22] Cua AB, Wilhelm K-P, Maibach HI. Skin surface lipid and skin friction: relation to age, sex, and anatomical region. Skin Pharm 1995; 8: 246-251.

[15.23] Manuskiatti W, Schwindt DA, Maibach HI. Influence of age, anatomic site and race on skin roughness and scaliness. Dermatology 1998; 196: 401-407.

[15.24] Sivamani, K. Raja, Godman, Jack, Gitis, V.Norm, Maiback, I.Howard, Review, Cofficient of Friction: Tribological Studies in Man–An Overview, Skin Research and Technology,200;9:227-234

16

WEAR BEHAVIOR AND DEBRIS DISTRIBUTION OF UHMWPE AGAINST Si_3N_4 BALL UNDER BI-DIRECTIONAL SLIDING

16.1 A General Review

The complex wear tracks of UHMWPE against Si_3N_4 ball were formed by UMT tester with the bi-directional sliding motion, which are similar to the wear tracks in hip joints. The wear behavior of UHMWPE was studied under the bi-directional sliding motion, lubricated with plasma solution. The results indicate that, the wear mass loss under unidirectional reciprocating motion is smaller than that under bi-directional sliding motion. The wear rates of UHMWPE under bi-directional sliding motion are in linear proportion to the defined parameter, which co-relates with the cross-shear theory. The main wear mechanism is ploughing under unidirectional reciprocating motion, while plastic deformation, adhesion and fatigue are the wear mechanism under bi-directional sliding motion. Under bi-directional sliding modes, the wear particle distribution range decreases with the rising of complexity of wear tracks, but the proportion of biological active wear particles increases. The particles size distribution follows lognormal distribution. The central size and the peak accumulation of UHMWPE particles linearly decreases and increases against the frequency ratio, respectively, besides the unidirectional reciprocating sliding.

16.2 Introduction

Ultra-high molecular weight polyethylene (UHMWPE) is a well-known biomaterial having low friction[16.1]. Owing to its superior mechanical toughness and wear resistance, UHMWPE has been used as acetabular cup in total artificial hip joints since early 1960s. The wear of UHMWPE components implanted in human body produces wear debris. As the service time of artificial joints prolongs, the aseptic losing and the osteolysis induced by UHMWPE wear becomes the main cause of long-term failure of hip joints replacement[16.2-16.4]. It has been known that the wear particles generated at the prosthetic surface will enter the peri-prosthetic tissue where they are phagocytosed by macrophages. The macrophages release pro-inflammatory cytokines and eventual loosening of the prosthesis, which results in the failure of total hip replacement. From tribological aspect, the goal of biotribological research of total hip replacement is to develop newer hip joint having "low wear and less harm" property, in order to reduce UHMWPE debris generation and control their bioactivity in human body[16.5].

Some researchers investigate the wear mechanism of UHMWPE in total joint replacements. Linear elastic stress analysis using finite element methods[16.6] shows that the maximum principal stress within the UHMWPE during normal walking is usually less than 10 MPa for the total joint replacement. But it has not led to any significant understanding of the wear mechanism of UHMWPE in total joint replacement for that the wear of UHMWPE is not an elastic process in the hip component. When acetabular cup/femoral head are put into contact the microscopic asperities are plastically deformed although the overall or nominal contact is elastic. An incremental residual plastic strain is built for that every contact asperity experiences repeating cyclic deformation during walking. Failure will occur when the ductility of material within each unit contact spot comes to limit value. Based on the critical strain criterion model, wear particle will be produced when accumulated plastic strain reaches the critical strain. Although this theory is validated by experimental results, the wear rate of UHMWPE in this experiment is lower than clinical results[16.2]. The low-wear phenomenon is normally encountered when the artificial joint material is measured with the conventional simple wear tester[16.7-16.8]. However, the joint simulators have received limit success in reproducing clinical wear rates[16.9-16.12].

Some authors have attempted to explain the low-wear phenomenon on simple wear test machines. Mckellop[16.13] proposes that the conventional wear testers is overly simplistic in terms of motion and loading configurations, and low UHMWPE wear may be closely associated with the linear motion of conventional wear tester. Bragdon *et al.*[16.14] further proposes that higher clinical wear rates in UHMWPE acetabular cups may be associated with multi-directional motion of the hip-joint. Wang *et al.*[16.7] suggests that the motion-dependent behavior of UHMWPE wear is attributed to the unique molecular structure of UHMWPE that molecules orient preferentially in the direction of sliding[16.15, 16.16, 16.18]. In linear-tracking motion, molecular orientation leads to strain – hardening of the wear surface, which results in wear resistance enhancement as sliding proceeds. In multi-directional motion, the UHMWPE wear surface experiences both shear and tensile stresses in multiple directions. Strengthening in one particular direction will result in weakening in the perpendicular direction – a phenomenon that is often observed in oriented linear polymers[16.16, 16.19]. Recent computer simulation of the human joint kinematics has indeed indicated that stresses experienced by the surface in both the hip and the knee joint are multi-directional[16.16, 16.17, 16.20].

It has been proven that there are different modes of multi-directional sliding in not only real hip but also various simulators[16.21, 16.22]. Most of the gait slide tracks are oval figures, but there are also tracks with very high aspect ratio and small tracks similar to HUT-3 simulators. The tracks on acetabular cup, produced by BRM simulator, include figures of eight line, straight line, nonsysmmetric oval and elliptic figures. Because the wear of the most common acetabular cup material, UHMWPE, has been found to be highly sensitive to the motion modes[16.10, 16.16, 16.23], it is important to investigate the influence motion patterns on the wear of UHMWPE. Up till now, we know little about the effect of these sliding tracks on the wear of UHMWPE. Turell *et al.*[16.24] studied the effect of elongated and closed rectangular motion patterns on the wear of UHMWPE. Their results obey the orientation softening theory.

In this chapter, four kinds of sliding patterns including straight line, oval shape, eight (butterfly-like) figure and double butterfly figure to conduct wear tests of UHMWPE against Si_3N_4 ceramic ball on a ball-sliding-on-disc machine has been designed, which represents the sliding tracks of a single contact point between the femoral head of an orthopaedic implant and the acetabular cup during testing. The wear behavior of UHMWPE under these sliding modes and their quantitative characterization are studied. Naturally, there is the infinite number of tracks on hip cup, and wear of the hip cup is the wear summation of all tracks. The research of single type of wear track will be helpful for a better understanding of the wear mechanism of acetabular cups.

16.3 Experimental details

16.3.1 *Test materials*

The UHMWPE samples were prepared by using the hot-press molding method. The molecular weight of UHMWPE is 5,000,000 g/mol. UHMWPE powder was pre-pressed in designed molder under 5 MPa at room temperature for 15 minutes. Then the molder was heated to 190-200° C without applied pressure for two hours. Afterwars, UHMWPE sample was pressed under 15 MPa load until it cooled to 50° C atmosphere. The UHMWPE sample was prepared in disc shape with diameter of 30 mm and thickness of 10 mm. The friction surface of UHMWPE sample was polished to having roughness of R_a= 0.2-0.4 μm. Before testing, UHMWPE samples were cleaned in acetone in ultrasonic bath. Si_3N_4 ball was selected as the counterpart of UHMWPE to simulate the wear of artificial joints consisting of UHMWPE cup and ceramic head. The diameter of Si_3N_4 ball is 4 mm.

16.3.2 Wear tests

The ball-on-disc wear tests were performed on an UMT tester, developed at CETR (Campbell, California). This tester is capable to perform multiple friction and wear tests with in-situ monitoring of many tribological parameters including friction force, in-situ wear depth, contact acoustic emission, *etc.* The schematic of contact between Si_3N_4 ball and UHMWPE is shown in Fig. 16.1. The UHMWPE disc reciprocates in Y direction, which is driven by an eccentric rotator. The Si_3N_4 ball is mounted on a holder connecting to a three-dimension force sensor. The carriage reciprocates in X direction. The sliding distance of disc and ball reciprocating are 12 mm. During the wear tests, the UHMWPE disc and Si_3N_4 ball slide at different frequencies, which result in different wear track patterns. Table 16.1 shows four track patterns plotted by X and Y position data recording during wear tests. The sliding distance, *l*, in one cycle is approximately calculated as

$$l = \sum_{i=1}^{n} \sqrt{(X_i - X_{i+1})^2 + (Y_i - Y_{i+1})^2} \qquad (16.1)$$

where X_i and Y_i is the position data of X and Y coordinates, n represents the number of data in one cycle.

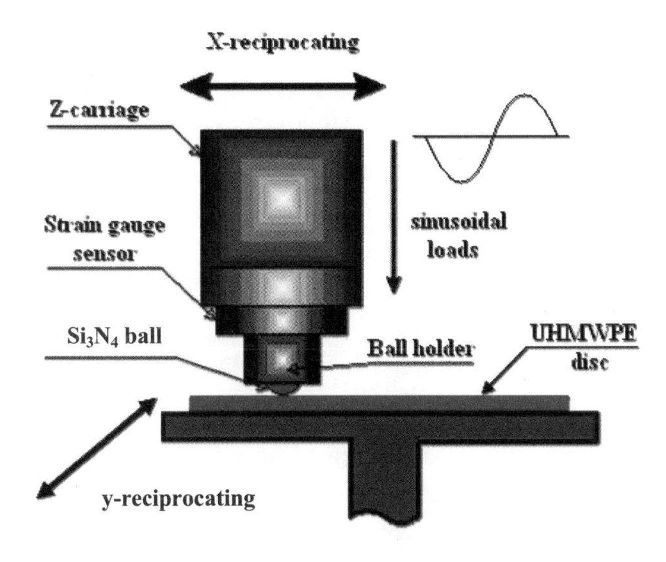

Fig. 16.1 Schematic diagram of ball-on-disc wear test rig with cross-sliding function to form bi-directional sliding modes, simulating complex wear tracks of UHMWPE against Si_3N_4 ball.

Table 16.1 Reciprocating frequency and sliding tracks with different frequency ratio F_r

Sliding modes	**Reciprocating**	**Oval sliding**	**Double-elliptical sliding**	**Triple-elliptical sliding**
Sliding frequency	$F_x = 0, F_y = 0.5$	$F_x = 0.5, F_y = 0.5$	$F_x = 1, F_y = 0.5$	$F_x = 1.5, F_y = 0.5$
$F_r = F_x / F_y$	0	1	2	3
Track pattern		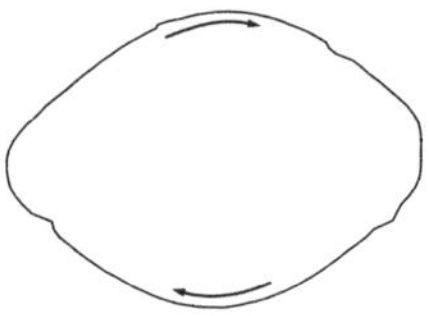	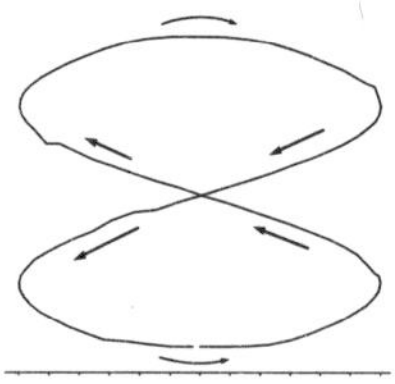	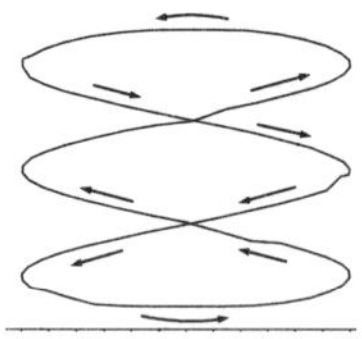
Sliding distance per cycle/mm	24	36.39	56.08	78.81

Wear tests were run in 25% plasma solution for 10,000 reciprocating cycles. A sinusoidal dynamic load changed from 20 N to 25 N was applied normally on the Si_3N_4 ball, the oscillating frequency is 0.5 Hz. After wear tests terminated, the tested UHMWPE sample was cleaned in an acetone-filled ultrasonic bath and dried at 80°C in an attemperator. Wear mass loss of UHMWPE disc was measured by using of an electronic scale with 0.01 mg accuracy. The specific wear rate of UHMWPE, M_s is determined as

$$M_s = \frac{M_w}{nl} \tag{16.2}$$

where M_w represents the wear mass loss of UHMWPE in wear tests, and n is testing cycles.

The wear particles of UHMWPE in plasma lubricant were collected after wear tests. They were analyzed by using of laser particle size analyzer. The particles size distribution and average diameter of particles in different sliding patterns were compared. In order to investigate the wear mechanism of UHMWPE in bi-directional sliding, the worn surface of UHMWPE disc was observed by using of scanning electric microcopy.

16.4 Results and discussion

16.4.1 *Wear of UHMWPE*

The wear mass loss of UNMWPE disc under four sliding patterns with various bi-directional frequency ratios is shown in Fig. 16.2. It is suggested that the changes in sliding patterns influence the wear amounts of UHMWPE. The wear mass loss of UHMWPE in bi-directional sliding patterns is higher than that under uni-directional reciprocating sliding. Among three bi-directional sliding cases, the wear mass loss of UHMWPE decreases with increasing frequency ratio. Comparing the wear mass loss for four sliding modes, uni-directional reciprocating sliding ($F_r = 0$) results in the lowest wear mass loss of 0.18 mg. The highest wear mass loss of 1.98 mg occurs in oval sliding case ($F_r = 1$), similar to rotating sliding of ball-on-disc, which is more than ten times of the wear mass loss in uni-directional reciprocating. The wear mass loss under double-elliptical motion ($F_r = 2$) or triple-elliptical motion ($F_r = 3$) has close values of 0.93 mg and 0.88 mg, respectively.

A frequency parameter has been defined to identify the sliding patterns for different bi-directional motions, the parameter is formulated as;

$$\delta = \frac{F_x}{\sqrt{F_x^2 + F_y^2}} = \frac{F_r}{\sqrt{1 + F_r^2}} \tag{16.3}$$

where $F_r = F_x / F_y$, F_x and F_y represents the reciprocating frequency in X and Y direction, respectively.

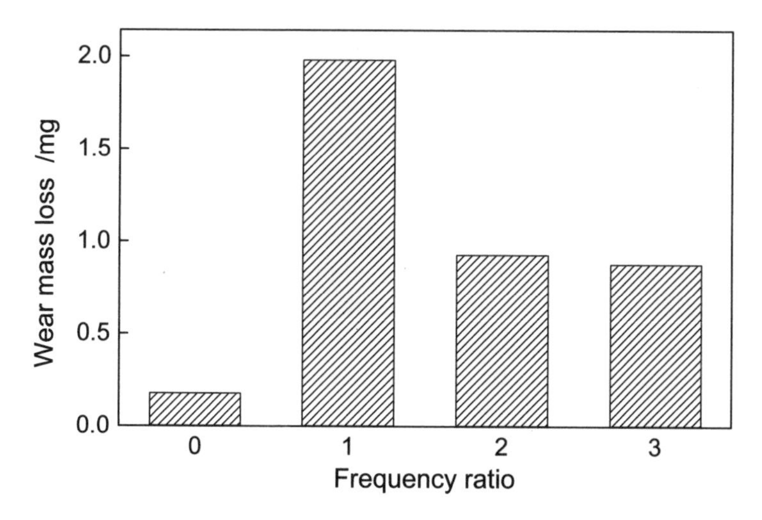

Fig. 16.2 The wear mass loss of UHMWPE disc under various frequency ratio, applied with sinusoidal dynamic load from 20 N to 25 N, 10,000 test cycles in plasma lubrication.

The relation of the specific wear rate of UHMWPE to frequency parameter is shown in Fig. 16.3. It is shown that the specific wear rates of UHMWPE disc in bi-directional sliding conditions decrease linearly to the frequency parameter. However, the wear rate in uni-directional sliding has large diversity to linear region. This indicates the difference in wear mechanism between the bi-directional sliding wear and uni-directional sliding wear of UHMWPE.

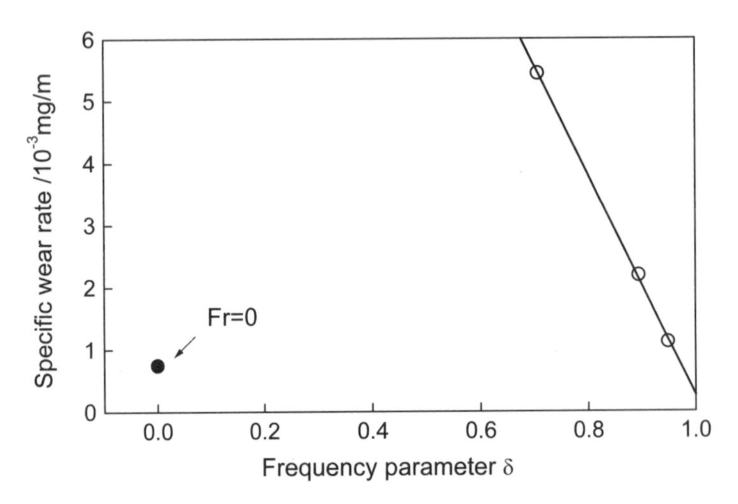

Fig. 16.3 UHMWPE wear rates against Si_3N_4 ball in relation to defined parameter $F_r/\sqrt{1+F_r^2}$, applied with sinusoidal dynamic load from 20 N to 25 N, 10,000 test cycles in plasma lubrication.

The specific wear rates of UHMWPE in bi-directional sliding modes proportional to frequency ratio are resulted from the increasing of friction energy. When Si_3N_4 ball slides against UHMWPE along the wear path defined by these four kinds of sliding modes, friction energy, which is proportional to sliding velocity, is dissipated in both X and Y directions. According the cross-shear theory, the sliding in principal direction, *Y*, leads to plastic deformation or macromolecular orientation, whereas the sliding in the secondary direction, X, leads to material removal by intermolecular splitting[16.13-16.16]. Therefore, the dissipative energy in *X* direction will control the wear of UHMWPE under bi-directional sliding, which can be described as

$$\frac{\Delta W_x}{\Delta W} \propto \frac{V_x}{\sqrt{V_x^2 + V_y^2}} \propto \frac{F_x}{\sqrt{F_x^2 + F_y^2}} \tag{16.4}$$

where ΔW_x represents the friction energy in *X* direction, ΔW is the total friction energy. Because uni-directional reciprocating has F_x=0, so it is evident that ΔW_x=0. Therefore, the wear rate of UHMWPE disc under reciprocating sliding results in the lowest values, which has been validated in literatures[16.21, 16.24].

16.4.2 *Wear mechanism*

In order to investigate the effect of sliding patterns on wear mechanism of UHMWPE, the worn surfaces of UHMWPE discs were observed with a scanning electric microcopy. Fig. 16.4 shows the SEM pictures of worn surface on UHMWPE in different sliding patterns. The topography in Fig. 16.4a represents the worn surface of UHMWPE in uni-directional reciprocating. The light ploughs indicate that abrasion is main wear mechanism. The surface in Fig. 16.4a looks much smoother than that under bi-directional sliding motion, so the lowest wear mass loss is resulted in. The worn surface under circumferential sliding ($F_r = 1$) is shown in Fig. 16.4b. The rougher topography of worn surface includes fibrous piling, shredding and cavity formation, all these features suggest the adhesion wear mechanism. In addition, fatigue cracks also appear on the worn surface of UHMWPE under circumferential sliding. Such complex wear mechanism is a possible reason of high wear mass loss in this sliding mode.

There are ripples perpendicular to the sliding direction formed on the worn tracks of UHMWPE under double-elliptical and triple-elliptical sliding. These surface topographies are formed due to plastic deformation in sliding contact area, which is unique feature on the intersection points of worn surfaces in $F_r = 2$ and $F_r = 3$ sliding modes, as shown in Fig. 16.4c and Fig. 16.4d. Because the ripples on UHMWPE are more easily broke off when they are subjected to cross directional shearing, more wear particles will be removed and higher wear mass loss of UHMWPE will be induced. The observations on the worn surfaces outside cross position of double-elliptical sliding ($F_r = 2$) and triple-elliptical sliding ($F_r = 3$) are shown in Fig.16.4e and Fig. 16.4f, where appears the feature of slight ploughing similar to that in uni-directional sliding. Some pits can be found on the worn surface at both cross position and sliding track under triple-elliptical sliding, which suggests that fatigue wear mechanism plays a certain part in the wear of UHMWPE in bi-directional sliding modes. Such phenomena are not found under uni-directional reciprocating sliding.

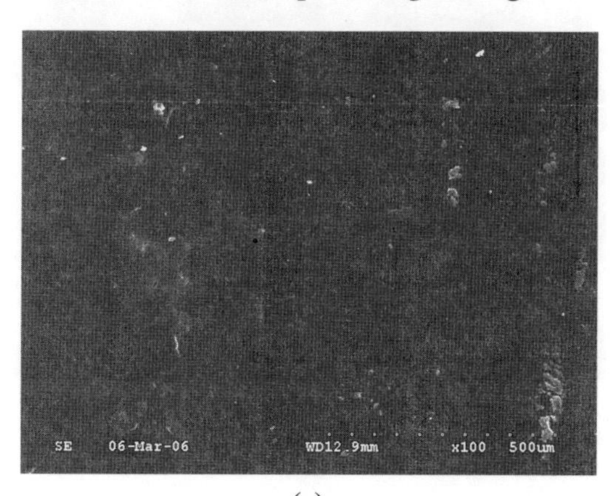

(a)

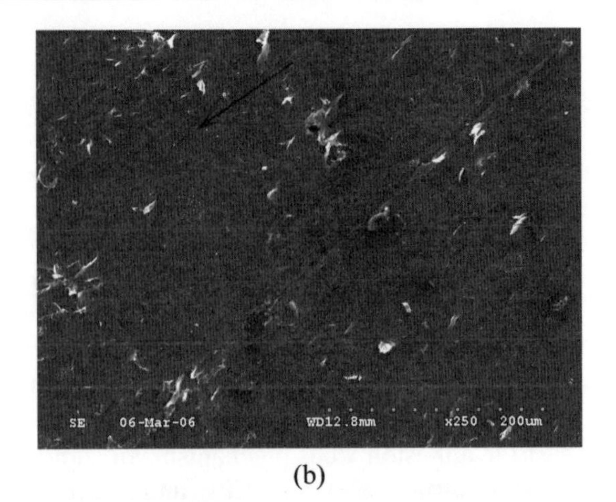
SE 06-Mar-06 WD12.8mm x250 200um

(b)

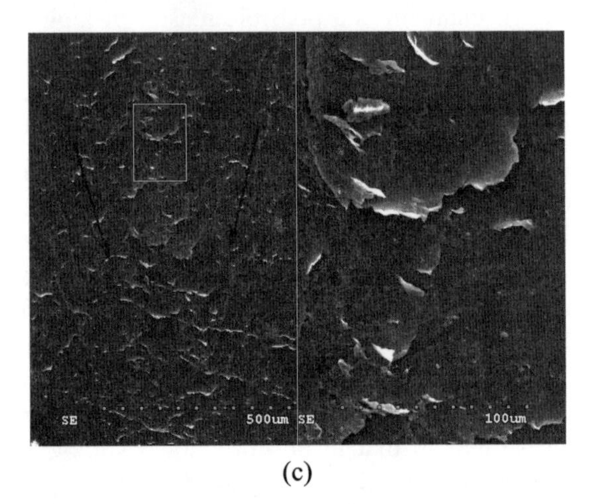
SE 500um
SE 100um

(c)

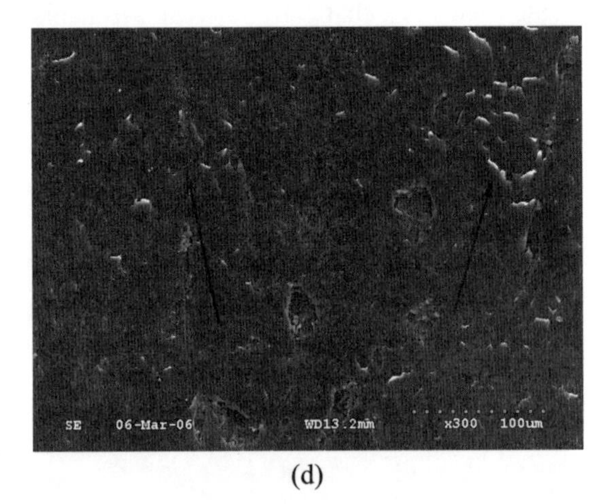
SE 06-Mar-06 WD13.2mm x300 100um

(d)

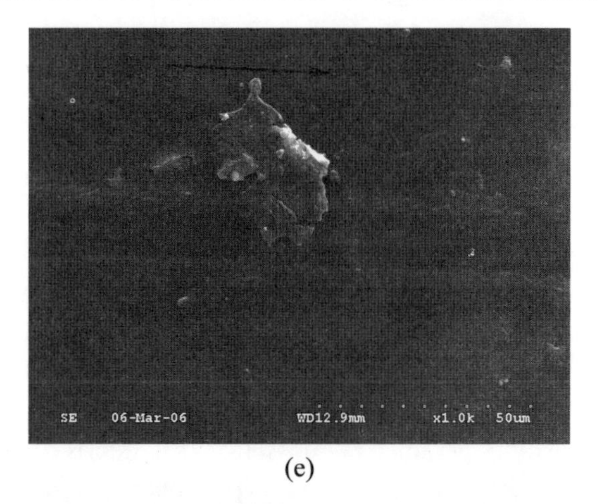

(e)

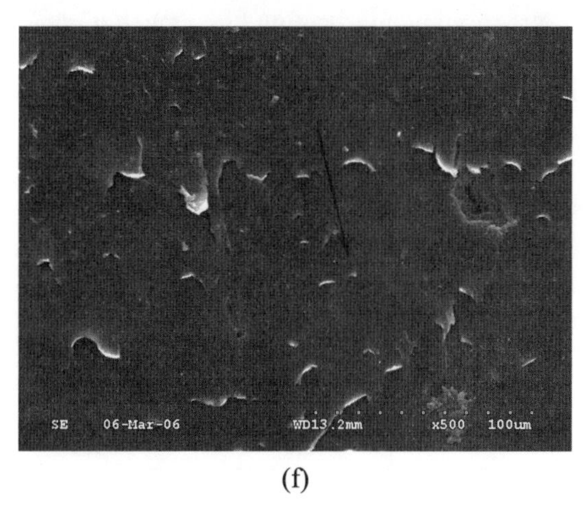

(f)

Fig. 16.4 SEM observation of UHMWPE worn surfaces on (a) reciprocating track (F_r=0), (b) oval sliding track (F_r=1), (c) cross position on double-elliptical sliding track (F_r=2), (d) cross position on triple-elliptical sliding (F_r=3), (e) double-elliptical sliding track and (f) triple-elliptical sliding track.

16.4.3 Size distribution of wear particles

It is known that the biological activity of UHMWPE wear debris is a function of particles size. The most biological active UHMWPE particle size ranges 0.1-1.0 μm. The particles in the 1.0-10 μm size range are five times less active than the particles in the 0.1-1.0 μm size range, and the particles in the 10-100 μm size range are 25-fold less active than the smallest particles[16.25]. In order to investigate the size distribution of UHMWPE wear particles, the UHMWPE particles generated in wear process were collected, they were then quantitatively analyzed by using of a laser size counter. The size distribution of UHMWPE debris generated in different sliding modes is shown in Fig. 16.5. By statistical analysis, it is obtained that the distribution of UHMWPE wear particle follows lognormal probability density function, that is

$$f = \frac{1}{\sqrt{2\pi}\sigma d}\exp\left[-\frac{(\ln d - \mu)^2}{2\sigma^2}\right] \tag{16.5}$$

Where f represents the accumulation distribution in percentage, d represents wear particle diameter, and σ is standard deviation of particle size, μ is logarithm of mean of particle size, they determine the shape and position of distribution curves. Table 16.2 gives the distribution parameter of σ and μ for UHMWPE wear particles generated in four sliding modes. It is described that the shape and position parameters of UHMWPE debris from bi-directional sliding have little difference, however, parameters describing the debris distribution of uni-directional reciprocating have much higher values.

Table 16.2 Distribution parameter of UHMWPE wear debris for four sliding modes

Frequency ratio	μ	σ
0	4.08	0.707
1	2.87	0.451
2	2.63	0.397
3	2.51	0.361

Curves in Fig. 16.5 indicate that the distribution shape and peak position of UHMWPE particles change due to bi-directional sliding modes. The uni-directional sliding mode results in a wide-range and low-peak distribution shape of the UHMWPE wear particles. However, bi-directional sliding modes result in a narrow-range and high-peak distribution shape. Among three bi-directional sliding cases, the size range of UHMWPE wear particles decrease against frequency ratio increasing. For example, the size range of UHMWPE wear particle generated under triple elliptical sliding ($F_r = 3$) double elliptical sliding ($F_r = 2$) and locates 4-35 μm and 4-55 μm, respectively. But the size range of UHMWPE wear particles produced in oval sliding mode ($F_r = 1$) is 5-110 μm. These results suggest that the size ranges of UHMWPE wear debris generated from bi-directional cross sliding (*i.e.* in $F_r = 2$ and $F_r = 3$ modes) are two times less than the size ranges of UHMWPE debris from oval sliding mode ($F_r = 1$), and six times less than the size ranges of those from reciprocating sliding ($F_r = 0$).

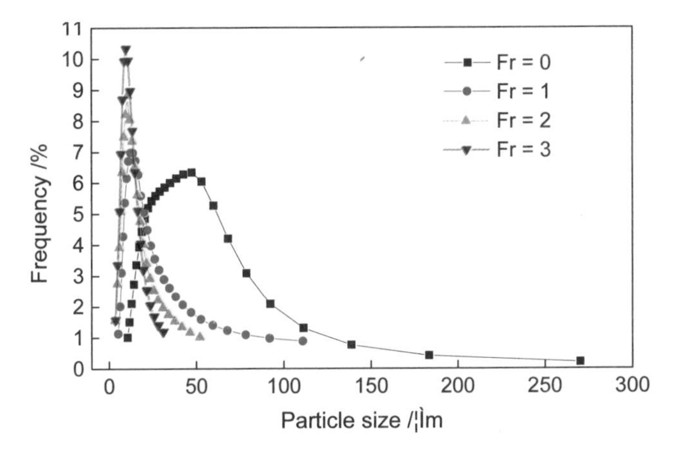

Fig. 16.5 The size distribution of UHMWPE wear particles generated in wear process with four sliding track patterns, applied with sinusoidal dynamic load from 20 N to 25 N, 10,000 test cycles in plasma lubrication.

The bar charts of the proportion of UHMWPE wear particles in two size ranges are drawn in Fig. 16.6. The particles in size range of 0.1-1 μm were not detected. It is seen that the UHMWPE particles generated in reciprocating sliding mode (F_r = 0) mainly locate at 10-100 μm, they occupies 97% of the total particles, but wear particles in 1-10 μm only have very little portion. With increasing frequency ratio, the proportion of UHMWPE wear particles in 1-10 μm size range increases, *i.e.* 15.8%, 27.2% and 35.6% respectively for F_r=1, 2 3. However, proportion of UHMWPE wear particle in 10-100 μm decreases, *i.e.* 84.2%, 72.8% and 64.4% respectively. This suggests that the percentage of biological active UHMWPE wear particles generated in bi-directional sliding decreases with increasing frequency ratio.

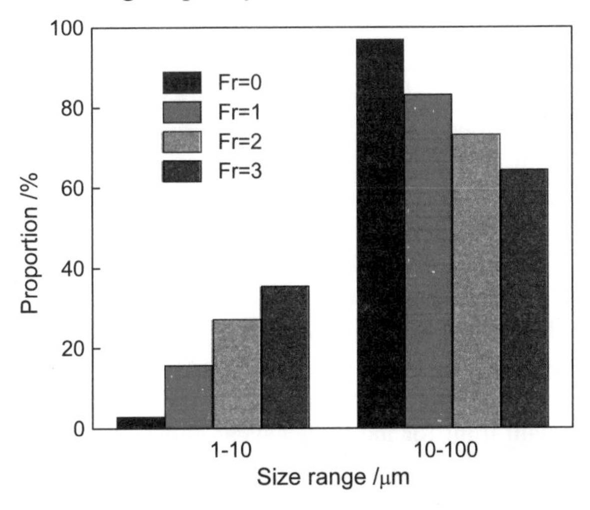

Fig. 16.6 The group proportion of UHMWPE wear particles generated in wear process with four sliding track patterns.

The central size corresponding to the peak accumulation of UHMWPE wear particles decreases to frequency ratios, for three bi-directional sliding modes, but the peak accumulation percentage of UHMWPE wear particles increases, as shown in Fig. 16.7. However, central size and peak accumulation of UHMWPE wear particles generated in uni-directional reciprocating have large deviation in comparison with the linear region of bi-directional sliding modes.

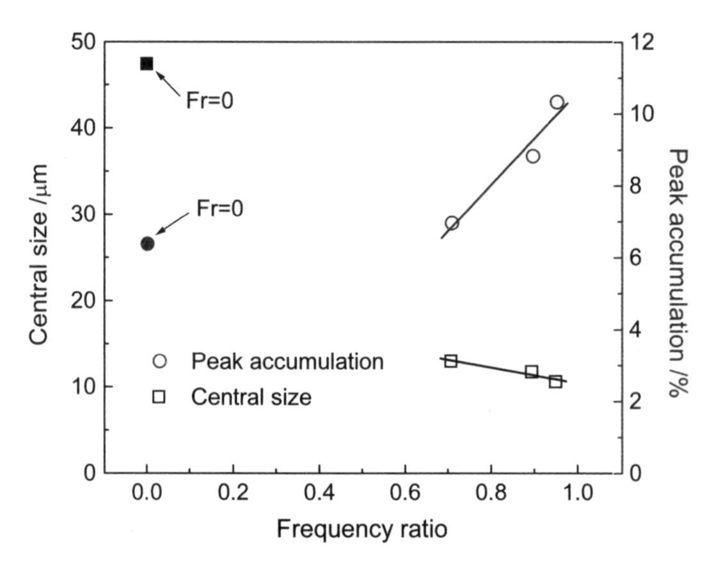

Fig. 16.7 Relation of central size and peak accumulation of UHMWPE wear particles to frequency ratios, representing four sliding track patterns.

16.5 Conclusions

From the above analysis and investigations, the following conclusions are drawn[15.26]:

The complex sliding tracks formed in hip joints were simulated by the UMT tester with well-defined synchronized programmable simple motion components. The aim is to investigate the effect of wear tracks on the wear of UHMWPE. The results indicate that the wear mass loss under uni-directional reciprocating sliding is smaller than that under bi-directional sliding modes. The wear rates of UHMWPE under bi-directional sliding modes are in linear proportion to the defined parameter, which co-relates with the cross-shear theory. Under bi-directional sliding modes, the wear particle distribution range decreases with the rising of complexity of wear tracks, but the proportion of biological active wear particles increases. So the complex wear tracks are harmful to the implant joint for the higher wear and more active wear particles. The particles size distribution follows lognormal distribution. The central size and the peak accumulation of UHMWPE particles linearly decreases and increase against the frequency ratio, respectively, besides the unidirectional reciprocating sliding. The main wear mechanism is ploughing under uni-directional reciprocating sliding, while plastic deformation, adhesion and fatigue are the wear mechanism under bi-directional sliding modes.

References

[16.1] N. Gitis, Plastic and Composite Materials for Sliding Guideways, (1987), Moscow, Engineering Press, (in Russian)

[16.2] A. Wang, D.C. Sun, C. Stark, J.H. Dumbleton, Wear mechanisms of UHMWPE in total joint replacements, Wear, 181-183 (1995) 241-249

[16.3] J. Fisher, D. Downson, H. Hamdzah, H.L. Lee, Effect of sliding velocity on the friction and wear of UHMWPE for use in total artificial joints, Wear, 175 (1994) 219-255

[16.4] D. Xiong, S. Ge, Friction and wear properties of UHMWPE/Al_2O_3 ceramic under different lubricating conditions, Wear, 250 (2001) 242-245

[16.5] S. Ge, C. Huang, Biotribological behavior of UHMWPE hip joints, Proceeding of the 4th China International Symposium on Tribology, Xi'an, China (2004): 68-74

[16.6] D. L. Bartel, V.L. Bicknell, T.W. Wrighte, J. biomed. Mater (Am), 68 (7) (1986) 1041-51

[16.7] A. Wang, V.K. Polineni, A. Essner et al. The significance of nonlinear motion in the wear screening of orthopaedic implant materials, J. Test. Evaluation (1975)

[16.8] M.J. Griffith, M.K. Seidenstein, D. Williams, et al. Socket wear in Chamley low friction arthroplasy of the hip, Clin. Orthopaed. Related Res., 137 (1978) 37-47

[16.9] I.C. Clarke, Wear of artificial joint materials, hip joint simulator studies, Engna Med., 10 (1981) 189-198

[16.10] H.A. Mckellop,vP. Campbell, S.-H. Park, et al. The origin of submicron polyethylene wear debris in total hip arthroplasty, Clin. Orthopaed. Related Res., 311 (1995) 3-20

[16.11] A. Essner, A. Wang, C. Stark, et al. A simulator for the evalutation of total knee replacement component wear, 5th world biomaterials Congress, May 28- June 2, 1996, Toronio, Canada, PP.I-580

[16.12] D.O. O'connor, C.R. Bragdon, D.W. Burke, et al. A 12 station upright hip simulator wear testing machine employing oscillating motion replicating the human gait cycle. Trans. 21st Annual Meeting of The society for biomaterials, The society for biomaterials, Minneapolis, MN, 1995, p.359

[16.13] H.A. Mckellop, Comparison between laboratory wear tests and clinical performance of past bearing materials, ASTM Workshop on Characterization and Performance of Articular Surface, ASTM, Denver, CO, 17 May 1995

[16.14] C.R. Bragdon, M.Jasty, J.D. Lowenstein, et al. Mechanism of wear of retrieved polyethylene acetabular components, 63rd Annual Meeting, American Academy of Orthopaedic surgeons, 22-26 February, 1996, Atlanta, GA, Paper no.376

[16.15] A. Wang, C. Stark, J.H. Dumbleton, Mechanistic and morphological origins of ultra-high molecular weight polyethylene wear debris in total joint replacement prostheses, Proc. Instn. Mech. Engrs. Part H: J. Eng. Med. 210 (1996) 141–155.

[16.16] A. Wang, D.C. Sun, B. Edwards, M. Sokol, A. Essner, V.K. Polineni, C. Stark, J.H. Dumbleton, Orientation softening in the deformation and wear of ultra-high polyethylene, Wear 203/204 (1997) 230–241.

[16.17] C.R. Bragdon, D.O. O'Connor, J.D. Lowenstein, M. Jasty, W.D. Syniuta, The importance of multidirectional motion on the wear of polyethylene, Proc. Instn. Mech. Engrs. Part H: J. Eng. Med. 210 (1996) 157–166.

[16.18] A. Wang, A. Essner, V.K. Polineni, C. Stark, J.H. Dumbleton, Lubrication and wear of ultra-high molecular weight polyethylene in total joint replacements, Tribol. Int. 31 (1-3) (1998) 17–33.

[16.19] C.C. Hsio, Flow orientation and fracture strength of a model linear hard polymer solid, J. Polym. Sci. 44 (1960) 71–79

[16.20] B.S. Ramamurti, C.R. Bragdon, D.O. O'Connor, J.D. Lowenstein, M. Jasty, D.M. Estok, W.H. Harris, Loci of movement of selected points on the femoral head during normal gait, J. Arthroplasty 11 (7) (1996) 852–855

[16.21] V. Saikko, O. Calonius. Slide track analysis of the relative motion between femoral head and acetabular cup in walking and in hip simulators. Journal of Biomechanics,2002,35(4):455-464.

[16.22] D. Bennett, J. Orr, R. Baker, Movement loci of selected points on the femoral head for individual total hip arthroplasty patients using three-dimensional computer simulation, J. arthroplasty 15 (7) (2000) 909-915

[16.23] Saikko, V., A multidirectional motion pin-on-disk wear test method for prosthetic joint materials. Journal of Biomedical Materials Research 41(1998), 58–64

[16.24] M. Turell, A. Wang, A. Bellare, Quantification of the effect of cross-path motion on the wear rate of ultra-high molecular weight polyethylene, Wear 255 (2003) 1034–1039

[16.25] J.L. Tipper, J.B. Matthews, E. Ingham, T.D. Stewart, J. Fisher, M.H. Stone, Wear and functional biological activity of wear debris generated from UHMWPE-on-zirconia ceramic, metal-on-metal, and alumina ceramic-on-ceramic hip prostheses during hip simulator testing. Friction, lubrication, and wear of artificial joints (edited by I.M. Hutchings), Professional Engineering Publishing, 2003.

[16.26] Gitis, N., Vinogradov, M., Prashad, H., Shirong Ge, Shibo, "Wear behaviour and debris distribution of UHMWPE against Si_3N_4 ball under bi-directional sliding" PFAM, 2008, New Delhi

17

MICRO/NANO SCALE MECHANICAL AND TRIBOLOGICAL CHARACTERIZATION OF SILICON CARBIDE FOR ORTHOPEDIC APPLICATIONS

17.1 A General Review

Micro/nano mechanical and tribological characterization of SiC has been carried out. For comparison, measurements on SiC, CoCrMo, Ti-6Al-4V, and stainless steel have also been made. Hardness and elastic modulus of these materials were measured by nano indentation using a nano indenter. The nano indentation impressions were imaged using an Atomic Force Microscope (AFM). Scratch, friction, and wear properties were measured using an accelerated micro-tribometer. Scratch and wear damages were studied using a Scanning Electron Microscope (SEM). It is found that SiC exhibits higher hardness, elastic modulus, scratch resistance as well as lower friction with fewer and smaller debris particles compared to other materials. These results show that SiC possesses superior mechanical and tribological properties that make it an ideal material for use in orthopedic and other biomedical applications.

17.2 Introduction

Silicon carbide (SiC) has great potential for overcoming the current inadequacies of orthopedic materials due to its inherent characteristic of being chemically inert and consequently resistant to harsh mechanical and chemical environments. While initial research has been performed on SiC for implantation and biocompatibility, including as a coating in orthopedics for the prosthetic-bearing surface and the un-cemented joint prosthetic, and SiC is currently in clinical use as a coating for stents to enhance hemocompatibility[17.1-17.8], SiC has not undergone the focused and comprehensive studies necessary to make it an accepted alternative to current orthopedic materials. Due to the inertness of SiC, this material would become permanently integrated into the new bone growth without the disadvantages of wear debris generation, oxidation, and metallosis that generally occurs with implanted materials. Prior to its development as an implant, SiC must be thoroughly evaluated to establish its safety for implantation and its material characteristics as compared to current materials. Once its advantages in relation to current materials are demonstrated, further studies can be undertaken to proceed with the development of the scaffold structures ultimately leading to clinical trials with the new material.

The major disadvantage with traditional orthopedic materials used for joint replacement, such as titanium, cobalt chrome, and stainless steel, is their propensity to wear. The particles produced as wear debris lead to bone loss and implant loosening[17.9]. Due to this, prostheses may need to be replaced every 10 to 15 years. Additionally, these materials oxidize over time creating a reactive implant surface, as well as local and systemic metallosis. Solid surfaces, such as metal implants, allow different levels of natural integration with bone or other tissues in the body, based on the degree of biocompatibility of the implants[17.10]. Thus, a new mechanically stable material that lasts longer, generates less wear, invites better tissue growth, and is stable and biocompatible would be a great advancement to this field. In addition, joint replacement surgery in children and young adults less than 30 is generally only reserved for patients with significant debilitation due to a greater risk of complications, such as increased implant wear and shorter implant life spans, even though they may experience substantial pain or activity loss[17.9, 17.11]. Ideally, orthopedic devices should meet the load bearing and movement requirements of an active population with an extended material lifetime devoid of degradation[17.12].

The continuous research on implantable biomaterials for use in orthopedic surgery is due to the limitations of the currently available biomaterials. As described above, periprosthetic osteolysis are caused by the generation of wear debris by current orthopedic biomaterials[17.13, 17.14]. In contrast, SiC possesses all of the properties lacking in the above situations: mechanical strength, superior biocompatibility, and an anti-abrasive nature.

SiC is expected to possess superior biocompatibility properties because it is chemically inert and exhibits properties that may indicate improved mechanical characteristics such as wear and hardness compared with standard orthopedic materials[17.5, 17.15-17.17]. There has been a variety of research on SiC for implantation and biocompatibility, including as a coating in orthopedics for the prosthetic-bearing surface and the un-cemented joint prosthetic, and SiC is currently in clinical use as a coating for stents to enhance hemocompatibility[17.1-17.8]. One study shows that SiC particles were well tolerated by rabbit bone, did not cause aggregation of macrophages, and were apparently harmless. This study found SiC particulate matter to be safe, and a possible stimulant of bone in-growth[17.1]. Despite the promising characteristics of this material as demonstrated from this research, few facilities have the capabilities to fabricate and further develop the material for orthopedic applications.

Reliable studies are the key for practical application and commercialization of today's advanced biomaterials. Mechanical and tribological aspects are of critical importance in determining long-term stability of medical implants and devices. Aseptic loosening of total joint replacements begins as a mechanical problem, which is initiated by joint articulation and other relative motion between surfaces. Wear debris can cause calcar resorption and implant loosening. Debris generated as a result invokes biological responses that accelerate loosening by propelling changes in the bone. One of the crucial factors is the release of wear debris particles that initiate the inflammatory cascade. An in-depth understanding of the mechanical and tribological properties of biomaterials is therefore fundamental to the control of fracture failure and wear in new implants.

The objectives of this study were to identify a new orthorpedic material, SiC, that exhibits superior mechanical and tribological properties, and to understand failure mechanisms of current orthorpedic materials. Nano indentation, micro-scratch, and wear experiments were conducted using a nano-indenter and a micro-tribometer. Indentation, scratch, and wear damage mechanisms were analyzed by atomic force microscopy (AFM) and scanning electron microscopy (SEM).

17.3 Experimental

17.3.1 *Test Samples*

Silicon carbide bulk crystals were grown using the seeded physical vapor transport (PVT) technique[17.18]. In this method, SiC powder is used as a source and heated to 2500 °C, resulting in extensive sublimation. The SiC species are then transported by diffusion to the seed on which SiC material crystallizes into bulk. Conventional PVT technique is the only method used to grow large diameter SiC single crystals suitable for electronic devices. This method allows control of the growth parameters, such as growth rate, thermal gradient, seeding, etc., that is used to produce SiC material with predictable crystallinity, polytype, doping concentration, etc.

6H-SiC bulk crystals were grown under the following conditions: The temperature of the seed was varied between 1900 °C and 2300 °C, controlled by a pyrometer. Resistive heating, as well as radiofrequency-heating furnaces, were used with the temperature gradient $\leq$ 30 °C/cm between the source and seed. The argon background pressure was maintained between 1 and 50 Torr. The grown SiC single crystals were oriented using X-ray diffractometer and sliced into wafers using a wire saw. The wafers were then lapped and polished to obtain a smooth surface (RMS $\leq$ 5 Å) and to perform the necessary studies. Finally, the SiC samples were cleaned using standard procedures including an ultrasonic bath.

Three commonly used orthopedic biomaterials were tested along with SiC, including cobalt-chromium-molybdenum (CoCrMo), titanium-6 aluminum-4 vanadium alloy, (Ti-6Al-4V), and stainless steel 316L (SS 316L). (ASTM F138) were kindly provided by Oncore Orthopaedics (Austin, TX). They were machined from metal bars into disks measuring 12.0 mm in diameter and 2.5 mm in thickness. Before testing, the surfaces of the disks were hand polished using a series of silicon carbide metallographic papers down to a grit size of 1200. Properties of the materials tested are listed in Table 17.1.

Table 17.1 Mechanical and Tribological Properties of SiC, CoCrMo, Ti-6Al-4V, and Stainless Steel

Material	Hardness (GPa)	Elastic modulus (GPa)	Critical load (m/V)	Wear Coefficient of Friction		Particle Size (μm)	Damage Index
				Beginning	End		
SiC	45.8	370	27	0.17	0.2	0.4	1
CoCrMo	10.1	300	13	0.17	0.23	1.3	2
Ti-6A1-4V	5.6	127	12	0.23	0.26	7.9	4
Stainless steel	6.2	223	9	0.45	0.40	8.1	3

17.3.2 *Mechanical and Tribological Characterization*

Nano-indentation tests were performed using a Troboscope nano-mechanical testing system (Hysitron Inc.) in conjunction with a Veeco Dimension 3100 AFM system (Veeco Metrology Group). The Hysitron nanoindenter monitors and records the load and displacement of the indenter, a diamond Berkovich three-sided pyramid, with a force resolution of about 1 nN and displacement resolution of about 0.1 nm. Hardness and elastic modulus were calculated from the recorded load–displacement curves. The indentation impressions were then imaged *in situ* using the indenter tip. A series of 10 indentations were performed for each sample.

Nano-indentation hardness is defined as the indentation load divided by the projected contact area of the indentation. It is the mean pressure that a material will support under load. From the load–displacement curve, hardness can be obtained at the peak load as

$$H = \frac{P_{\max}}{A} \tag{17.1}$$

where $P_{\max}$ is the peak load and A is the projected contact area. For an indenter with a known geometry such as the Berkovich tip used in this study, the projected contact area is a function of contact depth, which is measured by the nano-indenter *in situ* during indentation[17.19-17.21]. Therefore, the projected area A can be measured and calculated directly from the indentation displacement.

The elastic modulus was calculated using the Oliver–Pharr data analysis procedure, Oliver WC et al.[17.22] beginning by fitting the unloading curve to a power–law relation. The unloading stiffness can be obtained from the slope of the initial portion of the unloading curve, $S = dP/dh$. Based on relationships developed by Sneddon[17.23] for the indentation of an elastic half space by any punch that can be described as a solid of revolution of a smooth function, a geometry-independent relation involving contact stiffness, contact area, and elastic modulus can be derived as follows:

$$S = 2\beta\sqrt{\frac{A}{\pi}}\,E_r \tag{17.2}$$

where β is a constant which depends on the geometry of the indenter (β = 1.034 for a Berkovich indenter)[17.19-17.21], and E_r is the reduced elastic modulus which accounts for the fact that elastic deformation occurs in both the sample and the indenter. E_r is given by

$$\frac{1}{E_r} = \frac{1-\nu^2}{E} + \frac{1-\nu_i^2}{E_i} \tag{17.3}$$

where E and ν are the elastic modulus and Poisson's ratio for the sample, and E_i and ν_i are the same quantities for the indenter. For diamond, E_i = 1141 GPa and $\nu_i = 0.07$[17.19-17.21].

Micro-scratch as well as friction and wear tests were carried out using a CETR micro-tribometer (CETR Inc.). In the micro-scratch tests, a conical diamond indenter having a tip radius of 1.5 μm and an included angle of 60° was drawn over the sample surface, and the load was ramped up until substantial damage occurred. An acoustic emission sensor was used to detect crack formation during scratching. The 1 mm long scratches were made by translating the sample while ramping the loads on the conical tip over different loads ranging from 6 m*N* to 50 m*N*. The coefficient of friction, friction force, normal load, and acoustic emission signal were detected *in situ* during scratching. The friction and wear tests were conducted against a WC ball with a 4 mm diameter and surface finish of about 2 nm RMS in reciprocating mode. Typical test conditions were as follows: stroke length 7.0 mm, frequency 0.1 Hz, average linear speed 1.0 mm s^{-1}, normal load 20 m*N*, temperature 22 ± 1 °C, and relative humidity 45 ± 5%. The friction force was measured for a total sliding distance of 5 m.

Scratch and wear tracks were observed using a Philips XL30 field-emission environmental SEM.

17.4 Results and Discussion

17.4.1 *Hardness and Elastic Modulus*

Representative load-displacement curves and AFM images of indentations made at 5 m*N* peak indentation load on SiC, CoCrMo, Ti-6Al-4V, and stainless steel are compared in Figure 17.1. SiC exhibits the lowest indentation depth and highest slope of unloading curve, followed by CoCrMo, stainless steel and Ti-6Al-4V. The AFM images show the same trend as the load-displacement curves; SiC exhibits the smallest indentation mark, followed by CoCrMo, stainless steel, and Ti-6Al-4V. Pile-up (the indented material around the indenter above its original surface[17.20]) was found around the indentation impressions for all samples. A smaller pile-up was observed around the indentation impressions of SiC.

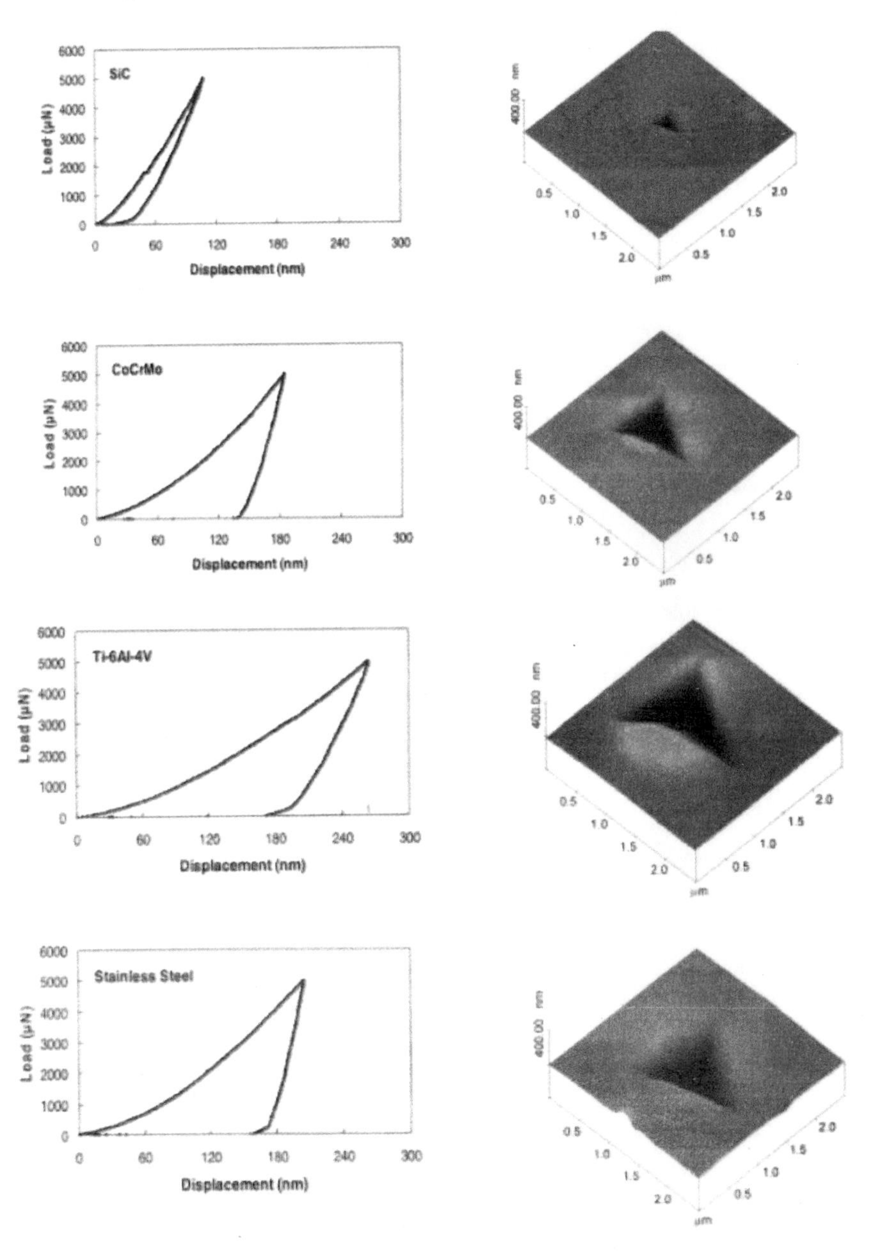

Fig. 17.1 (a) Representative load–displacement curves and (b) AFM images of indentations made at 5 m*N* peak indentation load on SiC, CoCrMo, Ti-6Al-4V, and stainless steel.

Multiple partial loading/unloading nano-indentations made on SiC, CoCrMo, Ti-6Al-4V, and stainless steel were used to determine their hardness and elastic modulus as a function of depth. The data collected in the tests are displayed in Fig. 17.2, along with hardness and elastic modulus values that were calculated from the load–displacement curves using standard techniques[17.19-17.21]. SiC exhibits the highest hardness of about 45.8 GPa and elastic modulus of about 370 GPa among the samples examined, followed by CoCrMo, stainless steel, and Ti-6Al-4V. The hardness values of stainless steel and Ti-6Al-4V are comparable while the elastic modulus of Ti-6Al-4V is higher than that of stainless steel.

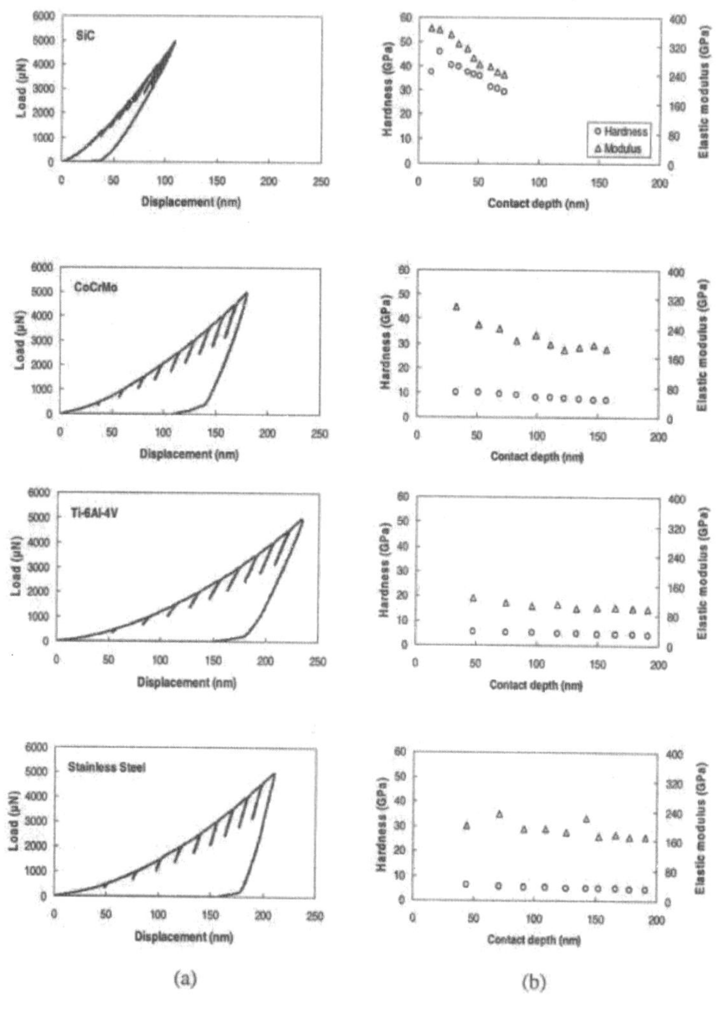

Fig. 17.2 Multiple partial loading/unloading nano-indentations made on SiC, CoCrMo, Ti-6Al-4V, and stainless steel for determining their hardness and elastic modulus as a function of depth.

17.4.2 *Scratch Resistance*

Figure 17.3 shows the coefficient of friction and acoustic emission profiles as a function of increasing normal load and SEM images of three regions over scratches: at the beginning of the scratch (indicated by A on the friction profile), at the point of initiation of damage at which the coefficient of friction increases to a high value or increases abruptly (indicated by B on the friction profile), and towards the end of the scratch (indicated by C on the friction profile), for SiC, CoCrMo, Ti-6Al-4V, and stainless steel. At the initial stages of scratching, SiC exhibits a lower coefficient of friction than other samples. With increasing normal load, the coefficient of friction of SiC remains a constant low value of 0.03 until the normal load reaches the critical load of about 22 m*N* at which its coefficient of friction rises to 0.1. A corresponding acoustic emission signal was detected at 22 m*N*, indicating crack formation. CoCrMo and stainless steel exhibit a continuous increase in the coefficient of friction with increasing normal load before significant damage occurred to the samples while Ti-6Al-4V exhibits a large variation in the coefficient of friction at the beginning of scratch.

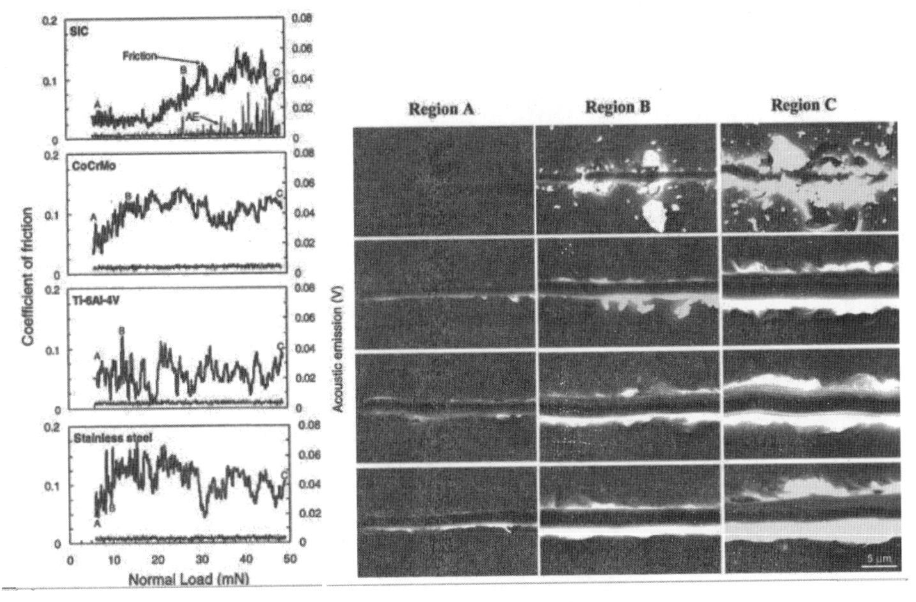

Fig. 17.3 Coefficient of friction and acoustic emission profiles as a function of increasing normal load and SEM images of three regions over scratches: at the beginning of the scratch (indicated by A on the friction profile), at the point of initiation of damage at which the coefficient of friction increases to a high value or increases abruptly (indicated by B on the friction profile), and towards the end of the scratch (indicated by C on the friction profile), for SiC, CoCrMo, Ti-6Al-4V, and stainless steel.

The SEM images show that below the critical loads all samples were damaged by plowing, associated with the plastic flow of material. In region A, for SiC, a shallower plowing scratch track was found without any debris on the side of the scratch, which is likely the cause for the lower coefficient of friction before the critical load. For all other samples, in addition to plowing scratch tracks, pile-up and debris particles were found on the sides of the scratches. SiC exhibits the smallest scratch track among the samples examined, followed by CoCrMo, stainless steel, and Ti-6Al-4V. After the critical load for SiC, cracks and debris particles were found on the side of the scratch from the critical load to the end. The crack formation at/after the critical load is responsible for the detected acoustic emission signals. For all other samples, deep plowing with plastic flow of material and pile-up was observed on the side of the scratch and a curly chip was found at the end of the scratch. This is a typical characteristic of ductile metal alloys. The acoustic emission sensor used in this study is not sensitive to the plastic deformation. This is why no acoustic signals were detected for CoCrMo, stainless steel, and Ti-6Al-4V.

17.4.3 *Friction and Wear Resistance*

The coefficient of friction as a function of sliding distance for various samples sliding against a WC ball is compared in Fig. 17.4. The SEM images of wear tracks and debris formed on all samples when sliding against a WC ball after a sliding distance of 5 m are compared in Fig. 17.5 and Fig. 17.6. Among the samples examined, SiC exhibits the lowest coefficient of friction, followed by CoCrMo, Ti-6Al-4V, and stainless steel. SiC shows a constant low coefficient of friction of about 0.17 during the sliding. CoCrMo shows an abrupt increase in the coefficient of friction at a sliding distance of 2 m. For Ti-6Al-4V and stainless steel, an abrupt increase in the coefficient of friction was observed right from the beginning of sliding, and then the coefficient of friction remains at higher values compared to SiC and CoCrMo.

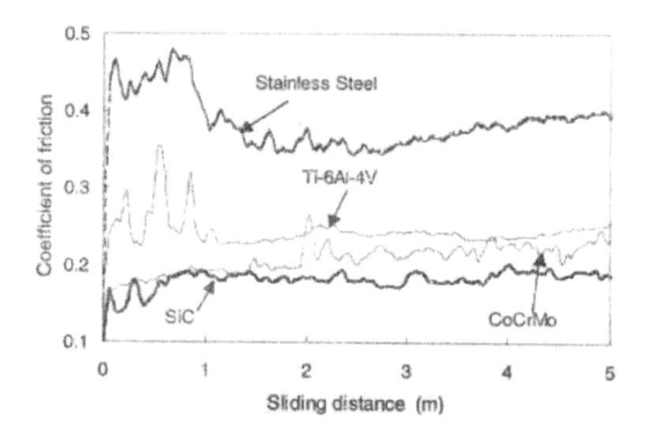

Fig. 17.4 Coefficient of friction as a function of sliding distance for SiC, CoCrMo, Ti-6Al-4V, and stainless steel when slid against a WC ball.

The SEM images Fig. 17.5 show that SiC has the smallest wear track and the least amount of debris, followed by CoCrMo, stainless steel, and Ti-6Al-4V. For Ti-6Al-4V and stainless steel, scratches with plastic flow of material were found within the wear tracks, which is believed to be attributed to the sudden increase in the coefficient of friction at the beginning of sliding. The amount of debris particles accumulated at the end and on the sides of the wear track of Ti-6Al-4V is at least several times more than those of other samples. SEM images of the debris particles Fig. 17.6 show that the debris particle size of SiC is the smallest, followed by CoCrMo, Ti-6Al-4V, and stainless steel. The debris particle size is relatively uniform for SiC and CoCrMo. For Ti-6Al-4V and stainless steel, in addition to small debris particles, many large chip-like debris particles were found. The edges of these large chip-like debris particles exhibit plastic deformation tearing characteristics. The large debris particles were, in turn, pulverized into small debris particles by the WC ball during subsequent sliding. These debris particles were severely strain hardened with high contact and shear stresses by the WC ball at the tribological interface. In addition, the debris particles, in particular the large chip-like debris particles, scratched the sample, generating additional sample damage with a higher coefficient of friction.

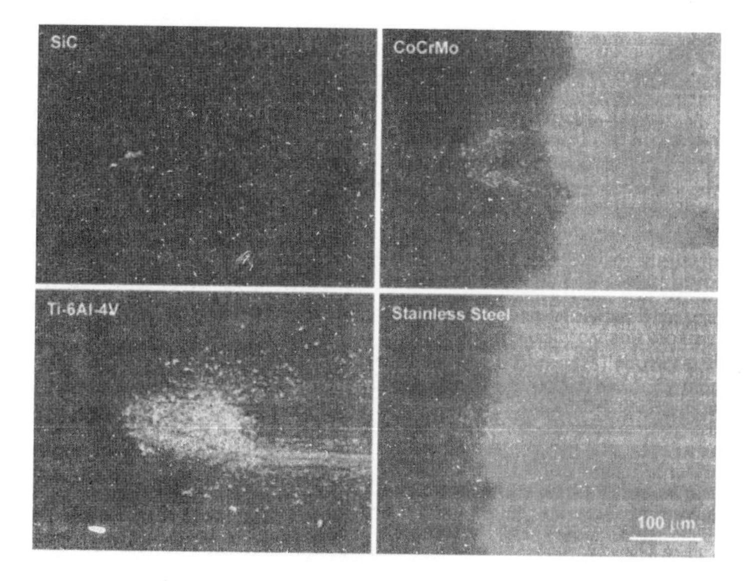

Fig. 17.5 SEM images of wear tracks and debris formed on SiC, CoCrMo, Ti-6Al-4V, and stainless steel when slid against a WC ball after a sliding distance of 5 m. The end of the wear track is on the left-hand side of the image.

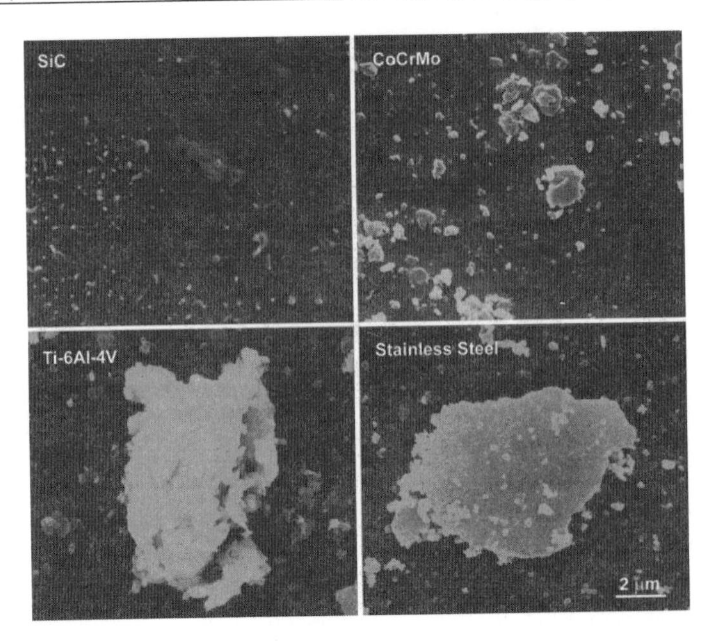

Fig. 17.6 SEM images of wear debris particles formed on SiC, CoCrMo, Ti-6Al-4V, and stainless steel when slid against a WC ball.

During wear tests, sample surface experiences shear stress. The maximum wear shear stress locates beneath the sample surface rather than on the sample surface[17.24]. This maximum wear shear stress can result in yielding (plastic deformation) for low strength materials and, consequently, delamination and bulking of the surface layer. Among the four materials tested in this study, the relatively low strength materials are Ti-6Al-4V and stainless steel. This may be why large chip like debris particles were found with Ti-6Al-4V and stainless steel. The higher hardness and elastic modulus of SiC makes plastic deformation and material flow more difficult. It is well known that wear surfaces of metals and alloys are easily oxidized and contaminated through reactions with environmental elements at the tribological interface. Oxidation and contamination layers are brittle and contribute to the generation of debris particles at the tribological interface. It has been proven that SiC is chemically inert and resistant to harsh mechanical and chemical environments. Without surface oxidation or a contamination layer, SiC retained the superior mechanical properties throughout the process. This is believed to be another key contribution to the high wear resistance of SiC.

Fig. 17.7 summarizes the hardness, elastic modulus, scratch, and wear results of SiC, CoCrMo, Ti-6Al-4V, and stainless steel. The damage bar chart in Fig. 17.7 was plotted based on optical image examination of the scratch and wear tracks, as well as, debris after tests: 0 represents no apparent damage; 1, low damage; 2, moderate damage; 3, heavy damage; and 4, severe damage.

Good correlation exists between mechanical properties and scratch/wear damage. Higher mechanical properties result in less scratch/wear damage. The currently used metals and alloys such as CoCrMo, Ti-6Al-4V, and stainless steel cannot meet the ever-increasing requirement for better tribological performance in orthopedic applications. SiC has been identified as the superior material due to its better mechanical and tribological properties, and should find more applications in biomedical applications. The advanced coating technology has made it possible to coat SiC on various substrates and retain the same high mechanical and tribological properties[17.15-17.17, 17.21].

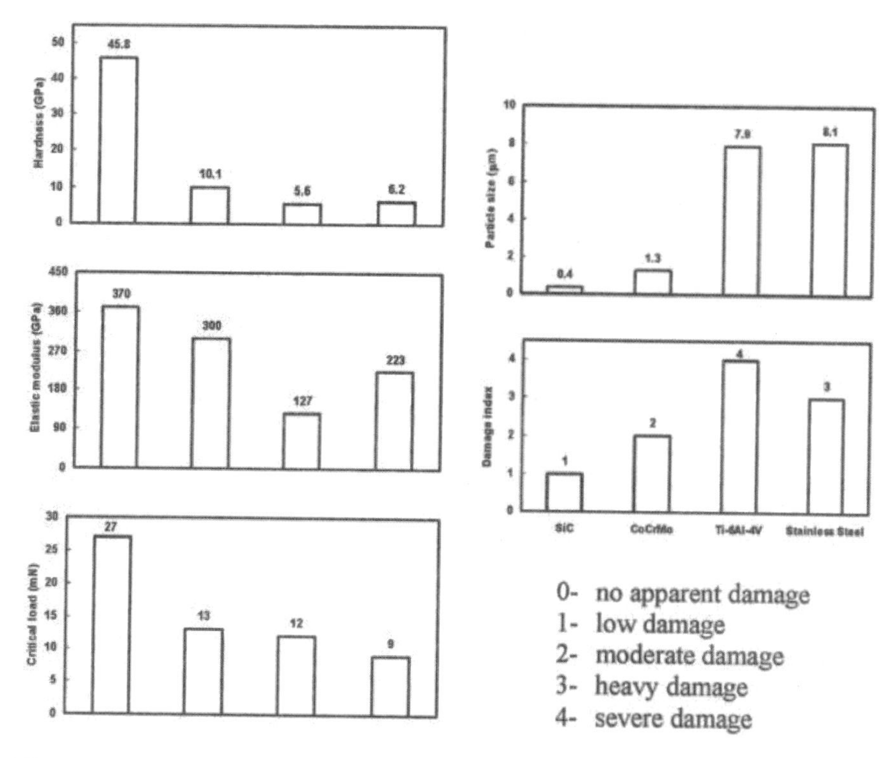

Fig. 17.7 Bar charts summarizing the hardness, elastic modulus, scratch, and wear results of SiC, CoCrMo, Ti-6Al-4V, and stainless steel.

17.5 Conclusions

From the above studies the following conclusions are drawn [17.25]:

SiC shows higher hardness, elastic modulus, and scratch resistance, as well as, lower friction than the other materials currently used in orthopedic applications. Compared with CoCrMo, Ti-6Al-4V, and stainless steel, SiC exhibits a low coefficient of friction with fewer and smaller debris particles. A good correlation between mechanical properties and scratch/wear damage was found. Higher hardness and elastic modulus together with the inertness of SiC are attributed to its superior tribological properties. Severe plastic ploughing damage, observed at the beginning of wear sliding, indicates that the currently used metallic orthopedic materials do not provide adequate load-carrying capabilities. The nano-indentation technique together with accelerated scratch, friction, and wear tests used in this article can be satisfactorily used for the characterization of orthopedic materials. SiC has proven to be an exceptional choice as a material in orthopedic applications.

References

[17.1] Aspenberg P, Anttila A, Konttinen YT, Lappalainen R, Goodman S, Nordsletten L, Santavirta S. Benign response to particles of diamond and SiC: bone chamber studies of new joint replacement coating materials in rabbits. Biomaterials 1996;17:807–812.

[17.2] Santavirta S, Takagi M, Nordsletten L, Anttila A, Lappalainen R, Konttinen YT. Biocompatibility of silicon carbide in colony formation test *in vitro*: a promising new ceramic THR implant coating material. Arch Orthop Trauma Surg 1998;118:89–91.

[17.3] Allen M, Butter R, Chandra L, Lettington A, Rushton N. Toxicity of particulate silicon carbide for macrophages, fibroblasts and osteoblast-like cells *in vitro*. Biomed Mater Eng 1995;5:151–159.

[17.4] Kotzar G, Freas M, Abel P, Fleischman A, Roy S, Zorman C, Moran J, Melzak J. Evaluation of MEMS materials of construction for implantable medical devices. Biomaterials 2002;23:2737–2750.

[17.5] McLaughlin J, Harris G. SiC for subretinal applications. Research Accomplishments 2000, National Nanofabrication Users Network; 2000. p 34–35.

[17.6] Nordsletten L, Hogasen A, Konttinen Y, Santavirta S, Aspenberg P, Aasen A. Human monocytes stimulation by particles of hydroxyapatite, silicon carbide, and diamond: *in vitro* studies of new prosthesis coatings. Biomaterials 1996;17:1521–1527.

[17.7] Monnink S, van Boven A, Peels H, Tigchelaar I, de Kam P, Crijns H, van Oeveren W. Silicon carbide coated coronary stents have low platelet and leukocyte adhesion during platelet activation. J Invest Med 1999;47:304 –310.

[17.8] Kalnins U, Erglis A, Dinne I, Kumsars I, Jegere S. Clinical outcomes of silicon carbide coated stents in patients with coronary artery disease. Med Sci Monit 2002;8:PI16 –PI20.

[17.9] McGee MA, Howie DW, Costi K, Haynes DR, Wildenauer CI, Pearcy MJ, McLean JD. Implant retrieval studies of the wear and loosening of prosthetic joints: a review. Wear 2000;241:158–165.

[17.10] Friedman RJ, An YH, Jiang M, Draughn RA, Bauer TW. *In vivo* mechanical and histological evaluation of bone ingrowth and apposition to metal implants of different surface textures in the rabbit femur. J Orthop Res 1996;14:455– 464.

[17.11] Bessette B, Fassler F, Tanzer M, Brooks C. Total hip arthroplasty in patients younger than 21 years: a minimum 10 year follow-up. Can J Surg 2003;46:257–262.

[17.12] Lemons J, Freese H. Metallic biomaterials for surgical implant devices. BONEZone 2002; Fall.

[17.13] Goodman SB, Lind M, Song Y, Smith RL. *In vitro, in vivo*, and tissue retrieval studies on particulate debris. Clin Orthop 1998;352:25–34.

[17.14] Kadoya Y, Kobayashi A, Ohashi H. Wear and osteolysis in total joint replacements. Acta Orthop Scand Suppl 1998;278:1–16.

[17.15] Li X, Bhushan B. Micro/nanomechanical characterization of ceramic films for microdevices. Thin Solid Films 1999;340:210–217.

[17.16] Bhushan B, Sundararajan S, Li X, Zorman CA, Mehregany M. Micro/nanotribological studies of single-crystal silicon and polysilicon and SiC films for use in MEMS devices. In: Bhushan B, editor. Tribology issues and opportunities in MEMS. Netherlands: Kluwer Academic; 1998. p 407–430.

[17.17] Li X, Bhushan B, Takashima K, Baek CW, Kim YK. Mechanical characterization of micro/nanoscale structures for MEMS/NEMS applications using nanoindentation techniques. Ultramicroscopy 2003;97:481– 494.

[17.18] Cherednichenko DI, Drachev RV, Sudarshan TS. Self-congruent process of SiC growth by physical vapor transport. J Crystal Growth 2004;262:175–181.

[17.19] Bhushan B, Li X. Nanomechanical characterization of solid surfaces and thin films (invited). Int Mater Rev 2003;48:125–164.

[17.20] Li X, Bhushan B. A review of nanoindentation continuous stiffness measurement technique and its applications. Mater Character 2002;48:11–36.

[17.21] Bhushan B, Li X. Nanomechanical characterization of ceramic materials. In: Gogotsi Y, Domnich V, editors. High pressure surface science and engineering. Bristol: IOP Publishing; 2003.p 321–348.

[17.22] Oliver WC; Pharr GM. An improved technique for determining hardness and elastic-modulus using load and displacement sensing indentation experiments. J Mater Res 1992;7:1564 –1583.

[17.23] Sneddon IN. The relation between load and penetration in the axisymmetric boussinesq problem for a punch of arbitrary pro-file. Int J Eng Sci 1965;3:47–56.

[17.24] Bhushan B. Principles and applications of tribology. New York: Wiley; 1999.

[17.25] Li Xiaodong, Wang Xinnan, Bandokov Robert, Morris Julie, An H. Yuehuel, Sudarshan S. Tangali, "Micro/Nanoscale Mechanical and Tribological Characterisation of SiC for orthopedic Application", Wiley interscience, DOI: 10.1002/Jbm.b.30168.

18 The Frictional Coefficient of Bovine Knee Articular Cartilage

18.1 A General Review

The normal displacement of articular cartilage was measured under load and in sliding, and the coefficient of friction during sliding was measured using a UMT-2 Multi-Specimen Test System of CETR. The maximum normal displacement under load and the start-up frictional coefficient have similar tendency of variation with loading time. The sliding speed does not significantly influence the frictional coefficient of articular cartilage.

18.2 Introduction

Cartilage has excellent biomechanical and tribological properties with low friction and minimum wear in diarthrodial joints throughout the lifetime of most people, and the lifetime of articular cartilage can be 40 years or longer. This has inspired material and bionic scientists to study the mechanism of such excellent tribological characteristics in order to develop artificial joints. Various mechanisms have been proposed to explain the remarkable low friction behavior of articular cartilage, such as fluid film, mixed and boundary lubrication. In the fluid film lubrication regime, both cartilage surfaces are completely separated by a layer of synovial fluid, which results in minimal friction[18.1, 18.2]. However, this lubrication theory fails to explain the low friction in sensorial joints under conditions of little motion such as start-up after a long period of standing. Therefore, various mixed lubrication mechanisms have also been proposed, including boosted lubrication[18.3], weeping lubrication, McCutchen C W[18.4] and biphasic lubrication[18.5]. In particular, the load carried by the fluid phase in articular cartilage is responsible for low friction, and the loading time prior to sliding is an important factor[18.6-18.8]. In addition to these mixed lubrication theories, boundary lubrication is extremely important in ensuring low friction and protecting the surfaces when two cartilages are in direct contact[18.9].

In this study, investigations of the frictional behaviour of articular cartilage of bovine knee with mixed/boundary lubrication regimes in bovine serum, by measuring normal displacement under load and the coefficient of friction after periods of stationary loading varying from 5 sec to 60 min have been carried out. The influence of sliding speed up on friction is also investigated.

18.3 Materials and Methods

18.3.1 *Materials*

Articular cartilages were collected from bovine femoral condyles and from the moral-patella articulating surfaces, with the underlying subchondral and cancellous bone retained, then were cut to the square form of 7 mm × 7 mm bone and rectangular form of 30 mm × 10 mm bone which were called upper cartilage and lower cartilage respectively. The average thickness of the cartilage layer was 1.3 mm, and that of subchondral bone was 2 mm. The surface of cartilage was kept horizontal by polishing the underside of the underlying subchondral bone, so as to ensure that the contact areas were horizontal on all specimens. Articular cartilage specimens were stored frozen at –20 °C in saline. Prior to the tests the cartilage specimens were defrosted and immersed in bovine serum for at least one hour before use.

Bovine serum with concentration of 99% was bought from Jianghai Bioengineering Limited Company of Heilongjiang Province. A UMT-2 Multi-Specimen Test System of CETR was used to measure the normal deformation, tangential force, normal force and sliding length simultaneously.

18.3.2 *Methods*

18.3.2.1 *Normal displacement test under load*

A loading test was carried out to investigate the normal displacement of cartilages under load. The upper cartilage was stuck to a stiff 8 mm × 8 mm metal fixture, and the fixture was connected to the tool holder of the Test System with a pin. The tool holder, pin and fixture were in line. The lower cartilage was fixed to the bottom of a bath made of Ultra High Molecular Weight Polyethylene (UHMWP) filled with bovine serum (Fig. 18.1). When the two cartilages contacted, a strain gauge measured changes in the Z-carriage caused by the displacement of the cartilages. The upper cartilage was loaded to 40 N under programmed control, and the computer automatically recorded the displacement. Normal displacement was the total displacement of upper and lower cartilages. Loading time ranged from 5 sec to 60 min.

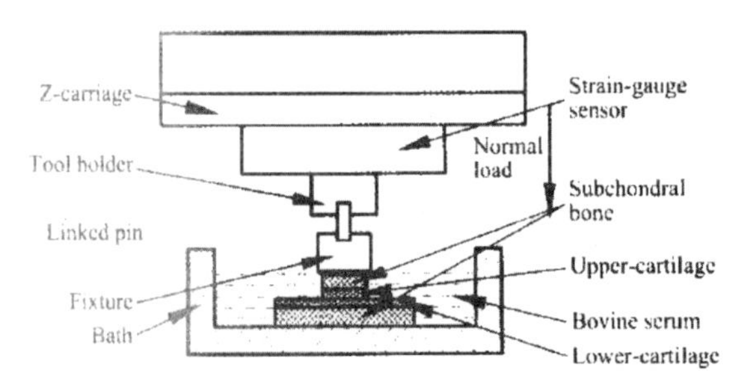

Fig. 18.1 Experimental methods of loading test of articular cartilage.

18.3.2.2 *Frictional test in sliding*

Sliding tests were carried out to investigate the frictional behavior of cartilages in sliding. Prior to sliding test, the cartilage was loaded vertically. When the upper cartilage was sliding the lower cartilage remained at rest (Fig. 18.2). The friction force in sliding direction (F_x) and the normal force (F_z) were measured by force transducers in the test system. The frictional coefficient was calculated as F_x/F_z. The sliding was in one direction and at a constant speed. Each loading and sliding test was controlled by computer. The normal load of 40 N applied to the upper cartilage produced a stress in the range of 0.5 MPa to 4 MPa (depending on the contact area) representative of physiological loading. Lower sliding speeds were employed to ensure that the contact operated in mixed or boundary lubrication[18.9]. Different sliding speeds were used to investigate their influence on frictional behavior of the articular cartilages. After the sliding test the cartilages were unloaded and immersed in bovine serum for several minutes allowing them to rehydrate again and to recover their former structure. The time for which the cartilage was unloaded and immersed in bovine serum was twice that under load.

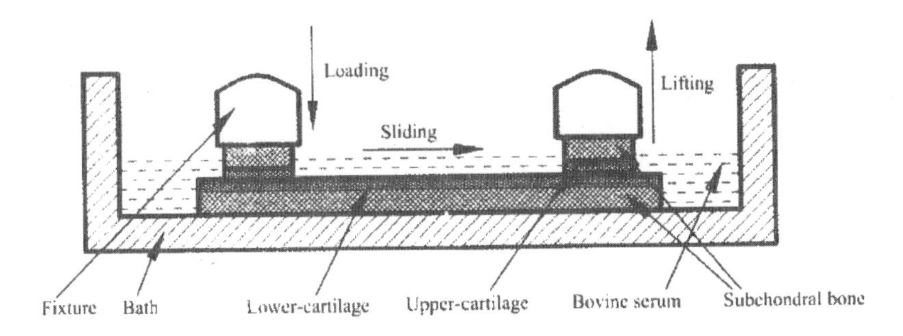

Fig. 18.2 Experimental methods of sliding test of articular cartilage.

The experiments were conducted with loading times of 5 sec, 30 sec, 1 min, 2 min, 4 min, 7 min, 12 min, 20 min, 30 min, 45 min and 60 min, and under sliding speeds of 2 mm.s^{-1}, 4 mm.s^{-1} and 7 mm.s^{-1} respectively. The sliding distances were 10 mm, 12 mm and 14 mm, respectively. Each experiment was repeated three times under the similar conditions, hence 11×3 ×3 experiments were performed. The normal displacement and the coefficient of friction were the mean value of three experiments.

18.4 Experiment Results

18.4.1 *Normal displacement*

Fig. 18.3 shows the normal displacement as function of the loading time. The normal displacement initially ascended sharply, and then rose slowly after about 12 min. Fig. 18.3 shows that the repeatability of experiments is good, that the articular cartilage has good deformation recovery, and that the normal displacement depends strongly on the loading time. Fig. 18.4 shows how the normal displacement varies with loading time. The slope initially declines sharply, and decreases slowly after about 12 min. Regression analysis gives the relationship between normal displacement and loading time as

$$S = \frac{1}{0.392 + 2.51t^{-0.5}} \quad (t > 0) \tag{18.1}$$

where **S** is normal displacement and t loading time.

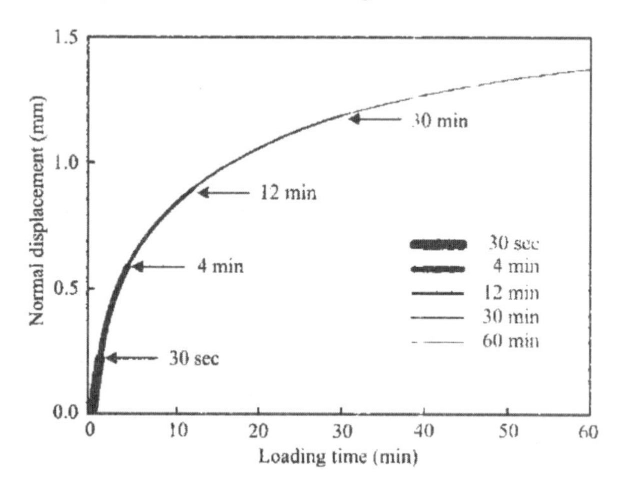

Fig. 18.3 The normal displacement as function of loading time.

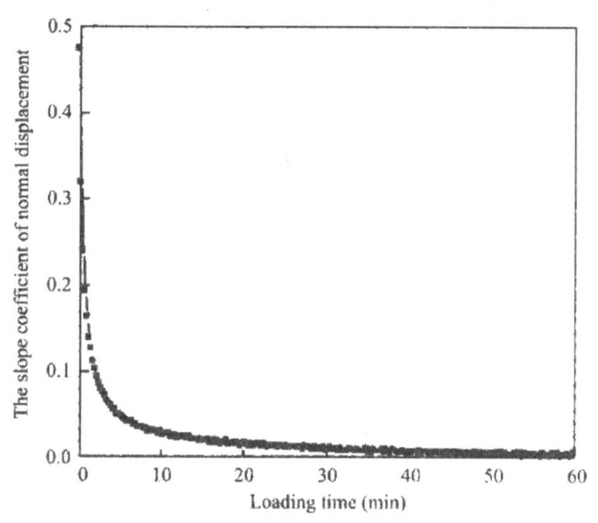

Fig. 18.4 The slope coefficient of normal displacement varied with loading time.

Fig. 18.5 shows excellent agreement of Eq. (18.1) with experimental results. Therefore, Eq. (18.1) can describe normal displacement of articular cartilage with loading time.

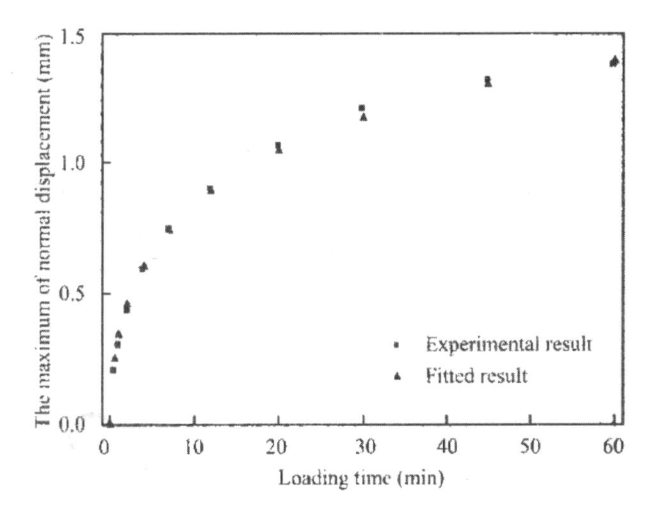

Fig. 18.5 Comparison of maximums of normal displacement and fitted results under different loading times.

18.4.2 *Frictional coefficient*

Fig. 18.6 shows how the coefficient of friction varied with sliding speed when the loading time was 5 sec. The coefficient of friction initially ascended sharply, then descended and ascended gradually. The influence of sliding speed on the coefficient of friction is insignificant. These results indicate that the articular cartilage was in the boundary or mixed lubrication regimes when loading time was 5 sec. It can be speculated that, with loading time between 5 sec and 60 min, articular cartilages were all in the boundary or mixed lubrication regimes.

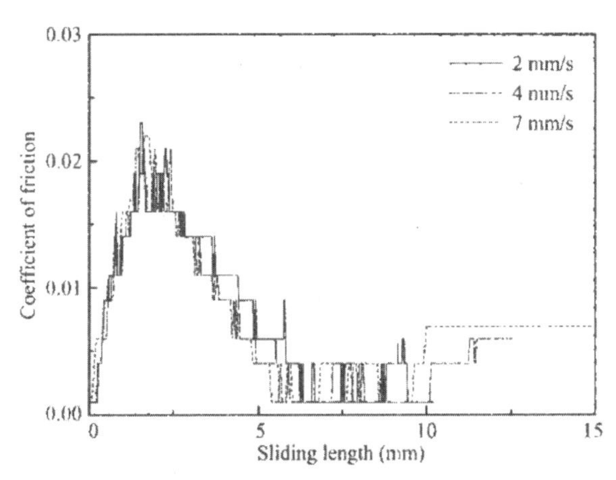

Fig. 18.6 The coefficient of friction as function of sliding length with different sliding speeds, loading time was 5 sec.

Fig. 18.7 shows the coefficient of friction and normal displacement as functions of sliding time, in which the sliding speed was 4 mm.s^{-1} and loading time was 4 min. We can divide the coefficient of friction into four stages, in stage 1 from *a* to *b*, the coefficient ascended sharply; in stage 2 from *b* to *c*, the coefficient declined sharply; in stage 3 from *c* to *d*, the coefficient descended slowly; in the stage 4 from *d* to *e*, the coefficient rose slowly. The coefficient approached its maximal value at point 6, which is called the coefficient of start-up friction. We can also divide the normal displacement into four stages, in stage 1 from *u* to *v*, normal displacement stayed constant; in stage 2 from *v* to *w*, normal displacement descended slightly; in stage 3 from *w* to *x*, normal displacement declined sharply, and the upper cartilage slid up the slope of lower cartilage's groove, therefore, this stage is also referred to as the climbing stage; in stage 4 from *x* to *y*, normal displacement rose slowly.

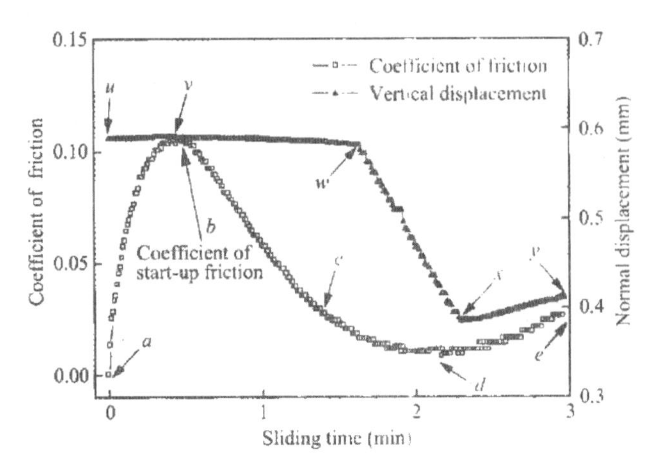

Fig. 18.7 The coefficient of friction and normal displacement varying with sliding time, the sliding speed was 4 mms^{-1} and loading time was 4 min.

Fig. 18.8 shows coefficient of start-up friction as function of loading time when the sliding speed was 4 mm/s. The coefficient initially ascended sharply, and rose tardily after about 12 min. Thereby, we can obtain the relationship between coefficient of start-up friction and loading time as

$$\mu_m = \frac{1}{0.455 + 20.090\,t^{-0.5}} \quad (t > 0), \tag{18.2}$$

where μ_m is the coefficient of start-up friction and t loading time.

Eq. 18.2 is also plotted on Fig. 18.8, which shows excellent agreement with the experimental results. Comparing Eq. 18.1 and Eq. 18.2, we may speculate that the coefficient of start-up friction is probably influenced by normal displacements prior to sliding, and the maximum normal displacement may be used to quantify the coefficient of start-up friction.

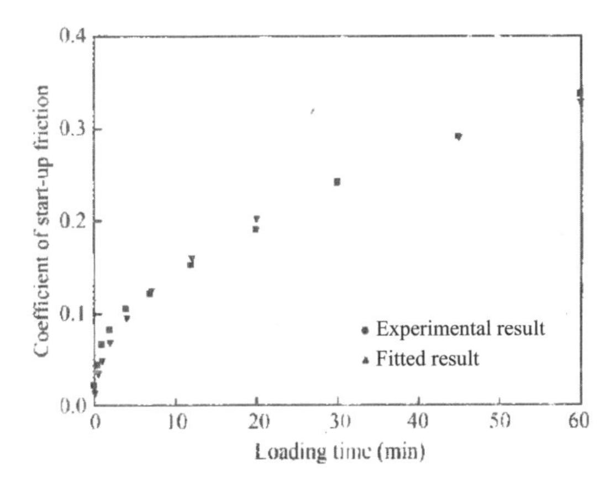

Fig. 18.8 The coefficient of start-up friction and fitted results varying with loading time, sliding speed was 4 mms^{-1}.

18.4.3 *Frictional property influenced by sliding speeds*

Fig. 18.9a shows the relationship between the coefficient of friction and sliding length, and Fig. 18.9b shows the relationship between coefficient of friction and sliding time with different sliding speeds and a loading time of 4 min. The trend of change of the coefficient is similar to that in Fig. 18.6 and Fig. 18.7. The sliding speed has almost no influence on the coefficient of start-up friction. At sliding speeds of 2, 4 and 7 $mm.s^{-1}$, the coefficient of start-up friction was 0.107, 0.106 and 0.107, respectively. However, the higher the sliding speeds, the less the time in achieving the coefficient of start-up friction. We may conclude that sliding speed does not significantly influence the frictional coefficient of articular cartilage.

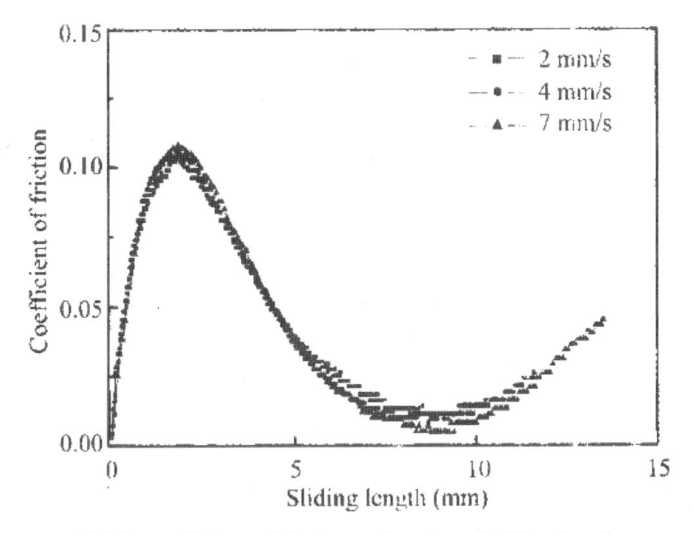

(a) The coefficient of friction as function of sliding length

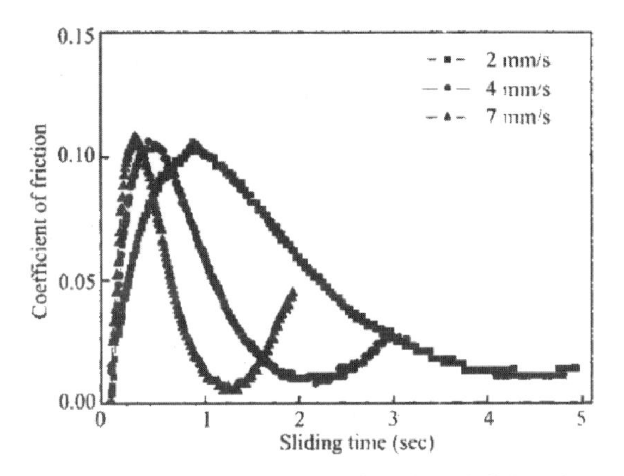

(b) The coefficient of friction as function of sliding time

Fig. 18.9 The coefficient of friction as function of sliding length and sliding time.

18.5 Discussion

Articular cartilage is composed of a network of fine collagen fibrils within which a network of hydrophilic proteoglycan aggregate molecules is immobilized and restrained. Collagen fibrils and proteoglycans are the structural components transmitting the internal mechanical stresses resulted from the loads applied to the cartilage. The structural components, together with water, determine the mechanical behavior of this tissue[18.10]. For this reason, it was often modeled as a biphasic material[18.11].

In loading, normal displacement could be referred as an indication of the fluid flow in the cartilage and the load carried by the fluid phase. An initial instantaneous elastic deformation in response to load was primarily due to the change in shape of the cartilage. At this point the fluid flow reached its maximum and the fluid phase carried the largest portion of the load. As the loading time increased, the slope of normal displacement with loading time decreased (Fig. 18.4). As the rate of flow of fluid in the cartilage declined, the load carried by the fluid phase decreased and the load carried by the solid phase increased. At 45-60 min the contact between upper cartilage and lower cartilage was close to equilibrium, if little further deformation occurred then the load carried by the fluid was close to zero. Thus, normal displacement is significantly affected by loading time. These results are consistent with previous studies[18.6, 18.9]. Of course, it is easy to understand that normal displacement functioned in a passive correlation with the load carried by the solid phase. After an equal period of load removal to that of previous loading, the loading test produced a similar displacement when reloaded, indicating a full recovery of fluid content (Fig. 18.3). This is consistent with previous studies[18.12]. This shows conclusively that articular cartilage is viscoelastic and can recover automatically under certain conditions.

On initial loading, a large amount of the load is carried by the fluid phase of the cartilage. As the static loading time increases, the load carried by the solid phase increases and that carried by the fluid phase decreases. Hence, the overall frictional force increases which can be simply expressed as

$$F_{\mathrm{r}}(t) = \mu_r(t)W, \tag{18.3}$$

where $F_{\mathrm{r}}(t)$ is overall friction force, $\mu_{\mathrm{r}}(t)$ overall or aggregate friction coefficient, and W total load.

As analysed above, the load W carried by the cartilage is composed of two parts carried by the solid phase and by the fluid phase, when Eq. (18.3) can be rewritten as

$$F_{\mathrm{r}}(t) = \mu_s(t)\, W_s(t) + \mu_f(t)\, W_f(t), \tag{18.4}$$

where μ_s(t) is the effective coefficient of friction attributed to the solid phase, $\mu_{\mathrm{f}}(t)$ is the effective coefficient of friction attributed to the fluid phase, $W_{\mathrm{s}}(t)$ is the load carried by the solid phase and $W_f(t)$ is the load carried by the fluid phase. Since μ_s is significantly greater than μ_f, the second term of Eq. (18.4) can be ignored and we have

$$F_{\mathrm{r}}(t) = \mu_s(t)\, W_s(t). \tag{18.5}$$

Eq. (18.5) indicates that the frictional force is mainly determined by the load carried by the solid phase. According to the conclusion above, the friction force is mainly associated with normal displacement under load. When the total load W is a fixed value, the coefficient of friction is also related to normal displacement.

Prior to initial sliding, the upper cartilage is loaded upon the lower cartilage, and the deformed area of lower cartilage is larger than the contacted area of upper cartilage. Of course, the deformation of the area around the contact of the lower cartilage is less than the contacted area, and was strongly related to loading time. Fig. 18.10 sketches the dynamic variations of the normal displacement of upper cartilage and lower cartilage with sliding length.

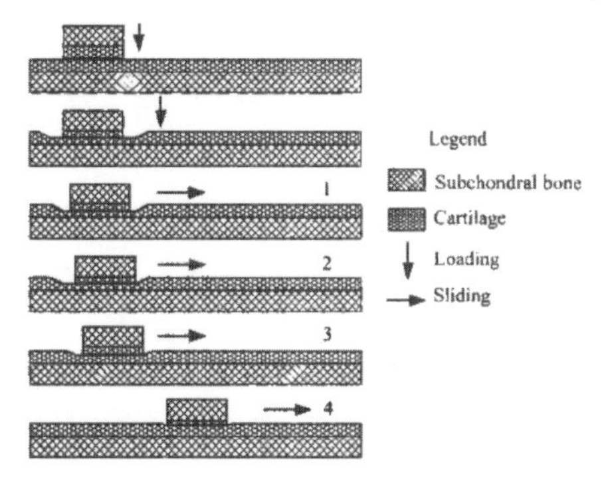

Fig. 18.10 Normal displacements of upper cartilage and lower cartilage dynamically vary with sliding length.

In stage 1, the coefficient ascended sharply in a short period. At initial instantaneous sliding, the normal displacement was maximum, and the load carried by the solid was also maximum. The coefficient of friction ascended rapidly from zero, and shortly arrived at the maximum. The result is in agreement with previous studies[18.13]. Normal displacement kept almost unchanged, which showed the upper cartilage just started to slide.

In stage 2, the coefficient of friction declined sharply. When the upper cartilage slid on the surrounding area of lower cartilage, the load carried by the solid in this area was obviously less than that on the contact area, which made the coefficient of friction reduce quickly. Normal displacement went down slightly, which showed the upper cartilage was on the surrounding area of lower cartilage.

In stage 3, the coefficient of friction reduced slowly. When the upper cartilage passed through the circumjacent area of lower cartilage, it bore the trend of climbing up the slope of the lower cartilage's groove, which caused a sharp reduction of normal displacement. But the load carried by the solid phase did not decline rapidly.

In stage 4, the coefficient of friction ascended slowly. When the upper cartilage completely passed through the slope of the groove in the lower cartilage, normal displacement of the upper cartilage was larger than that of the lower cartilage, because the lower cartilage was not completely loaded. In the following sliding, the upper cartilage transmitted compressive stress to the lower cartilage, which resulted in increase of load carried by the lower cartilage. Normal displacement then began to ascend gradually until the sliding stopped.

18.6 Conclusions

The following is the outcome of the study carried out[18.14]:

The frictional behaviour of articular cartilage of the bovine knee with mixed boundary lubrication regimes in bovine serum has been studied by measuring the normal displacement under load and the coefficient of friction during sliding. The articular cartilage is highly viscoelastic and is able to recover fully after the load is released. Prior to sliding, the coefficient of start-up friction is mainly determined by the normal displacement, and they have the similar tendency of variation with loading time. The sliding speed does not significantly influence frictional coefficient of the articular cartilage. The results of this study can be used as useful information in the development of bionic joints.

References

[18.1] Dowson D. Models of lubrication of human joints, in: Lubrication and wear in living and artificial human joints. *Proceeding of Institute Mechanical Engineers,* 1967, **181**, 45–54.

[18.2] Dowson D, Jin Z M. Micro-elastohydrodynamic lubrication of synovial joints. *Engineering in Medicine,* 1986, **15**, 63–65.

[18.3] Walker P S, Unsworth A, Dowson D, Sikorski J,-Wright V. Mode of aggregation of hyaluronic acid protein complex in the surface of articular cartilage. *Annals of the Rheumatic Diseases,* 1970, **29**, 591–602.

[18.4] McCutchen C W. The frictional properties of animal joints. *Wear,* 1962, 5, 1–17.

[18.5] Mow V C, Lai W M. Recent developments in synovial joint biomechanics. *SIAM Rex,* 1980, **22**, 275–317.

[18.6] Forster H, Fisher J. The influence of loading time and lubricant on the friction of articular cartilage. *Proceeding of Institute Mechanical Engineers, H: Journal of Engineering in Medicine,* 1996, **210**, 109–1 19.

[18.7] Mukherjee N, Wayne J S. Load sharing between solid and fluid phases in articular cartilage: I-Experimental determination of in situ mechanical conditions in a porcine knee. *Journal* of *Biomechanical Engineering,* 1998, **120**, 614–619.

[18.8] Mukherjee N, Wayne J S. Load sharing between solid and fluid phases in articular cartilage: 11-Comparison of experimental results and u-p finite element predictions. *Journal of Biomechanical Engineering,* 1998, **120**, 620–624.

[18.9] Jin *Z* M, Pickard J E, Forster H. Fritional behaviour of bovine articular cartilage. *Biorheology,* 2000, **37**, 57–63.

[18.10] Mow V C, Proctor Ch S, Kelly MA. Biomechanics of articular cartilage. *Basic Biomechanics* of *the Musculoskeletal System,* 2nd ed, (editors) Nordin M A, Frankel V H. Lea and Febiger, Philadelphia, 1989, 3 1–58.

[18.11] Spilker R L, Suh J K, Mow V C. A finite analysis of the indentation stress-relaxation response of linear biphasic articular cartilage. *Journal of Biomechanical Engineering,* 1992, **114**, 191–201.

[18.12] Edwards J. Physical characteristics of articular cartilage. *Proceedings of the Institution of Mechanical Engineers,* 1967, **181**, 16–24.

[18.13] Mabuchi K, Ujihira M, Sasada T. Influence of loading duration on the start-up friction in synovial joints: Measurements using a robotic system. *Clinical Biomechanics.* 1998, **13**, 492–494.

[18.14] Shan-hua Qian, Shi-rong Ge, Qing-liang Wang, The Frictional Coefficient of Baline Knee Articular Cartilage. Journal of Bionic engineering 3 (2006), 079-085.

Section - V

Tribology in Engines

19

INVESTIGATIONS OF LOW FRICTION RING PACK FOR GASOLINE ENGINES

19.1 A General Review

Lower emissions, reduced friction and low lubricant oil consumption are the main drivers for new gasoline engines. In terms of piston ring pack, the trend is to reduce ring tangential load and width. On the other hand, the main concern is to have proper ring conformability and lube oil control. This chapter presents the comparison of a baseline ring pack with a low friction pack in terms of friction, blow-by control and lube oil consumption. Besides ring width and tangential load reductions, evaluations of ring materials are also carried out.

Narrow compression rings, 1.0 and 0.8 mm, were engine tested. PVD top ring was also tested and showed about 10% friction reduction compared to the usual Gas Nitrided one. 3-piece 1.5 mm oil rings were compared with the usual 2.0 mm ones. Being more flexible, the narrower oil rings can have same conformability with reduced tangential load.

Friction was measured in the mono-cylinder SI Floating Liner engine in 5 operational conditions. Effect of cylinder roughness on friction is discussed by reciprocating bench tests.

Compared with a typical 1.2/1.2/2.0 mm SI ring pack, the proposed 1.0/1.0/1.5 mm pack brought about 28% reduction in ring friction in the tested conditions, which would mean in about 1% of fuel savings in urban use.

19.2 Introduction

Mechanical losses in an internal combustion engine account for approximately 10% of the total energy of the consumed fuel. This amount represents around 25% of the effective power at full load, more at part loads. At idle or no-load, 100% of the indicated power is consumed by friction. The piston and the piston rings are the largest contributors to the mechanical losses, but the relative share varies with engine type and load condition. Fig. 19.1 shows the energy distribution for a 2.OL SI engine at full load/5000 rpm.

The search for reducing friction is continuous, but the interest for low friction components has increased recently, especially sparked by the fuel price increase and more rigorous emissions legislation. A common question during engine design is "how much friction reduction / fuel economy can be expected from a given design change". Fig. 19.2 shows a rough estimation based on total friction, but it remains difficult to estimate the benefits of a given design change, e.g. lower ring Ft or reduced piston skirt area. Simulations can be done to address single design changes, that later can be engine tested in a complete design pack. An exercise of this approach was made in[19.2], where simple equations to estimate friction change were compared with more complex simulations and finally with engine tests.

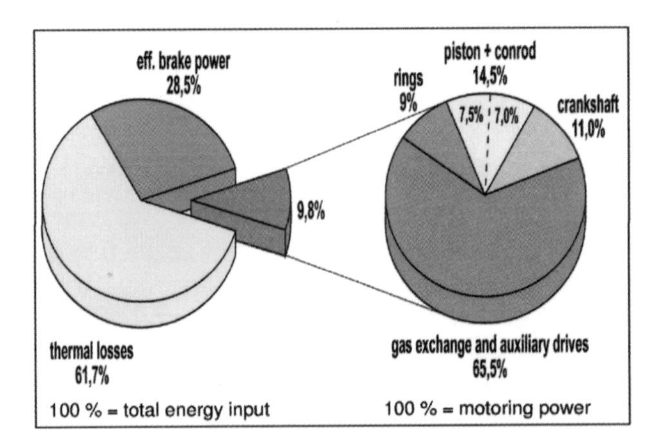

Fig. 19.1 Breakdown of total energy distribution and engine mechanical losses[19.1].

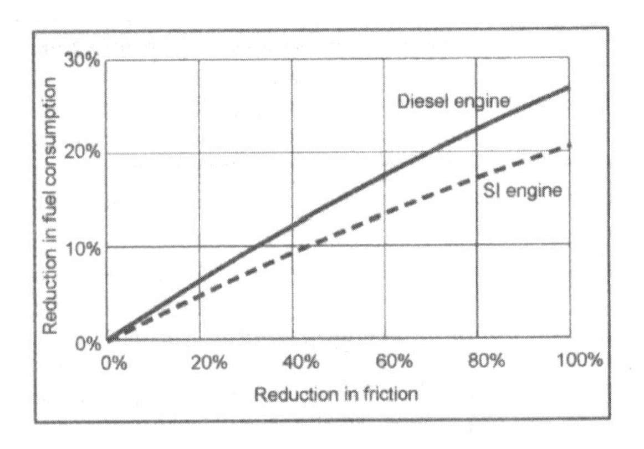

Fig. 19.2 Estimation of fuel economy due to engine friction reduction (2000 rpm)[19.3].

Although the specific friction share among the components depends on engine design, operation condition and other factors, it is generally accepted that the main friction contributor is the piston-cylinder system and that among the largest potentials to reduce friction are:

(a) Liner finish
(b) Oil control ring tangential load
(c) Ring width
(d) Materials with lower friction coefficient

Those items will be discussed ahead

19.2.1 *Liner Finish*

Friction on lubricated sliding surfaces may be considered as the sum of the boundary and hydrodynamic regime contributions. See Fig. 19.3. The boundary, asperity, friction coefficient is much higher than the hydrodynamic one, so transferring the load supported by the asperities to the oil film greatly reduces friction[19.4]. Smoother surfaces usually presents lower asperity contact and lower oil film thickness, not only reducing friction but also lube oil consumption.

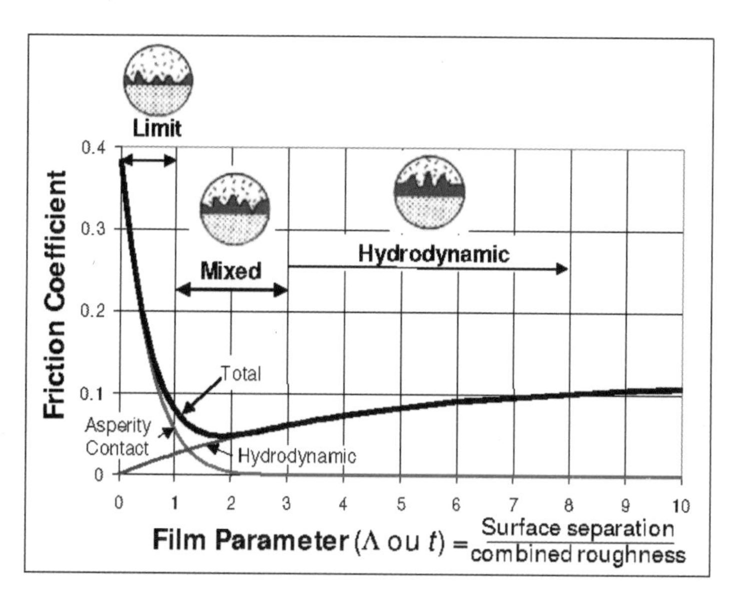

Fig. 19.3 Stribeck Curve.

As an example of friction reduction due to smoother bore finishes, fig. 19.4 shows the average cycle friction coefficient of 2 different liner finishes against Physical Vapor Deposition PVD coated rings. Ring and liner specimens removed from actual engine parts were used. The test was conducted in a CETR UMT-2 reciprocating tester. For this study, the specimens were tested at different speeds and 50 N. Reciprocating stroke was 10 mm and the liner specimens were flooded with SAE 30 lubricant at ambient temperature. A similar, previous, but more detailed study on surface finish is described in[19.5].

	Plateau Honed	Slide Honed
R_q [μm]	0.84	0.49
R_{pk} [μm]	0.25	0.28
R_k [μm]	0.74	0.39
R_{vk} [μm]	1.85	1.47

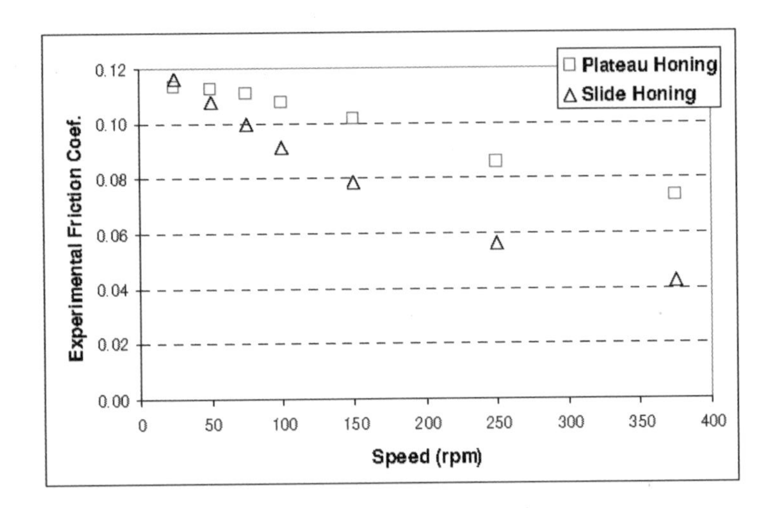

Fig. 19.4 Friction reciprocating bench test.

At low speed conditions, friction is dominated by asperity contact and defined mostly by the reciprocating material properties. See Fig.19.4. At 25 rpm, the 2 surface finishes presented same friction coefficient. With increase in speed, the lubrication regime moves in direction to the hydrodynamic one. The Slide, smoother, finish presented lower friction, e.g., the measured friction reduction with the Slide honing was around 40% at 375 rpm.

19.2.2 *Oil Control Ring*

Due to its relative high load, necessary for good lube oil control, the OCR is the major contributor to the ring pack friction. The tangential load of OCR is basically determined by ring unitary load, P_0. P_0 is the tangential load, F_t, divided by the ring contact area. For 3-piece rings, the typical P_0 values are from 0.6 to 1.2 N/mm^2. Reducing the OCR tangential load is an obvious way to reduce friction, but one important drawback is the correspondent reduction of ring conformability. Conformability is the ring capacity to conform to a deformed bore and usually quantified by the k ring parameter, called "conformability coefficient", used in ring calculations[19.6].

$$\mathrm{K} = \frac{F_t\, D_n^2}{4\, E\, J} \tag{19.1}$$

where

F_t : ring tangential load
D_n : ring nominal diameter
E : Modulus of Elasticity (N/mm^2)
J : Moment of Inertia of the ring cross section (mm^4)

One way to preserve the necessary ring conformability while reducing its load, is to use more flexible ring cross sections. For the usual OCR designs, this means to reduce the rail radial thickness, usually followed with a reduction of its axial width. Fig. 19.5 compares oil ring designs in terms of conformability. 2-piece OCR is usually applied to high loaded applications due to its higher durability and conformability. On the other hand, 3-piece design is mostly applied to SI engines due to its side sealing that provides better lube oil control at partial loads[19.9].

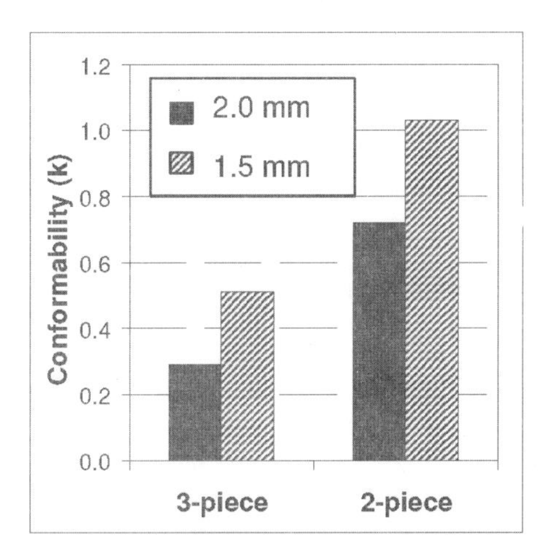

Fig. 19.5 Conformability comparison of different OCRs.

Low width 3-piece rings usually have the ISO ES2 expander, where the stress on the expander is proportional to its wire width (w), ring Ft and the distance between the segments (IS). See Fig.19.6.

$$\text{Stress}_{\text{Expander}} \propto \text{Ft} \,.\, \text{IS} \,.\, \text{W}^3 \tag{19.2}$$

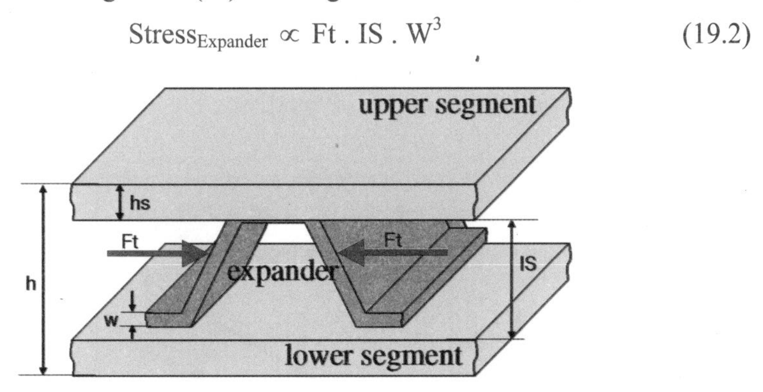

Fig. 19.6 Scheme of 3-piece loading.

Figure 19.7 shows a thermal test, where closed rings are kept in an oven at 300 °C to investigate the loss of tangential load due to thermal relaxation of the expander. Narrow segments allow lowering the Ft for the same P_0. But the distance IS will increase and usually a wider expander wire is needed due to assembling restrictions. As consequence, expander stress is increased, which leads to higher thermal relaxation and Ft loss under operation. Compare curves A and B in Fig. 19.7. On the other hand, a narrow OCR, not only allows the use of narrow segments, but also has an expander with lower stress. Even this happens with the same Ft and segment width, due to its lower distance between segments. Compare curves B and C in Fig. 19.7. The less stressed 1.5 mm has lower Ft loss than the 2.0 mm designs.

Ring	Width (mm)	P0 (N/mm^2)	Segment Width (mm)	Ft (N)
A	2.0	0.8	0.46	36.4
B	2.0	0.8	0.35	23.1
C	1.5	0.8	0.35	23.1

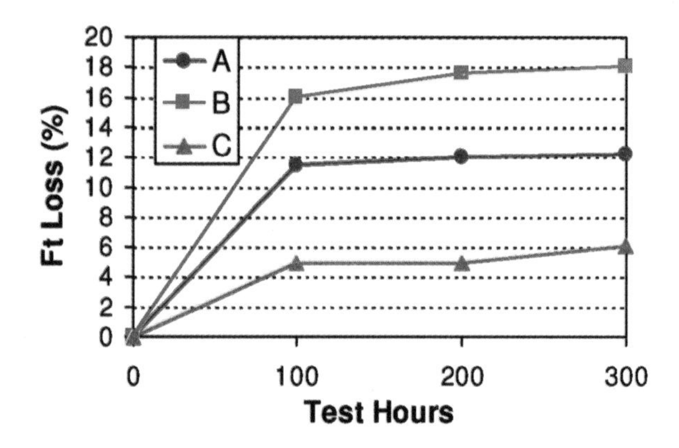

Fig. 19.7 Thermal stability tests on 3-piece OCR.

Compared with the usual 2 mm, 3 piece OCR, the 1.5 mm design enables two design paths: similar conformability with reduced tangential load or higher conformability with similar tangential load. See Fig. 19.8.

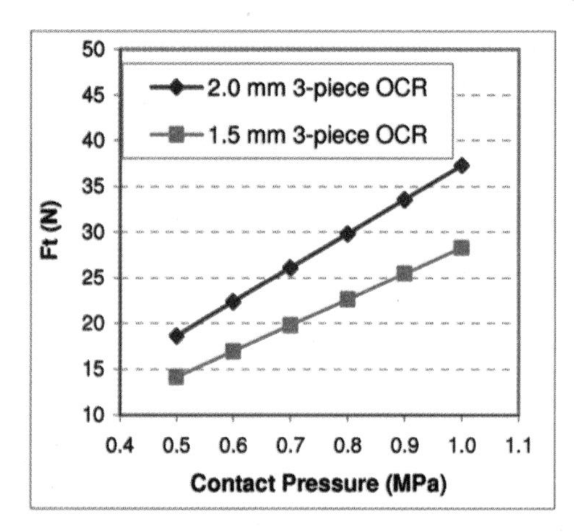

Fig 19.8 3-piece OCR Ft comparison for different widths.

19.3 Floating Liner Experiments

In the next items, influence of ring coating, OCR tangential load and ring width will be presented using measurements in the floating liner device. Floating liner characteristics are given in Table 19.1. The Musashi Institute developed floating liner has been extensively used for engine friction measurements, e.g.[19.7], [19.8]. Basically, the floating liner consists in a modified mono-cylinder engine, where the liner has vertical freedom and a load cell measures the piston/piston ring vertical load applied to the liner. See Fig. 19.9. Friction was measured in 5 operation conditions: 1500 rpm @ BMEP of 380, 500 and 630 kPa; 2500 rpm @ 500 kPa and 2500 rpm @ 500 kPa.

Table 19.1 Floating Liner Characteristics

Engine Type	Single Cylinder, 4 stroke SI gasoline
Displacement (liter)	0.499
Bore × Stroke (mm)	86 × 86
Compression rate	10 : 1
Crank ratio (L/R)	3.5
Operation Conditions (Speed and BMEP)	1500 rpm @ 380, 500, 630 kPa 2000 rpm @ 500 kPa 2500 rpm @ 500 kPa
Oil Type	SAE 5W-30 SL/GF-3 class
Cylinder Temp. [°C]	100 (at middle stroke]
Oil Temperature [°C]	85 (at main gallery]

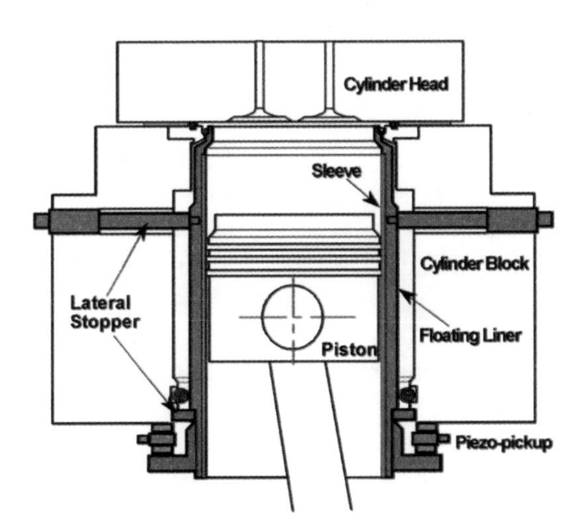

Fig. 19.9 Floating Liner Device.

The piston had a special large, 7.5 mm, top land due to the floating liner characteristics, but the same piston design, except for different groove dimensions when necessary, was used in the described tests. Pin offset is

1.0 mm. The same liner was also used for all tests. The liner was run for 10 h before the tests for break-in. The first chronological test had six test cycles because of the break-in for the new piston and new piston ring. The other tests had 4 test cycles. The first measurement cycle was just after assembly. Each successive cycle was after 15 min. Each cycle consisted of the 5 operation conditions. Each measurement data is the average of 128 combustion cycles. For each operation condition, combustion pressure and friction was measured at each crank angle. Fig. 19.10 shows, as example, a typical measurement at 1500 rpm, BMEP = 380 kPa.

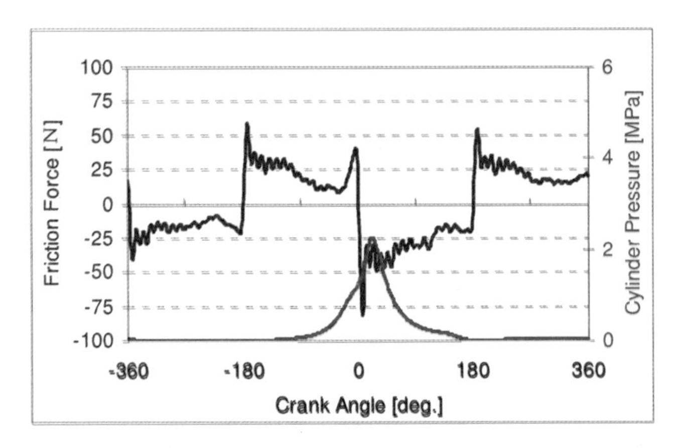

Fig. 19.10 Typical measurement at 1500 rpm, BMEP = 380 kPa.

For this chapter, the measured total Friction Mean Effective Power, FMEP, of the 3 last cycles of each pack/operation condition was averaged and is presented. The baseline pack (1.2/1.2/2.0 mm width) was the last tested to minimize the eventual break-in effects of the reduced friction packs.

As expected, for the same speed, friction increased with load, and for the same load, friction increased with speed. Table 9.2 shows the average measured values for the baseline ring pack. Despite some differences, the friction ranking among the tested ring packs was the same for the 5 operation conditions.

Table 19.2 Floating Liner Measurements/Baseline Ring Pack

Rpm	BMEP [kPa]	FMEP [kPa]	Friction %
1500	380	15.3	4.0
	500	16.1	3.2
	630	17.4	2.0
2000	500	16.6	3.3
2500	500	19.5	3.9

19.3.1 *OCR Tangential Load*

Figure 19.11 compares 3-piece OCRs with $P_0 = 0.6$ and $0.8\ N/mm^2$, Ft = 27.3 and 32.3 N, respectively. The figures shows the measured friction force at 1500 rpm, 380 kPa and the FMEP for the 5 tested conditions. The ring pack with OCR with $P_0 = 0.6$ presented about 32% lower friction at 1500 rpm, 21% at 2500 rpm.

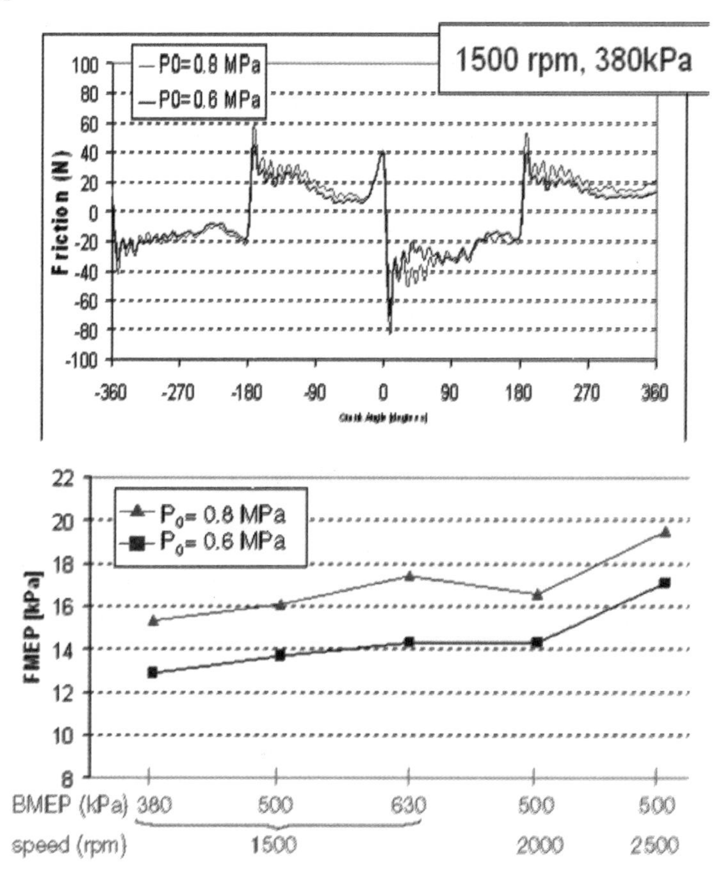

Fig. 19.11 Friction Force and FMEP with different OCR P_0.

19.3.2 *Low Width Ring Pack*

Lower friction can be achieved by the use of narrower rings. Especially for the top ring, its width defines the radial load applied to the bore, because the ring is loaded mostly by the combustion gas pressure. For 2nd rings, the tangential load is basically proportional to its width. For OCR, the use of narrower ring and segments, allowed a good conformability even with lower tangential loads. As example, the tested 1.5 mm ring has 40% more conformability than the 2.0 mm wide, even with 25% lower the tangential load. See Table 9.3. Friction of a 1.2/1.2/2.0 mm ring pack was compared with a 1.0/1.0/1.5 mm.

Table 19.3 Ring Pack Characteristics

	1.2/1.2/2.0 mm	1.0/1.0/1.5 mm
Top	1.2 mm GNS Asymmetrical Barrel Face Ft = 8.9 N	1.0 mm GNS Symmetrical Barrel Face Ft = 6.6 N
2nd	1.2 mm Taper face Gray Cast Iron Ft = 8.2 N	1.0 mm Taper face Ductile Cast Iron Ft = 7.0 N
OCR	2.0 mm 3-piece GNS P_0 = 1.0 MPa Ft = 33.3 N	1.5 mm 3-piece GNS P_0 = 0.6 MPa Ft = 15.7 N
Σ Ft	50.4 N	29.3 N

The low width ring pack presented approx. 28% less friction, the reduction was slightly more at low speed/low load, decreasing at higher speeds/loads. See Fig. 19.12. This reduction would mean a friction reduction ~1% of the engine BMEP. See Table 19.4. Similar fuel saving can be expected.

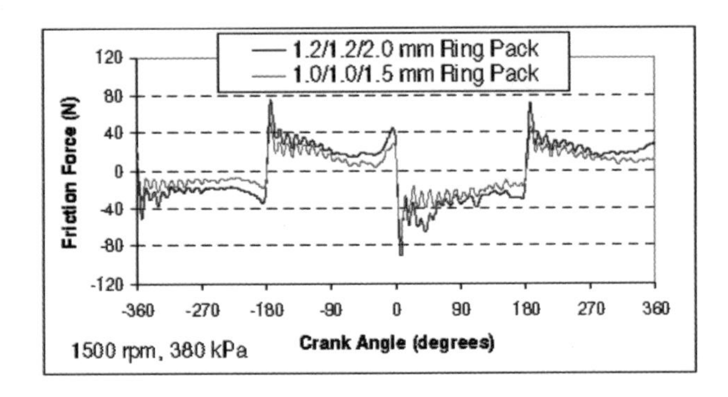

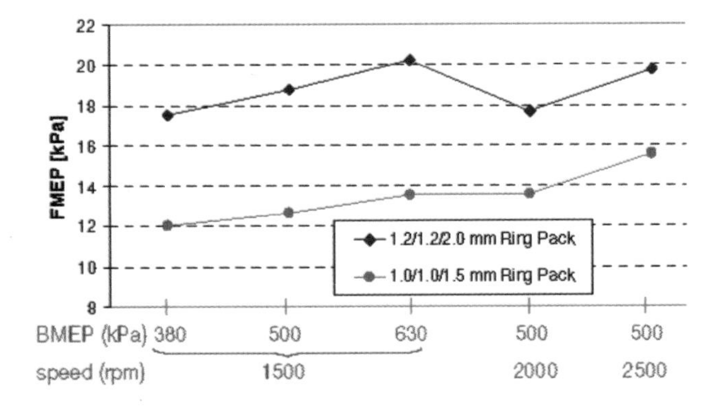

Fig. 19.12 Friction Force and FMEP comparison between 1.2/1.2/2.0 and 1.0/1.0/1.5 mm ring packs.

Table 19.4 Friction Reduction and Potential Fuel Saving

Speed	BMEP (kPa)	FMEP (kPa)		Friction Ratio (FMEP/BMEP)	BMEP gain
		Test 1	Test 2		
1500	380	17.5	12.0	4.6%	1.4%
	500	18.7	12.7	3.7%	1.2%
	630	20.2	13.5	3.2%	1.1%
2000	500	17.7	13.6	3.5%	0.8%
2500	500	19.8	15.6	4.0%	0.8%
Average				3.8%	1.1%

19.3.3 PVD Top Ring

The top and oil rings are usually coated with a wear resistant coating or surface treatment, usually Moly based, Chromium coated or Gas Nitrided Steel. CrN PVD is being introduced in the SI market due its exceptional wear resistance and lower friction. Fig.19.13 compares the friction coefficient of these coatings.

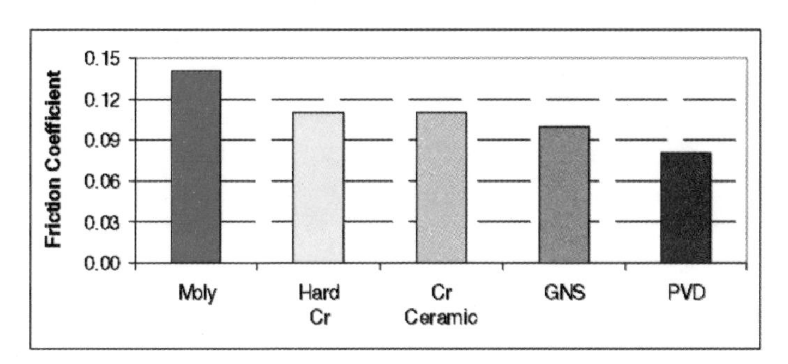

Fig. 19.13 Friction Coefficient of the different ring coatings.

GNS and PVD were compared in the Floating Liner device. An even narrower ring pack was used for this comparison. The ring pack width was 0.8/0.8/1.5 mm. Except for the top ring coating, ring characteristics are kept the same in the 2 ring packs. Fig.19.14 shows the measured friction force at 1500 rpm, 380 kPa and FMEP for the 5 tested conditions. The PVD top ring presented 10% less friction at 1500 rpm, 5% at 2500 rpm.

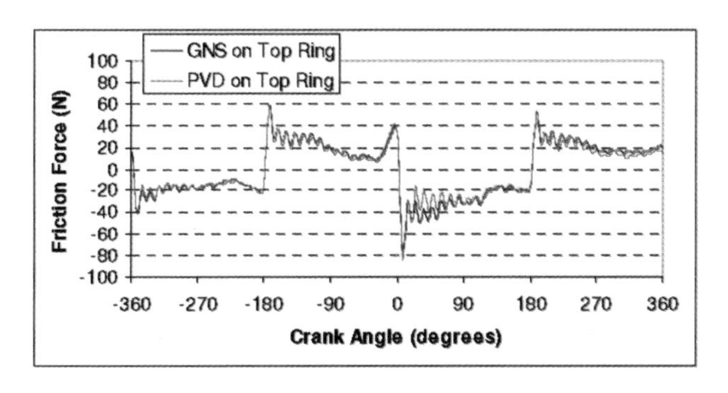

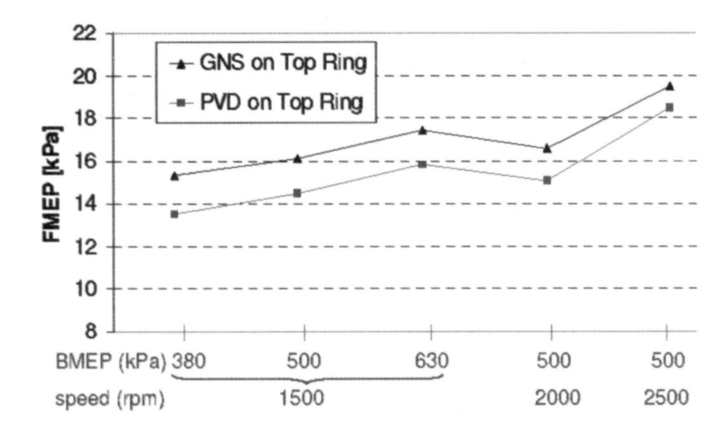

Fig. 19.14 Friction Force and FMEP comparison of the PVD coated top ring against the usual Gas Nitrided Steel, GNS.

19.4 Blow-By and Lube Oil Consumption

Due to the reduction of inertia, narrower compression rings are more axially stable, especially under high rpm, low load operation conditions. Spikes of Blow-by, ring fluttering, may occur in such operation conditions. Fig. 19.15 shows the improvement of Blow-By control with narrow ring packs. The baseline ring pack was 1.2/1.5/2.0 mm wide with 3-piece OCR with P_0 = 0.6 MPa. The low friction ring pack was 0.8/1.0/1.5 mm wide with 3-piece OCR with P_0 = 0.6 MPa.

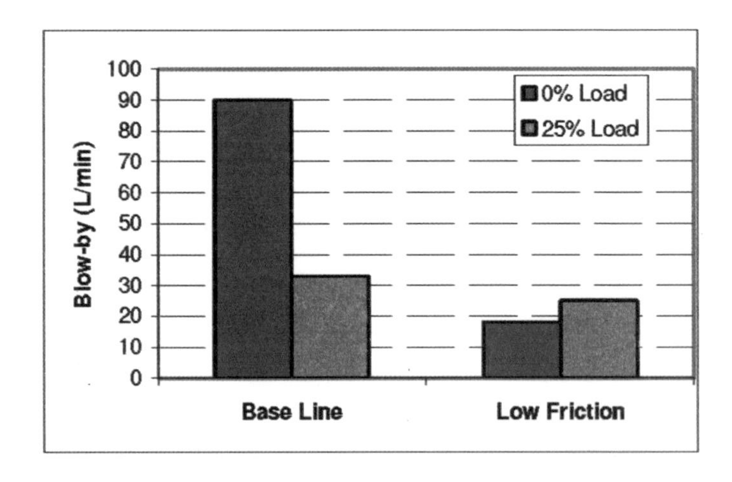

Fig. 19.15 Blow-by at 6000 rpm with Gasoline V6, 3.2L engine.

As discussed earlier, the 1.5 mm 3-piece OCR enables reduced Ft with similar contact pressure and similar conformability. Fig. 19.16 compares 2.0 and 1.5 mm 3-piece OCR in a SI 2.0L, 83 kW, engine. Top and 2nd rings were the same in the 2 tests. LOC was similar at full load and part load.

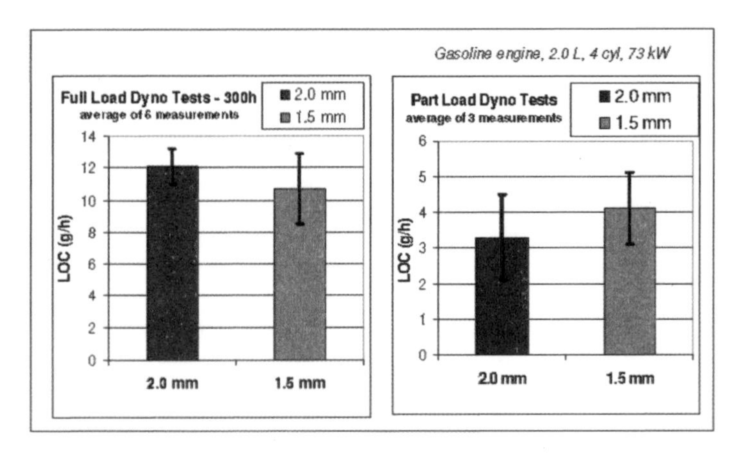

Fig. 19.16 2.0 and 1.5 mm 3-piece OCR comparison

The use of low friction ring packs with regular or rougher bore finishes may jeopardize lube oil consumption due to the increase of break-in time and thicker lube films. As example, fig. 19.17 compares a baseline ring pack (1.2/1.5/2.0 mm width) and a low friction pack (1.0/1.2/1.5 mm). The baseline 2-piece OCR had P_0 = 1.7 MPa, while the low friction had P_0 = 1.2 MPa.

	Bore Roughness	
	Baseline	Smooth
Rpk (μm)	0.3	0.1
Rk (μm)	1.4	0.5
Rvk (μm)	1.5	0.7

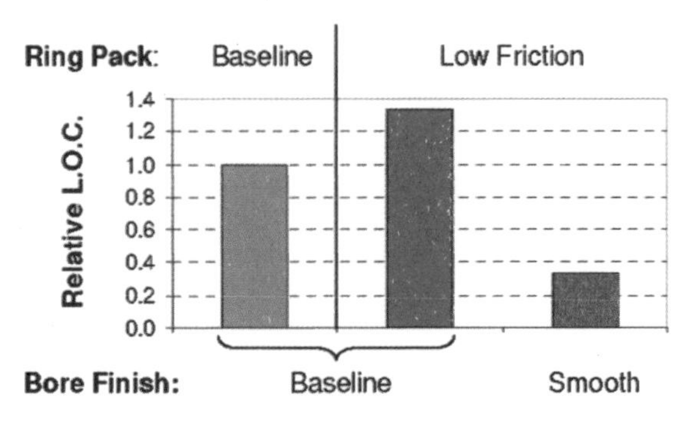

Fig. 19.17 L.O.C. at rated power, 2.0L GDI engine.

19.5 Conclusions

Based on the studies reported in this chapter, the following conclusions are drawn[19.10]:

Reducing the ring pack load, especially the OCR Ft, reduced the friction. Reducing the OCR P_0 from 0.8 to 0.6 MPa reduced in 15% the friction.

Changing the ring pack from 1.2/1.2/2.0 to 1.0/1.0/1.5 mm, with simultaneous reduction of the P_0 from 1.0 to 0.6 MPa, reduced the friction to 28%, which would mean about ~1% fuel saving.

The use of PVD on top ring reduced the measured friction by ~10%.

Low friction ring pack may require improved bore finishing to better lube oil control.

References

[19.1] Schelling, Freier, "Factors influencing the friction power of pistons" - MAHLE publication, circa 1995.

[19.2] Tomanik et. al., "Reduced Friction Power Cell Components" - paper SAE2000-01 -3321.

[19.3] Basshuysen, R., Schafer, R, "Internal combustion Engine Handbook", SAE International.

[19.4] Jocsak J. et. al., "The effects of cylinder liner finish on piston ring-pack friction" – ASME paper ICEF 2004-952.

[19.5] Jocsak I. et. al., "The Characterization and Simulation of Cylinder Liner Surface Finishes" – ASME paper ICES 2005-1080.

[19.6] Tomanik, E., "Piston Ring Conformability in a Distorted Bore" – paper SAE 960356

[19.7] Sato, O. et. al., "Improvement of Piston Lubrication in a Diesel Engine by Means of Cylinder Surface Roughness" – SAE paper 2004-01-0604

[19.8] Nakayama, K. et. al., "The Effect of Crankshaft Offset on Piston Friction Force in a Gasoline Engine" SAE paper 2000-01-0922.

[19.9] Ferrarese, A.; Rovai, F., "Oil Ring Design Influence on Lube Oil Consumption of SI Engines", ASME paper ICEF 2004-868

[19.10] Tomanic, Eduardo, Ferrarese, Andre "Low Friction Pack for Gasoline Engines", ASME paper ICEF 2006-1566.

Nomenclature

BMEP: Brake Mean Effective Pressure [kPa]
FMEP: Friction Mean Effective Pressure [kPa]
Ft: Ring Tangential Load [N]
h: Ring width [mm]
hs: segment width [mm]
IS: distance between segments [mm]
OCR: Oil Control Ring
P_0: Ring Unitary Pressure [MPa]
PVD: Physical Vapor Deposition
w: Expander Wire Width [mm]

20

Investigations into Characterization and Simulation of Cylinder Linear Surface Finishes

20.1 A General Review

This chapter presents an investigation into the characterization and performance prediction of different cylinder liner surfaces commonly used in modern internal combustion engines. The topography of liner specimens was studied, and the friction and wear between a piston ring and each liner surface was measured using a horizontal reciprocating bench tester. The load, speed, and lubricant supply during testing were chosen to ensure that the piston ring and liner operated primarily in a mixed lubrication regime.

A computer program was developed to model the performance of the piston ring and liner specimens under the conditions observed during the reciprocating bench test. The Greenwood and Tripp statistical asperity contact model was employed to describe the rough surface contact behavior between the liner specimen and piston ring. Two different methods of characterizing the liner specimen surface roughness and determining the inputs required for the Greenwood and Tripp model from the surface measurements were considered. The friction observed experimentally was compared to the friction predicted by the model, and the ability of the model to predict the absolute friction for a given surface and the relative difference in friction between two different surfaces was investigated.

20.2 Introduction

With an increasing demand for greater durability and decreased oil consumption in internal combustion engines, it has become necessary to optimize cylinder liner surface finish design. New cylinder liner surface finishes are being used in production engines that are specifically designed to reduce lube oil consumption, friction, and wear[20.1]. In general, these surface finishes tend to be smoother than conventional cylinder liner finishes. Fig. 20.1 shows typical fax films and *Rk* values of different surface finish currently in production in diesel engines[20.2]. Referring to Fig. 20.1, it can be seen that new surface finishes, such as those produced by slide and plateau honing, are smoother than a standard honed surface but still contain relatively deep valleys. In order to study the influence of the new surface finishes on piston ring-pack performance, it is necessary to develop methods to characterize and model the surface interactions under mixed lubrication.

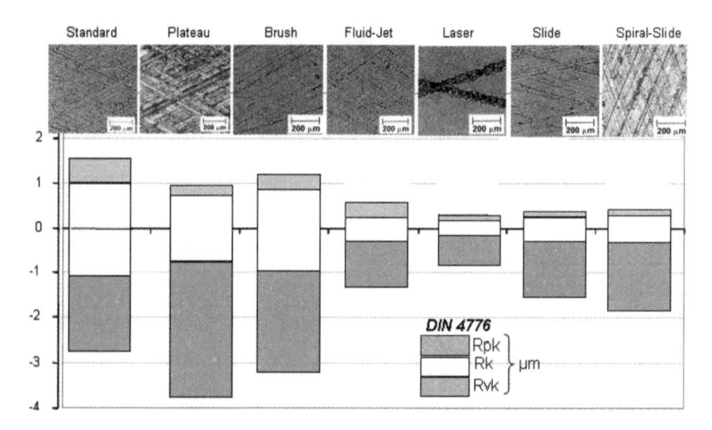

Fig. 20.1 Fax film and *Rk* values of the different cylinder liner surface finishes currently in production

There have been numerous computer simulations developed to study the behavior of the piston ring pack[20.3-20.5]. Most of the models rely on a statistical description of surface roughness as a practical means of incorporating its effects on ring-pack performance. The Greenwood and Tripp model is a statistical asperity contact model that has been commonly used in ring-pack simulations to describe rough surface contact[20.6]. The effect of surface roughness on oil flow is often addressed by the use of an average Reynolds equation, which adjusts the smooth surface solution of the Reynolds equation for the effects of surface roughness using statistical flow factors[20.7]. However, simple assumptions are usually made to describe liner surface roughness, and there is limited information in the literature on the statistical characterization of actual cylinder liner surfaces[20.8-20.10].

This study attempts to develop and evaluate two methods of determining surface roughness parameters using actual surface data from two common cylinder liner finishes. Although in this study, the methods were not applied to study ring-pack performance within an engine, both methods of rough surface characterization are readily adaptable to axis-symmetric piston ring-pack simulations that employ a stochastic description of surface roughness. Therefore, both the experimental data and the methods of surface characterization will potentially enable better prediction of the effects of ring-pack performance under engine operating conditions.

20.3 Statistical Parameters

Three statistical parameters that can be used to describe the nature of a probability distribution are its standard deviation (σ), skewness (*Sk*), and kurtosis (*Ku*).

$$\sigma = \sqrt{\int_{-\infty}^{\infty} y^2(z)\phi(z)dz} \tag{201}$$

$$Sk = \frac{1}{\sigma^3} \int_{-\infty}^{\infty} y^3(z)\phi(z)dz \tag{20.2}$$

$$Ku = \frac{1}{\sigma^4} \int_{-\infty}^{\infty} y^4(z)\phi(z)dz \tag{20.3}$$

In Eqs. (20.1), (20.2), and (20.3), $y(z)$ is the distance from the mean of the distribution, and $\phi(z)$ is the probability distribution function of the distribution. Skewness is a measure of the distance between the mean and mode, the asymmetry, of a distribution. Kurtosis is a measure of the "peaked ness" of a distribution. A Gaussian probability density function is symmetric about its mean (zero skewness), and has a kurtosis value of three.

20.4 Surface Characterization

The Greenwood and Tripp model was chosen to describe asperity contact between the piston ring and cylinder liner surface. This model calculates micro-contact and pressures that arise when two rough surfaces approach each other. According to the Greenwood and Tripp model, nominal asperity contact pressure between two rough surfaces can be expressed using Eqs. (20.4) - (20.6).

$$P_c\left(\frac{d}{\sigma}\right) = k' E' \int_{\frac{d}{\sigma}}^{\infty} \left[z - \frac{d}{\sigma}\right]^{2.5} \phi(z)dz \tag{20.4}$$

Where,

$$k' = \frac{16\sqrt{2}\pi}{15}(\eta\beta\sigma)^2\sqrt{\frac{\sigma}{\beta}} \tag{20.5}$$

In Eqs. (20.4) and (20.5), P_c is the nominal asperity contact pressure between the two surfaces, d is the mean separation of the two surfaces, η is the asperity density per unit area, β is the asperity peak radius of curvature, $\phi(z)$ is the probability distribution of asperity heights, and z_s is defined as the offset between the asperity height mean and the surface height mean. The composite Young's modulus and the composite standard deviation of asperity heights used by the Greenwood and Tripp model are defined by eqs. (20.6) and (20.7), respectively.

$$E' = \frac{1}{\frac{1-\nu_1^2}{E_1} + \frac{1-\nu_2^2}{E_2}} \tag{20.6}$$

$$\sigma = \sqrt{\sigma_1^2 + \sigma_2^2} \tag{20.7}$$

In equations (20.6) and (20.7), E_1 and E_2 and v_1 and v_2 are the respective Young's modulus and Poisson's ratio of cylinder the two contacting surfaces, and σ_1 and σ_2 are the standard deviation of asperity heights of the two surfaces. The Greenwood and Tripp model assumes that contact is elastic, and the asperities are parabolic in shape and identical on the contacting surfaces.

In reality, contact between the ring and cylinder liner in an engine may include some plastic deformation, especially during the initial break-in period. As a result, ring to liner contact may enter an elastic-plastic regime[20.11]. However, as concluded by other researchers, it is assumed that contact between the cylinder liner and ring can be treated as elastic even though plastic deformation may be occurring[20.12]. When an asperity begins to deform plastically, the load carried by that asperity will be distributed to other asperities at a similar height that are still undergoing elastic deformation. In addition, Greenwood and Tripp have shown that the nominal asperity pressure calculated for elastic contact is very similar to that calculated for plastic deformation. It should be noted that the effect of oil film and oxide layers on the surfaces of the ring and liner may also play an important role in asperity contact, and have not been considered in this study.

It was also assumed that probability distribution of the cylinder liner surface roughness is representative of the combined distribution of the cylinder liner and ring surfaces, since it is generally the rougher and softer of the two surfaces. Therefore, $\phi(z)$ is estimated using the cylinder liner probability distribution.

Although the Greenwood and Tripp model has been used extensively in the literature, there is limited experimental data on the values of the Greenwood and Tripp asperity contact parameters. It is usually assumed that the standard deviation of asperity heights is equivalent to the roughness of the surface, and the product $(\sigma\beta\eta)$ takes on values between 0.03 and 0.05.

In the present study, two different methods of characterizing surface asperities and determining the Greenwood and Tripp asperity contact parameters in Eqs. (20.4) and (20.5) are investigated. Both methods determine the Greenwood and Tripp asperity contact parameters through the analysis of a 2-D profilometer trace of the liner surface, an example of which is shown in Fig. 20.2. Waviness was removed from the profile traces considered in the study with the use of an Rk filter. The Rk filter, as defined by DIN 4776, is designed to reduce the overshoot in the roughness profile that occurs on both sides of a deep valley. The Rk filter was preferred over the use of other filtering methods because it generally gives a better representation of roughness for cylinder liner surfaces, which usually contain deep valleys, as is shown in Fig. 20.1. An underlying assumption is that the surface roughness trace provided is a good representation of the liner surface.

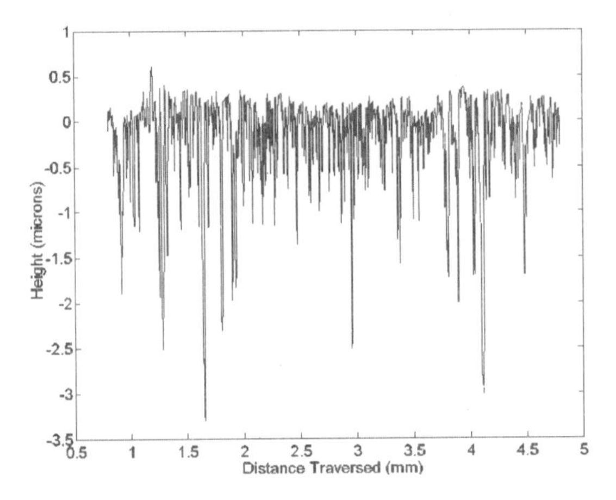

Fig. 20.2 Surface profilometer trace

In both methods, an asperity peak is defined as a data point from the surface roughness trace that is higher than either of its adjacent points, as is stated in Eq. (20.8).

$$asp(i) \equiv h(i) > (h(i-1), h(i+1)) \tag{20.8}$$

In Equation (20.8), $h(i)$ is the height of the data point i in the profilometer trace, and $asp(i)$ is the height of an asperity peak. This definition of an asperity is influenced by the distance between data points in the profilometer trace. Therefore, it is important to choose a sampling rate high enough that asperity peaks can be properly identified. It is also necessary to estimate the number of asperities per unit area, or asperity summit density, from the asperity peak density per unit length obtained from the 2-D profilometer trace. Based on both experimental and analytical analysis, the 3-D density of asperity summits is related approximately to the 2-D density of asperity peaks by Eq. (20.9)[20.13, 20.14].

$$\eta \approx 8\left(\frac{2D\ asperity\ density}{length}\right)^2 \tag{20.9}$$

The two methods of determining the Greenwood and Tripp asperity contact parameters considered in this study differ only in their treatment of the probability density of asperity heights.

20.5 Gaussian Asperity Height Distribution

The first method employed to obtain the distribution of asperity heights from actual surfaces, hereafter referred to as the Gaussian Distribution (GD) method, was previously proposed by one of the current authors and has also been extended to 3-D measurements[20.10, 20.13]. If only the asperities above the mean of the surface are considered, the GD method assumes that the distribution of asperity heights is Gaussian in nature. An asperity above the mean plane of the surface in a profilometer trace is identified using Eq. (20.8). Combined with the assumption of a Gaussian distribution, only the mean height and the standard deviation of the asperities above the mean plane are required to uniquely define the asperity height distribution. The Gaussian distribution is defined in Eq. (20.10).

$$F(z) = \frac{1}{\sqrt{2\pi}\sigma} exp\left(\frac{-(z - z_s)^2}{2\sigma^2}\right) \tag{20.10}$$

In Eq. (20.10), all parameters pertain to the distribution of asperities with peak heights above the mean plane of the surface. Since the Gaussian distribution can be described analytically, this method has the inherent advantage of being less computationally involved. For liner surface finishes considered in this study, the distribution of asperities above the surface mean was not strictly Gaussian in nature, but more closely approximated Gaussian behavior than the complete asperity height distribution.

20.6 Asperity Height Distribution

The distribution of asperity heights above the mean of the surface is not necessarily Gaussian in nature. With the availability of an accurate surface profilometer trace, it is possible to construct the complete distribution of asperity heights using the actual surface roughness data. Thus, an alternative method proposed by the authors to determine the distribution of asperity heights, referred to hereafter as the Actual Distribution (AD) method, employs the measured surface data to infer the shape of the asperity height distribution. The methodology behind the construction of the asperity height distribution is described below.

The population of asperities in a surface roughness trace, and the respective height of each asperity peak, is identified based on the definition given in Eq. (20.8). The total range of asperity peak heights is subsequently divided into equal height subintervals. The number of subintervals used is dependent on the number of asperities present within the trace. In the present study, the number of subintervals was set equal to the square root of the total number of asperities, rounded to the nearest integer[20.15].

Each asperity is assigned to the appropriate subinterval based on its height, and a frequency distribution of asperity heights is computed. The frequency distribution is assumed to be a discrete representation of the actual probability distribution of asperity heights. Fig. 20.3 displays the surface height and asperity height distributions calculated using the AD method for a typical production cylinder liner finish. The noise present in the asperity height frequency distribution in Fig. 20.3 does not affect the calculation of asperity contact because the Greenwood and Tripp model integrates the frequency distribution when calculating asperity contact pressure, as is illustrated in Equation (20.4). Although the AD method is more computationally involved, it removes the need to assume the shape of the distribution of asperity heights.

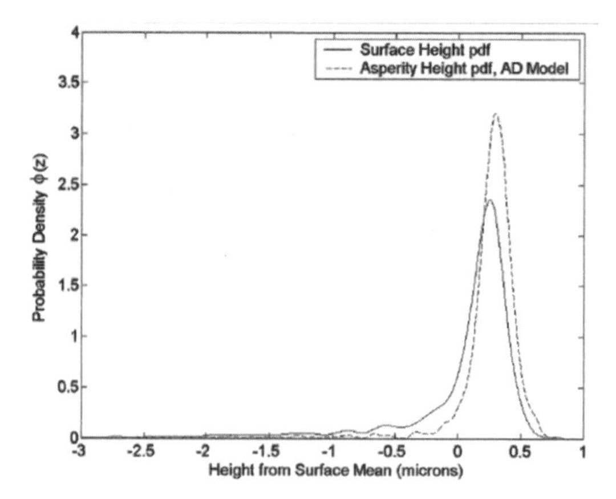

Fig. 20.3 Comparison of surface and asperity height probability density functions

In analyzing typical cylinder bore finishes, a similarity between the actual asperity height and surface height probability distributions was observed. In general, the behavior of the two probability distributions was very similar when a wide variety of cylinder bore finishes were analyzed. Figs 20.4 to 20.6 display the relationship between the standard deviation, skewness, and kurtosis of the surface height and asperity height distributions for a variety for SI cylinder bore finishes in new condition, and after a 15 hours dynamometer test

From examining Figs 20.4-20.6, there appears to be a strong correlation between the nature of the surface height distribution and the asperity height distribution in the case of typical SI cylinder bore finishes. This finding is significant in that it does not agree with the assumption made by Greenwood and Tripp, and other researchers, that the complete distribution of asperity heights tends towards a Gaussian distribution regardless of the shape of the underlying surface height distribution[20.6, 20.16].

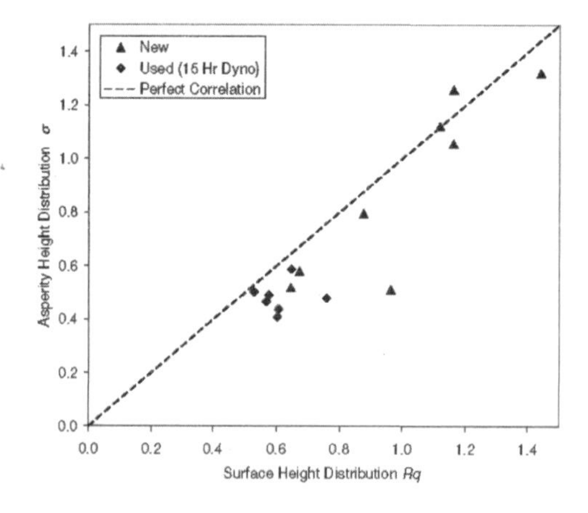

Fig. 20.4 Comparison of surface height and asperity height standard deviations of SI cylinder bore finishes

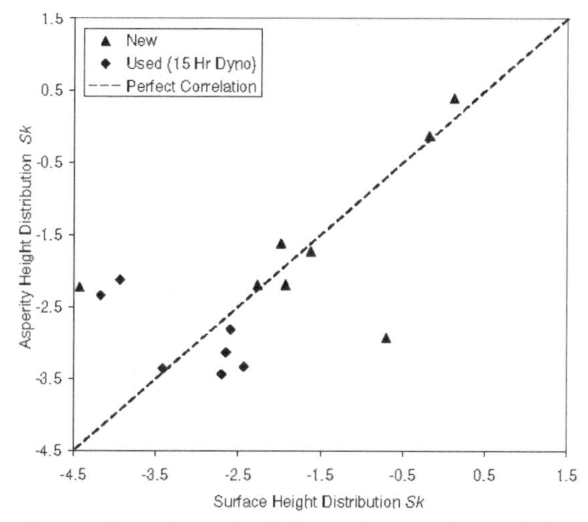

Fig. 20.5 Comparison of surface height and asperity height skewness of SI cylinder bore finishes

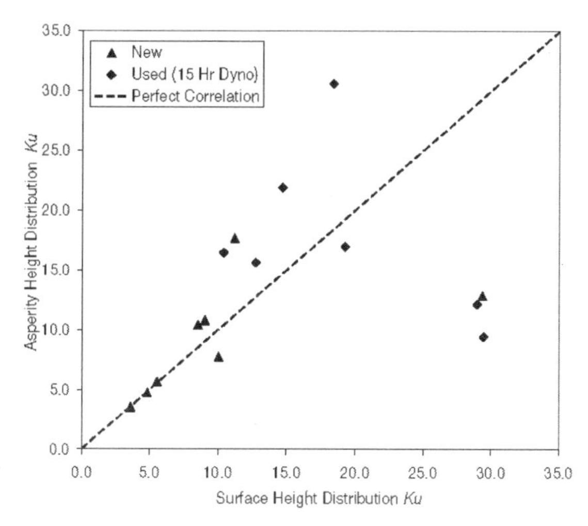

Fig. 20.6 Comparison of surface height and asperity height kurtosis of SI cylinder bore finishes

20.7 Cylinder Liner and Piston Ring Surfaces

The intention of this study was to examine surface finishes that are representative of production cylinder bores. Thus, a typical plateau honed liner and a typical slide honed liner were chosen. A steel ring specimen, representative of a production compression ring, was used in conjunction with both production cylinder bores. Table 20.1 summarizes the finishes of the two liner surfaces used in the experiment. Table 20.2 summarizes the input parameters predicted for the Greenwood and Tripp model for the two surfaces considered. The standard deviation of asperity heights for the piston ring was taken as 0.15 μm.

Table 20.1 Characteristics of liner surfaces considered

	Plateau Honed	Slide Honed
R_q [μm]	0.84	0.49
R_{pk} [μm]	0.25	0.28
R_k [μm]	0.74	0.39
R_{vk} [μm]	1.85	1.47
Sk	–2.79	–3.85
Ku	13.82	25.99
Honing angle	35° – 45°	35° – 45°

Table 20.2 Calculated Greenwood and Tripp Parameters

	Plateau		Slide	
	GD Method	AD Method	GD Method	AD Method
σ [μm]	0.18	0.64	0.11	0.29
β [μm]	29.2	29.2	27.6	27.6
$\eta \times 10^{10}$ [m^{-2}]	1.05	2.10	1.38	2.50
Z_s [μm]	0.29	0.01	0.16	0.06
Sk	1.11	–2.95	1.04	–5.35
Ku	7.82	17.16	4.41	52.53

20.8 Asperity Contact Pressure Comparison

A comparison was made between the contact pressure predicted by the GD method and AD method. Figs 20.7 and 20.8 show the contact pressure predicted by the Greenwood and Tripp model using both methods for a slide honed and plateau honed surface, respectively.

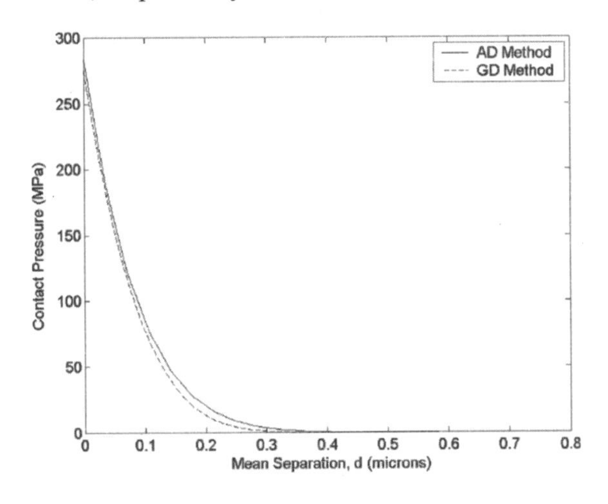

Fig. 20.7 Comparison of predicted asperity contact pressure for slide honed surface

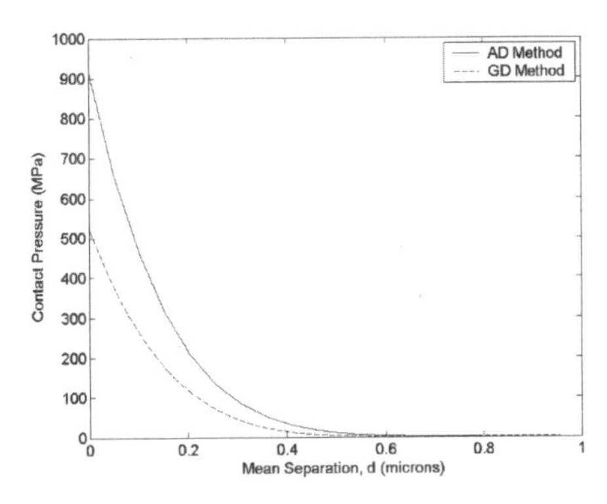

Fig. 20.8 Comparison of predicted asperity contact pressure for plateau honed surface

By examining Figs 20.7 and 20.8, it is seen that the asperity contact pressure predicted by the Greenwood and Tripp model using the AD and GD methods are different, especially for the plateau honed surface. The differences seen in the predicted contact pressure are a direct result of the way in which the asperity height probability distribution is calculated in the two methods. It should be stressed that it is not the shape of the probability distribution alone that is affecting predicted contact pressure. The underlying populations of asperities used in the GD and AD methods are different, leading to different values of asperity density, asperity peak radius of curvature, and the standard deviation of asperity heights. By examining Eq. (20.5), it can be seen that these values in turn affect predicted contact pressure.

20.9 Experiment

Friction and wear measurements were obtained with the use of a ring/liner reciprocating tester. A CETR UMT-2 reciprocating tester was used in the current study. Normal load is applied using a closed-loop servo mechanism, and normal load and friction forces are measured with strain-gages. Details of the experimental setup are shown in Fig. 20.9, and listed in Table 20.3. The tests were conducted using a procedure developed by MAHLE to accelerate ring wear and facilitate its evaluation. The accelerated ring wear test is designed to simulate the extreme ring conditions of heavy duty diesel (HDD) engines.

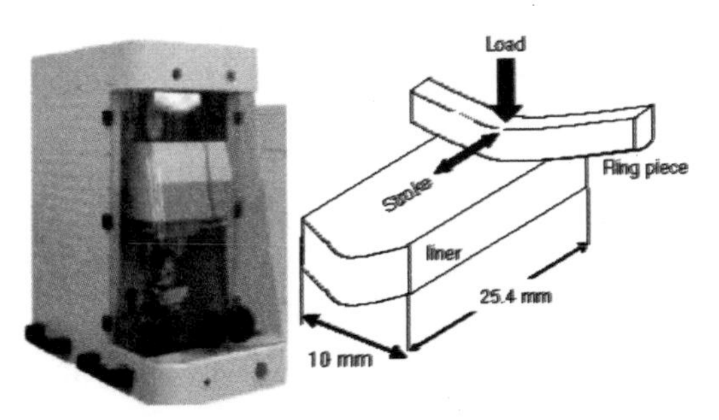

Fig. 20.9 CETR UMT-2 reciprocating tester and schematic of reciprocating test

Table 20.3 Operating conditions and geometric parameters

Parameter	Value	Unit
Ring Stroke	10	mm
Ring Specimen Length	10	mm
Ring Face Profile	Skewed Barrel	-
Ring Young's Modulus	350	GPa
Liner Young's Modulus	100	GPa
Lubricant	SAE 30W	-
Lubricant Temperature	25	°C
Applied Pressure, P	120	bar
Reciprocating Sped, N	900	rpm
a_{bc}	0.1	-

The duration of the test is four hours with an applied ring pressure of 120 bar, a rotational speed of 900 rpm, and the liner specimen flooded with 20 ml of SAE 30 Texaco Regal oil. To accelerate ring wear, the oil is doped with 0.57 g of 0.05 μm diameter Alumina particles and 1.9 g of quartz 600 mesh per liter of oil. The testing procedure is as follows.

For the first five seconds the ring and liner are broken-in by varying the applied load from 75 to 360 N, at which point a constant load of 360 N is maintained for the next four hours. The accelerated wear test is used to rank the wear behavior between ring coatings and liner materials and finishes, and a minimum of six replications is done for each ring/liner combination. For this study, the liner specimens were cut from regular production Pearlitic cast iron cylinder liners from an HDD engine with a 130 mm diameter bore. The top ring specimens were cut from 3 mm wide production gas nitride steel rings with a CrN PVD coating.

20.10 Lubrication Model

A computer model was developed to predict the friction produced by the ring specimen during the reciprocating tests.

The model employs the Greenwood and Tripp model to calculate asperity contact pressure, with both the GD and AD method for determining asperity height distribution included. The model is based on MIT's axis-symmetric ring-pack friction model, which simultaneously solves the Reynold's equation in the lubricant flow direction between the ring and liner, and a quasi-steady force balance on the piston ring. The model considers unsteady inlet and exit conditions and hydrodynamic, mixed, and boundary lubrication regimes[20.3].

The model layout for a cross-section of the ring specimen is detailed in Fig. 20.10.

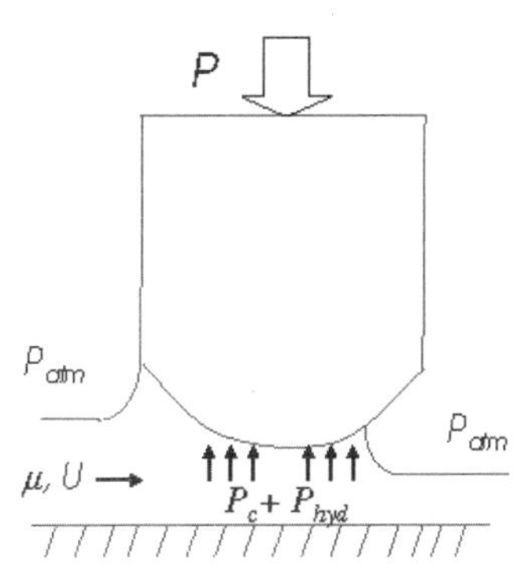

Fig. 20.10 Schematic of model layout for ring specimen cross- section

In Fig. 20.10, P is the unit pressure applied to the ring specimen, U is the instantaneous sliding velocity of the ring, μ is the lubricant dynamic viscosity, P_{atm} is atmospheric pressure, and P_c and P_{hyd} are the pressures generated by asperity contact and hydrodynamic shear, respectively. In the bench test the oil covers the entire liner specimen, so it was assumed in the analysis that there was sufficient lubricant supply to flood the inlet of the ring specimen throughout the entire stroke of the ring. The total friction force experienced by the ring specimen is the summation of hydrodynamic and boundary contact friction, as is shown in Eq. (20.11).

$$F_f = \int_{\text{ring face}} a_{bc} P_c dA + \int_{\substack{\text{ring} \\ \text{wetted area}}} \left[C_1 \frac{\mu U}{h} - C_2 \frac{h}{2} \frac{dP}{dx} \right] dA \qquad (20.11)$$

In Eq. (20.11), F_f is the total ring friction, P_c is the nominal asperity contact pressure predicted by the Greenwood and Tripp model, a_{bc} is the coefficient of friction for boundary lubrication, h is the local oil film thickness, U is the piston sliding speed, and dP/dx is the pressure gradient in the flow direction. C_1 and C_2 are functions involving average flow factors developed by Patir and Cheng that account for the effect of surface roughness on lubricant flow[20.7].

The statistically derived flow factors developed by Patir and Cheng adjust the solution of the Reynolds equation for smooth surfaces for the effect of surface roughness. However, these flow factors are for isotropic surfaces with a Gaussian distribution of surface roughness, and plateau and slide honed liners are anisotropic and display highly non-Gaussian roughness. A detailed analysis of the effects of cylinder bore surface texture, including honing grooves, on oil flow is beyond the scope of this study. As has been done in the literature, a simple truncation procedure was used to determine the effective Gaussian roughness of a non-Gaussian surface[20.17, 20.18]. The truncation procedure assumes that the deep valleys created by honing contain stagnant fluid and can be neglected for the purposes of flow resistance.

The instantaneous total coefficient of friction at time t, $a(t)$, is defined in Eq. (20.12).

$$a(t) = \frac{F_f(t)}{L} \tag{20.12}$$

The average coefficient of friction observed over one cycle of ring motion, a_T, is defined by Eq. (20.13).

$$a_T \equiv \frac{N}{60} \int_0^{60/N} a(t)dt \tag{20.13}$$

In Eq. (20.13), N refers to the rotational speed of the crankshaft in revolutions per minute.

The operating conditions and geometric parameters of the reciprocating tester required as model inputs for this study are summarized in Table 20.3. Both the lubricant temperature and coefficient of boundary contact friction values were estimated. It may be noted that the effects of the alumina and quartz on oil viscosity were also neglected.

A subtlety that underlies the modeling of rough surfaces in a mixed lubrication regime is the choice of the surface reference height at which the fluid is assumed stagnant relative to the surface, and a no-slip condition is valid. The fluid film height between the piston ring and cylinder is defined as the separation between the rough surface reference heights of two rough surfaces. In the following study, the mean height of the rough surface was used, and assumed to be the appropriate reference height.

20.11 Results

20.11.1 *Friction Evolution and Wear*

In addition to data on friction, the accelerated wear test also provided information on the wear rates experienced by both the ring and liner specimen. As the ring and liner specimens wear during the test, the measured friction will typically reduce. Fig. 20.11 shows the friction variation for both the plateau and slide honed surfaces for the duration of the test.

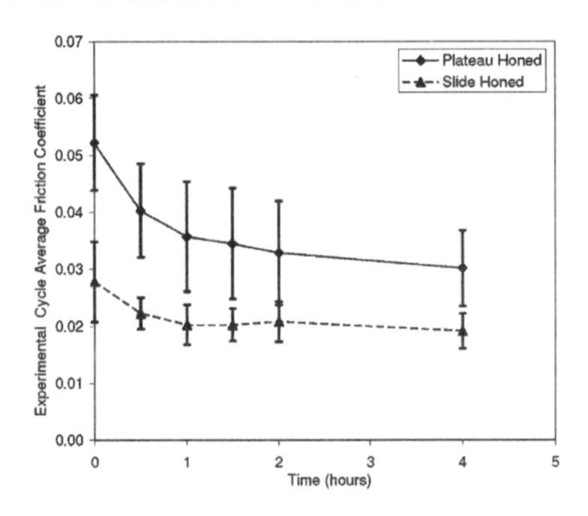

Fig. 20.11 Variation of friction coefficient along test

Both ring and liner wear at the conclusion of the four hours test, as measured by profile variation, was very low. This is thought to be a consequence of the PVD coating used on the ring specimens. The ring wear volume per unit length is listed in Table 20.4. Fig. 20.12 shows typical ring profiles, new and after test, and the ring and liner maximum wear thickness.

Table 20.4 Measured ring wear volume

	Plateau		**Slide**	
Wear Volume [μm^2]	30	-	9	-

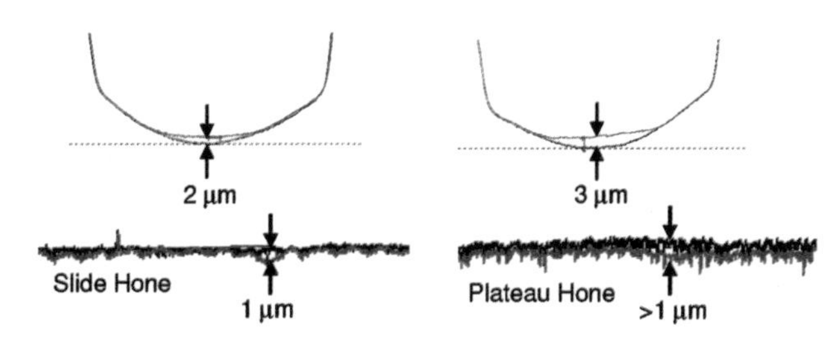

Fig. 20.12 Typical new versus worn ring/liner profiles

The difference in ring wears predicted by the simulation for the two surface finishes was estimated using the empirical model developed by Archard[20.19]. Archard's wear model is given in Equation (20.14).

$$W = \left[\frac{K}{HV}\right] L\,.\,\Delta s \tag{20.14}$$

In Eq. (20.14), W is the amount of wear, HV is the Vickers surface hardness, L is the normal load, Δs is the distance slid, and K is the wear coefficient. It is assumed that only the normal load carried by asperity contact, and not the load carried by hydrodynamic pressure, contributes to wear[20.20]. Further, since the ring and liner materials were kept constant in the current study, it was also assumed that both K and HV were the same for all experiments. The total load carried by asperity contact was used to estimate L at each point during the cycle. The total wear per cycle can be estimated using Eq. (20.15).

$$W = \left[\frac{K}{HV}\right] \int_{cycle} \left[\int_{ring\ face} P_c dx \right] ds \tag{20.15}$$

In Eq. (20.15), W is the total wear per revolution. Determining predicted wear for the test requires the determination of the unknown constant K, and including the wear evolution of the ring profile in the simulation. As the ring wears its running face will generally flatten, as is seen in Fig. 20.11, leading to lower asperity contact pressures as the total ring load is carried more uniformly along the ring running face. This analysis is beyond the scope of this study.

However, a rough estimate of the ratio of initial wear predicted by the model between the plateau and slide honed surfaces can be obtained. Table 20.5 shows the wear per cycle divided by (*K/HV*) predicted by the model for both surfaces, and the standard deviation refer to the range of predictions for three different surface roughness traces of the same surface.

Table 20.5 Predicted initial wear

	Plateau		**Slide**	
	GD Method	AD Method	GD Method	AD Method
(HV/K).W_{cycle} [Nm]	2.70	3.87	0.73	0.88
Std. Dev. [nm]	$\pm$0.04	$\pm$0.52	$\pm$0.09	$\pm$0.04

The ratio of observed ring wear volume for the plateau and slide honed surfaces, and the ratio of the initial wear predicted by the model for the two surfaces are shown in Eqs. (20.16) and (20.17), respectively.

$$\frac{W_{T,PH}}{W_{T,SH}} = 3.33 \tag{20.16}$$

$$\left[\frac{\mathrm{K/HV}}{\mathrm{K/HV}}\right]\frac{\mathrm{W_{P,PH}}}{\mathrm{W_{P,SH}}} = \frac{\mathrm{W_{P,PH}}}{\mathrm{W_{P,SH}}} = 3.7, 4.4 \tag{20.17}$$

In Eqs. (20.16) and (20.17), the subscripts *T* and *P* refer to test and predicted, respectively, and *PH* and *SH* refer to the plateau and slide honed surfaces, respectively. The two values given in Eq. (20.17) are the predicted wear ratios using the GD and AD methods, respectively, in the model. By comparing Eqs. (20.16) and (20.17), it can be seen that the both the model predictions and experimental results associate higher wear with the plateau honed surface. The ratios of wear predicted by the model for the two surfaces are slightly higher than the ratio of ring wear observed during the experiment. This may, at least partially, be a result of only considering the unworn profiles in determining the model wear predictions.

To minimize the influence on measured friction of both the change in the ring profile and liner finish during the test, only the first five seconds (75 cycles) were used to compare the friction measured experimentally with the friction predicted through simulation.

20.11.2 *Friction*

The lubrication model results were compared with the experimental results obtained from the reciprocating tester. The experimental data was obtained with an acquisition rate of 345 Hz, which proved to be too low to obtain a good representation of the instantaneous friction evolution along the stroke. Further experiments, without oil doping and at slower speeds, are planned in order to have a better friction mapping along the stroke.

As a result, only the average friction coefficients obtained from simulation and experimentally from the reciprocating tester were compared. Both the GD and AD methods of asperity height distribution characterization were considered. Three separate surface roughness traces from different locations on the liner specimen of both the slide honed and plateau honed surface were analyzed. Thus, the sensitivity of the model results to different measurements of the same surface was considered. In Fig. 20.13, the cycle average friction coefficient measured experimentally is compared with the model predicted average friction coefficient. In Fig. 20.13, error bars are included to illustrate the variation in the predicted cycle average friction coefficient using different traces of the same surface.

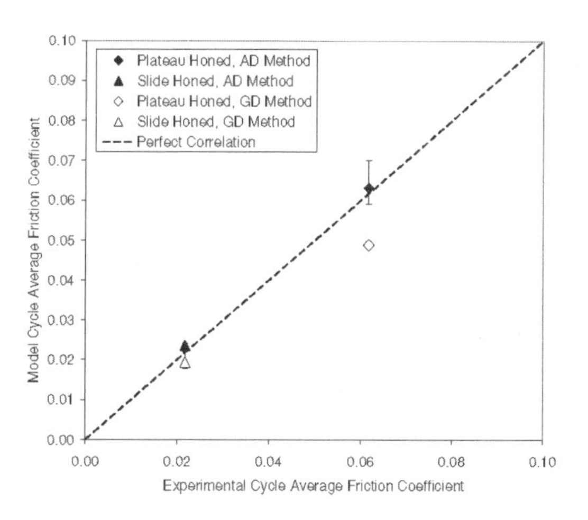

Fig. 20.13 Comparison of cycle average friction coefficient

Examining Fig. 20.13, it is evident that both the GD and AD methods accurately predict the cycle average friction coefficient for the slides honed surfaces. However, the AD method also accurately predicted the cycle average friction coefficient for the plateau honed surfaces, while the GD method underestimated it. The difference in the predicted coefficients of friction between the GD and AD methods correlate directly with the difference in predicted asperity contact pressures shown in Figs 20.7-20.8. It should also be noted that the variation in the predicted cycle average friction coefficient is significantly larger for the AD method when analyzing different surface traces from the plateau honed liners.

In general, the AD method is much more sensitive to variations in the surface roughness trace than is the GD method. This is a direct result of the difference in how each method calculates the distribution of asperity heights. The AD method constructs the asperity height frequency distribution from the measured profile, while the GD method only extracts aggregate statistical parameters. As a result, one outlying asperity peak height from a population of many asperity peaks obtained from a surface roughness trace can significantly affect the shape of the asperity height probability distribution, but will not have much effect on the standard deviation of the total population of asperities. Thus, an inherent advantage of the GD method compared to the AD method is its lower sensitivity to random deep valleys or high peaks that may be present in a particular surface roughness trace.

20.11.3 *Oil Film Thickness*

The reciprocating tester is not capable of measuring oil film thickness between the ring and liner specimen. Therefore, only the predicted oil film thickness is presented in this study. The predicted ring minimum oil film thickness for the slide honed and plateau honed surfaces is shown in Figs 20.14 and 20.15, respectively.

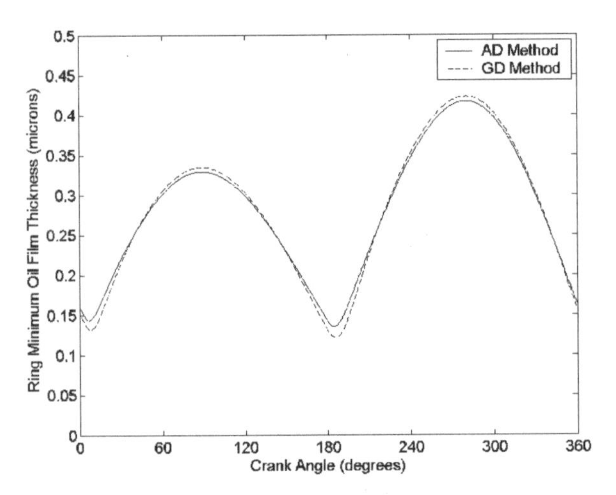

Fig. 20.14 Ring specimen minimum predicted oil film thickness with slide honed surface

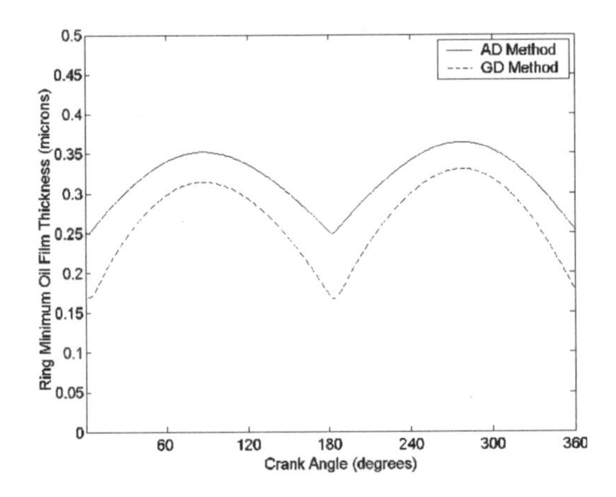

Fig. 20.15 Ring specimen minimum predicted oil film thickness with plateau honed surface

By examining Figs 20.14 and 20.15, it is evident that the oil film thicknesses predicted by the two methods are in much better agreement for the slide honed surface than for the plateau honed surface. This is a result of the better agreement of predicted asperity contact pressure for the slide honed surface than for the plateau honed surface, as is shown in Figs 20.7 and 20.8. It is not meaningful to directly compare oil film thicknesses predicted for the slide honed and plateau honed surfaces because the effective ring profiles were slightly different in each experiment. This difference was a result of both manufacturing variations in the ring profile and slight installation misalignment in the apparatus.

20.11.4 *Effect of Surface Finish on Friction*

To isolate the effect of plateau and slide honed surfaces on the cycle average friction, a simulation was run in which the same ring profile was assumed to interact with both surfaces. The simulation results using both the AD and GD methods are given in Table 20.6. In Table 20.6, the standard deviation gives the range of predicted cycle average friction coefficients for three different surface roughness traces of the same surface.

Table 20.6 Predicted effect of surface finish on cycle average friction coefficient

	Plateau		Slide	
	GD Method	AD Method	GD Method	AD Method
a_r	0.0376	0.0519	0.0234	0.0193
Std. Dev.	$\pm$ 0.0008	$\pm$ 0.01	$\pm$ 0.0006	$\pm$ 0.0024

By examining Table 20.6, it is evident that both the GD and AD methods predict lower cycle average friction for the slide-honed surface. Since the coefficient of boundary contact friction was assumed to be the same for both surfaces, the lower predicted friction is a result of lower nominal asperity contact pressures per cycle with the slide- honed surface.

20.11.5 *Applicability to Engine Running Conditions*

Significant ranges of lubrication conditions seen in an engine also occur during a cycle of the bench tester. Lubrication conditions ranging from primarily boundary lubrication to primarily hydrodynamic lubrication are predicted to occur despite the high ring load and low sliding speeds during the bench test. This is due to offsetting effect of high oil viscosity at the low operating temperatures of the test, and the large ring wetted area due to the flooded inlet condition. The wide range of lubrication conditions is illustrated in Fig. 20.16, in which the predicted proportions of total ring load carried by asperity contact and hydrodynamic pressure over a cycle are plotted for a slide-honed surface. During ring reversal, boundary lubrication is almost reached as the majority of total ring load is supported by asperity contact. During mid-stroke conditions, pure hydrodynamic lubrication is almost attained with very little asperity contact present.

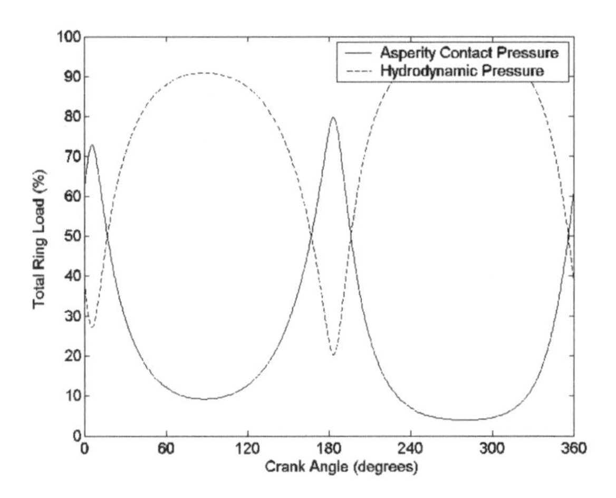

Fig. 20.16 Proportion of total ring supported by asperity contact and hydrodynamic pressure

The coefficient of boundary contact friction was assumed to be similar to that found in an engine. A rough comparison can be made between the average hydrodynamic coefficient of friction during the bench test and during an engine cycle using the Reynolds equation, as is shown in Equations (20.18) and (20.19).

$$f \approx \sqrt{\frac{\mu U}{PB}} \tag{20.18}$$

$$\frac{f_E^2}{f_T^2} \approx \left(\frac{\mu_E}{\mu_T}\right) \cdot \left(\frac{U_E}{U_T}\right) \cdot \left(\frac{P_T}{P_E}\right) \cdot \left(\frac{B_T}{B_E}\right) \tag{20.19}$$

In Equations (20.18) and (20.19), f is the coefficient of hydrodynamic friction, U is the sliding velocity, P is the applied ring pressure, B is the ring wetted width, and the subscripts E and T refer to engine and test, respectively. Table 20.7 provides an estimate of the cycle average values of these parameters for the bench test, and for typical diesel engine operating conditions. It should be noted that f is also the dimensionless duty parameter for the Stribeck curve that defines the lubrication regime under which the ring is operating.

Table 20.7 Estimated bench tester and diesel engine operating parameters

Parameters	Bench Tester	Diesel Engine
Liner Temp, [°C]	25	130
μ, [Pa.s]	0.2	0.005
U, [m/s]	0.5	10
P, [bar]	120	20
B, [mm]	1.5*	0.5*

* Wetting widths based on ring width of 3 mm

By combining the estimates given in Table 20.7 and Eq. (20.19), it is seen that the average coefficient of hydrodynamic friction estimate for a diesel engine is the same order of magnitude to that of the bench tester, as is shown in Eq. (20.20).

$$\frac{f_E}{f_T} \approx \sqrt{\left(\frac{0.005}{0.2}\right) \cdot \left(\frac{10}{0.5}\right) \cdot \left(\frac{120}{20}\right) \cdot \left(\frac{3}{1}\right)} \approx 3 \qquad (20.20)$$

Since the coefficients of boundary contact friction and hydrodynamic friction are similar for the bench tester and a diesel engine, the reciprocating tester operates in similar range of lubrication conditions, or covers a similar range on the Stribeck diagram, as a diesel engine. Note that the absolute friction over the cycle experienced by the bench tester is higher than typically observed in an engine, as illustrated by the low oil film thickness during the cycle in Figs 20.14 and 20.15. Nonetheless, the results obtained from the bench tester are not incomparable to the conditions seen by a piston ring in a diesel engine.

20.12 Conclusions

The followings are the concluding remarks based on the above studies[20.21]:

This study attempts to develop and validate two stochastic methods that describe cylinder liner surface roughness. These methods are combined with the Greenwood and Tipp statistical asperity contact model to allow prediction of asperity contact pressure, and subsequently friction and wear.

The accelerated ring wears bench test procedure allowed an investigation into the relative time evolution of friction, and relative wear behavior of the plateau and slide honed surfaces. Higher friction was observed for the plateau honed surface throughout the four hours test, and the plateau honed surface appeared to settle to a higher steady-state level of friction. Both the ring and liner wear measured at the end of the test were higher for the plateau honed surface, and preliminary wear estimations from the model simulation predicts higher wear for the plateau honed surface.

In general, both methods proposed are capable of distinguishing between the plateau and slide honed surfaces considered in terms of the friction generated by a piston ring. However, the AD method, which uses the actual distribution of asperities obtained from a surface roughness trace, shows stronger agreement with experimental results than the GD method. The AD method is capable of accurately predicting both the absolute average friction for the plateau and slide honed surfaces, as well as the relative difference in cycle average friction between the two surfaces. The AD method proved to be very sensitive to outlying peaks and deep valleys in the surface roughness trace.

The GD method, which assumes a Gaussian distribution of asperities lie above the mean of the surface, accurately predicted the average friction coefficient for the slide honed surface, but under-estimated it for the plateau honed surface. However, the GD method was unaffected by outliers in the surface roughness trace because it derives only aggregate statistical parameters, the standard deviation and mean height, from the trace. Thus, the two methods considered appear to present a trade-off between accuracy and robustness.

The experimental data was obtained with an acquisition rate of 345 Hz, which proved to be too low to obtain a good representation of the instantaneous friction variation along the stroke. Thus, validating the instantaneous friction predicted by the simulation for the GD and AD methods is outside the scope of this study. To gain a better understanding of the ability of both methods in predicting instantaneous friction, further experiments are planned with higher data acquisition rates.

A detailed description of the effects of 3-D surface roughness topography on ring pack friction, and inclusion of a detailed model of elastic-plastic contact are outside the scope of this study. The effect on ring pack performance of anisotropic cylinder bore surface texture with particular attention given to honing grooves, and a more comprehensive study on the effects of elastic-plastic contact are topics for future study.

References

[20.1] Lemke, H., 2003, "Characteristic Parameters of the Abbot Curve", MTZ 5/2003, **64**, pp. 438-444.

[20.2] MAHLE, 2000, "Recommendation for the Specification of Cast Iron Cylinder Bore Surfaces" Technical Information. Publication.

[20.3] Tian, T., Wong, V. W., Heywood, J. B., 1996, "A Piston Ring-Pack Film Thickness and Friction Model for Multigrade Oils and Rough Surfaces", SAE paper 962032 and SAE Transactions, Journal of Fuels and Lubricants, **105**, pp. 1783-1795.

[20.4] Radcliffe, C. D., Dowson, D., 1995, "Analysis of Friction in a Modern Automotive Piston Ring Pack", Elsevier Lubricants and Lubrication, pp. 355-365.

[20.5] Hu, Y., Cheng, H.S., Arai, T., Kobayashi, Y. and Aoyama, S., 1993, "Numerical Simulation of Piston Ring in Mixed Lubrication – A Nonaxisymmetrical Analysis", ASME Journal of Tribology, 93-Trib-9.

[20.6] Greenwood, J. A. and Tripp, J., 1971, "The Contact of Two Nominally Flat Rough Surfaces", Proc. Inst. Mech. Engrs., **185**, pp. 625-633.

[20.7] Patir, N. and Cheng, H. S., 1979, "Application of Average Flow Model to Lubrication between Rough Sliding Surfaces", ASME Journal of Lubrication Technology, **101**, pp. 220-230.

[20.8] Bolander, N. W., Steenwyk, B. D., Kumar, A., Sadeghi, F., "Film Thickness and Friction Measurement of Piston Ring Cylinder Liner Contact with Corresponding Modeling including Mixed Lubrication", ICED Fall Technical Conference Proceedings, ASME Paper ICEF2004-903.

[20.9] Jocsak, J., Wong, V., Tian, T., 2004, "The Effects of Cylinder Liner Finish on Piston Ring-pack Friction", ASMEICED Fall Technical Conference Proceedings, ASME Paper ICEF2004-952.

[20.10] Tomanik, E. et. al., 2003, "A simple numerical procedure to calculate the input data of Greenwood-Williamson model of Asperity Contact for Actual Engineering Surfaces" Leeds-Lyon Symposium on Tribology: Tribological research and design for engineering systems TRIBOLOGY SERIES, **41**, pp. 205-2 16.

[20.11] Kogut, L., and Etsion, I., 2003, "A Finite Element Based Elastic Plastic Model for the Contact of Rough Surfaces", Tribology Transactions, 46, pp. 383-390.

[20.12] McCool, J.I., 1986, "Comparison of models for the contact of rough surfaces", Wear, **107.**

[20.13] Greenwood, J. A., 1984, "A Unified Theory of Surface Roughness", Proc. Roy. Soc. Lon., **393**, 133-157.

[20.14] Tomanik, E., 2005, "Modelling of the Asperity Contact Area on Actual 3D Surfaces", SAE paper 2005-01-1864.

[20.15] Montgomery, D. C., Runger, G.C., Norma, F. H., 2001, *Engineering Statistics, Second Edition,* John Wiley and Sons Inc., New York.

[20.16] Johnson, K. L., 1985, Contact Mechanics, Cambridge University Press, London.

[20.17] Visscher, M., Dowson, D. and Taylor, C. M., 1995 (Elsevier, Amsterdam, 1996), "Surface Roughness Modeling for Piston Ring Lubrication: Solving the Problems" in The Third Body Concept: Interpretation of Tribological Phenomenon, Proc. of the 22nd Leeds-Lyon Symposium on Tribology.

[20.18] Tian, T., 2002, "Dynamic behaviors of piston rings and their practical impact. Part 2: oil transport, friction and wear of ring/liner interface and the effects of piston and ring dynamics", Proc. Instn Mech Engrs, **216** Part J.

[20.19] Hutchings, I. M., 1992, *Tribology: Friction and Wear of Engineering Materials*, Eduard Arnold, London.

[20.20] Tomanik, E. Nigro, F., 2001, "Piston Ring Pack and Cylinder Wear Modelling" SAE paper 2001-01-0572.

[20.21] Jocsak,J., Wong, W. Victor, Tomanik, E., Tian, T. 2005, "The Characterization and Simulation of Cylinder Liner Surface Finishes", ASME, Proc. of ICES 2005, Paper ICES2005-1080.

Nomenclature

$a(t)$	Instantaneous total coefficient of friction
a_{bc}	Boundary lubrication coefficient of friction
a_r	Cycle average total coefficient of friction
d	Separation of mean surfaces, [m]
E	Elastic modulus, [Pa]
F_f	Total friction, [N]
h	Height from mean plane of surface, [m]
K_u	Kurtosis
N	Crank rotational speed, [RPM]
P	Nominal asperity contact pressure, [Pa]
P_{hyd}	Hydrodynamic pressure, [Pa]
R_q	Root mean square (RMS) surface roughness, [μm]
R_k	Core roughness depth (DIN 4776), [m]
R_{pk}	Reduced peak height (DIN 4776), [μm]
R_{vk}	Reduced valley height (DIN 4776), [μm]
Sk	Skewness
Z_s	Asperity height mean-plane offset, [μm]
W	Wear, [μm]
$\phi(z)$	Probability density function
σ	Standard deviation of asperity height distribution, [m]
η	Asperity density, [asperities/m^2]
β	Asperity peak radius of curvature, [m]
ν	Poisson's ratio

Section - VI

Films and Coatings

21
Multi-Sensor Testing of Thin and Thick Coatings for Adhesion and Delamination

21.1 A General Review

There are numerous techniques known for adhesion and delamination testing, some of the most common being a tape test, stud-pull test, scratch test and an indentation test. In the tape test, a tape is pulled off the surface containing a scratch, which provides the failure initiation. In the stud pull test, a stud held with thermosetting epoxy is pulled off the film surface. The indentation test, wherein a ball is pressed into the surface, is used for hard coatings, and the failure pattern indicates acceptable behavior. In the scratch test, where an indenter moves in both vertical (loading) and horizontal (sliding) directions, and an acoustic emission sensor allows for detection of the initiation of fracture, while the scratch pattern indicates the type of failure.

The UMT series Micro-Tribometer has been developed to perform all the common adhesion tests. During any of them, it can simultaneously measure contact or surface electrical resistance, displacement or deformation or depth of penetration, contact acoustic emission, temperature, forces in all three directions and digital video of the contact area. This report covers evaluation of the adhesion and delamination properties of coatings by the scratch test.

21.2 Instrumentation

The fully computerized Micro-Tribometer mod. UMT-2 provides a combination of precision linear and rotational, including reciprocating, motions to the specimens, with programmable speeds ranging from 0.5 μm/s to 50 m/s. A load is applied via a closed-loop servo-mechanism and is programmed to be kept constant or linearly increasing, ranging from 0.1 mN to 1 kN. The environmental temperature and pressure can also be controlled. Friction force (F_x), normal load (F_z), electrical contact resistance (ECR), and contact acoustic emission (AE) are all measured at a total sampling rate of 20 kHz, displayed in real time and recorded for further analysis.

Specimens in a wide variety of shapes and dimensions can be accommodated. For scratch-adhesion tests various tool geometries and materials can be used, including metal or ceramic ball or needle, diamond stylus or CETR patented micro-blade.

21.3 Experiment

Two types of coatings were tested in this work, 50-micron thick soft elastomer coatings used on ink-jet cartridges and few-nanometer thin hard diamond-like coatings used on hard magnetic disks. These two types have been chosen since they are very different in both hardness and thickness and cover a wide spectrum of practical applications.

The effects of scratching tools were studied on three types of tools, 1.6 mm balls from stainless steel and tungsten carbide, 10 micron sharp diamond stylus and a novel tungsten carbide micro-blade, which is as sharp as the stylus (10 micron), but very wide (0.8 mm).

To achieve effective results, the multi-sensor technology was utilized, with the simultaneous F_z, F_x, ECR and AE measurements.

21.4 Results on Thick Soft Coatings

Neither steel nor tungsten carbide balls produced useful delamination data. The sharp diamond stylus scratched the coating, but did not delaminate it. The micro-blade produced repeatable both scratch-resistance and delamination/adhesion results.

The micro-blade test consisted of linearly increasing the applied load from 1 cN to 100 cN, while slowly sliding the micro-blade at 1 mm/s, with continuous multi-sensor process monitoring.

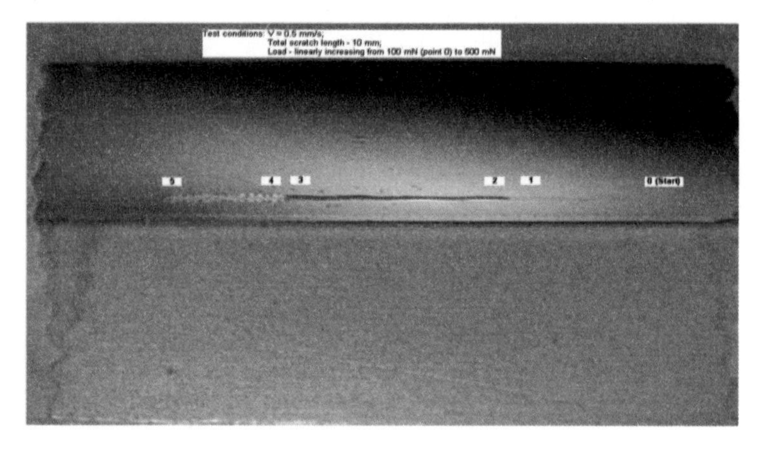

Fig. 21.1 Coated Surface Tested With Micro-Blade

The optical photo of the tested surface is shown in Fig. 21.1, the corresponding friction force plot is presented in Fig. 21.2. One can clearly see three zones, namely: deformation with no debris formation at very low loads (right part of Fig. 21.1, left part of Fig. 21.2), micro-scratching with production of a lot of micro-debris (middle parts of Figs 21.1 and 21.2), followed by delamination with chunks of debris formed at higher loads (left part of Fig. 21.1, right part of Fig. 21.2). The critical loads (or times) of scratching (on the borderline of the deformation and micro-scratching zones) and delamination (on the borderline of the scratching and delamination zones) can be found easily in these tests, and tend to be recurring.

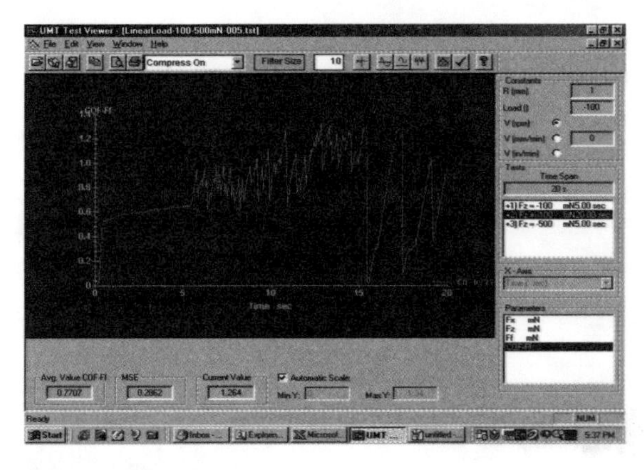

Fig. 21.2 Friction Force in Test with Micro-Blade

Though both the optical and force plots show all the three zones, the borderlines between them, defining the critical loads (times) or micro-scratching and delamination, can be determined with higher accuracy by utilizing additional electrical and acoustic measurements.

If a coating is non-conductive and a substrate is conductive, then electrical contact resistance is measured. If a coating is conductive, the electrical surface resistance is monitored. In either case, the electrical measurements typically show both critical thresholds of the onset of scratching (when electrical resistance begins to change) and breaking through the coating (when electrical resistance reaches the level of the substrate resistance). An example of a sharp threshold is given in Fig. 21.3.

The high-frequency contact acoustic emission reflects the scratching and delamination processes of solid coatings when their structure is relatively ordered and they are relatively hard. For instance, most metal and ceramic coatings emit substantial acoustic waves during such tests, while most soft polymers do not emit measurable acoustics. The acoustic emission resolution in terms of surface defects is defined by its frequency. For example, to detect a 10 nm scratch or delamination defect in the coating at a test speed of 1 cm/s (taking into account that 1 cm = 10,000 nm and a signal has to be several times higher than the process), one needs the signal frequency of at least 5 kHz.

In fact, we use the range of up to 5 MHz, which allows for detection of the very beginning of the tiniest micro-scratches and micro-delamination.

The effective use of the contact acoustic emission signal is illustrated in Fig. 21.3 below. In this example, the critical threshold of breaking through the coating can be determined by the sharp drop in electrical resistance, and is fully supported by the sharp increases in friction and acoustic signals. It is interesting that both friction and acoustic show a substantial pre-failure rise at the same lower load, corresponding to the beginning of coating damage. Moreover, the ultra-sensitive acoustic emission started even earlier, which reflected the onset of tiny coating damage, undetectable yet by the electrical and force signals.

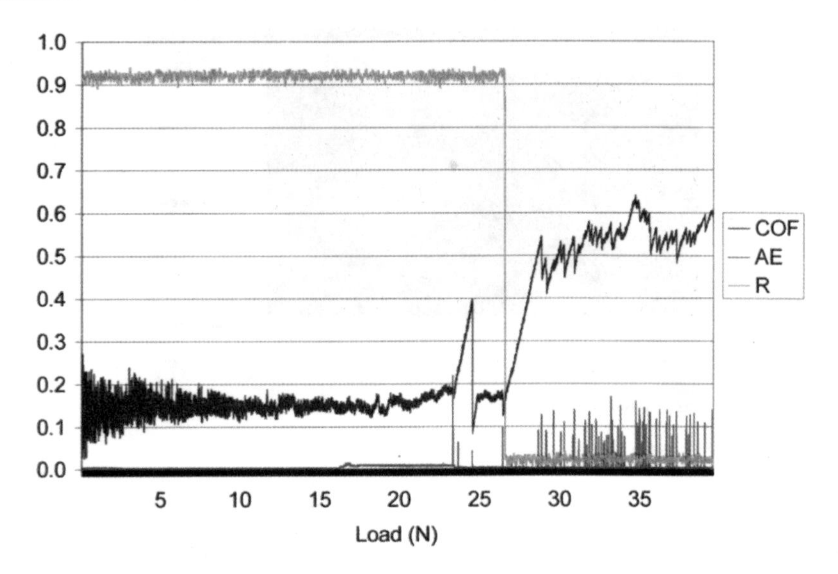

Fig. 21.3 Multi-Sensor Determination of Coating Failure

21.5 Results on Thin DLC Coatings

DLC, or diamond-like carbon coatings, are used in various industries. Their deposition technologies and so properties vary from application to application. For example, we have observed in numerous tests that the durability of very thin DLC coatings on hard magnetic disks and heads is the highest, while the durability of the DLC coatings on window glass and razor blades is much lower. Correspondingly, the scratch adhesion test procedure should be different.

Soft copper or gold-plated 4 mm or 6 mm balls have produced the most repeatable test data on the softest DLC coatings. Hard 1.6-mm tungsten carbide balls produced quite repeatable results on mid-range DLC coatings. Sharp diamond styluses scratched the coating, but did not delaminate it. Also, the sharp stylus data had limited repeatability due to the surface morphology differences from one scratch to another. The tungsten carbide micro-blade, averaging the scratch data over their substantial width, produced the most repeatable results for the hardest and thinnest coatings. Again, in all these tests the critical thresholds were determined by simultaneous measurements of forces, electrical and acoustic signals (see Fig. 21.3), though the latter ones were not informative for very thin coating films. The electrical resistance of DLC coatings was always measurable, though the harder and more diamond-like was a coating, the less conductive it was.

The thinnest coating tested was a 1.5-nm DLC film on a magnetic head wafer, the durability of which was found to be dependent on the rate of deposition. The Fig. 21.4 includes some of the results for thin hard DLC coatings on magnetic disks. For each DLC thickness four disks were tested, and the high data repeatability between them is obvious from this figure. Reduction in film thickness by a factor of 4, from 10 nm to 2.5 nm, decreased the critical load by a factor of 5, from 125 to 25 cN. Lubricating the disk with a thin 2-nm layer of a topical Fomblin lubricant, common for the magnetic disks, increased the critical load by a factor of 4.5 for the thin 2.5 nm coating.

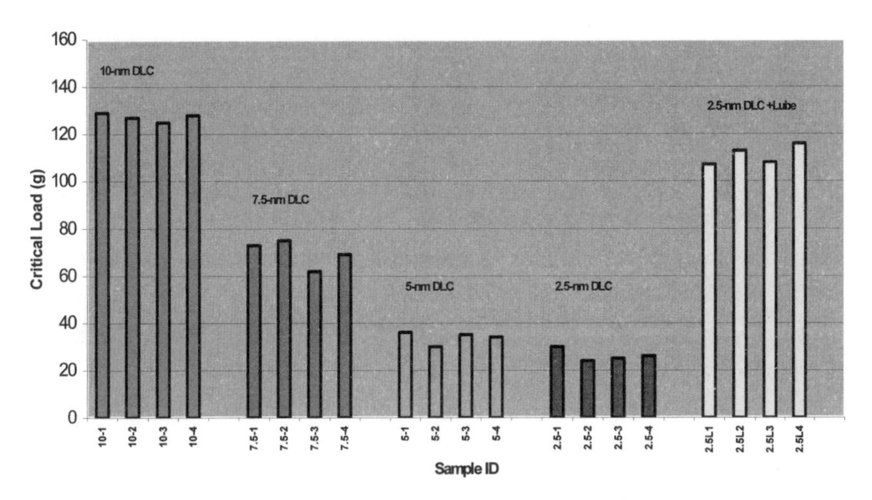

Fig. 21.4 Micro-Scratch Test Data for Thin Hard Coatings

21.6 Conclusions

1. The multi-sensing technology, based on simultaneous high-resolution force, electrical and acoustic (when possible) measurements, allows for very accurate determination of both delamination/adhesion and scratch resistance of both thin and thick coatings. Using the test tools with higher contact area (or length), like micro-blade or ball, may allow for more repeatable results than those obtained with sharp styluses.
2. The micro-tribometer mod. UMT-2 provides a useful platform for all common types of adhesion and delamination tests, with the complete utilization of all the advantages of the multi-sensing technology.

22

Comprehensive Tribo-Mechanical Testing of Hard Coatings

22.1 A General Review

The following test procedures were used for the comprehensive evaluation of hard-coated metal samples:

Reciprocating wear tests for evaluation of coating friction and durability;

Scratch-hardness tests under constant load for scratch-resistance and micro-hardness;

Scratch-adhesion tests under increasing load for coating adhesion and toughness;

Nano-indentation tests for coating nano-hardness and elastic modulus evaluation.

All the tests were performed on the same tester, re-configured for each type of test within a minute by installing corresponding replaceable attachments. The Universal Nano & Micro Tester with multiple sensors (nano-indentation, friction, contact electrical resistance, acoustics, temperature, wear depth) and integrated both atomic force and optical microscopes has been utilized to characterize physical and mechanical properties of thin films, with in-situ monitoring their changes during micro and nano indentation, scratching, reciprocating, rotating and other tests.

Clear differences between ductile and brittle coatings of diamond-like carbon, tungsten carbide and titanium carbide, as well as between different deposition technologies, have been observed within a wide range of thicknesses, from ultra-thin (3 to 5 nm) to thick (50 to 75 microns).

22.2 Description of Test Equipment

A number of tests were carried out on the Universal Nano and Micro Tester models. UNMT-1 and UMT. The general appearance of the tester is shown in Fig. 22.1. It is a highly precision instrument for nano and micro mechanical and tribological testing of practically all types of thin films and coatings, including metals, ceramics, composites, polymers, etc. It provides a combination of rotary and linear, including reciprocating, motions in both vertical and horizontal directions, and measures numerous parameters with high precision, such as normal and lateral forces and torques in two to six axes in ranges from micrograms to kilograms, wear and deformation, contact acoustic emission, contact electrical resistance, temperature, etc. A normal-load sensor provides feedback to the vertical motion controller, actively adjusting the sample position to ensure the user-programmed either constant or changing load during testing. The tester has fully computerized motor-control and data-acquisition system, with test data acquired, calculated and displayed in real time, as well as stored for future retrieval. An optional nano-indentation head measures nano-hardness and elastic modulus of thin films. The unique modular design provides the capability of fast switching between multiple testing modes (such as friction, wear, micro-scratch, nano-indentation, AFM imaging, etc.) by simply swapping the easily-replaceable drive stages and sensors. A high-frequency multi-channel data-acquisition system, with data sampling at thousands times per second, allows for detection of almost instantaneous tiny sub-micro-contact and sub-micro-failure events in sophisticated test sequences. Integrated optical microscopy allows for precision sample micro-positioning, digital video of the in-situ dynamics of surface failure, and micro-images of wear tracks, indents and scratches. Integrated atomic force microscope provides nano-imaging of test surfaces, wear tracks, indents and scratches, both periodically during testing and post-test.

Fig. 22.1 Photo of the UMT tester

22.3 Description of Four Test Procedures

The following test procedures were used for the most comprehensive evaluation of the sample coatings properties:

(a) Reciprocating wear tests for evaluation of coating friction and durability.

(b) Scratch-hardness tests under constant load for scratch resistance and micro-hardness measurement.

(c) Scratch-adhesion tests under progressively increasing load for evaluation of the coating adhesion and scratch toughness properties.

(d) Nano-indentation tests for coating nano-hardness and elastic modulus evaluation.

All four types of tests were performed on the same tester, re-configured for each type of test by installing corresponding replaceable attachments.

(a) Friction and Wear Test

The friction and wear tests were carried out using a 1.6 mm sapphire ball and the ASTM G-133 standard test method for linearly reciprocating ball-on-flat sliding wear. The test sample was mounted on a table of the lower drive and was reciprocating in contact with a stationary ball, as shown in Fig. 22.2. The ball in the ball holder was mounted via suspension on a dual-axis force sensor, attached to the carriage, which is a part of the electro-mechanical loading mechanism. A lateral slider on the carriage was used for positioning of the ball on the coated sample. An acoustic emission sensor was attached to the ball holder for monitoring the intensity of the high-frequency vibrations generated in the ball-to-sample interface.

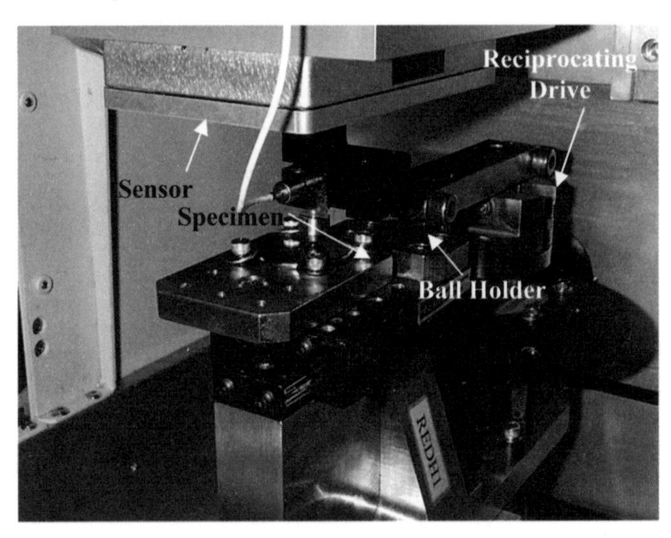

Fig. 22.2 UMT with upper stationary ball on lower reciprocating plate

The tests were conducted at a constant normal load of 2 N over a distance of 5 mm on the specimens, with a 3 Hz frequency of oscillation for 1 hour. The ball surface facing the specimen was changed before the start of each test by rotating the ball and inspecting under an optical microscope to make sure the new ball surface is fresh and smooth. The friction force Fx and normal load Fz were simultaneously recorded during the tests. The width W and depth H of the wear tracks were measured after the test at five different locations on the wear scar, and the average values were reported. Then a mean cross sectional area A (in mm^2) was calculated, and then specific wear rate WR was calculated:

$$\dot{WR} = \frac{A \times 5}{F_Z \times 3(2 \times 0.005) \times 3600} \text{ mm}^3/\text{N}\times\text{m}.$$

COF and AE data as a function of time are plotted in Fig. 22.3 for both samples.

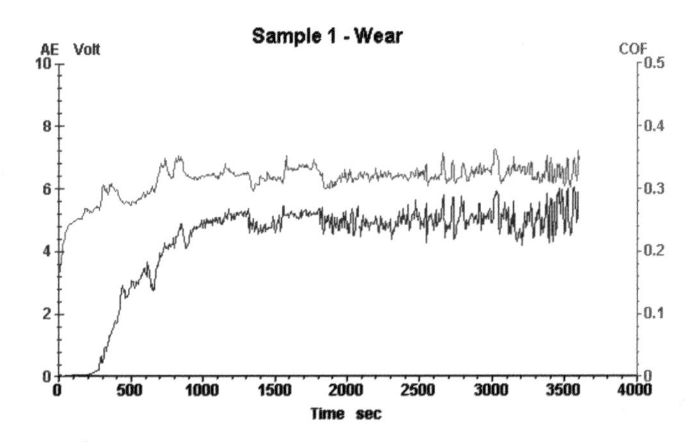

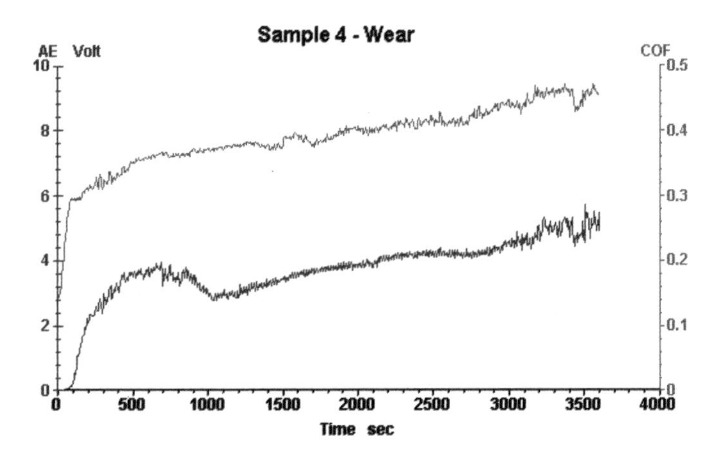

Fig. 22.3 Friction coefficient (pink) and AE (blue) signals during reciprocating wear tests

Table 22.1 shows the mean values of the coefficient of friction (COF) and AE signal after break-in period for both specimens, as well as wear scar width W, depth H, cross-sectional area A, and specific wear rate WR.

Table 22.1 Friction and Wear Data

Sample #	COF	AE	W µm	H µm	A $mm^3 \times 10^6$	$WR \times 10^6$
1	0.32	5.1	105	0.49	25.4	1.18
4	0.41	4.0	87	0.46	20.1	0.93

In the reciprocating wear test, Sample 1 had lower friction, while Sample 4 demonstrated lower wear and acoustic emission, indicating less surface damage.

(b) Scratch-Hardness Tests under Constant Load

Micro-scratch-hardness tests were performed on both specimens using a diamond stylus with the tip radius of 5 µm. The test was conducted with the reference to the ASTM G171-03 standard method for determination of scratch hardness of materials, modified for thin coatings.

The scratch hardness of a material can be determined by producing a scratch on the sample surface by a sharp hard (diamond) tool with known tip geometry, under the constant, controllable load. Measuring the scratch width, one can define the sample scratch hardness as:

$$HSp = k \times Fz / W^2$$

where HSp - scratch hardness number; k - constant; Fz - applied load; W - scratch width.

When the constant k is unknown, the scratch hardness can be determined by comparison of the scratch width on the sample and on a reference material with known hardness:

$$HS_{sample} = (HS_{ref} \times W_{ref}^2 / Fz_{ref}) \times Fz_{sample} / W_{sample}^2$$

where the indexes "sample" and "ref" refer to the test sample and reference material, correspondingly.

The diamond stylus in a stylus holder was mounted on the force sensor with a suspension (instead of and similar to the ball in the wear test Fig. 22.2) to provide automated programmable sample shifting for multiple scratches, the tested sample was mounted on a table of the lower linear drive. AE sensor was attached to the stylus holder to monitor the high-frequency signal, generated during scratching and indicating the intensity of material fracture. The normal load of 0.4 N was applied to the stylus and maintained constant with an active feedback from the force sensor. The scratch was produced by dragging the stylus along the sample surface with the upper lateral slider. The scratch length was 5 mm, the dragging speed – 0.5 mm/s. The test was repeated 3 times on each sample to verify the data consistency and repeatability. A polished fused quartz (with the hardness of 9.5 GPa) was used as the reference material for the scratch-hardness calculations.

COF and AE data as a function of time are plotted in Fig. 22.4.

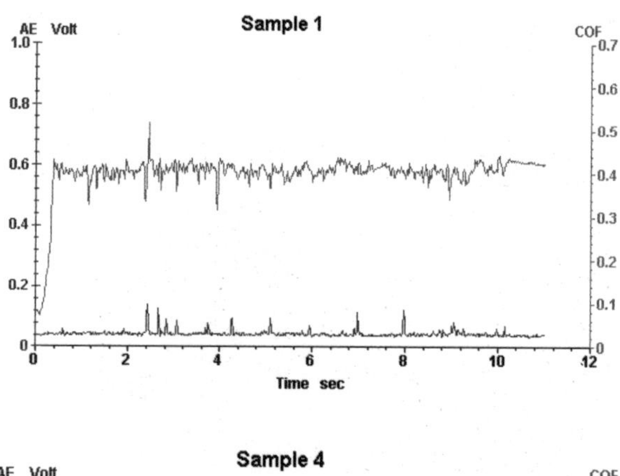

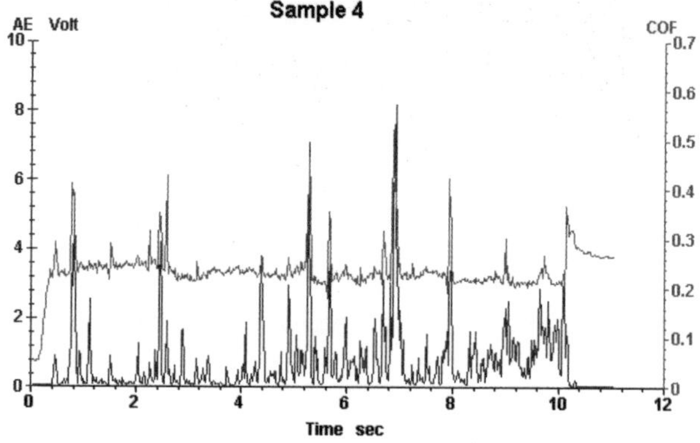

Fig. 22.4 Friction coefficient (pink) and AE (blue) signals during micro-scratch tests on Samples 1 and 4

The scratch width was measured on the same UNMT tester, by replacing the force sensor with an AFM head and taking the AFM images of the scratches. Both the force sensor with stylus and the AFM head were mounted on a fast-exchange fixture, which allows for their fast and easy swap (Fig. 22.5).

Table 22.2 shows the mean values of the coefficient of friction (COF) and acoustic emission (AE) for both specimens, along with the scratch width W and scratch micro-hardness HS. The raw data for the scratch width measurements are shown in the Appendix.

Table 22.2 Micro-Scratch Test Data with Diamond Stylus

Sample #	Mean COF	Mean AE	W µm	HS GPa
1	0.49	0.04	5.87	23.4
4	0.27	0.67	5.0	32.2

Fig. 22.5 UNMT-1 with the AFM head and optical digital microscope attachments

Fig. 22.6 represents 3-D AFM images of the scratches on both samples. Screenshots with the scratch width measurements are shown in Figs 22.7 and 22.8 for samples 1 and 4, respectively.

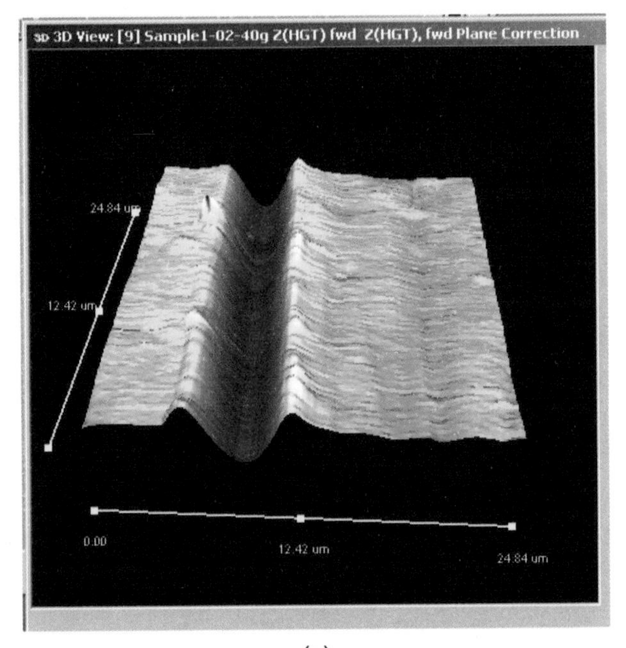

(a)

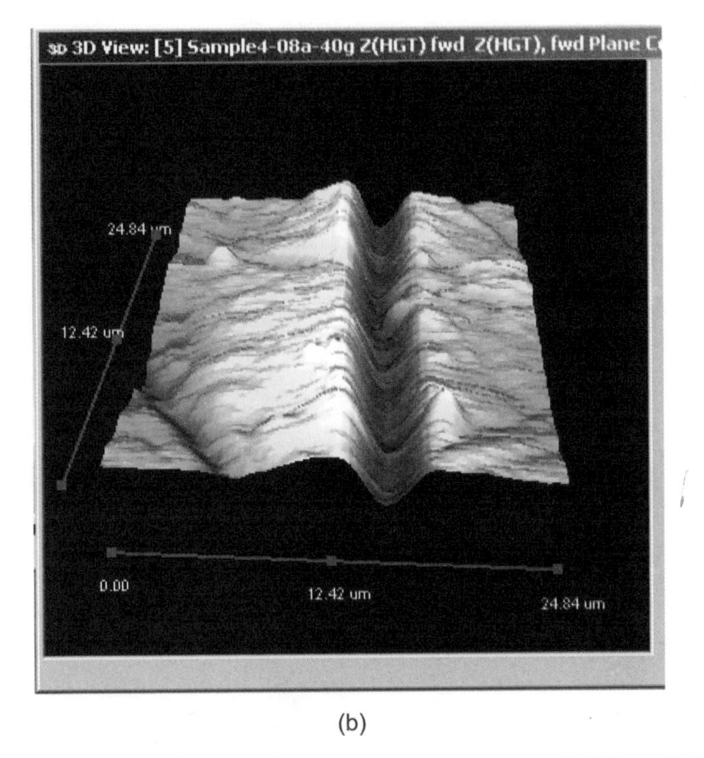

(b)

Fig. 22.6 3-D AFM images of scratches on Sample 1 (a) and Sample 4 (b) (scan size 25 x 25 μm)

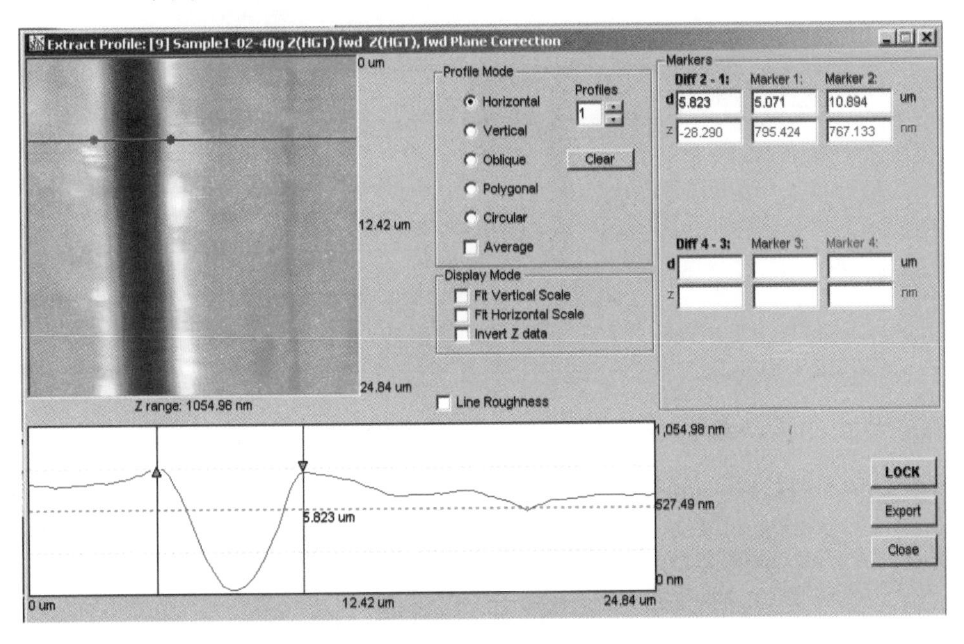

Fig. 22.7 Example of scratch width measurement for Sample 1

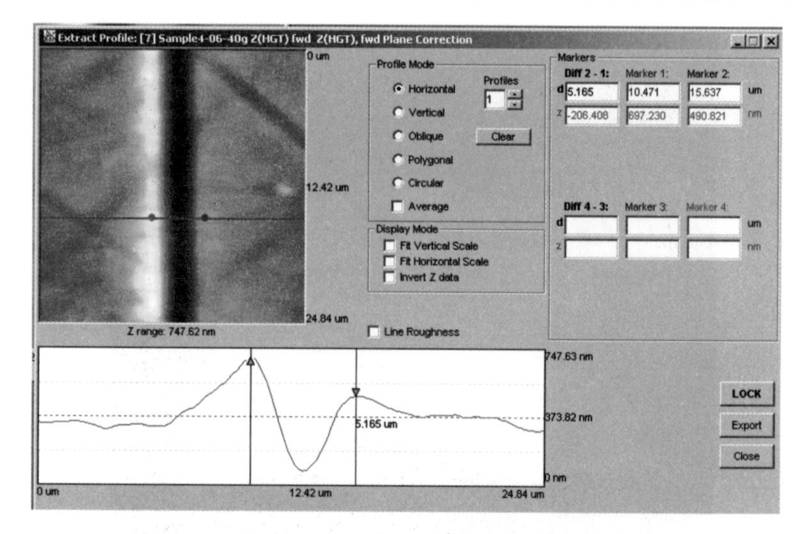

Fig. 22.8 Example of scratch width measurement for Sample 4

The AFM image analysis software allows for the sample surface roughness measurement on either the entire scanned image or a selected area. The example screenshot of the roughness measurement shown in Fig. 22.9.

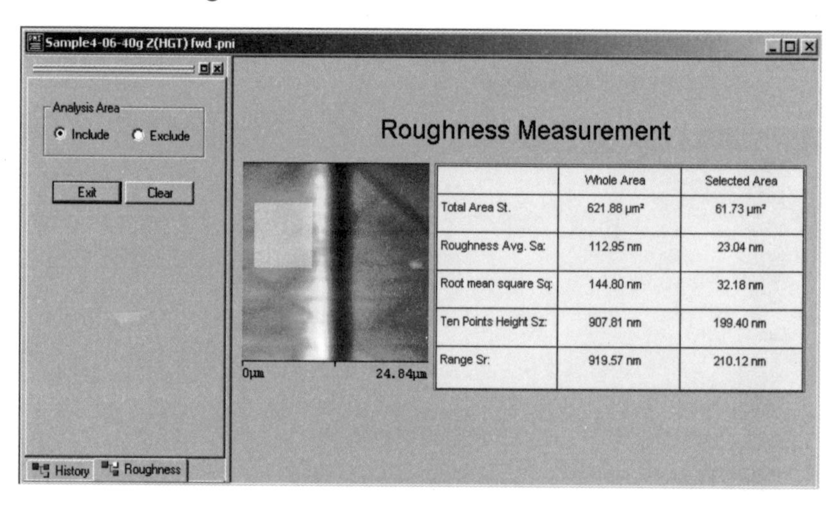

Fig. 22.9 Screenshot for surface roughness measurement (Sample 4)

In the micro-scratch-hardness test, Sample 1 exhibited lower hardness and acoustic emission, and higher dragging force than Sample 4. It indicates that the coating 1 is more ductile and is easier to deform plastically (deeper penetration of the stylus corresponds to higher dragging force), while the coating 4 is harder and more brittle (higher AE level corresponds to higher degree of material fracture).

(c) Scratch-Adhesion Tests under Increasing Load

The macro-scratch tests with progressively increasing load were conducted for evaluation of the coatings adhesion and scratch toughness using a diamond Rockwell indenter with the tip radius of 200 microns. As in the previous test, the indenter in its holder was mounted on the force sensor (of higher load range) with a suspension. The tested sample was mounted on the table of the lower linear drive to provide automated programmable sample shifting for multiple scratches. An AE sensor was attached to the indenter holder to monitor the high-frequency emission during scratching, indicating the intensity of material fracture.

The initial load of 1 N was applied to the indenter and stabilized for a few seconds. The scratch of the length of 10 mm was produced by dragging the indenter along the sample by the upper lateral slider with a speed of 1 mm/s. During the indenter dragging, the normal load was increased linearly from 1 to 60 N, according to the pre-programmed test script, while the load Fz, dragging force Fx and AE level were continuously monitored and saved. The tests were repeated several times on each sample to verify the data consistency and repeatability.

In the beginning of the scratch, at low loads, the indenter tip was mostly sliding on the sample surface, not causing coating damage. When the load reached a certain critical level, the coating started fracturing and/or delaminating from the substrate, since at higher load on the indenter having large tip radius, the depth of the mechanical stress distribution in the sample exceeded the coating thickness. This process reflected in sharp increases of both Fx and AE signals. The value of the critical load is used for the coating scratch toughness and adhesion evaluation.

An example of Fz, COF and AE plot, illustrating the critical load detection, is shown in Fig. 22.10.

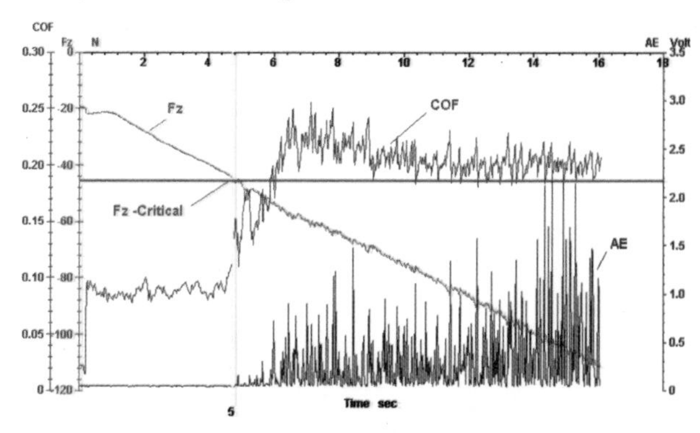

Fig. 22.10 Definition of the critical load (intersection of yellow and red lines) in scratch-adhesion test at progressively increasing load Fz.

Table 22.3 summarizes the COF and AE before and after coatings failure, and the critical loads.

Table 22.3 Scratch-Adhesion Test Data with Rockwell Indenter

Sample #	COF before Failure	COF after Failure	AE before Failure	AE after Failure	Critical Load, N
Sample 1	0.09	0.12	0.04	0.24	15.3
	0.08	0.14	0.04	0.92	20.6
	0.08	0.15	0.05	0.56	20.2
	0.08	0.17	0.04	0.81	18.6
	0.08	0.15	0.04	1.08	20.2
	0.08	0.12	0.05	0.89	18.8
Mean	**0.08**	**0.14**	**0.04**	**0.75**	**19.0**
Sample 4	0.09	0.18	0.05	0.46	28.9
	0.09	0.14	0.05	0.22	41.9
	0.08	0.15	0.05	0.29	27.6
	0.08	0.15	0.05	0.47	31.9
	0.08	0.14	0.04	0.46	29.7
	0.08	0.14	0.05	0.51	33
Mean	**0.08**	**0.15**	**0.05**	**0.40**	**32.2**

Typical graphs of COF and AE signals as functions of time are shown in Fig. 22.11 for both samples 1 and 4. In the macro-scratch-adhesion test, Sample 1 exhibited lower critical load and higher AE level after the coating failure than Sample 4, which indicates that the coating 1 has lower scratch fracture toughness at higher loads, or lower adhesion to the substrate. The dragging (friction) coefficient values were almost identical for both samples, which confirms that AE is more sensitive to the coating failures.

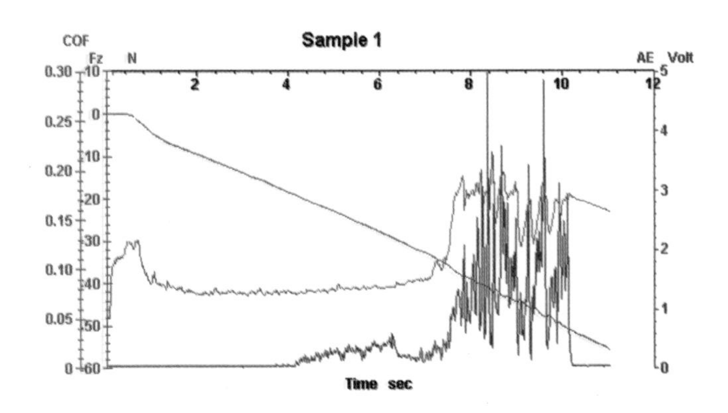

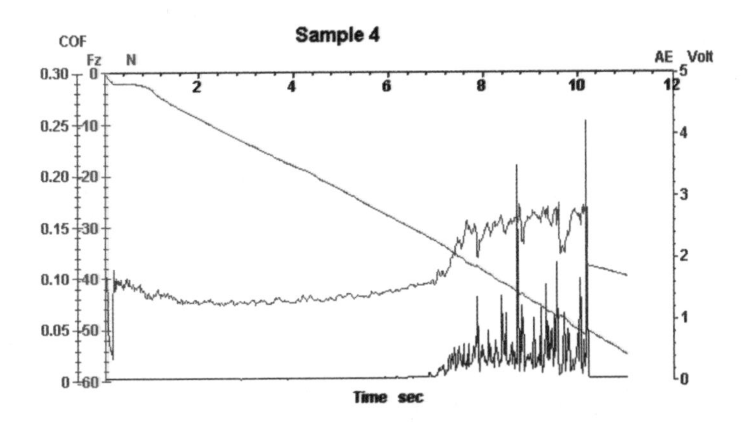

Fig. 22.11 Load Fz (red), dragging (friction) coefficient (pink), and AE (blue) signals during macro-scratch-adheison tests

(d) Nano-indentation Test

The tests were performed on the UNMT-1 with the nano-indentation head NH-1 (see photo in Fig. 22.12), attached to the same fast-exchange fixture as described above for an AFM head.

Fig. 22.12 Photo of Nano-head NH-1

Both samples were tested using a standard nano-indentation technique at the maximum loads of 70 mN. Nano-hardness (H) and Reduced Elastic Modulus (Er) measurement results are summarized in Table 22.4 for six measured points on each sample.

Table 22.4 Nano-indentation test data

	Sample # 1		Sample # 4	
Tests	H GPa	Er GPa	H GPa	Er GPa
#1	19.53	169.60	24.42	191.89
#2	20.81	174.55	26.61	179.80

Contd...

	Sample # 1		Sample # 4	
Tests	H GPa	Er GPa	H GPa	Er GPa
#4	22.55	193.90	26.87	193.04
#5	20.22	171.53	26.35	198.08
#6	20.91	176.81	26.72	184.54
Average	**20.79**	**173.45**	**26.21**	**187.72**
Deviation(+/-)	1.51	19.8	1.22	6.77

The typical loading-unloading curves are shown in Figs 22.13 and 22.14.

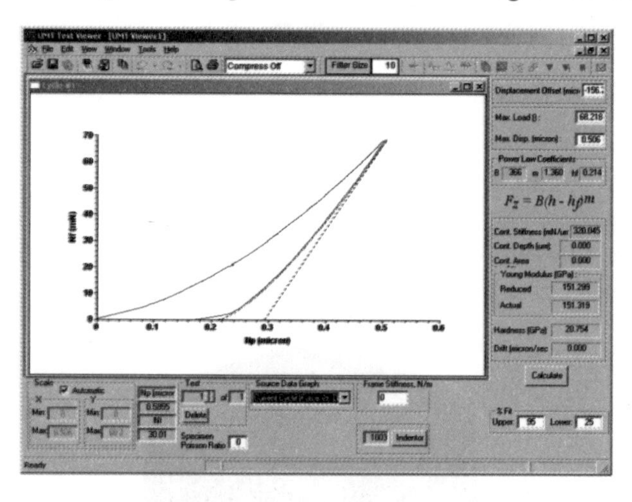

Fig. 22.13 Loading-unloading curve on Sample 1

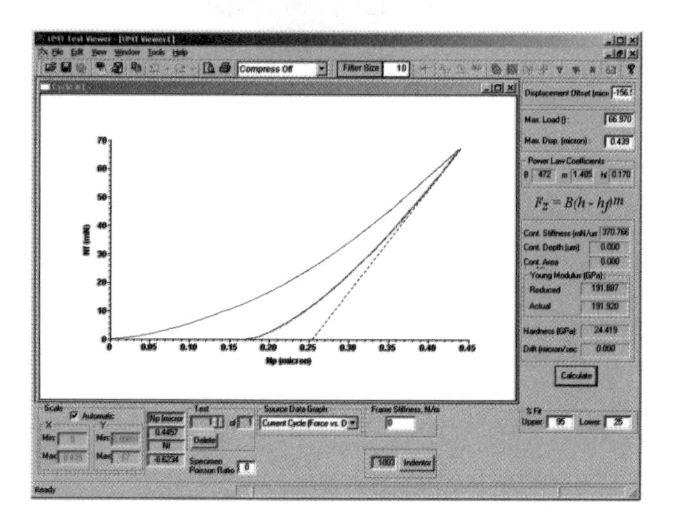

Fig. 22.14 Loading-unloading curve on Sample 4

In the nano-indentation test, Sample 4 showed higher values of both H and Er, while Sample 1 showed lower values and higher data scattering of Er. The scattering can be attributed to the surface roughness induced effects.

22.4 Conclusions

Based on the test results reported in this chapter, the following conclusions are drawn:

1. The Universal Nano and Micro Tester model UNMT-1 is uniquely capable of performing multiple tests for comprehensive precision evaluation of mechanical properties of hard coatings, including wear resistance, micro- and macro- scratch resistance, adhesion, micro- and nano-hardness, and elastic modulus. This capability is based on the modular design of the tester, allowing for inter-changeable attachments, precision servo-control of normal load, advanced sensors with simultaneous monitoring of normal and lateral forces, micro and nanc wear and deformation, high-frequency acoustic signal and electrical resistance, as well as integrated digital optical microscope and AFM.
2. The test procedures on the UNMT-1 are proven to be effective for accurate quantitative characterization of durability and other mechanical properties of hard coatings.

23

Quantitative Nano & Micro Scale Metrology of Thin Films and Coatings

23.1 A General Review

A novel test procedure has been developed to study mechanical properties of thin films and coatings, based on simultaneous measurements of a servo-controlled normal load, friction force, contact acoustic emission and electrical resistance. This multi-sensing methodology has been realized on the Nano & Micro Tribometer mod. UNMT, which can perform all common mechanical tests of indentation, scratching, reciprocating wear, rotating wear, with integrated both atomic-force and optical microscopy that allow for very sensitive detailed imaging of the coatings, with periodic assessment of the properties and surface topography at various stages of the tests. Experiments were performed on various films used in liquid crystal displays and copper interconnects in semiconductor chips.

23.2 Introduction

Quantitative nanometer resolution metrology tools have become a standard in semiconductor, data storage, display and other hi-tech industries, where products have to be tested for thin-film properties. Latest advances in nano and micro tribology have led to the development of integrated instrumentation. An innovative optical in-situ surface characterization experimental system, combined with the Raman spectroscopy instrument, was reported by Singer et al.[23.1, 23.2] for investigation of chemical-mechanical properties of debris during tribological testing. Recently, co-focal Raman spectroscopy was combined with atomic force microscopy for nondestructive chemical characterization at nanometer scale[23.3]. A combinatorial approach in micro-tribology tests[23.4] indicated the need for in-situ characterization instruments. A recent nano-scale tribometer combination with AFM has been reported[23.5]. Therefore, the number of novel tribometer applications is growing fast and covering fast changing industries, including biomedical[23.6].

In the present study, micro-scratch and micro-wear tests were performed on flat panel displays, coated with thin conductive films. Performance of a quantitative nano & micro-tribometer, integrated with SPM and high resolution optical microscope imaging, is also demonstrated on Cu coated Si wafer samples where quantitative material properties were derived at several intermediate characterization steps by means of nano-indentation.

23.3 Experiment

A picture of the micro-tribometer used for display testing is shown in Fig. 23.1. Each LCD sample is cut into 10 x 50 mm test specimens and clamped between two electrical connectors on a lower stage. A normal load is applied via a closed-loop servo control.

In scratch and adhesion tests, the lower stage is stationary, the upper tool is Rockwell-C diamond indenter making unidirectional linear motions. In wear tests, the lower stage is either stationary or linearly reciprocating, the upper tool is either same diamond indenter or sapphire ball, either linearly reciprocating (when stage is stationary) or stationary (when stage is reciprocating). During testing, normal load and electrical surface resistance ESR are monitored and recorded. When the conductive coating is present, ESR is at its low level reflecting the conductive coating; when the coating is completely cut (worn) through, ESR is at its maximum level corresponding to dielectric properties of the substrate. In our scratch tests, the indenter moved slowly on the coating, causing some material removal (see schematics in Fig. 23.2). A series of runs with progressively increasing normal loads, though constant within each run, was performed in each test. The normal load started from 1 or 1.5 N in the 1st run and was increased by 0.5 N each run until the coating was cut through. The critical load characterizing the coating scratch resistance was defined as the minimum load to cut through the coating completely.

A 5-micron thick Cu coated Si wafer was placed into the UNMT tester and compressed by 5 N load. A chemical-mechanical planarization process was carried out by a commercially available machine mod. CP-4 (CETR, Campbell, USA)[23.7]. A 5 mm work-hardened steel sphere used in tribology tests (ASTM G99-90) was utilized for this experiment. The 5 mN load nano-indentation tests were performed before and after work-hardening of the Cu layer. A cube corner diamond pyramid tool of 40 nm radius was used for all 56 indents. The nano-indentation ramp loading profile consists of 10 s loading, 3 s holding and 10 s unloading segments.

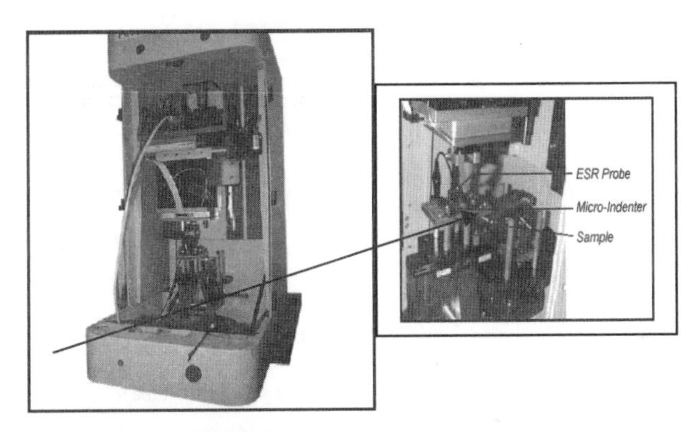

Fig. 23.1 Photo of UNMT for Coating Evaluation; Inset Shows the Close-View Photo of UNMT Set-Up for Display Testing.

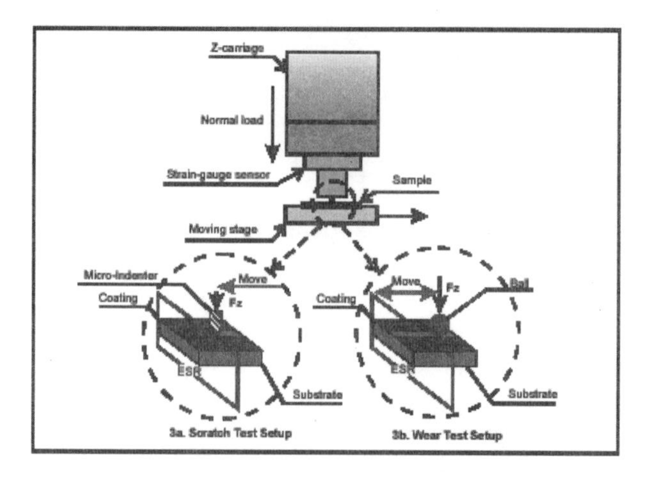

Fig. 23.2 Schematic of Scratch and Wear Test Set-Up.

Reduced elastic modulus and hardness values were derived by following the Oliver-Pharr methodology[23.9, 23.10]. Pre and post indentation AFM images (utilizing the UNMT with integrated AFM) were taken and analyzed at the selected locations. Optical images prior to each indentation were taken to locate the surface areas modified by compression.

23.4 Results and Discussion

23.4.1 *LCD Display Coating*

The results for three different LCD samples, each tested three times, (see Table 23.1) show repeatable differences between the coatings.

Table 23.1 Results of scratch and wear testing on LCD display coatings

SCRATCH TESTING			
Sample ID	**Critical Load (N)**		
	1st Test	2nd Test	3rd Test
Sample1	5	5.5	5.5
Sample2	4	4	4
Sample3	2.5	3	2.5
WEAR TESTING			
Sample ID	**Critical # of Cycles, thousands**		
	1st Test	2nd Test	3rd Test
Sample1	2.7	2.5	2.6
Sample2	2.1	2.2	2.0
Sample3	1.1	1.2	1.1

Typical scratch raw data is presented in Figs. 23.3 to 23.5, illustrating various stages of the process of cutting through the coatings with the load increase.

In the wear test, schematics of which is shown in Fig. 23.2, the indenter was stationary, while the sample stage was reciprocating, causing coating wear. A constant load of 1 N was chosen, under which there was no complete failure for all samples in scratch tests. A series of reciprocating cycles was run until the coating was worn through. The critical number of cycles characterizing the coating wear resistance was defined as the minimum number of cycles to wear through the coating completely. The results for three different LCD samples, each tested three times, summarized in Table 23.1, shows the repeated differences between the coatings, correlated with the scratch data.

23.4.2 Electroplated Cu Coating on Si

Typical loading-unloading curves for compressed (1) and not compressed (2) samples are shown in Fig. 23.6. Here, compressed area curve (1) indicates more pronounced elastic response typical for the contact work-hardening phenomenon. Greater contact stiffness yields the higher value of reduced elastic modulus and Martens hardness, and was found to be in the range of 135 ± 10GPa and 1.45 ± 0.17 GPa, respectively. A load-displacement curve for the not compressed area (2) exhibits more plastic-ductile type response and has a lower contact stiffness value for reduced elastic modulus and hardness. Experimentally derived reduced elastic modulus and hardness values were 112 ± 8GPa and 1.30 ± 0.15 GPa, respectively.

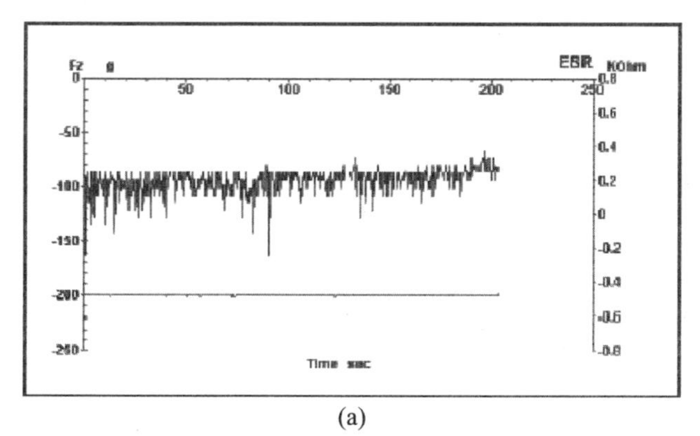

(a)

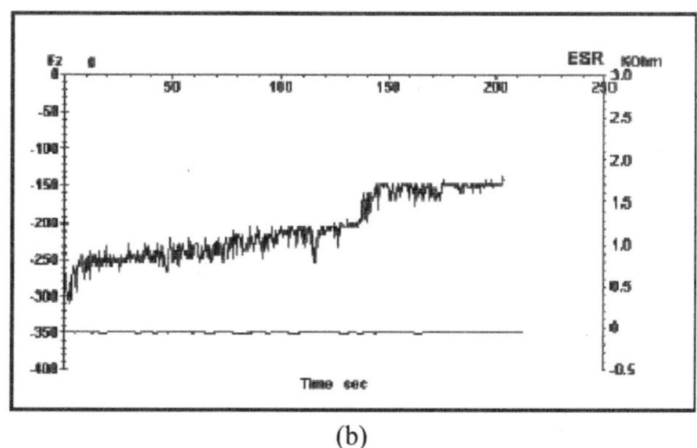

(b)

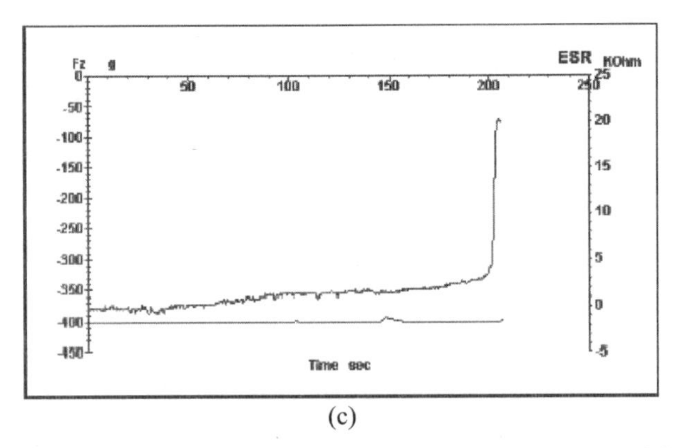

(c)

Fig. 23.3 Scratch raw plots for sample 1 at different loads: (a) 200 g (2N); coating was not cut (ESR remained low), (b) 350 g (3.5N); coating started to break (ESR increased slightly) and (c) 400 g (4 N); coating was broken but not totally cut through (ESR increased).

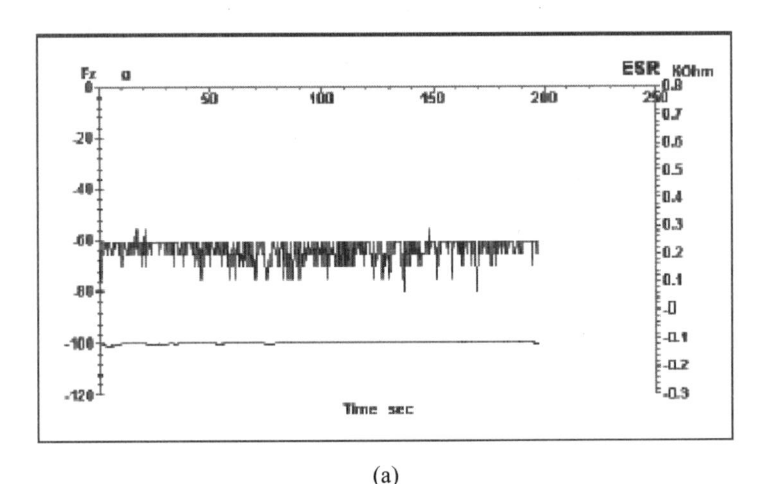

(a)

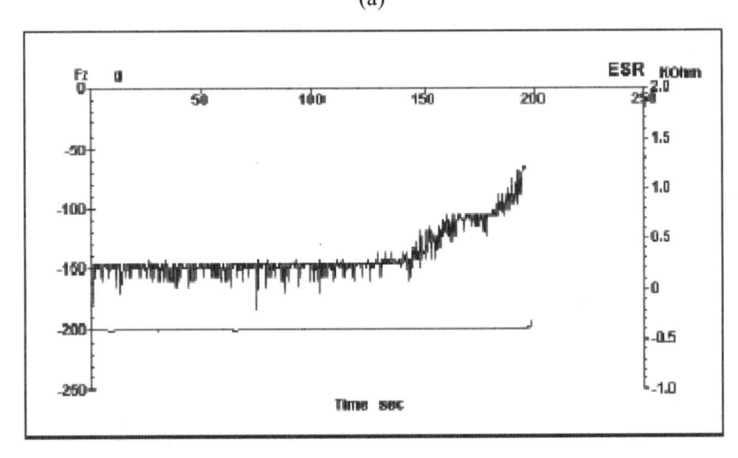

(b)

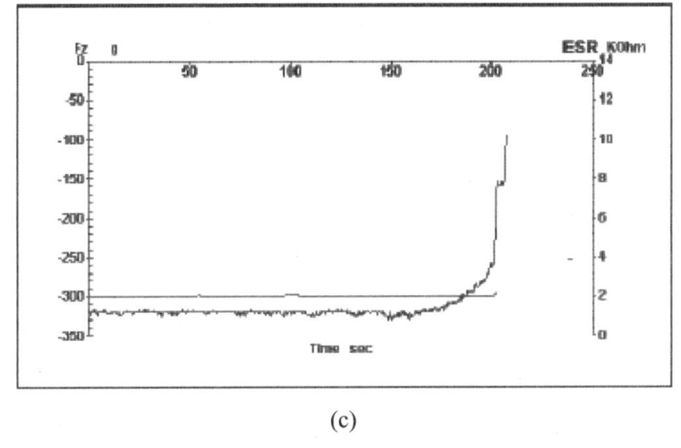

(c)

Fig. 23.4 Scratch raw plots for sample 2 at different loads: (a) 100 g (1N); coating was not broken (b) 200 g (2N); coating started to break and (c) 300 g (3N); coating was broken but not totally cut through.

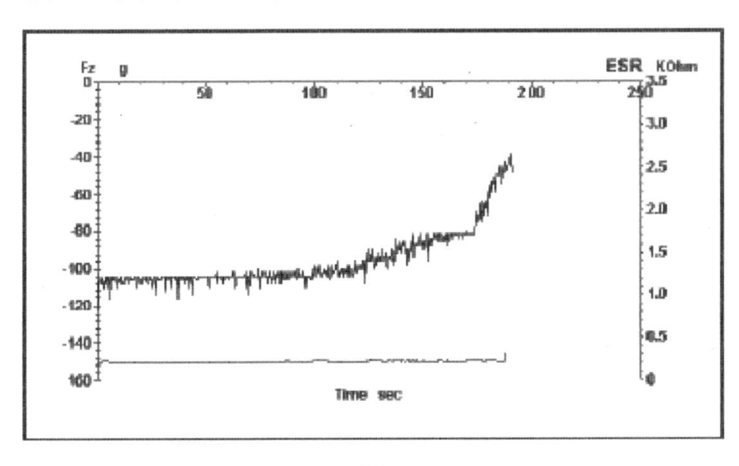

(a)

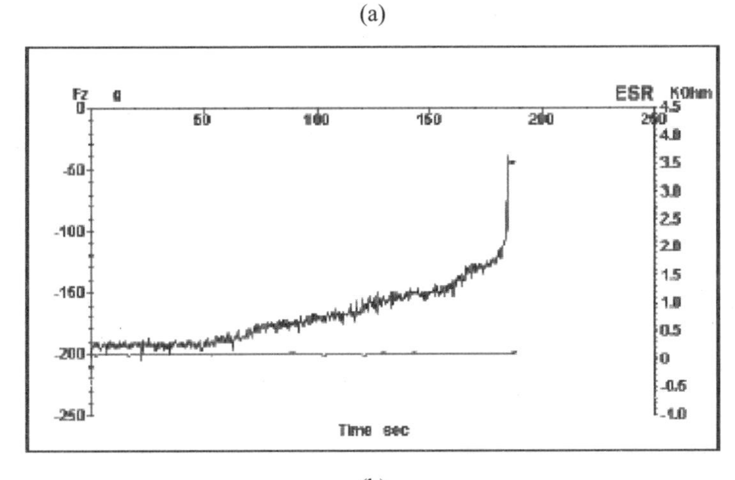

(b)

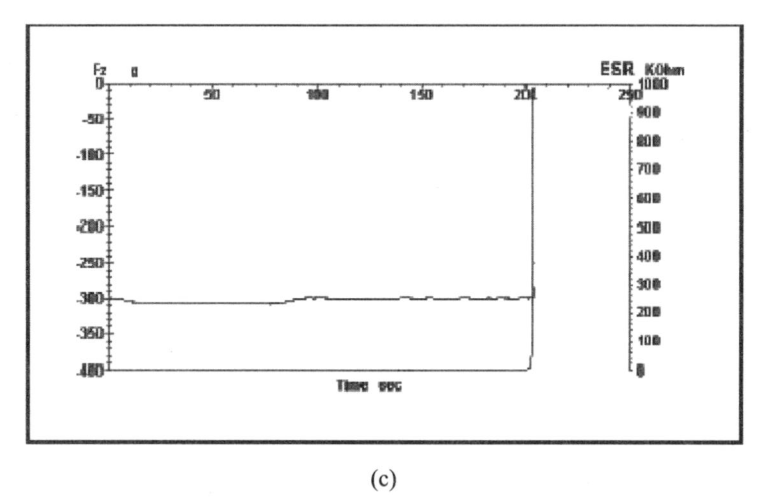

(c)

Fig. 23.5 Scratch raw plots for sample 1 at different loads: (a) 150 g (1.5N); coating was broken but not totally cut through, (b) 200 g (2N); coating was broken but not totally cut through, and (c) 300 g (3N); coating was totally cut through as ESR jumped to its maximum level.

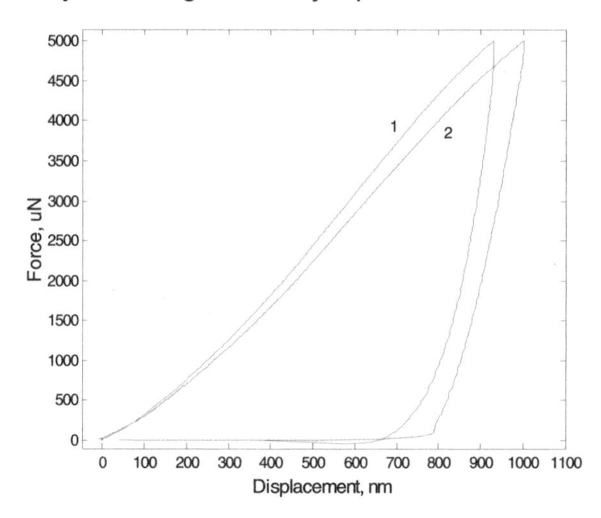

Fig. 23.6 Loading-unloading curves for treated (1) and not altered (2) Cu coated Si wafer sample.

A similar Cu work-hardening phenomena can be observed by testing nano-mechanical properties of Cu coated wafers during CMP process or by tribological tests and is reported in the literature[23.11].

The experimental results were consistent with the AFM images. An image of a typical indent is shown in Fig. 23.7. Here, a triangle shaped indent and small amount of associated material pile-up is seen in the center of the scan.

An optical images of the sample indicated boundary where a compressed surface can be observed.

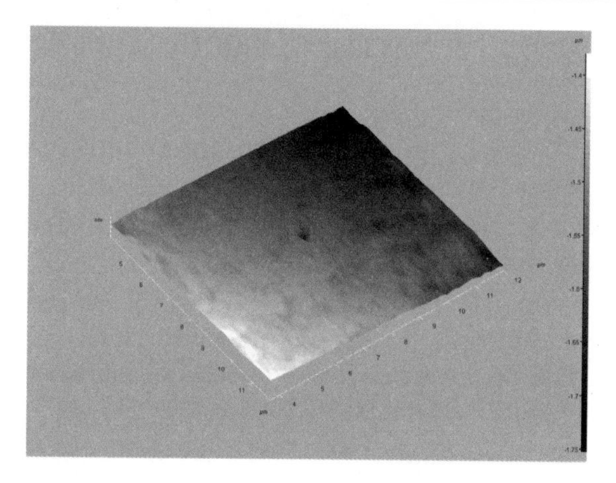

Fig. 23.7 AFM image of indent.

23.5 Conclusions

The novel test procedures on the Nano & Micro-Tribometer mod. UNMT, based on the precision servo-control of normal load and simultaneous measurements of electrical resistance, allow for accurate and repeated quantitative evaluation of scratch, adhesion and wear of conductive coatings. The multi-sensing test procedure allows for accurate and recurring quantitative evaluation of durability of thin films and thick coatings.

Performance of the nano-indenter is demonstrated by the work-hardening effect observed on very thin copper films.

A powerful combination of a micro-tribometer, nano-indenter and AFM is filing the gap between micro-scale tribology and nano-scale metrology.

References

[23.1] Singer, I.L., Dvorak, S.D., Wahl, K.J., 2000, "Investigation of Third Body Processes by *In Vivo* Raman Tribometry," NordTrib2000 Conf. Proc. held in Porvo, Finland, June 11-14, 2000, pp. 1-6.

[23.2] Singer, I.L., Dvorak, S.D., Wahl. K.J. and Scharf, T.W., 2003, "Role of third bodies in friction and wear of protective coatings," *J. Vac. Sci. Technol. A* **21** pp.232-240.

[23.3] Fisher, H., Jauss A., Hollrichter O., 2004, "Combining Cofocal Raman Microscopy and AFM in a Modular Microscopy System," GIT Imaging & Microscopy, **6**, 1, pp. 48-49.

[23.4] Eglin, M., Rossi, A., Spenser, N.D., 2003, "A Combinatorial Approach to Elicidating Tribological Mechanisms," Tribol. Letters, **15**, 3, pp. 193-198.

[23.5] Liu, X., Gao, F., 2004, "A novel multi-function tribological probe microscope for mapping surface properties, "*Meas. Sci. Technol.,* **15**, pp. 91-102.

[23.6] Gitis, N., 2004, "Tribometrological Studies in Bioengineering," Proc. of SEM X Int. Congress on Experimental Mechanics held in Costa Mesa on June 7 – 10, 2004, pp. 1 – 11.

[23.7] Sikder, A.K., Kumar, A., Thagella, S., 2004, "Effects of Properties and Growth Parameters of Doped and Undoped Silicon Oxide Films on Wear Behavior during Chemical Mechanical Planarization Process," J. Mater. Res., **19**, 2004, pp. 996-1010.

[23.8] Daugela, A., Gitis, N., and Meyman, A. "Contact Mechanical Impedance Method In Nanoindentation", Proc. TRIB2004: 2004 ASME/STLE International Joint Tribology Conference, Long Beach, California USA, October 24-27, 2004.

[23.9] Oliver W.C., Pharr, G.M., 1992, J. Mater. Res., **7**, pp. 601 – 608.

[23.10] ISO 14577-1, Metallic materials - Instrumented indentation test for hardness and materials parameters - Part1: Test method.

[23.11] Sikder, A. K., Kumar A., Shukla P., Zantye P.B. and Sanganaria M., "Effect of Multistep Annealing on the Mechanical and Surface Properties of Electroplated Cu Thin Films", J. Elec. Mater. **32**, (2003) pp. 1028.

24 Integrated Multi-Sensor Tribo-SPM (Scanning Probe Microscope) Testing for Nano/Macro Tribo-Metrology

24.1 A General Review

A novel quantitative nano and micro-tribometer with integrated SPM (Scanning Probe Microscope) and optical microscope imaging has been developed to characterize numerous physical and mechanical properties of liquid and solid thin films and coatings, with in-situ monitoring their changes during micro and nano indentation, scratching, reciprocating, rotating and other tribology tests. Both the material properties and surface topography can be assessed periodically during the tests. The integrated multi-sensing tribo-metrology is illustrated with two examples, of nano-indentation characterization of silicon wafer based coatings and micro-scratch characterization of diamond-like carbon coatings on magnetic media on the same instrument.

24.2 Introduction

Quantitative nanometer resolution metrology tools have become a standard in semiconductor, data storage and other hi-tech industries, where products have to be tested for thin-film properties. Though it is critical to characterize advanced thin films and coatings, today's off-line nano-scale metrology tools can capture only a limited number of parameters. There is an immediate need for process control instruments capable of in-situ nanometer scale quantitative characterization at different stages of manufacturing process. Latest advances in nano and micro tribology forced the development of integrated instrumentation. A new generation of innovative optical/SPM instruments for chemical and mechanical characterization of surfaces is being developed for tribological testing applications. A combinatorial approach in micro-scale tribology tests Eglin, M., et al.[24.1] indicated the need for in-situ characterization instruments integrated into a tribometer. A recent nano-scale tribometer combination with AFM (Atomic Force Microscope) instrument has been reported[24.2]. Therefore, the number of novel tribometer applications is growing fast and covering fast changing industries, such as biomedical[24.3].

Performance of a quantitative nano and micro-tribometer[24.4], integrated with AFM and high resolution optical microscope, is demonstrated in two examples, on silicon wafer coatings, where quantitative material properties were derived at several intermediate characterization steps with nano-indentation, and on diamond-like carbon coatings on magnetic disks, where disk durability was characterized by micro-scratching with a patented variable-angle blade micro-scratcher.

24.3 Nano-indentation

24.3.1 *Instrumentation*

Photo of the newly developed instrument is shown in Fig. 24.1. The Universal Nano and Micro Tribo tester (UNMT) system consists of a fully automated tribometer[24.4], capable of performing numerous multi-axial linear and rotary tribological tests, 3 μm resolution optical microscope, closed-loop interchangeable SPM scanner and a nano-indentation instrument Nanohead-1.

Fig. 24.1 Photo of UNMT with integrated AFM and optical microscope.

The UNMT has easily interchangeable rotary and linear, including fast oscillations, lower and upper drives, that provide a speed range from 0.001 (0.1 μ/s) to 10,000 rpm (50 m/s). It can apply a servo-controlled load that can be programmed to be constant or variable, in several ranges from 0.1 μN to 1 kN. The AFM and Nano-head can be installed onto the motorized Z-stage using a quick-release connector. All the UNMT motions and signals are controlled with a dedicated PC and sophisticated control software package. The system is placed on an active anti-vibration table to ensure precision measurements of the order of nano-meters. Both the Nano-head and the AFM have sub-nano-meter resolution in terms of displacement noise floor. The Nano-head has a sub-micro-Newton force noise-floor, maximum ranges of Z–displacement and force of 300 μm and 500 mN, respectively. Instrument calibration and mechanical properties extraction for the Nano-head are performed automatically according to the ISO 14577 standard for instrumented nano-indentation[24.5]. A closed-loop SPM scanner has up to a 100 × 100 μm scanning area.

24.3.2 *Methodology*

The Nanohead–1 shown in Fig. 22.12 of Chapter 22 is a sub-nanometer resolution quantitative nano-mechanical test instrument integrated into the Universal Nano and Micro Mechanical Tester and capable of performing nano-indentation tests, where the applied load F_z and penetration depth h are continuously monitored. The load versus depth plots are generated and processed from the collected data. The sample hardness H and the reduced elastic modulus E_r are calculated from the unloading segment of the curve as below. The reduced modulus is defined as follows:

$$E_r = S\frac{\sqrt{\pi}}{2\sqrt{A}} \tag{24.1}$$

where S is the unloading stiffness $\left(\frac{dF_Z}{dh}\right)$ and A is the projected contact area.

The stiffness S is calculated by fitting the unloading curve to the power law curve, i.e.:

$$F_Z = B(h - h_f)^m \tag{24.2}$$

Here, B, h_f, and m are arbitrary fitting parameters. The stiffness is calculated from the following derivative:

$$S = \frac{dF_Z}{dh}(h_{\max}) = mB(h_{\max} - h_f)^{m-1} \tag{24.3}$$

The nano-indentation hardness follows the classical hardness description and is defined by the ratio of the maximum load to the projected contact area:

$$H = \frac{F_{Z\max}}{A} \tag{24.4}$$

The contact area is determined via a tip calibration function $A(h_c)$ where h_c, the contact depth, is found by using the following equation:

$$h_c = h_{\max} - \varepsilon\frac{F_{Z\max}}{S} \tag{24.5}$$

Edge effects caused by surface deflection at the contact perimeter are compensated by taking the geometric constant ε = 0.75 as discussed in [24.6-24.7].

24.3.3 *Experiment*

Eight silica nitride and titanium nitride coated samples were blindly tested and characterized using the methodology and the instrument described above. Over 120 indentation tests were performed on all the samples.

The samples were individually mounted on AFM specimen stubs and tested using a Berkovich tip, a 3-sided tip with an included angle of 142°, with a tip radius of approximately 80 nm, as seen in Fig. 24.2. The samples were tested using a trapezoidal loading profile that loaded in 5 s, held the maximum load for 5 s to allow for creep, and unloaded in 5 s. Several tests were conducted at various loading rates to ensure that the sample was not strain-rate dependant and so was not causing hardening at higher loading rates. The majority of the tests were performed at 3000 μN, but additional tests were performed at the higher and lower loads to ensure that there was no change in the properties as a function of the depth into the sample. Here, 1500, 3000, and 4500 μN loads were chosen for nano-indentation tests to ensure that resulting indentation depth was in the range of 150 nm, that is less than 5% of the thickness of coatings (4 μm). For the silica-base coatings, substrate effects are observed on the data when indentation depth is approaching 30-40 % of the total coating thickness. Overall, different types of samples produced different reduced elastic modulus and nano-hardness values, indicating that both the nano-indentation technique and the instrument can distinguish different samples. Scattering of the data is due to the non-homogeneity of the coatings and potential instrumentation errors.

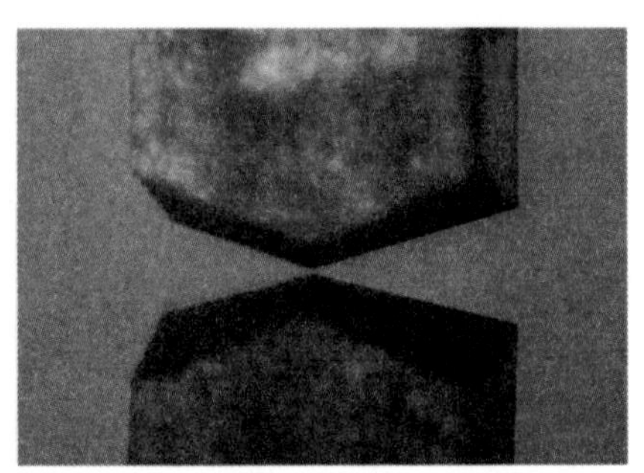

Fig. 24.2 Optical microscope side view of indentation process. (Reflection of the tip is seen on the sample, at distance to the surface of 5-7 μm.)

The results are summarized in the Table 24.1, yielding the following observations:

1. Similarly patterned TiN coated samples #5 and #6 have *Er* and *H* differences of 10%. Indentation into the silicon substrate of the sample #5 gave $Er = 72$ GPa and $H = 10.5$ GPa values, that are in the range of the typical values for silicon (e.g., for a fused quartz $Er = 69.6$ GPa and $H = 9.5$ GPa).

2. Mechanical properties of the pair of similarly patterned SiN coated samples #7 and #8 were found to be almost the same, suggesting that the same type of coating process was used.
3. TiN coated sample #1 had the highest hardness, though its *Er* was in the same range as for #7 and #8.
4. The softest was Cu coating #2, which had *Er* and *H* values of the order of magnitude lower than the rest of the samples.
5. Sample #4 had *Er* and *H* similar to the bare silicon.

Table 24.1 Mean values of measured REM and hardness with corresponding data scattering

Samples	#1 (SiN)	#2 (Cu)	#3 (SiN)	#4 (Si)	#5 (TiN)	#6 (TiN)	#7 (SiN)	#8 (SiN)
REM, Er, [GPa]	105.4 ± 5.1	14.3 ± 0.7	89.1 ± 2.28	74.2 ± 1.78	110.26 ± 4.8	128.9 ± 6.9	95.58 ± 5.1	94.59 ± 2.4
Hardness, H, [GPa]	16.64 ± 0.43	0.87 ± 0.04	13.71 ± 0.36	12.5 ± 0.55	15.21 ± 0.78	16.01 ± 0.4	14.33 ± 0.72	13.83 ± 0.3

A loading-unloading curve for the SiN coated wafer indent is shown in Fig 24.3. The sample had a well-pronounced elastic component. The AFM images were taken with the UNMT's SPM module in between the tests in order to image surface of the sample. A 16 × 16 μm AFM image taken between the nano-indentation tests of the TiN coated sample #2 is shown in Fig. 24.4.

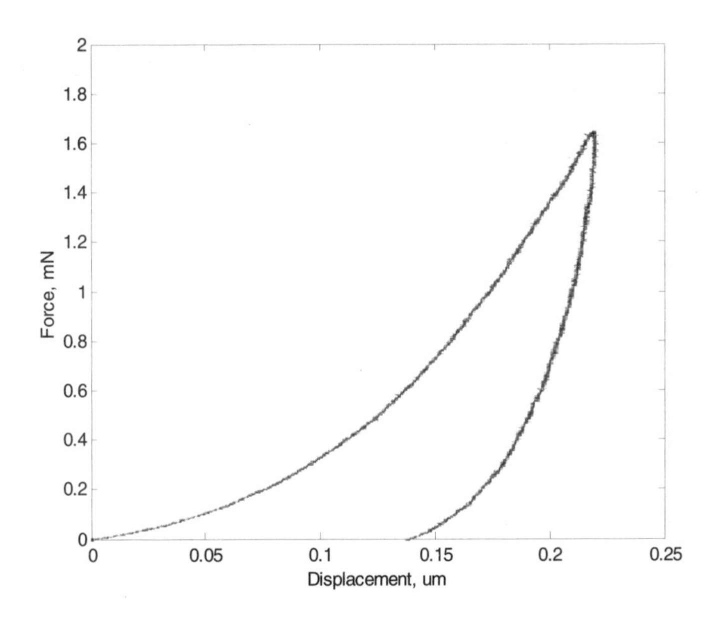

Fig. 24.3 Loading-unloading curve on Si wafer.

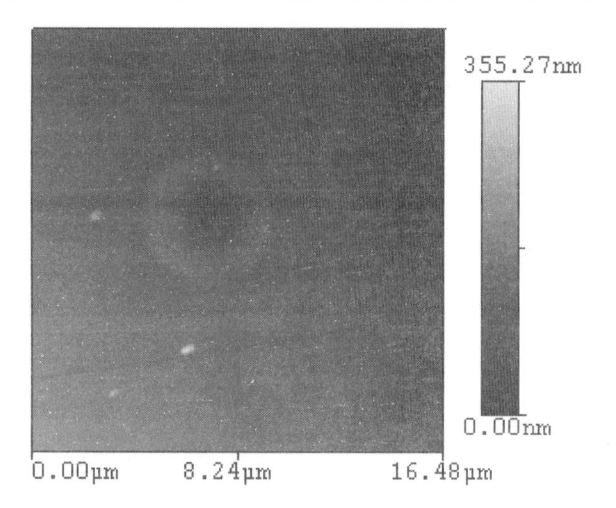

Fig. 24.4 AFM image of TiN coated sample.

24.4 Micro scratch

24.4.1 *Instrumentation*

Micro/nano scratch testing is frequently used to investigate the behavior of thin films under various loading conditions. In particular, scratch testing can be used to investigate coating adhesion and failure modes. Many types of scratch probes are utilized for different types of coatings such as the DLC (diamond-like carbon) coating on magnetic disks. A tungsten carbide micro-blade with adjustable-angle holder has been used to evaluate durability of both freshly deposited bare overcoats and finished (lubricated and burnished) disk surfaces. The 0.8 mm micro-blade was chosen based on the contact stress analysis. For spherical or cylindrical contact geometry, the contact stresses are distributed deeper than a few nanometers of film thickness. For evaluation of thin films, however, the contact stresses should be concentrated within or near the films, which is achieved with the special micro-blade geometry. The schematic of the UNMT used in this study is shown in Fig. 24.5, the close-up view of the micro-blade with holder is shown in Fig. 24.6. The test procedure involved servo-controlled loading, multiple sensors, and precision motion.

24.4.2 *Experiment*

The fast and quantitative micro-scratch method was applied for ultra-thin coatings of 4.5 nm, covered with a mono-layer 1.5-nm lubricant. Three types of lubricated disks, two samples of each, were tested, named as Bias Carbon at 300 V (samples 300_1 and 300_2), Bias Carbon at 150 V (samples 150_1 and 150_2), and Bias Carbon at 50 V (samples 50_1 and 50_2). Here, bias means film deposition voltage. The tests were performed at both A and B sides of the disk, three tests on each side at three different locations, OD (disk radius 47.5 mm), MD (disk radius 32 mm), and ID (disk radius 16 mm). During the test, while the micro-blade moved slowly against the film coating, progressive materials removal occurred.

As shown in Fig. 24.7, the measured coefficient of friction COF, normal force Fz, and electric contact resistance ECR were monitored with time. The normal force was gradually increased from 20 mN to 1000 mN. The coating film was cut through in 36 s, which corresponded to a critical load of about 300 mN. At that critical load, COF shifted to a higher value with a different slope, while ECR dropped to practically zero, because the micro-blade made contact with the conductive magnetic film after cutting through the coating.

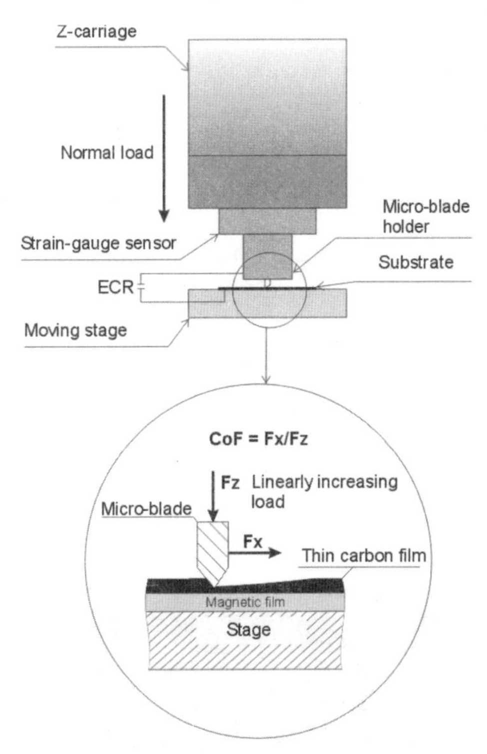

Fig. 24.5 Schematics of UNMT micro-scratch module.

Fig. 24.6 Photo of UNMT micro-blade module.

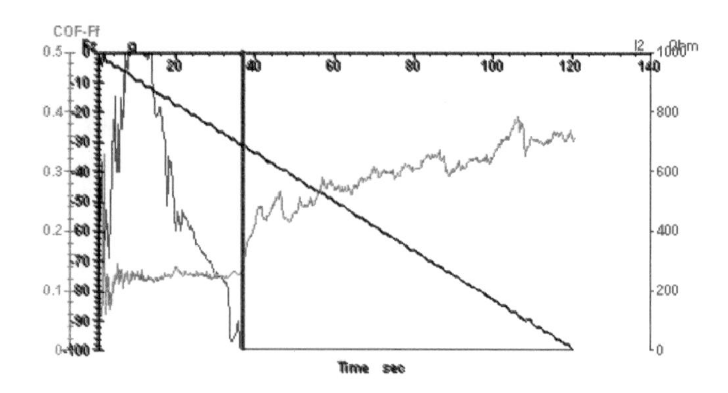

Fig. 24.7 Coefficient of friction, contact electrical resistance and ramping vertical force versus time.

The fact of the ultra-thin coatings break-through at the observed critical loads was confirmed by the integrated AFM images.

The UNMT tribometer allows for mapping of the tested surfaces, with 3-dimensional maps of coating durability, with a statistical analysis data for both the radial and the circumferential directions.

Wear tracks were examined after the test, using the Optical Surface Analyzer (made by Candela Instruments). The results are shown in Fig. 24.8. They confirmed that wear tracks were very recurring in both track pattern and track depth.

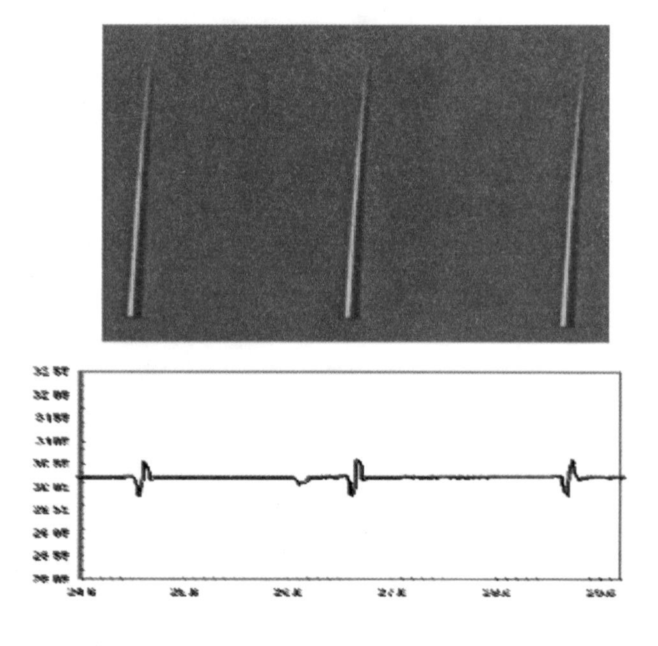

Fig. 24.8 Optical reflectometry image and profile.

Table 24.2 summarizes the critical loads for the disk overcoats at different disk radii. The scratch resistance was the highest for the disks with the 300 V-bias carbon, slightly lower for the disks with 150 V-bias, and the lowest for the disks with 50 V bias.

The Table 24.2 shows some radial non-uniformity of disk durability, with the scratch-resistance being the highest at the MD of the disks, lower at the ID and the lowest at the OD; this was then related to a sputtering chamber fixture, which was re-designed based on this durability data.

Table 24.2 Micro Scratch Test Results for DLC Coating on Magnetic Discs

Sample	Critical Load (g) at Disk Radius						
	16 mm		32 mm		47.5 mm		Average
	Side A	Side B	Side A	Side B	Side A	Side B	
300_1	32	35	44	45	28	31	37.3
300_2	35	36	46	40	34	41	
150_1	37	39	41	49	25	35	34.8
150_2	28	29	36	44	25	30	
50_1	30	20	25	30	18	20	23.8
50_2	19	24	25	18	26	30	
Average	30.3		36.9		28.6		

24.5 Conclusions

The above studies concludes the followings:

The Universal Nano and Micro Tribometer with integrated nano-indenter, micro-scratcher, optical microscope, AFM and many other tribology modules is a powerful tool for nano and micro mechanical and tribology characterization of thin films and other nano-technology specimens.

In the example with nano-indentation, both hardness and reduced elastic modulus indicated differences in mechanical properties between different SiN and TiN coated blindly selected samples. In terms of the data scattering, the reduced elastic modulus measurement results were more pronounced than the nano-hardness ones.

Micro-scratch experiments resulted in finding critical loads to breakthrough the ultra-thin films. Simultaneous coefficient of friction and contact electrical resistance measurements helped to "fingerprint" the samples.

References

[24.1] Eglin, M., Rossi, A., Spenser, N.D., 2003. A Combinatorial Approach to Elicidating Tribological Mechanisms, Tribology Letters, **15**, 3, pp. 193-198.

[24.2] Liu, X., Gao, F., 2004, A Novel Multi-function Tribological Probe Microscope for Mapping Surface Properties, *Meas. Sci. Technol.,* **15**, pp. 91-102.

[24.3] Gitis, N., 2004, Tribometrological Studies in Bioengineering, Proceed. SEM X Congress on Experim. Mechanics, Costa Mesa, June 7–10, pp. 1 – 11.

[24.4] Gitis N., Daugela A., Vinogradov M., Meyman A., 2004. In-situ Quantitative Integrated Tribo-SPM Nano/Micro Metrology. Proceed. ASME/STLE Tribology Conf., Long Beach, October 24-27.

[24.5] ISO Standard 14577, 2002. Metallic Materials - Instrumented Nano-indentation for Hardness and Materials Parameters. Parts: 1, 2, 3.

[24.6] Oliver W.C., Pharr G.M., 1992, An Improved Method for Instrumented Indentation: advances in understanding and refinements to methodology. J. Mater. Res., **7**, pp. 1547-53.

[24.7] Oliver W.C., Pharr G.M., 2004, Measurements of Hardness and Elastic Modulus by Instrumented Nano-indentation: advances in understanding and refinement to the methodology. J. Mater. Res., **19**, No.1, pp. 3 - 21.

25
Evaluation of the Bond Strength of Thermal Sprayed Coatings

25.1 A General Review

A novel lateral force-sensing micro-indentation method, reported in this chapter was applied to evaluate the interfacial bond strengths of regularly grained and nano-grained Al_2O_3/TiO_2 composite coatings. The interfacial bond strength, i.e., the bond between a coating and its substrate (steel), was determined by performing micro-indentation tests, on a cross section near the interface. By monitoring changes in lateral force during indentation, the critical indentation force at interfacial de-bonding was determined. The interfacial bond strength was then determined based on this critical indentation force and interfacial stress analysis using the finite element method. The results were compared to those obtained from a pull-off test and showed consistency with them.

25.2 Introduction

Evaluation of the interfacial bond strength for coatings has been an important and difficult task in surface engineering. Many testing methods, such as the pull-off test [1, 2], bending test [3, 4], and peel-off test [5, 6], have been employed to evaluate the bond strength. However, these methods only provide average bond strength and test results may not correctly represent the intrinsic bond strength and reflect associated interfacial failure mode [7]. Application of these methods is also limited by other factors. For example, for the pull-off test, the adhesive used to glue a sample to the sample holder is required to have its bond strength higher than that of the interface. It is also difficult to use bending and peel-off tests to evaluate hard and brittle coatings [8]. Scratch testing [9, 10] is another technique for evaluating coating's adherence to substrate by moving a small diamond tip over the sample surface under a progressively increasing load. The initiation of interfacial de-bonding is detected from acoustic emission signals or the local load-displacement curve; the corresponding critical scratching load could thus be determined. However, cracking of coating may also produce acoustic signals or unusual changes on the load- displacement curve, so that the detected failure events may not be always related to de-bonding of the coating substrate interface. In addition, the critical scratching load is also affected by the shape of the stylus, coating thickness, mechanical properties of the coating and substrate, etc. These factors make the scratching method only semi-quantitative. Micro-indentation is a promising technique for interfacial bond evaluation. Micro-indentation may be performed on the cross section of a sample, either directly at interface between coating and substrate [11] or at the substrate near the interface [12], to evaluate the interfacial bond. The length of crack at interface caused by indentation is used to evaluate the interfacial toughness or strength. This method, however, only provides qualitative information on the interfacial bond; furthermore, at this stage, the method cannot be used to determine the initiation of interfacial de-bonding. There are also other methods for evaluating interfacial bonds, such as the cavitation test [9] and the laser spallation test [13]. However, these methods are only suitable for specific coatings, and their applications are limited.

Recently, a new method has been used applying lateral force-sensing indentation technique to evaluate the interfacial bond strength [14, 15]. During the test, indentation is performed on the cross section at the substrate side near the interface. Due to the asymmetrical constraint on the indenter tip when the tip is pressed near the interface, the indenter tip bears a non-zero lateral force whose magnitude increases as the indentation force is increased. When interfacial de-bonding occurs, the constraint from the interface is partially released. As a result, the tip will move in the opposite direction and result in a change in sign of the slope of the lateral force time curve as Fig. 25.1 shows. The critical indentation load corresponding to initiation of interfacial de-bonding can thus be determined. Based on the critical indentation load, the interfacial bond strength can be calculated using the finite-element method in combination with a Quadratic Delamination Criterion [14-16]. In this work, this method was applied to determine interfacial bond strength of a commercial thermal sprayed Al_2O_3/TiO_2 coating (Mecto 130) and a nano-structured Al_2O_3/TiO_2 coating on a mild carbon steel substrate. The obtained results were compared to results of a pull-off test. It was demonstrated that the lateral force-sensing indentation method was effective for the determination of interfacial bond strength.

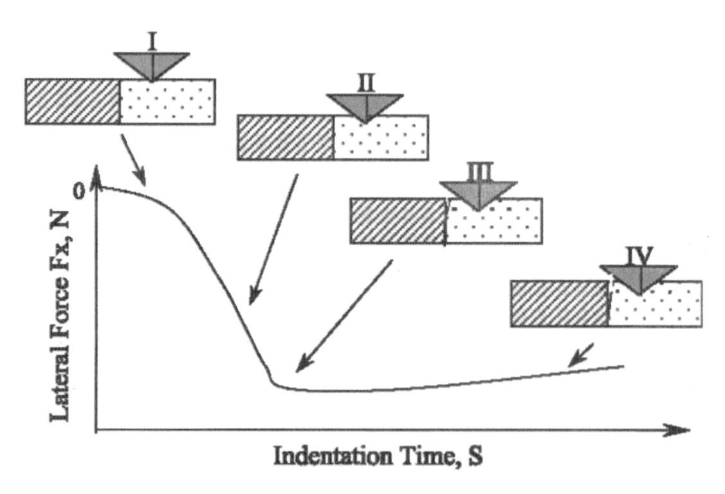

Fig.25.1 Variation of Lateral Force with Time

25.3 Experimental Method

In this work, interfacial bonds of thermal-sprayed $A1_2O_3/TiO_2$ composite coatings on steel substrate were evaluated. Compositions of the coatings were 87 wt.% $A1_2O_3$ and 13 wt.% TiO_2. One type of the coating had regular microstructure of melted splats. The plasma spray condition has been given in Ref. [17]. Another type of coating was deposited using nano-crystalline powers, which was composed of two different parts; one had the microstructure similar to that of Metco 130 with fully melted features, and the other had partially melted domains embedded in the melted splats [18]. Details about the fabrication of the $A1_2O_3/TiO_2$ nano-coating as well as microstructures and properties of these two types of coating have been reported in literature [17-19]. A schematic illustration of micro-structures of these two coatings is shown in Fig. 25.2. Both coatings were deposited on a mild carbon steel, which was blasted to remove rust and cleaned [17]. The coated steel was sectioned into 13 × 12 mm rectangular plates consisting of 3-mm-thick substrate and 100-μm-thick coating. The cross section of the samples was polished and the final surface roughness was about 0.05 μm.

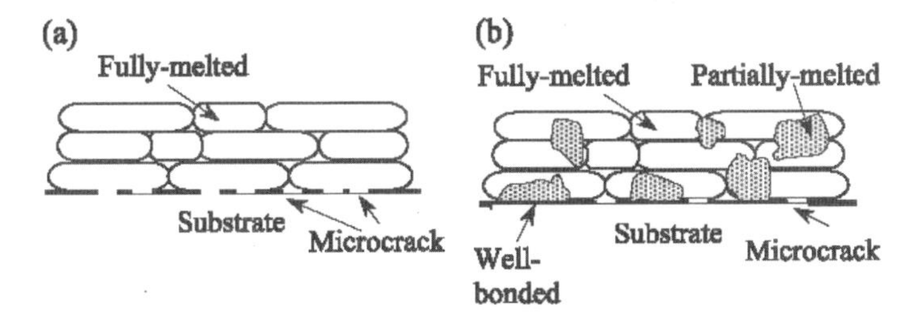

Fig. 25.2 Schematic illustrations of microstructures of two coatings. (a) Metco I30 and (b) nano-coating.

Micro-indentation experiments were performed using a Micro-Tribometer, developed by the Center for Tribology (California, USA). For the present study, a cone-shaped tungsten carbide indenter with a tip radius of 0.2 mm and tip angle of 30° was used. The indentation tests were performed on the sample cross-section at steel side near the interface at a distance of 30-80 μm as shown in Fig. 25.3. During indentation, the load increased linearly from 0 to 30 N at a speed of 0.013 N/s. During the test, the normal load, lateral force and time were recorded. The indentation position was determined using an optical microscope. The critical load corresponding to interfacial de-bonding was determined by averaging at least five measurements.

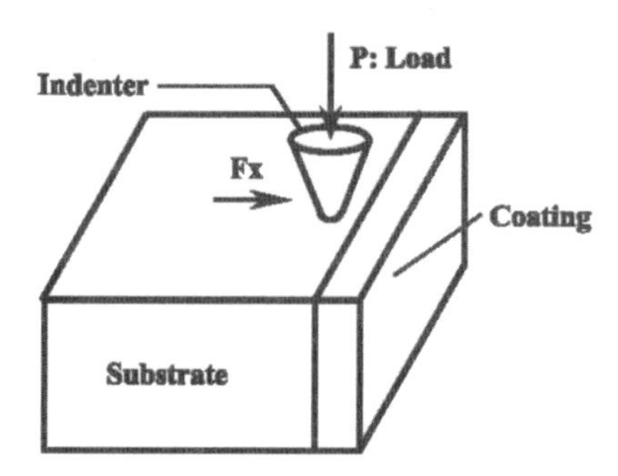

Fig. 25.3 Schematic illustration of the indentation test.

25.4 Determination of the Interfacial Bond strength

Typical normal load (L)-time curve and lateral force (Fx)-time curve are illustrated in Fig. 25.4. As shown, when the normal load was increased, the lateral force changed correspondingly. At the beginning of indentation, the absolute value of Fx increased gradually until the normal load reached a certain value, and then the absolute value of the lateral force decreased. Such a change corresponded to a change in slope sign of the Fx-t curve, indicating the occurrence of interfacial de-bonding as demonstrated in Refs. [14, 15]. An example of a resultant interfàcial crack is shown in Fig. 25.5.

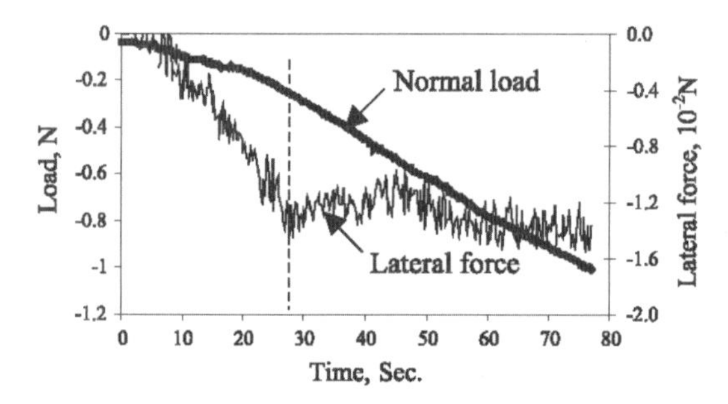

Fig. 25.4 Typical changes in load and lateral force vs. time for the nano-structured coating at distance of 40 micron from interface.

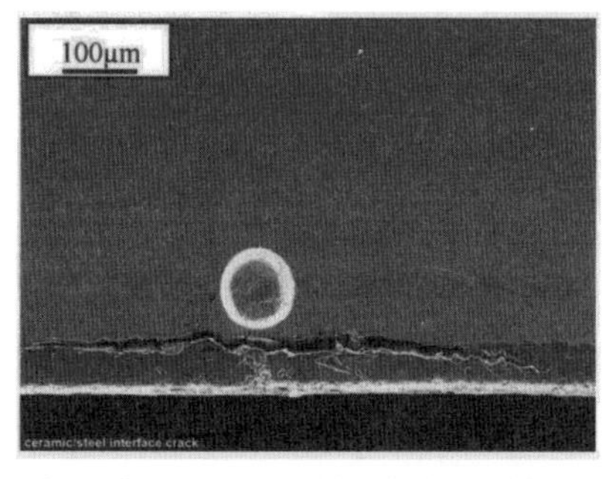

Fig. 25.5 Cracking at interface caused by indentation.

As mentioned above, the reason for the occurrence of lateral force is the asymmetric constraint on the indenter. When the stress at the interface was larger than the interfacial bond strength, de-bonding was initiated at the interface. As a result, the asymmetric constraint changed and this resulted in a change in sign of lateral force slope. The corresponding normal load is the critical load (Fc) at interfacial de-bonding.

The critical loads corresponding to the interfacial de-bonding at different indentation positions for both Metco 130 and nano-structured coatings are shown in Fig. 25.6. One can see that when indentation was applied near the interface, a smaller critical load is required to cause the interfacial de-bonding. Compared to the Metco 130 coating, the nano-coating needed a higher critical load to cause interfacial de-bonding at the similar indentation position. This means that the nano-coating had higher interfacial bond strength.

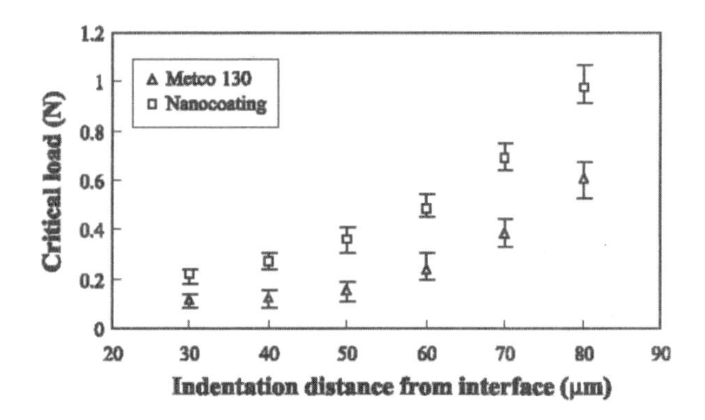

Fig. 25.6 The critical Loads vs. indentation distance from interface for the coatings.

To determine the local interfacial stress distribution for calculating the interfacial bond strength, a finite element model was employed to analyze the interfacial stress that corresponded to the measured critical load (Fc). The FEM analysis was made using ANSYS (version 7.0) software. All parameters used for the FEM analysis had the same values as those for the indentation experiment, such as the geometry and location of indenter and material properties.

The sample substrate was modeled using isotropic eight-node solid elements with elastic-plastic properties, while the ceramic coating was frilly elastic. The mesh near indenter and interface was refined to adequately reflect the stress gradient with sufficient accuracy. The indenter was assumed to be a non-deformable body. Mechanical properties of the substrate and coatings are given in Table 25.1. The bottom boundary of a specimen was constrained in all directions. Load was applied on the indenter until the critical load was reached. Using this model, the stress distribution at interface was calculated.

Table 25.1 Mechanical characteristics of involved materials

Materials	Young's modulus	Poisson's ratio	Yield stress	Tangential modulus
	E (GPa)	ν	σ_y (MPa)	E_t (GPa)
Carbon steel	200	0.25	540	10
Metco 130 coating	168.8	0.23	–	–
Nono-coating	158.2	0.23	–	–

Interfacial debonding generally occurs at free edge of interface, where the singularity could exist, as demonstrated in literature [20-22]. This may make the calculated stress at a particular point in the vicinity of the free edge meaningless. Thus, an average stress approach was adopted here. The average of stress component σ_{ij} is defined as [23, 24]:

$$\overline{\sigma_{ij}} = \frac{1}{x_{ave}} \int_0^{x_{ave}} \sigma_{ij} dx \tag{25.1}$$

where x_{ave} is a characteristic length along the interface, starting at the free edge, over which the integration is calculated. X_{ave} is treated here as an unknown parameter.

To determine the interfacial bond strength, a Quadratic Delamination Criterion [16] was adopted. This criterion takes into account both normal and shear stresses at the interface. The interfacial failure criterion between a coating and a substrate is defined as

$$\left(\frac{\overline{\sigma_{yy}}}{Z}\right)^2 + \left(\frac{\overline{\sigma_{yx}}}{S}\right)^2 = 1 \tag{25.2}$$

Where Z and S are the normal and shear interfacial bond strengths, respectively. $\overline{\sigma_{yy}}$ and $\overline{\sigma_{yx}}$ are respective average normal and shear stress over the average length x_{ave}. This criterion was used for the present calculation; the three unknowns, Z, S, and x_{ave} need to be determined.

In order to determine Z, S, and x_{ave}, at least three indentation tests are needed. In this work, five tests at different positions for a particular coating to find corresponding critical loads at interfacial de-bonding were performed. The following trial-and-error procedure was used to determine Z, S, and x_{ave} [16]. For each given trial set (Z, S, and x_{ave}), the predicted critical load could be obtained by finite element analysis. When eq. 25.2 is satisfied, the corresponding applied load is considered as the critical load, $Lc(i)$ ($i = 1, 2 \ldots, n$), where i represents each indentation position. The predicted critical load ($Lc(i)$) is then compared to experimentally determined critical load ($Le(i)$) using a ratio $Q(i) = Lc(i)/Le(i)$. The variation of coefficient, CV, is defined as

$$\mathrm{CV} = \frac{\sqrt{\frac{1}{n-1}\Sigma(Q(i)-\overline{Q})^2}}{\overline{Q}} \qquad (25.3)$$

where $\overline{Q}$ is the mean value of $Q(i)$. The best-fit set (Z, S, x_{ave}) could be determined by giving a mean value $\overline{Q} = 1$ while maintaining the minimum CV. Using this approach, we determined interfacial bond strengths for both the coatings, which are tabulated in Table 25.2.

Table 25.2 Interfacial bond strength determined using the lateral-force indentation method

Materials	Tensile strength (MPa)	Shear strength (MPa)	x_{ave} (mm)	Coefficient of variation (CV) (%)
Metco 130 coating	26.2	11.23	0.121	5.66
Nonocoating	44.01	29.45	0.120	3.94

25.5 Comparison of the Results Obtained using Different Methods

The result of the lateral force measurements (Table 25.2) indicates that the nono-structured coating has higher tensile and shear interfacial bond strengths, 44.01 and 29.45 MPa, than the commercial Metco 130 coating whose corresponding strengths are 26.2 and 11.23 MPa, respectively. The bond strength of the coating was also measured using a modified ASTM direct pull-off test [19], in which a coated sample is glued to another bulk material and a tensile stress is applied to cause interfacial de-bonding. The maximum stress at interfacial de-bond is the bond strength. Result of the pull-off test is given in Table 25.3 [19]. The results obtained using these two different methods are consistent. It is worth noting that the micro-indentation method yields slightly higher values than those from the pull-off test. As a matter of fact, due to the difference in mechanical properties between the coating and the substrate, normal stress and shear stress coexist at interface even under a uni-axial tension load. Therefore, it is not accurate or adequate to quantitatively evaluate interfacial failure using a pure tension test. Because the multi-axial stress state at interface has been taken into account, the lateral force micro-indentation should provide information that is closer to reality.

Table 25.3 Pull-off test results[25.19]

Materials	Tensile strength (MPa)	Coefficient of variation (CV) (%)
Metco 130	16.65	5.17
Nano-coating	39.30	3.95

Another reason for higher interfacial bond strength determined using the micro-indentation method is that some defects such as micro voids or micro cracks could be introduced at the interface during coating, sample cutting, and preparation. These defects may act as stress raisers, which facilitate interfacial de-lamination. Because the pull-off test only provides information on the overall performance of an interface, the stress concentration could result in interfacial failure under a smaller load than expected. For micro-indentation test, the defects are treated as one of interfacial micro-structural features. In this case, the influence of micro voids or micro cracks is, however, smaller, because of the lower probability of existence of the micro defects in a local area where micro-indentation test is performed. As a result, the determined bond strength should be higher. If micro-indentation test is performed in a defect-free region, the determined bond strength will be the intrinsic interfacial bond without influence from the interfacial defects.

It was demonstrated that the nano-structured coating had a higher interfacial bond than the Metco 130 coating. The commercial Metco 130 coating is a typical plasma-sprayed coating. Micro cracks were observed at its interface [18]. In the case of the nano coating, two different interfacial zones existed, one was similar to that observed in Metco 130 coating with micro cracks, the other was a partially melted zone as illustrated in Fig. 25.2, which is believed to be highly adherent to the substrate [18]. However, investigation of the mechanism responsible for higher interfacial bond of the nano-structured coating is beyond the scope of the present study.

From the above discussion, one might draw the conclusion that the lateral force-sensing micro-indentation technique can provide information on the local interfacial bond strength, including both normal and shear strength components. This technique is therefore not only useful to evaluation of interfacial bond strength but also suitable for fundamental investigation of defects' influences on interfacial strength if micro-indentation tests are performed in selected regions where interfacial defects, impurity segregation, and precipitates are present. Because the indentation can be carried out on both micro-level and nano-level, this technique would be effective for characterization and evaluation of a wide range of interfaces, including those in composites, coatings, and thin films.

25.6 Conclusions

From the above studies, the following conclusions are drawn :[25.25]

A newly developed lateral force-sensing micro-indentation technique was applied to evaluate interfacial bond strengths of regular and nano-structured Al_2O_3/TiO_2 coatings. Results of the test were compared to those obtained from a pull-off test. It was demonstrated that results of these two types of test were consistent. However, the interfacial bond strengths determined using the micro-indentation technique were higher than those from the pull-off test. Such difference could be attributed to the fact that what the micro-indentation test determined was closer to the intrinsic interfacial bond strength while the pull-off test only gave the average interfacial bond strength that was affected by interfacial defects such as micro cracks. Furthermore, the latter did not take account of the possible effect of singularity at free edge where the influence of shear stress might also exist, which could negatively affect the accuracy of the test. This study has also demonstrated that the lateral force-sensing micro-indentation technique is effective and feasible not only for evaluation interfacial bond strength but also suitable for fundamental investigation of effects of interfacial defects on the interfacial strength for a variety of interfaces in composites, coatings, and thin films.

References

[25.1] X.Q. Ma, F. Borit, V. Guipont, M. Jeandin, Journal of Advanced Materials 34 (2002) 52.

[25.2] Y.Z. Yang, Z.Y Liu, C.P. Luo, Yu.Z. Chuang, Surface and Coatings Technology 89 (1997) 97.

[25.3] G. Burkle, F. Banhart, A. Sagel, C. Wanke, G. Croopnick, H.J. Fecht, Materials Science Forum 386-388 (2002) 571.

[25.4] R. Beydon, G. Bernhart, Y. Segui, Surface and Coatings Technology 126 (2000) 39.

[25.5] J.Y. Sener, D.F. Van, F. Delannay, Journal of Adhesion Science and Technology 15 (2001) 1165.

[25.6] T. Lux, Surface and Coatings Technology 133-134 (2000) 425.

[25.7] D. Chicot, P. Démarécaux, J. Lesage, Thin Solid Films 283 (1996) 151.

[25.8] A.G. Evans, Adhesion Measurement of Films and Coatings 2 (2001) 1.

[25.9] H. Ollendorf D. Schneider, Surface and Coatings Technology 113 (1999) 86.

[25.10] M. Toparli, S. Sasaki, Philosophical Magazine. A 82 (2002) 2191.

[25.11] P.H. Demarecaus, D. Chicot, J. Lesage, Journal of Materials Science Letters 15 (1996) 1377.

[25.12] J. Bystrzycki, J. Paszula, R. Rrebinski, Journal of Materials Science 29 (1994) 6221.

[25.13] V. Gupta, J. Yuan, Journal of Applied Physics 74 (1993) 2397.

[25.14] D.Y. Li, Materials Science Forum 426-432 (2003) 2053.

[25.15] H. Zhang, Q. Chen, D.Y. Li, Acta Materialia 52 (2004) 2037.

[25.16] J.C. Brewer, P.A. Lagace, Journal of Composite Materials 11(1988) 1141.

[25.17] Y. Wang, S. Jiang, M. Wang, S. Wang, T.D. Xiao, P.R. Strutt, Wear 237 (2000) 176.

[25.18] P. Bansal, N.P. Padture, A. Vaailiev, Acta Materialia 51(2003) 2959.

[25.19] E.H. Jordan, M. Gell, Y.H. Sohn, D. Goberman, L. Shaw, S. Jiang, M. Wang, T.D. Xiao, Y. Wang, P. Strut, Materials Science and Engineering A301 (2001) 80.

[25.20] D. Munz, Y.Y. Yang, Journal of the European Ceramic Society 13 (1994) 453.

[25.21] M.A. Sckuhr, A. Brueckner-foit, D. Munz, Y.Y. Yang, International Journal of Fracture 77 (1996) 263.

[25.22] A.R. Akisanya, International Journal of Solids and Structures 34 (1997) 1645.

[25.23] R.Y. Kim, SR. Soni, Journal of Composite Materials 18(1984)70.

[25.24] S. Yi, L.X. Shen, J.K. Kim, C.Y. Yue, Journal of Adhesion Science and Technology 14 (2000) 93.

[25.25] H. Zhang, D.Y. Li, Surface and Coatings Technology 197 (2005) 137-141.

26

Friction and Wear of a Rubber Coating in Fretting

26.1 General

The friction characteristic of a rubber coating undergoing fretting-like oscillatory motion against a steel slider has been experimentally investigated. The variation of friction coefficient and rubber coating life as a function of load, velocity and displacement amplitude has been studied. Results of experiments on the effect of load, velocity and displacement amplitude on the coating life are discussed in this chapter.

26.2 Introduction

Fretting describes a type of contact failure in systems that undergo very small vibration amplitudes. This type of a failure is often associated with a combination of oxidation corrosion and abrasive wear between two contacting metallic bodies. Fretting is commonly found between the riveted joints of an airplane, tools of a pressing machine, in the contact of electrical switches, and so on. The oscillatory sliding motion causes not only interfacial heat generation but also stress concentration on the mating surfaces. The generation of heat contributes to oxidation corrosion while stress concentration leads to the formation of surface cracks and wear. As a result, a component undergoing fretting motion ultimately fails to perform its intended task.

Fretting research works have been devoted to different objectives ranging from the fundamental understanding of its mechanisms[26.1, 26.2] to establish relationships between fretting and fatigue[26.3-26.5] to assess cyclic variation of the friction coefficient[26.6, 26.7]. While these factors are intimately related, it is generally believed that the increase in the cyclic variation of friction is a major driving force behind the rapid growth in the contact stress, fatigue and wear.

Research aiming to combat fretting failure by means of application of a suitable coating to reduce friction are the subject of several published papers[26.8, 26.9]. Rubber coating is an example of a useful material often utilized to prevent direct contact between surfaces, to reduce leakage, to reduce stress concentration and to provide resistance to fretting by minimizing the vibration amplitude. Rivin has looked at the use of thin layers of rubber to accommodate relative displacements between the surfaces of meshing gears[26.10]. However, rubber tends to easily wear out and has a finite life. As a result, its protection of surfaces is limited.

A literature survey of research works dealing with wear and friction characteristics of rubber reveals that the majority of the open publications are limited to applications involving tires and belts. In these applications, the interest lies in the bulk properties of rubber undergoing continuous sliding motion[26.11-26.19, 26.22, 26.23]. The tribological behavior of a thin rubber coating as a means of protecting the metal substrate has thus far received little attention in the open literature.

In this chapter, results of experiments devoted to the understanding of the friction and wear characteristics of a relatively thin rubber coating stamped on steel plates have been brought out. The experiments were performed under different loads, oscillating velocities and displacement amplitudes.

Furthermore, a series of laboratory tests were also performed to determine the coating life as a function of load, displacement amplitude and velocity.

26.3 Experimental Procedure

The apparatus used for measurement of friction and wear was a universal micro-tribometer of Center for Tribology equipped with a computerized data acquisition system. A crank mechanism produced the necessary oscillatory motion. It was designed to accommodate changing the displacement amplitude. A variable-speed servomotor provided sinusoidal reciprocating motion. The schematic of the device is shown in Fig. 26.1. A piezo-force sensor was used to measure the normal force and tangential force. The resolution of the force sensor was 0.01 N. The load axis is controlled within 1 micron by the servomotor and the load error is within 0.2 N. The displacement and the rotating revolution per minute are controlled within 1 micron and 0.02 rpm by the servomotor, respectively.

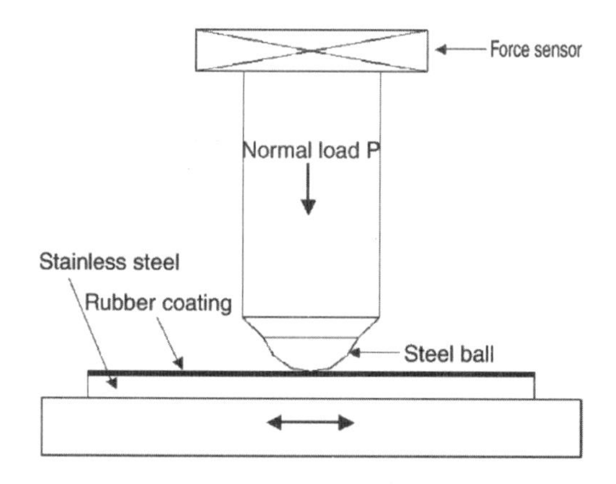

Fig. 26.1 Experimental apparatus.

The type of rubber was fluoropolymer. The typical properties of this material based on 1.90 mm thick slab were: shore hardness (HSA) of 66 MPa and tensile strength of 5.17 MPa. A screen-printing technique was used to have the coatings on the flat metal sheets (0.5 mm thickness). The screen was a wire mesh with only the designed area being open to allow the flow of coating. Once the sheets were coated, an oven cured the coatings.

All tests reported in this chapter were performed using a stainless steel ball that was of the same material as the plates. Its diameter and surface roughness (Ra) were 10 mm and 0.02 micron, respectively.

The normal force, the velocity and the displacement amplitude were varied to investigate the friction and wear of rubber coatings. All experimental tests were conducted under ambient conditions with a constant temperature of 20°C and a relative humidity of 40-50%.

26.4 Results and Discussion

In this section, series of experiments conducted to measure the friction coefficient of a rubber coating in contact with a stainless steel ball in reciprocating motion are reported. The normal loads were 2, 4, 10, 16 and 25 N. The displacement amplitude was 4.8 mm and the maximum velocity in the middle of the stroke was 0.6 mm/s.

26.4.1 *Friction exposed to oscillatory motion*

In a sinusoidally oscillatory motion, the velocity changes continuously from nil at each end of the stroke to a maximum at the middle of the stroke. Therefore, the friction coefficient changes continuously with velocity[26.11]. An example of experiments with fresh rubber is shown in Fig. 26.2. When the velocity is nil, the friction coefficient is small but finite. The maximum coefficient is seen to occur in the middle of the stroke and falls into its lowest value at the both ends of the stroke. This figure reveals that the friction coefficient is drastically affected by changes in the velocity as it oscillates.

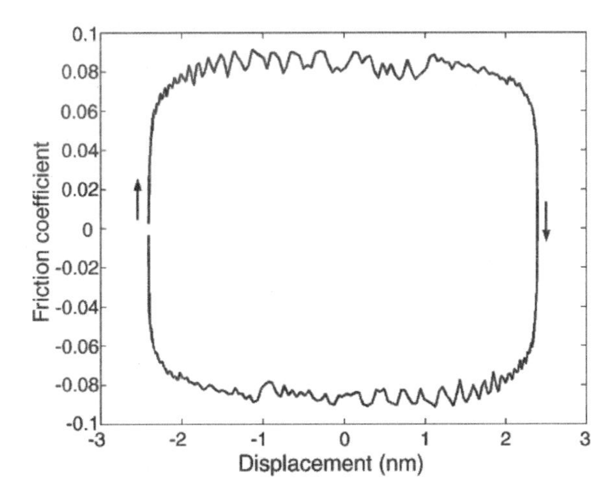

Fig. 26.2 Variation of friction coefficient with displacement under a normal load of 2 N.

The friction coefficient of the fresh rubber coating as a function of load is shown in Fig. 26.3. These results correspond to the friction at the middle of the stroke. It is shown that the friction coefficient of fresh rubber decreases in a non-linear fashion as the load increases.

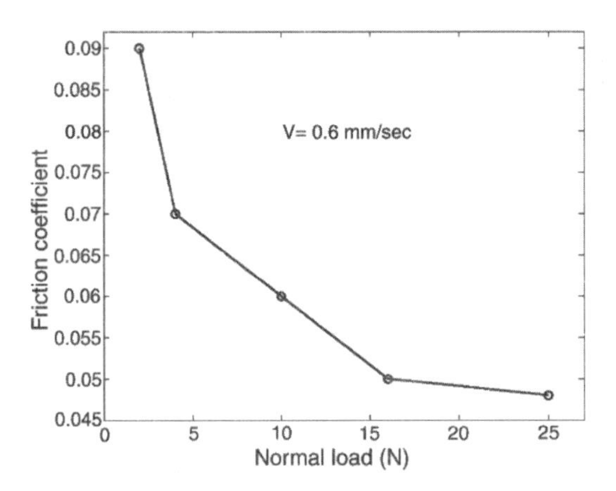

Fig. 26.3 Friction coefficient of fresh rubber as a function of load.

26.4.2 *Effect of load on friction coefficient and its relationship to coating life*

A series of experiments was continued until the rubber coating was fully worn off and the friction coefficient was measured under each one of the load values. The experiments were conducted in reciprocating motion, and so one cycle means one passage on the contact area. The results for coating life as a function of number of cycles are shown in Figs. 26.4 and 26.5. The friction coefficient of the rubber coating is seen to increase with the number of cycles as the rubber wears out. For example, the friction coefficient with 2N starts from 0.09 and reaches 0.15 after about 3600 cycles as shown in Fig. 26.4. When the load is higher, the friction coefficient varies much more rapidly. For example, the friction coefficient with 25 N is roughly 0.05 at the beginning but reaches 0.15 just after 100 cycles as shown in Fig. 26.4. The end cycle where the rubber coating is fully worn out is marked in Fig.26.4. At the beginning, the friction coefficient of fresh rubber decreases as load increases as mentioned earlier, and it also decreases as load increases when rubber is fully worn out. When the load is large the rubber quickly tears off and the friction coefficient sharply increases. As shown in Fig. 26.5, the higher the load, the shorter is the rubber coating life.

It has been reported that the wear volume of bulk rubber in continuous sliding increases as the normal load increases[26.12, 26.13], which agrees with the results of investigations with oscillatory motion. However, contrary to these results, Barquins[26.14] reported that the friction coefficient decreased as the number of cycles increased. In his research, the load and the sliding velocity were 20 mN and 500 micron/s, respectively. This disparity can be attributed to the lighter loading condition and the use of a glass ball in Barquins' experiments which resulted in a much less rubber wear.

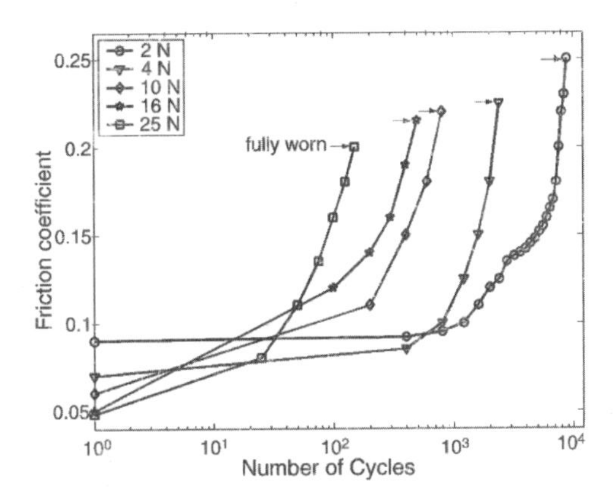

Fig. 26.4 Friction coefficient of rubber as a function of number of cycles.

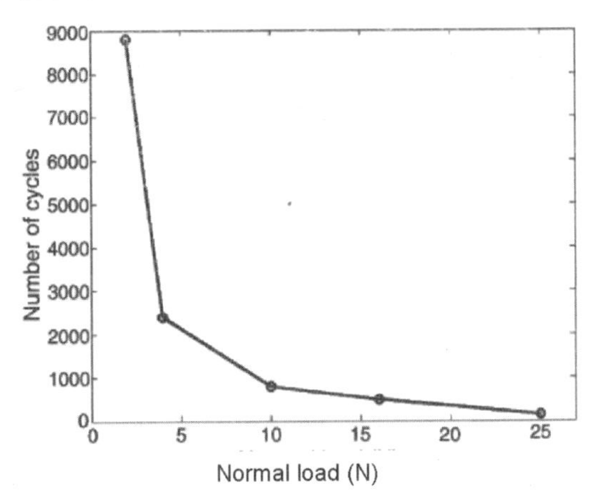

Fig. 26.5 Rubber coating life as a function of load.

26.4.3 *Effect of velocity on friction coefficient and its relationship to coating life*

A series of experiments were performed to measure the friction coefficient of rubber under various oscillating velocities and was continued until the rubber coating was fully worn off. The imposed load was 10 N and the other operating parameters were maintained as the same as the previous test in Sections 26.4.1 and 26.4.2. The results of the friction coefficient reported in this section correspond to the middle of the stroke where the maximum friction coefficient occurs. The friction coefficient of the rubber coating is found to increase as the sliding velocity is faster as shown in Fig. 26.6. It increases by four-fold from 0.03 at 0.025 mm/s to 0.12 at 100 mm/s.

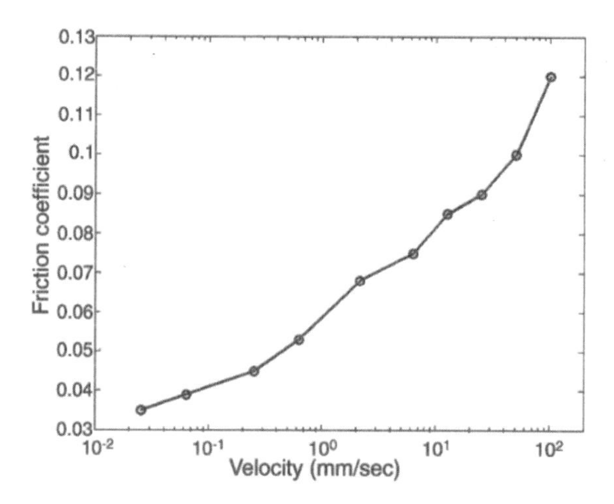

Fig. 26.6 Friction coefficient vs. velocity.

Fig. 26.7 shows how long the rubber coating lasts at different operating velocities. The coating life is measured in terms of the number of cycles before it is worn out. It is interesting to note that although the friction coefficient increases with increasing oscillating velocities, the coating life tends to be short at low velocity. As shown in Fig. 26.7, the rubber coating life is heavily dependent upon the velocity. Coating life is very short at very low oscillatory velocities, typically it lasts only 250 cycles at 0.025 mm/s. As the velocity is increased up to 2.5 mm/s, more cycles are needed for the rubber to completely wear off. At higher velocities, the coating life begins to deteriorate with increasing velocity. Further discussion on the effect of velocity on wear is given in Section 26.5.

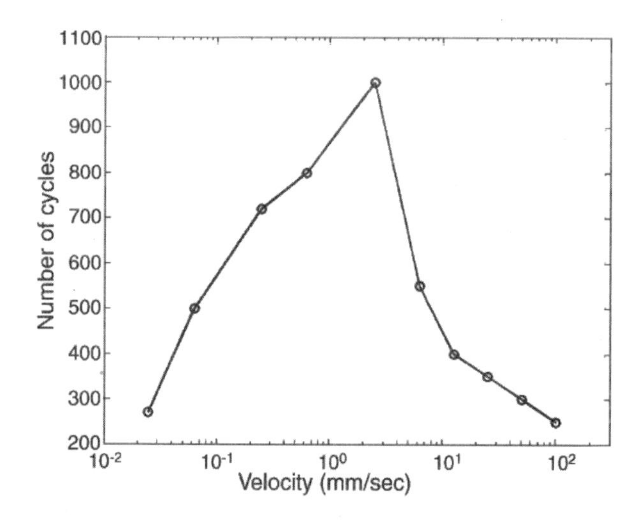

Fig. 26.7 Rubber coating life vs. velocity.

26.4.4 *Effect of displacement amplitude on coefficient of friction and its relationship to coating life*

The displacement amplitude of the reciprocating motion was varied from 0.076 to 4.8 mm, while the load was maintained at 10 N. The results of the friction coefficient obtained in the middle of the stroke with various velocities are presented in Fig. 26.8. As shown, the friction coefficient does not change much on varying the displacement amplitude.

The rubber coating life lasts longer when the displacement amplitude is shorter as shown in Fig. 26.9. Wear of rubber is a strong function of the displacement amplitude in oscillatory motion. The trend of life in the small displacement amplitude is similar to that in the big amplitude. That is to say, the coating life is short at low sliding velocities and at high velocities, in terms of the number of cycles necessary for the rubber to completely wear off.

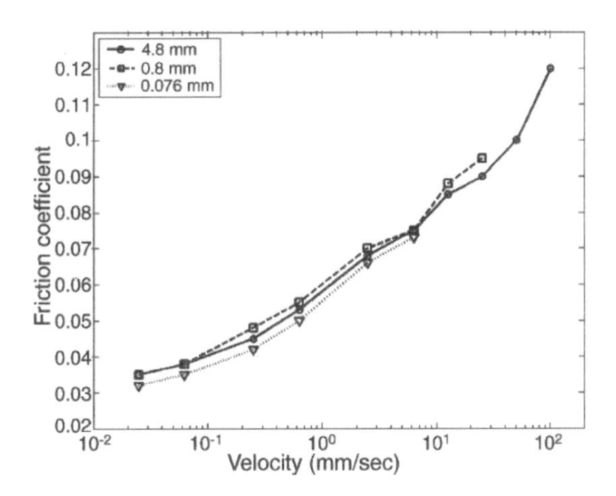

Fig. 26.8 Friction coefficient vs. velocity for various displacement amplitudes.

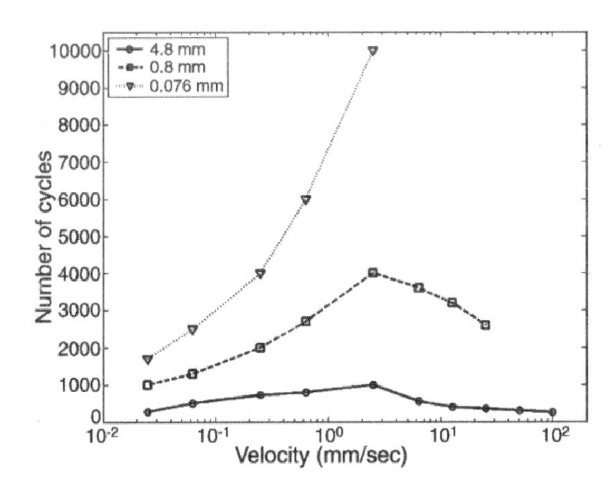

Fig. 26.9 Rubber coating life vs. velocity for various displacement amplitudes.

26.5 Discussion

Rubber is a viscoelastic material and its friction behavior is governed by viscoelastic material properties. In addition, rubber friction force depends on the operating speed and temperature. Further, viscoelastic losses and dependence on frequency can affect the tribological behavior of rubber.

M. Barquins[26.15, 26.16] obtained the following "master curve" formula with the functional form of $\mu = f(\log a_t\ v)$ to describe the dependence of friction coefficient on velocity, v and temperature:

$$\log a_t = -\frac{8.86(T - T_0)}{101.5 + T - T_0} \qquad (26.1)$$

where T represents the operating temperature and T_0 is a reference temperature. Barquins et al.[26.17] reported that friction force increased as the velocity increased but their experiments were conducted with a low velocity from 0.01 to 3.4 mm/s. B. Persson[26.18] reported that friction force decreased as the velocity increased in the high velocity range depending on the nature of the substrate surface roughness and on the mechanical properties of rubber. Persson's attention was focused on examining the behavior of bulk rubber undergoing continuous motion.

In this work, the temperature is constant and the velocity is sinusoidal, where in each stroke the rider undergoes an acceleration and deceleration. As shown in Fig. 26.2, the friction coefficient of the rubber coating increases in each stroke as the velocity increases. The maximum friction coefficient in the middle of the stroke increases as the velocity increases as shown in Fig. 26.8. The interaction between the varying velocity as well as acceleration/declaration within a stroke and the viscoelastic nature of rubber is quite complex and worth an additional detailed study.

Barquins and Courtel[26.15] and Barquins[26.16] studies revealed that Schallamach waves depend on temperature and the radius of the glass slider. In their studies, the bulk rubber and the glass slider were exposed to a relatively low load from 0 to 50 mN. The behavior of a rubber when used as a coating to avoid fretting, or when used for avoiding leakage, is also affected by the velocity and surface roughness of the slider[26.18]. All of the tests that detected a Schllamach's wave were conducted with a unidirectional and constant speed slider. In the experiments reported in the present chapter, no such waves were observed probably because of the oscillatory nature of the sliding motion, the high loads and the surface roughness. The Scallamach wave has been studied for an understanding of the friction mechanism of rubber[26.15, 26.16]. However, the relation, if any, between the Schallamach wave and rubber wear is still not clearly understood.

From a mechanical point of view, either a sphere on a plane or a cylinder on a plane is used in fretting studies. Theoretical analyses K. Johnson, A. Elkholy[26.20, 26.3] and experimental observations V. Lamacq[26.4] reveal that gross slip occurs at the end of the contact region between a ball (or a cylinder) and a plane. The maximum tensile stress is also found at the contact end [26.20, 26.3]. Given that the magnitude of the friction coefficient in metallic contact is inversely proportional to the sliding velocity[26.21], in tests involving reciprocating motion, the maximum friction is registered at the end of the stroke. These are the locations where fretting failure is likely to initiate. On the other hand, the friction coefficient of rubber is proportional to the velocity in oscillatory motion and varies from nil in the end of the stroke to the maximum at the middle of the stroke.

In the absence of coating protection, increasing the load on two contacting metals that undergo fretting, tends to slightly reduce the coefficient of friction[26.21]. Introduction of a suitable rubber coating would be useful since increasing the normal load reduces the coefficient of friction as shown in Fig. 26.3. A rubber coating can reduce the friction force on the whole contact path, especially at the extreme ends of oscillation amplitudes, where fretting fatigue is likely to initiate (see Fig. 26.2).

Many attempts have been made to find a relationship between the rate of abrasion loss and the physical and the mechanical properties of rubber[26.12, 26.13, 26.22, 26.23]. Y. Uchiyama[26.23] found the following relationship for determining the wear volume of rubber abrasion:

$$\mathrm{V} = k_1 \frac{\mu P}{\sigma_B} L \tag{26.2}$$

where V is the wear volume, μ represents the friction coefficient, P denotes the normal load, L the length of rubbing distance and k_1 is a constant The parameter σ_B is expressed by the following equation:

$$\sigma_\mathrm{B} = \sigma N \tag{26.3}$$

where σ is maximum amplitude of tensile stress and N is the number of cycles.

Fukahori[26.12] published the following relationship for the rate of abrasion with the mean strain amplitude, ε^* and the crack growth per cycle ($\mathrm{d}c/\mathrm{d}n$):

$$D = \frac{\mathrm{d}c(\varepsilon^*)}{\mathrm{d}n} \tag{26.4}$$

where $\varepsilon^* = \mu P/ES$, E is Young's modulus and S is the cross- sectional area. They proposed that abrasive wear of rubber is strongly influenced by the fracture resistance of the material undergoing repeated deformation as a result of two kinds of periodic motions: stick-slip oscillation and the micro-vibration generated during frictional sliding of rubber.

In this study, the rubber coating life is found to be short as the normal load increases as shown in Figs. 26.4 and 26.5, which is coincident with the prediction of Eqs. (26.2) and (26.4). But even though the friction coefficient at low velocity is very low as shown in Figs. 26.6 and 26.8 and other parameters are the same, coating life in low velocity is short as shown in Figs. 26.7 and

26.9. Even when rubber is exposed to a relatively low normal load, it is worn out after a small number of cycles at low velocity. Rubber wear seems to be also related to running time. Total running time is calculated from Fig. 26.7 and is shown in Fig. 26.10 with the coating life. As shown, the number of cycles in the positions marked by an arrow at low and high velocity are similar but the total running time at low velocity is very long. Rubber seems to be worn out easily in a small number of cycles under low frequency and long running time. It can be also concluded that rubber wear mechanism changes as sliding velocity changes.

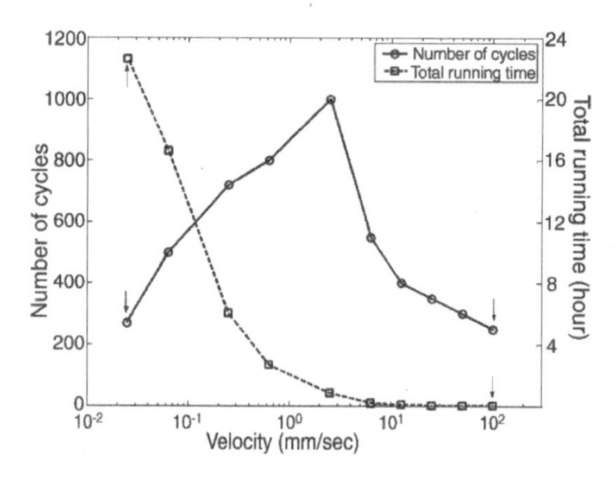

Fig. 26.10 Rubber coating life and total running time.

Fig. 26.11 shows a picture of fully worn rubber in the displacement amplitude 0.8 mm, in which all experimental conditions were the same except the velocity. The fully worn region of rubber at the velocity, 21.27 mm/s occurred in the middle of the stroke as expected. The worn region is small and especially very narrow in the direction perpendicular to the velocity as shown in Fig. 26.11(a). The fully worn area of rubber when the velocity is 0.025 mm/s is large and very wide in the direction perpendicular to the velocity, because the friction force is small and the total running time is very long. It is likely that the wear of rubber at low velocity oscillatory motion is related not to abrasion but to adhesion-type wear.

Fretting of a metal is divided into three regimes related to slip motion. They are: partial slip, mixed slip and gross slip conditions according to the displacement amplitude of sliding[26.6, 26.24]. Fouvry et al.[26.24] reported that the wear volume of metal in fretting increased as the displacement amplitude is increased. Liu and Zhou[26.6] showed that the friction coefficient and the depth of wear scar in dry sliding increased as the displacement amplitude increased. The present study of rubber coating revealed a similar result: the coating life is shorter as the displacement amplitude is increased. The difference between their metallic and our rubber contact can be explained by noting that in S. Fouvry[26.24], the fretting involved metallic contacts where the initiation of the wear scars started at the contact end points whereas rubber wear started at the center of the stroke as shown in Fig. 26.11. When a rubber coating is adopted in the contact the region of the maximum friction force varies from the middle of the stroke to the end of the contact as the number of cycle increases because rubber contact occurs in the beginning and metallic contact occurs after the rubber coating is fully worn out This relieves stress concentration on the contact.

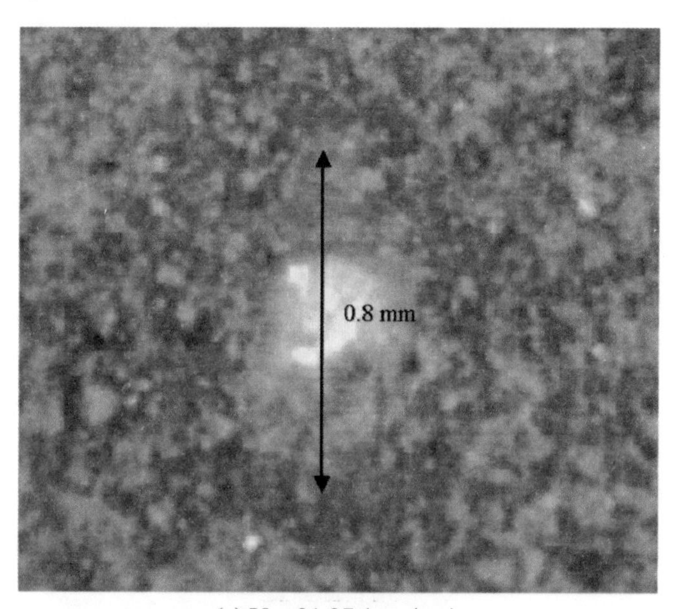

(a) V = 21.27 (mm/sec)

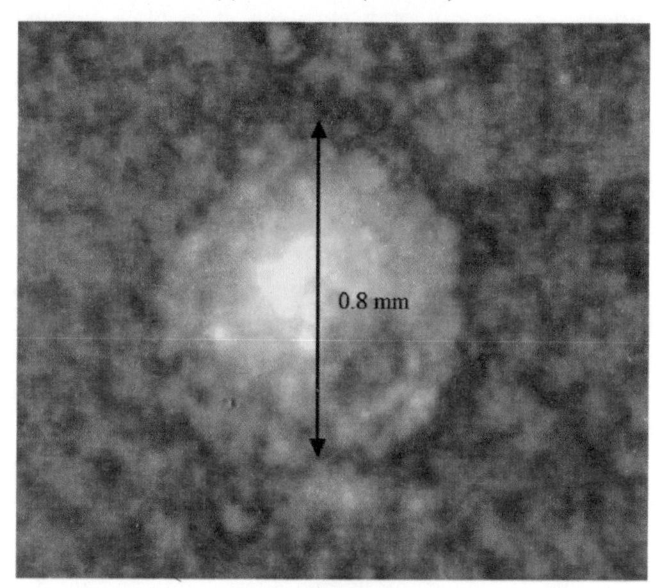

(b) V = 0.025 (mm/sec)

Fig. 26.11 Pictures of fully worn rubber (10 N): (a) v = 21.27 mm/s; (b) v = 0.025 mm/s.

The friction coefficient of the rubber coating is also less than the metallic contact as the number of cycles increases. An example of measured friction coefficient of metal against metal is shown in Fig. 26.12. The stainless steel ball slides on a stainless steel plate, and the friction coefficient increases rapidly as the number of cycles increases[26.4, 26.9, 26.13]. It reaches 0.8 after about 800 cycles in Fig. 26.12 but it reaches only 0.23 after 1200 cycles under the same operating condition as shown in Fig. 26.4. Therefore, it can be concluded that a rubber coating can protect the contact and prevent fretting failure.

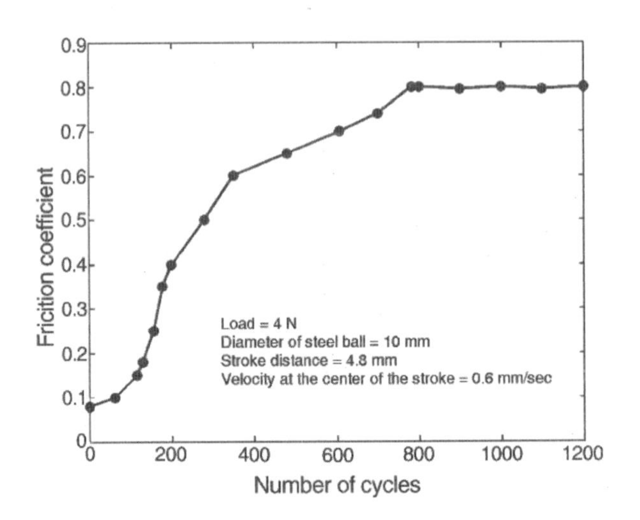

Fig. 26.12 Friction coefficient of a stainless steel plate against a stainless steel ball in reciprocating motion.

26.6 Concluding Remarks

The above studies draw the following conclusions[26.25]:

The friction characteristics of a rubber coating were investigated, particularly as related to fretting. The focus of this chapter was on the relationship between rubber friction and pertinent operating parameters such as the load, velocity and displacement amplitude. It was found that the friction coefficient of the rubber coating decreases as the load increases and increases as the velocity increases. However, the friction coefficient was found to be relatively insensitive to change of the displacement amplitude.

The rubber coating life is related to running conditions, i.e. the load, the velocity and the displacement amplitude.

The humidity and temperature may also play a role, but these factors were not considered in the present study as they were held constant. It is found that the rubber coating life decreases significantly as load is increased and as the displacement amplitude is longer. However, the coating life was found to be very short when the system undergoes oscillatory motion in a low velocity regime. Even though the friction coefficient may be low, the wear of rubber seems to be related to the total running time at low velocity and to be related to associated adhesion.

The friction force of the rubber coating is less than that of the metallic contact and the maximum friction force of the rubber coating is in the middle of the stroke but that of metallic contact is at the contact end at the end of the stroke. This can protect the contact and increase the life of a machine when operating in an environment susceptible to fretting.

References

[26.1] O. Vingsbo, M. Odfalk, N. Shen, Fretting maps and fretting behavior of some F.C.C. metal alloys, Wear 138 (1990) 153-167.

[26.2] E. Sauger, L. Ponsonnet, J. Martin, L. Vincent, Study of the tribologically transformed structure created during fretting tests, Tribol. Tnt. 33 (2000) 743-750.

[26.3] A. Elkholy, Fretting fatigue in elastic contacts due to tangential macro-motion, Tribol. Tnt. 29 (1996) 265-273.

[26.4] V. Lamacq, M. Dubourg, L. Vincent, A theoretical model for the prediction of initial growth angles and sites of fretting fatigue cracks, Tribol. Tnt. 30 (1997) 391-400.

[26.5] D. Swalla, R. Neu, Influence of coefficient of friction on fretting fatigue crack nucleation prediction, Tribol. Tnt. 34 (2001) 493-503.

[26.6] Q. Liu, Z. Zhou, Effect of displacement amplitude in oil-lubricated fretting, Wear 239 (2000) 237-243.

[26.7] M. Huq, J. Celis, Expressing wear rate in sliding contacts based on dissipated energy, Wear 252 (2002) 375-383.

[26.8] J. Braza, R. Furst, Reciprocating sliding wear evaluation of a polymeric tribological system, Wear 162—164 (1993) 748-756.

[26.9] C. Langlade, B. Vannes, M. Taillandier, M. Pierantoni, Fretting behavior of low-friction coatings: contribution to industrial selection, Tribol. Tnt. 34 (2001) 49-56.

[26.10] E. Rivin, Properties, prospective applications of ultra thin layered rubber-metal laminates for limited travel bearings, Tribol. mt. 16 (1983) 17-25.

[26.11] M. Barquins, R. Courtel, Adherence, friction and wear of rubber-like materials, Wear 158 (1992) 87-117.

[26.12] Y. Fukahori, Mechanism of rubber abrasion. Part 3: how is friction linked to fracture in rubber abrasion? Wear 188 (1995) 19-26.

[26.13] S. Schallamach, Abrasion of rubber, Prog. Rubber Technol. 46 (1984) 107-142.

[26.14] M. Barquins, Friction and wear of rubber-like materials, Wear 160 (1993) 1-11.

[26.15] M. Barquins, R. Courtel, Rubber friction and the rheology of viscoelastic contact, Wear 32 (1975) 133-150.

[26.16] M. Barquins, Sliding friction of rubber and Schallamach waves — a review, Mater. Sci. Eng. 73 (1985) 45-63.

[26.17] M. Barquins, D. Maugis, J. Blouct, R. Courtel, Contact area of a ball rolling on an adhesive viscoelastic material, Wear 51 (1978) 375-384.

[26.18] B. Persson, Theory of rubber friction and contact mechanics, J. Chem. Phys. 115 (2001) 3840-3860.

[26.19] A. Schallamach, How does rubber slide? Wear 17 (1971) 301-312.

[26.20] K. Johnson, Contact Mechanics, CUP, Cambridge, 1985.

[26.21] V. Dunaevsky, Friction and wear equations, in: E. Booser (Ed.), Tribol. Data Handbook, CRC Press, New York, 1997, pp. 445-454.

[26.22] A. Muhr, A. Roberts, Rubber abrasion and wear, Wear 158 (1992) 213-228.

[26.23] Y. Uchiyama, Studies on the friction and wear of rubbers. Part I: influence of mechanical properties on the abrasive wear of rubbers, mt. Polym. Sci.Technol. 11(1984) 74-80.

[26.24] S. Fouvry, P. Kapsa, H. Zahouani, L. Vincent, Wear analysis in fretting of hard coatings through a dissipated energy concept, Wear 203-204 (1997) 393-403.

[26.25] D. Kyun Baek, M.M. Khonsari, Friction and Wear of a Rubber in Fretting, Wear 258 (2005) 898-905.

27

Friction Coefficient Studies on Hold-Down Systems for Space Mechanisms

27.1 A General Review

Hydrogenated diamond-like carbon, DLC:H, would seem to be a promising coating for hold-down and other space devices due to its chameleon-like properties. It means that DLC film exhibit different friction coefficient in air and in vacuum. This chapter reports the results of the DLC:H properties in terms of residual stress, adherence, friction coefficient and morphology on different titanium samples with both textured and smooth surfaces simulated in air and in vacuum. The coatings were obtained by using an enhanced pulsed DC PECVD discharge asymmetrical capacitive-coupled deposition system with silane and methane as precursor gases, for interlayer and DLC film, respectively. Friction coefficients, total stress, critical load and Raman scattering analysis of DLC:H films were done in order to observe their changing adhesive and sliding properties as required in satellite devices. Despite concerns of eventual cold welding, DLC:H films produced significant results, showing capacity to lubricate within air atmosphere with lower friction coefficient, and yet help to hold surfaces together under vacuum due to higher friction coefficient conditions. Textural deformation of the samples in V form during lock tests suggests further studies as necessary to rule out a soldering effect.

27.2 Introduction

Solid lubricant coatings with chameleon-like properties have been studied for satellite devices where surfaces must stick and slide[27.1, 27.2]. MoS_2, the most popular solid lubricant in space applications, showed cold welding due to chemical degradation on the Galileo satellite[27.3]. In space applications, undesirable cold welding from metallic diffusion can occur between clean metallic surfaces due to extreme conditions of contact force and lack of oxidation typical of the ultra-high vacuum environment of space, or due to wear fretting or impact force caused by the launching sequence[27.4]. Since Galileo studies, some researchers have been sought intensely materials with lower friction coefficients and cold welding prevention in both terrestrial and space environments[27.5, 27.6]

DLC:H films have been studied as a protective anti-cold welding layer in addition to their use as a solid lubricant.[27.7]

Some studies have revealed interesting relationships between the surface protective coating and decreased cold welding in a vacuum environment [27.8, 27.9]. Some researchers have sought to minimize lubrication failure keeping low friction coefficient by improving DLC and MoS_2 films with others chemical elements [27.8, 27.9].

To obtain very low friction coefficients, the best results have been reached with high hydrogenated DLC films in vacuum with a super lubricity model[27.10]. This model has been proposed in order to explain how the absence of carbon dangling bonds and of the high density of C-H bonds influence friction coefficient. The carbon-hydrogen affinity is very strong and, hence, the DLC:H films become chemically very passive. It is known that when binding hydrogen constitutes over 40% of film content on carbon surfaces lower friction coefficients are obtained in high vacuum or inert environments [27.10]. In other words, in space, when at least 40% of the carbon atoms are bound with hydrogen, the DLC surfaces become chemically very inert, reducing friction between facing surfaces during sliding[27.11].

To obtain high friction coefficients for hold down in vacuum, DLC:H films have provided the best results. Their carbon atoms have not been bound to as many hydrogen atoms during deposition, thus maintaining the surface of free carbon dangling bonds that produce C-C bonds between counter faces once in space. The surface gas desorption, can increased the density of σ-bonds on DLC:H surfaces and increases their friction coefficients[27.3].Yet, in air environment the DLC:H films present low friction coefficients because other gases such as oxygen, water vapor, or hydroxyl ions, keep the surfaces passsivated, allowing only weak Van der Walls interactions. Thus, DLC:H functions as a lubricant at air environment and as a chemically sticky surface at high vacuum. DLC:H coating change its surface properties according to the atmosphere like a chameleon, to preserve its life.

The questions that arise are whether films can be deposit with higher adherence quality to protect against cold welding. How film integrity may be preserved during the vibration of the launch sequence and during deployment sliding? Has it been proved that the surface contact shape a key factor in film integrity? Is the chemical structure, the hardness, or the elasticity modulus of the base material the key to determine the risk of deformation, and lack of adherence? To preserve the DLC:H film integrity is necessary to choose, a metallic substrate with low elastic modulus and a transient interface.

The present chapter reports the studies concerning the best DLC:H films parameters as chameleon coatings deposited on Ti6Al4V for working as solid lubricant in air and as a stick coating in vacuum. Also, studies were carried out to know which combinations of coatings would provide the best locking effect between hold down pairs devices: DLC:H on DLC:H, or WC (tungsten carbide) on WC, or DLC:H on WC.

Raman scattering spectroscopy was used to analyze chemical structure and hydrogen content in DLC:H films. The scanning electron microscopy (SEM) and atomic force microscopy (AFM) were used to analyze the samples morphology. The profilometry was used to measure the total stress and a tribology system was used to evaluate the adherence on Ti6Al4V substrate surface and friction coefficient measurements. The experimental procedures and the results are presented below.

A new tribo-system was developed through cooperation with Fibraforte Ltda., Brazil, to test locking capacity.

Two inch diameters DLC:H samples were used as witness bodies to test friction coefficient in air atmosphere and high vacuum. These samples were submitted to do these tests to Center for Tribology INC (CETR), USA, and tests were carried out on UMT.

27.3 Experimental Procedures

A water-cooled 60 mm diameter cathode was mounted inside of a 20 liters high vacuum chamber. The cathode was placed 50 mm from a 60 mm diameter anode. An enhanced dual DC pulsed PECVD discharge was assembled. The pulse frequency of the DC pulsed system was kept constant at 25 kHz with a 50% duty cycle. During the deposition, the voltage waveform consisted of a fixed positive pulse amplitude of 30 V followed by a variable negative pulse of up to – 700V. Films with higher hydrogen concentration were obtained by r.f. 13.56 MHz technique described elsewhere[27.12].

DLC films were deposited up to a thickness of approximately 3μm on polished 3 mm thick Ti6Al4V disk with a 51.4 mm diameter and on 200 μm laser grooves periodic textured Ti6Al4V alloy using methane as the hydrocarbon source. Prior to methane deposition, amorphous 10 nm silicon film as an interlayer was deposited in order to improve the adhesion between DLC:H films and substrates. For the both films, the total flow and the working pressure were kept constant at 5 sc cm and 10 Pa, respectively, under 100°C. In addition, DLC:H films were deposited on Si(100) substrates to measure the total stress.

The substrates was cleaned ultrasonically in an isopropyl alcohol bath for 5 minutes before putting them into a vacuum chamber, after which they were additionally cleaned in a 10 Pa pressure of argon discharge during 30 minutes prior to deposition.

The smooth DLC:H film thickness was determined by stylus profilometry. Also, by using the stylus profilometry the total stress was determined by measuring the radius of curvature of the silicon substrates before and after the DLC film deposition and using the Stoney's equation, which is described in detail in the literature[27.10, 27.11].

The atomic arrangement of the films with low and high hydrogen concentration was analyzed by Raman scattering spectroscopy by using a Renishaw 2000 system, with an Ar^+ ion laser (λ = 514 nm) in backscattering geometry. The laser power on the sample was around 0.6 mW and the laser spot had a diameter of about 2.5 μm. The Raman shift was calibrated in relation to the diamond pick at 1332 cm^{-1}. All measurements were carried out in air at room temperature. SEM and AFM analyses were also used in order to evaluate the surfaces morphologies.

Friction coefficient measurements were carried out keeping the load constant at 5 N, while for the critical load measurements were varied from 0.2 to 20 N. A pin with a 6 mm diameter Ti6Al4V ball coupled to its tip was used as a counter part.

The friction coefficients and critical loads in air environment were determined using a CETR pin-on-disk tribometer system under ambient conditions (20 °C, 55% RH). A pin was mounted on a stiff lever with a precisely known weight and pressed onto the test DLC:H sample disk. Additional but similar experiments were done to measure the friction coefficient in a vacuum environment of 10^{-5} mbar. In this case, the smooth samples were put inside of the vacuum chamber coupled to the CETR system. For both cases, the friction coefficient was measured as a function of the rotation speed due to the strybeck movements in which the speed of the disc sample decreases over time. The strybeck movements was to choose in order to simulate the best sequence of satellite launching procedure,

For the critical load, a micro-scratch test was conducted on samples surfaces, using a diamond stylus (Rockwell C 120°) with a 200 μm radius diamond tip in order to evaluate the adherence between the film and the substrate. According to standard methodology, the load at which the coating is stripped from the substrate is deemed the critical load.

For hold-down testing the lock capacity between laser textured surfaces and between sprayed WC surfaces were measured using a system developed by a local company that makes solar panels and other aerospace devices. A load of 5 kN as required by satellite assembly procedure was used.

27.4 Results and Discussion

In Fig. 27.1 an AFM image of the DLC:H films on Ti6Al4V surface is shown as a typical low roughness in high magnification on 1 μm^2 area as low as 5.12 nm Ra and 6.43 nm RMS. Inside SEM image, show the typical smoothness in low magnification.

Fig. 27.2 shows a Raman scattering spectroscopy from DLC:H films for different hydrogen contents. Films with hydrogen concentration up to 25%, obtained by pulsed DC discharge show low luminescence, while the films with hydrogen over than 25%, obtained by r.f. discharge shows high luminescence. Fig 27.3, shows stress measurements as a function of the self bias for both techniques. It is apparent that the stresses for all films obtained by r.f. discharge are higher than that obtained by Pulsed DC discharge technique.

The chosen film for friction coefficient measurements in air atmosphere and vacuum and for cold welding prevention testing depict to that with best adherence and lower stress, so films deposited by pulsed DC discharge with 18-20% of hydrogen contents were used for all following tribological measurements.

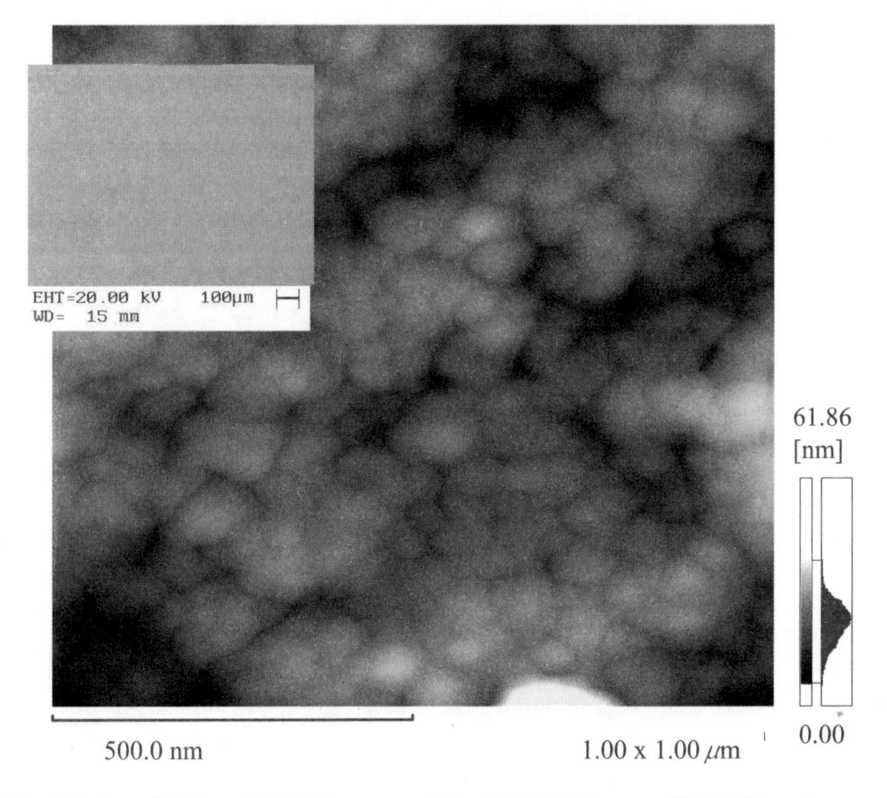

Fig. 27.1 AFM and SEM images of the DLC:H films on Ti6Al4V surface

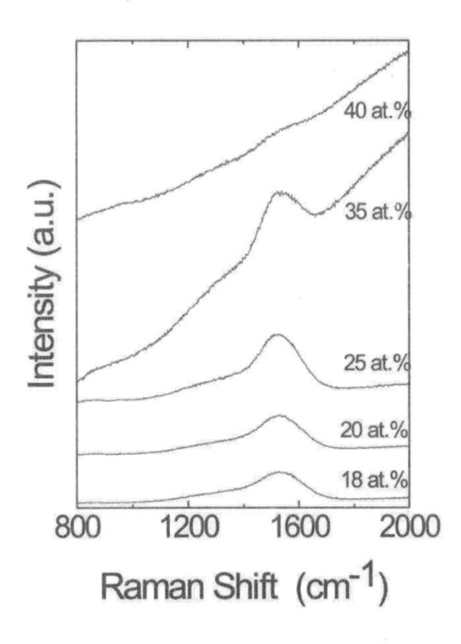

Fig. 27.2 Raman scattering spectroscopy from DLC:H films for different hydrogen contents

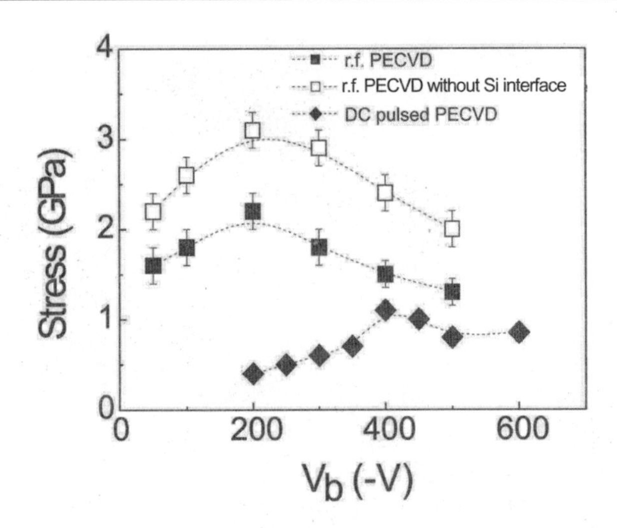

Fig. 27.3 Stress measurements as a function of the self bias for r.f. 13.56 MHz and Pulsed DC PECVD techniques.

Fig. 27.4, shows the critical load for DLC:H films with the highest hydrogenation, while Fig. 27.5 shows the critical load for DLC:H films with the lowest hydrogenation. A comparison between Fig. 27.4 and Fig. 27.5 shows that the lower critical load belongs to films with high hydrogenation. Fig 27.4 shows 18 N critical load, while Fig. 27.5 shows the critical load is over than 20 N. Low hydrogenated films are more adherent than high hydrogenated films.

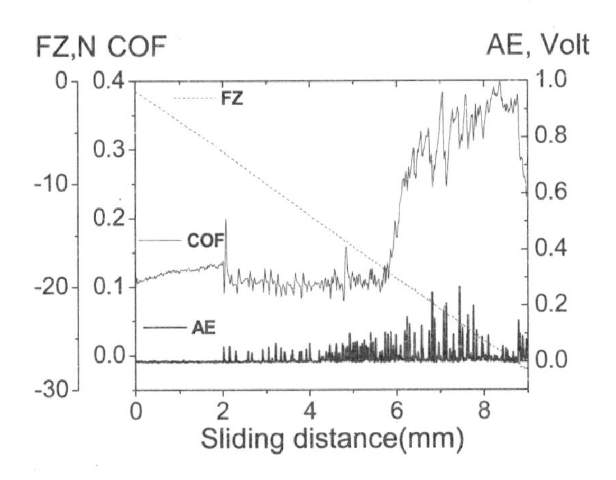

Fig. 27.4 Critical load for DLC:H films with the highest hydrogenation

Fig. 27.6 shows friction coefficient in strybeck test in three different speeds (0.001, 0.010, 0.100 m/s) in air atmosphere and in vacuum under down force of 5 N. The friction coefficient is almost constant for different sliding speed in air atmosphere and in vacuum. The friction coefficient in vacuum is twice higher than in air atmosphere showing their chameleon like property.

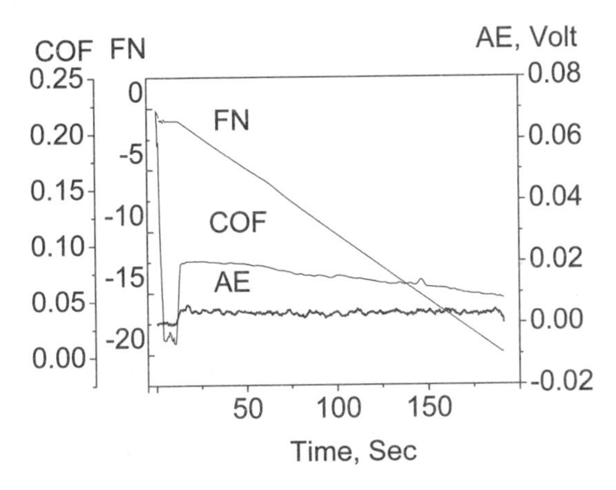

Fig. 27.5 Critical load for DLC:H films with the lowest hydrogenation

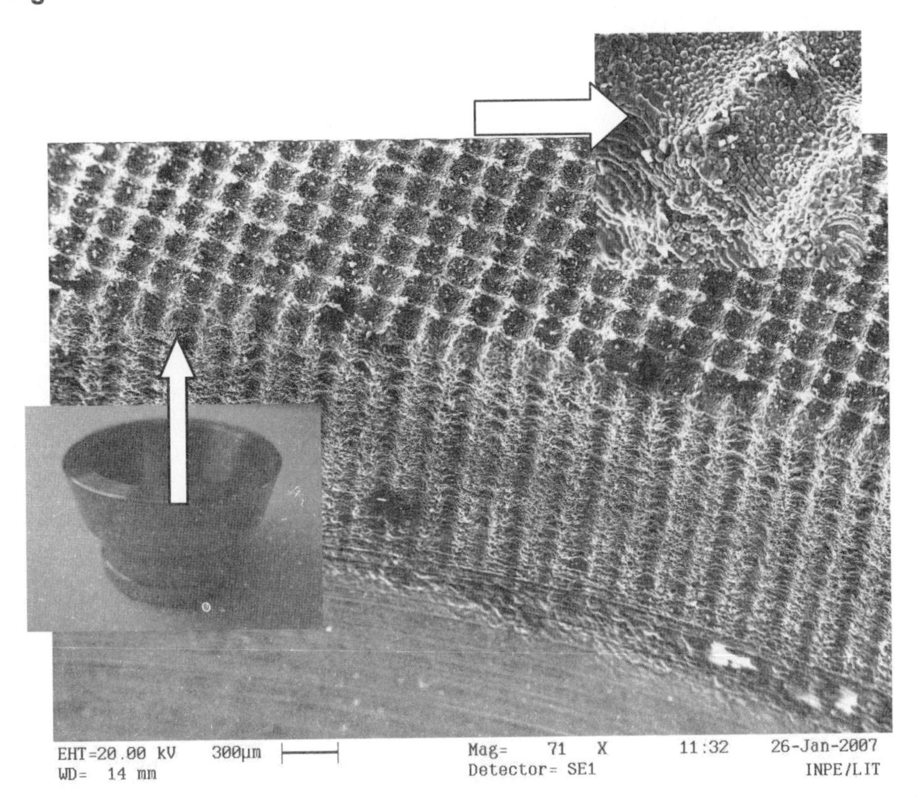

Fig. 27.6 SEM images of textured hold-down devices surface with DLC:H film, shown roughness surface detail (inside right up corner) after the locking test of DLC:H/DLC:H pair in air under 5 kN as normal force.

Fig 27.7 shows a schematic diagram system of friction coefficient measurement for locked test on hold-down pair. The normal force was kept constant at 5 kN. Fig. 27.8 shows the textured Ti6Al4V hold-down cylinder surface with WC submitted to the same locking test under down force of 5 kN. The friction coefficient results are summarized in Table 27.1. This table shows that, locking friction for the pair DCL:H/DLC:H is higher than WC/WC pair. An intermediary value was found for DCL:H/WC pair indicating a consistence of the measurements. Thus, DLC:H with 20% of hydrogen content showed superior locking capacity despite sample deformation.

Table 27.1 Locking friction of different hold-down pairs

Hold-down pairs	Locking friction	Normal load (kN)
DLC:H/ DLC:H	0.588 ± 0.007	5.0 ± 0.2
DLC:H/WC	0.488 ± 0.010	5.0 ± 0.2
WC/WC	0.354 ± 0.005	5.0 ± 0.2

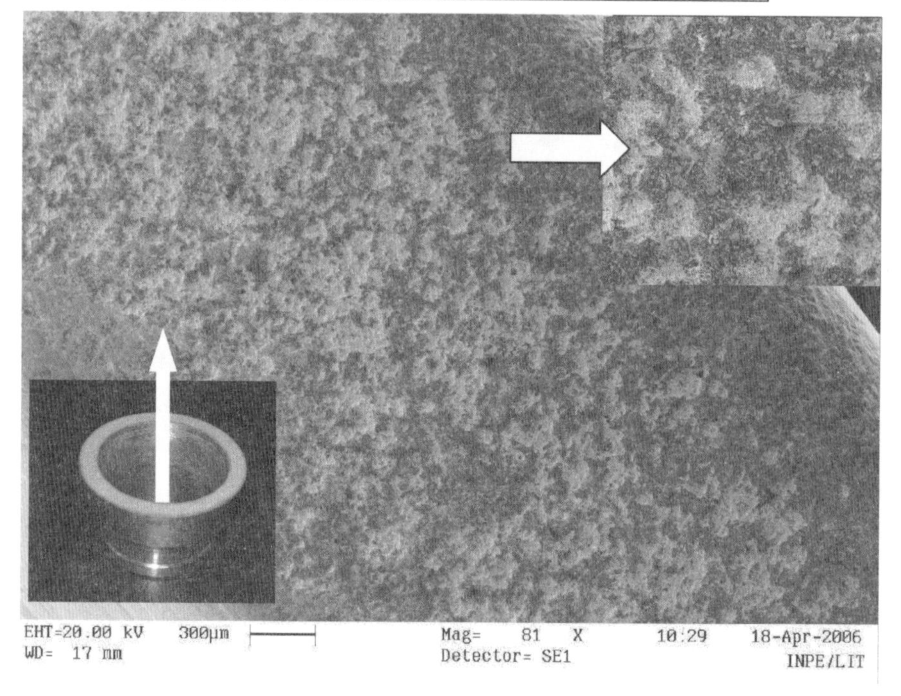

Fig. 27.7 SEM images of textured hold-down devices surface with WC film, shown roughness surface detail (inside right up corner) after the locking test of WC/WC pair in air under 5 kN as normal force.

Fig. 27.8 shows SEM image of textured surface with 3 μm thick DLC:H film of the hold-down devices (inside left down corner), and (inside right up corner) roughness surface detail after the locking test of DLC:H/DLC:H pair in air environment. The locking friction was 0.588 ± 0.007. The laser texturing method produced v-shaped grooves, which provided a strong locking surface after DLC deposition, but the peaks were susceptible to crushing. Therefore, some deformation of the v-shaped texture on Ti6Al4V hold-down cylinder appears as white patches. Such deformation on a space device would probably lead to peeling and cold welding can occur. Even though the locking friction is very high, it is apparent that this device deserves more development. A more rounded or block pattern may withstand pressure better than the v-shape used in this study.

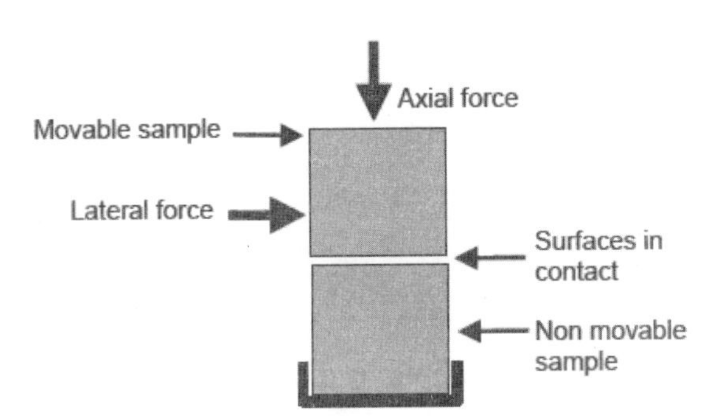

Fig. 27.8 Schematic diagram system of friction coefficient measurement for locked test on hold-down pair environment.

Fig. 27.9 shows SEM image of sprayed surface with 50 μm thick WC film of the hold-down devices (shown inside left down corner). The amplified SEM image (shown inside right up corner) shows surface roughness after the locking test of WC/WC pair in air environment. Locking friction was 0.354 ± 0.005, and some wear is visible.

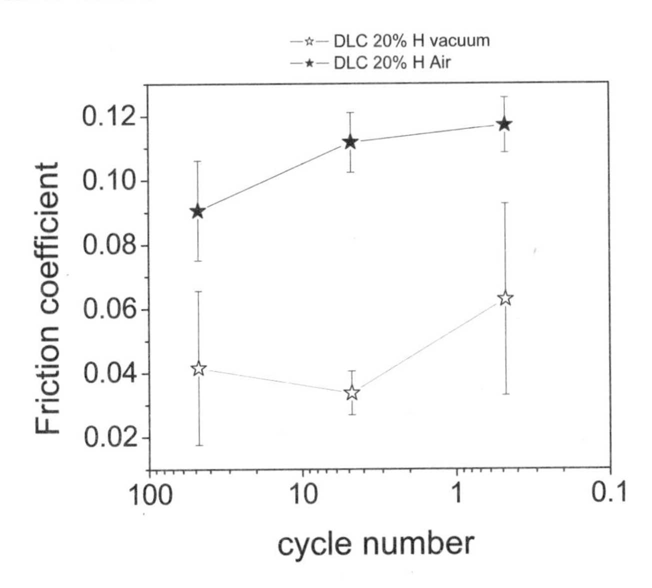

Fig. 27.9 Friction coefficient in strybeck test in three different speeds (0.001, 0.010, 0.100 m/s) in air atmosphere and in vacuum under 5N as normal force.

27.5 Summary and Conclusions

As demonstrated in this chapter DLC hydrogen containing around 20% films obtained by pulsed DC discharge presents: low friction coefficient in air atmospheric, high friction coefficient in vacuum, low luminescence in Raman spectroscopy analyses, low total stress in silicon wafer, good adherence in titanium alloy, good locked results on hold-down devices.

On the other hand it was observed that DLC films obtained by r.f. discharge has high luminescence in Raman spectroscopy analyses, increased total stress in silicon wafer, has reasonable adherence in titanium alloy resisted under a 200 µm radius diamond tip down 18.5 N.

After compared adherence, morphology and total stress results between two different DLC hydrogen concentrations were obtained. The results indicated that the DLC:H with low hydrogenation concentration could be used in hold-down surface devices to test locked performance in comparison with WC. Therefore, some deformation of the v-shaped texture on Ti6Al4V hold-down cylinder appears as white patches. Such deformation on a space device would probably lead to peeling and cold welding can occur. Even though the locking friction is very high, it is apparent that this device deserves more development. A more rounded or block pattern may withstand pressure better than the v-shape used in this study.

Finally it can be concluded that WC would currently seem to be the most promising coating for hold-down, because it may be produced by plasma spray on smooth surfaces with greater thickness that should prevent cold welding.

Since space devices are submitted to rapid acceleration during launch, followed by lower speeds in space orbit during deployment, such strybec test is a strong suggestion to simulate some satellite condition launch.

References

[27.1] C.C. Baker, R.R. Chromik, K.J. Wahl, J.J. Hu and A.A. Voevodin, Thin Solid Films. In Press, corrected Proof, available online, 12 February (2007).

[27.2] C.C.Baker, J.J.Hu and A.A.Voevodin. Surface and Coatings Technology. 201, Issue 7, 20, December (2006) 4224.

[27.3] S. Miyake, R. Kaneko, Thin Solids Films, 212 (1992) 256

[27.4] K. Miyoshi, Tribol. Int. 32 (1999) 673.

[27.5] Y. Lifshitz, Diamond Relat. Mater. 8 (1999) 1659.

[27.6] A. Grill, V. Patel, Diamond Relat. Mater. 2 (1993) 597.

[27.7] L.V.Santos, V.J.Trava-Airoldi, E.J.Corat, J.Nogueira, N.F. Leite, Surface & Coatings Technology, 200 (2006) 2593.

[27.8] N.A. Alcantar, C. Park, J.M. Pan, J. N. Israelachvili, Acta Materialia, 51 (2003) 31.

[27.9] A. A. Voevodin, J. S. Zabinski, Comp. Science and Tech. 65 (2005) 741.

[27.10] C. Donnet , J. Fontaine , T. Le Mogne, Belin M., C. Héau, J.P. Terrat, F.Vaux, G. Pont, Surf. and Coat. Tech. 120 (1999) 548.

[27.11] A. Erdemir, Tribology International 37 (2004) 577.

[27.12] Erdemir, A. and Donnet, C., J. Physical D: Appl. Phys. 39, (2006) 311.

[27.13] Bonetti L.F., Capote G., Santos L.V., Corat E.J. and Trava-Airoldi V.J. Thin Solid Films, Volume 515, Issue 1, 25 September (2006) 375.

28

Method of Characterizing Electrical Contact Properties of Carbon Nano-tube Coated Surfaces

28.1 A General Review

This chapter presents a method for electromechanical characterization of carbon nano tube (CNT) films grown on silicon substrates as potential electrical contacts. The method includes measuring the sheet resistance of a tangled CNT film, measuring the contact resistance between two tangled CNT films, and investigating the dependence on applied force and postgrowth annealing. Also, characterization of Au-CNT film contact resistance by simultaneous measurement of applied force and resistance has been discussed. A contact resistance as low as 0.024 Ω/mm^2 between two films of tangled single-wall carbon nano tubes grown on a polished silicon substrate have been measured and what was observed was an electromechanical behavior very similar to that predicted by classical contact theory.

28.2 Introduction

Power handling capability (low resistance) and heat dissipation are important issues for contact surfaces for many industrial applications including microsystems. Due to their excellent electrical and mechanical properties[28.1], carbon nano tubes appear to be promising contact materials which may also provide long life and wear resistance. However, while high current carrying capacity ($\sim 10^9$ A/cm^2) and near-ballistic conductance (R $\approx$ 6.5 kΩ) have been demonstrated along the length of an individual carbon nano tube (CNT)[28.2, 28.3], properties of CNT films at the micro to macroscales need further investigation to determine suitability for use as contact surfaces.

An important limitation for microsystems employing planar contact surfaces is the low forces generated by the actuation mechanisms which result in contact between only limited number of points [28.4, 28.5]. This problem may be addressed using CNT coatings, which provide highly dense nanoscale contact regions, enabling continuous and well-conformed contact surfaces. However, one must understand the reversible mechanical contact properties of nano tube – nano tube interfaces as well as nano tube – metal interfaces to fabricate high-performance contacts. Most of the literature on carbon nano tube-metal contacts reports on connections for applications such as transistors [28.6] and interconnects [28.7], where a permanent bond is formed between the CNTs and the metal. Theoretical studies have also been carried out to study carrier transport at carbon nano tube-metal heterojunctions[28.8, 28.9]. A recent study [28.10] demonstrated improvement in reversible contact resistance between a copper electrode and a copper surface by using an interfacial CNT layer. Reversible contact properties of a CNT coated metal wire and CNT coated electrode has also been reported [28.11]. But the studies on reversible contacts discussed above employ at least one electrode as one of the contact surfaces and do not represent the planar and parallel contact geometries required for both contact surfaces, such as in relays and switches. Here, a versatile method to characterize two contacting parallel planar CNT coated surfaces as well as metal-CNT surfaces is presented. The flexibility of the method allows changing of substrate material and the contact area, which are important parameters for the characterization of candidate surfaces for the specific application.

28.3 Experimental Method

Chemical vapor deposition is used to grow high quality tangled films (~ 1 μm thick) of primarily single-walled CNTs from a $Mo/Fe/Al_2O_3$ thin-film catalyst which is deposited on highly doped silicon by e-beam evaporation. The highest-yield films are grown using 85% /15% CH_4/H_2 at 825° and have a *G/D* ratio of over 25 as measured by Raman spectroscopy [28.12] (Fig. 28.1).

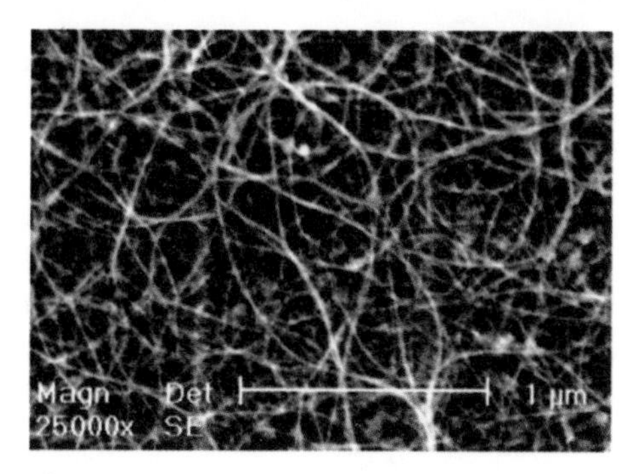

Fig. 28.1 Scanning electron microscopy (SEM) image of tangled CNT film grown on polished silicon surface using CH_4/H_2.

Firstly, the measurement the sheet resistance of the films using a standard four-point test is done. As control values, the resistance across samples of catalyst only (with no CNTs) is hundreds of kilo ohms, and in some cases there is no conduction. Next, two CNT coated rectangular substrates are brought together in cross configuration such that the CNT surfaces are in contact [Fig. 28.2(a)]. This four-point test configuration (Kelvin configuration) is used to measure the contact resistance between two CNT films, where current is sourced into one leg of a sample, passes through the contact area, and then flows out one leg of the other sample, while the voltage drop across the contact area is measured. The tests are performed with a CETR UMT microtribometer,[28.15] which allows simultaneous force/displacement and electrical contact resistance measurements. The samples are fixed on a glass slide, and electrical contacts to copper wires are made through colloidal silver paint and conductive epoxy [Fig. 28.2(b)]. As a control value, there is no measurable cross-configuration conductance across two catalyst only samples, and thus conduction is only through two contacting CNT surfaces. Since the substrate is highly doped and therefore conductive, in the Kelvin test configuration, the current flows through the silicon substrate in the contact region, instead of flowing through the CNT film. It is found that conduction occurs through the thickness of the thin aluminum oxide under layer supporting the catalyst film, which enables electrical connection between the CNT film and the conductive substrate.

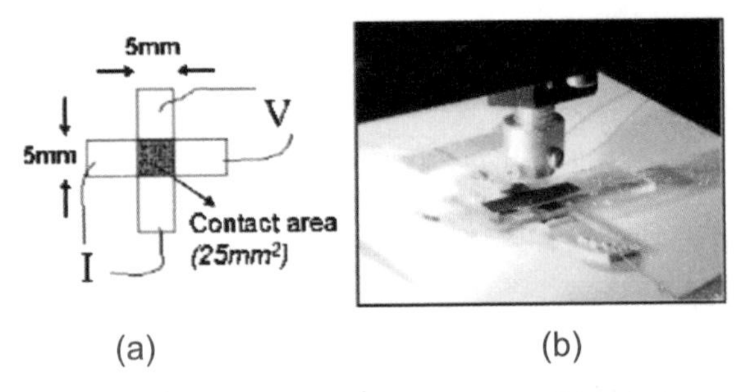

Fig. 28.2 (Color online) (a) Schematic of Kelvin structure setup used to measure contact resistance. (b) Measurement setup for measuring contact resistance between two CNT films. The substrates are brought together in a "cross" configuration, such that two surfaces covered with CNT films come into contact.

Next, the measurement of the contact resistance between a metal surface (Au) and the CNT film using a similar setup, as shown in Fig. 28.3, is carried out. This enables to determine the contact resistance between the two contacting surfaces by simultaneously measuring the current flowing trough the contact and the voltage drop at the contact. An approach is made to the CNT coated surface with a Au-plated ball (d ~ 4 mm) to record force and resistance simultaneously.

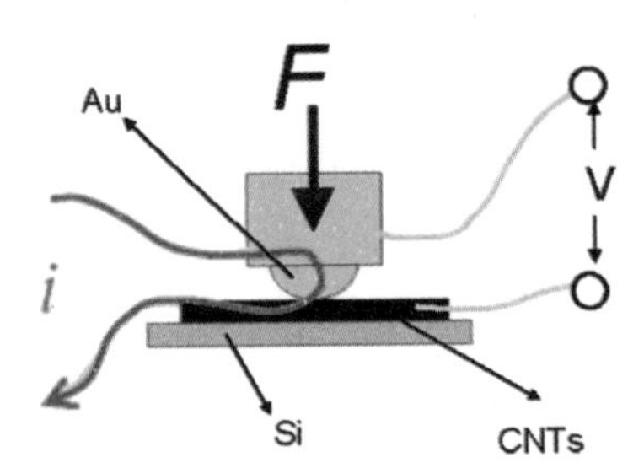

Fig. 28.3 (Color online) Measurement setup for CNT-Au contact resistance measurement. Current is sourced through the Au ball and the film, and the voltage drop between the Au ball and the film is measured.

28.4 Results and Discussion

The sheet resistance of the tangled CNT film is between 100 and 180 Ω/mm^2, corresponding to resistivities between 1×10^4 and 1.8×10^4 Ωm. This agrees with published values for single-wall carbon nano tube (SWCNT) films produced by other methods [28.13]. Data are taken across different lengths of CNT film. The trend is linear indicating bulk transport through the CNTs and resistance dominated by a large number of individual CNT-CNT contacts within the film. The samples are then annealed at different temperatures in Ar (99.9999%, Air gas) up to 500 °C and no significant change in sheet resistance is observed.

The lowest measured contact resistance between two contacting CNT films is 0.6 Ω (0.024 Ω /mm^2), and the values are typically between 0.6 Ω and 1.4 Ω with an average of 0.87 Ω. The dependence of the contact resistance on the contact force is shown in Fig. 28.4. Upon oscillation of the load from a light to a firm load (up to 5 kg), the resistance values oscillate; however, only the first few cycles are needed to reach a steady-state resistance. During these first cycles of compression, it is likely that increased CNT-CNT adhesion by van der Waals forces is modifying the nature of the CNT-CNT contacts. However, it is observed that compressing the CNT films in areas that are not in contact with another CNT surface does not change the cross-configuration resistance; therefore, this compression mechanism only affects the CNT-CNT contact resistance and not the sheet resistance. It is also observed that the contact resistance between two CNT films greatly increases (up to four times) after annealing at temperatures above 300 °C, possibly due to oxidation of the CNTs. The next measurement was the contact resistance between the Au-plated ball and the CNT film while recording applied force. Figure 28.5 compares this resistance with Au ball and Au-plated flat surface for reference.

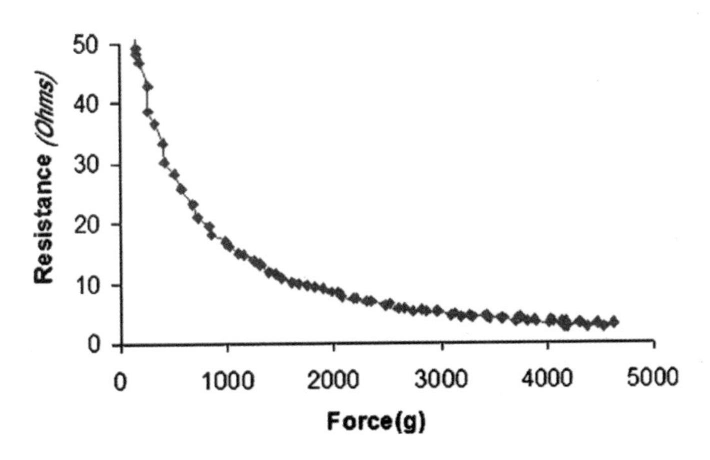

Fig. 28.4 (Color online) Contact resistance between two CNT films as a function of applied force. The contact area is 25 mm^2.

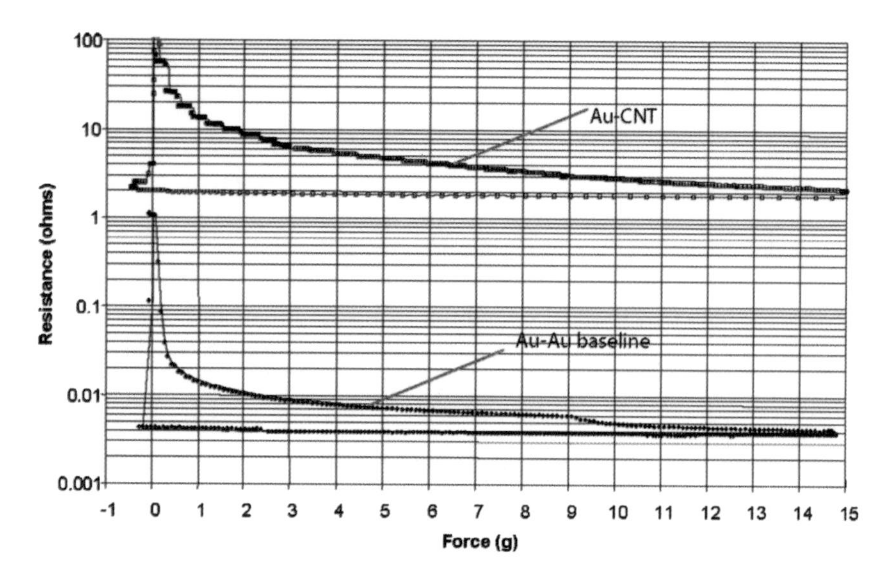

Fig. 28.5 Experimental results showing measured resistance values between two Au surfaces and between Au and CNT film as a function of the applied force.

The characteristics of both CNT-CNT and Au-CNT contact measurements are very similar to classical contact theory (Holm theory)[28.14], where the contact resistance decreases as the force increases. Classical contact theory states that contacting surfaces first establish contacts through asperities, and these asperities form "a" spots, which then become larger while new "a" spots are also formed as the contact force increases. According to Holm theory [28.14], at a given contact force, the effective contact area, i.e., the sum of the areas of all spots, is calculated using the yield stress of the contacting films. For the case of contacting CNT surfaces, however, the effective area calculation is not straightforward as the CNT film is made of "spaghettilike" carbon nano tubes (Fig. 28.1), and the theory is difficult to apply directly. It is believed that there are two mechanisms responsible for this behavior. First, as soon as the contact force increases, the contact forces between two given carbon nanotube sidewalls increase, which facilitates electron transfer. As the two CNT films are pressed together, the thickness of the film reduces and the effective CNT density increases due to bending of CNTs sideways, which results in more CNT-CNT sidewall contact points. Since the contact points act as parallel resistors, more contact points result in reduced overall resistance.

It is also observed that the contact force required to obtain low contact resistance is much higher for two contacting CNT films. This can be explained by the fact that carbon nano tubes have very large elastic modulus [28.1] (terapascal range), and large forces are needed to bend them and obtain more contact points. We conclude that a tangled CNT film against a Au coated surface works better than two contacting tangled films. It is believed that the contact resistance is a strong function of the surface density of the carbon nanotubes, which affects the effective contact area. Although the CNT film conforms well to the Au surface, the contact resistance is still higher than Au-Au resistance (Fig. 28.5) due to the low surface density.

A method and experimental setup is presented to characterize contact resistance between two CNT films.[28.16]

This method characterizes the contact resistance between metal surfaces and CNT films. The method is versatile and can be used to characterize CNT films of different properties and morphologies and can serve as a platform for future research and investigation of the contact properties of carbon nano tube films with different densities (number of CNTs per area), different qualities of carbon nano tubes (G/D ratio), different CNT film thicknesses, as well as contact properties of CNT films with different metals.

References

[28.1] M S. Dresselhaus, G. Dresselhaus, and P. Avouris, *Carbon Nanotubes: Synthesis, Structure, Properties, and Applications* (Springer, Berlin, 2001).

[28.2] S. D. Li, Z. Yu, C. Rutherglen, and P. J. Burke, Nano Lett. **4,** 2003 (2004).

[28.3] H. T. Soh, C. F Quate, A. F. Morpurgo, C. M. Marcus, J. Kong, and H. J. Dai, Appl. Phys. Lett. **75**, 627 (1999).

[28.4] J. Qiu, J. H. Lang, A. H. Slocum, and A. C. Weber, J. Microelectromech. Syst. **14**, 1099 (2005).

[28.5] K. L. Johnson, *Contact Mechanics* (Cambridge University Press, Cambridge, 1985).

[28.6] D Mann, A. Javey, J. Kong, Q. Wang, and H. J. Dai, Nano Lett. **3**, 1541 (2003).

[28.7] F Kreupl, A. P. Graham, G. S. Duesberg, W. Steinhogl, M. Liebau, E. Unger, and W. Honlein, Microelectron. Eng. **64**, 399 (2002).

[28.8] A. A. Farajian, H. Mizuseki, and Y. Kawazoe, Physica B (Amsterdam) **22,** 675 (2004).

[28.9] N. Nemec, D. Tomanek, and G. Cuniberti, Phys. Rev. Lett. **96**, 076802 (2006).

[28.10] M. Park, B. A. Cola, T. Siegmund, J. Xu, M. R. Maschmann, T. S. Fisher, and H. Kim, Nanotechnology **17**, 2294 (2006).

[28.11] Y. Tzeng, Y. Chen, and C. Liu, Diamond Relat. Mater. **12**, 774 (2003).

[28.12] A J. Hart, A. H. Slocum, and L. Royer, Carbon **44**, 348 (2006).

[28.13] L. Hu, D. S. Hecht, and G. Gruner, Nano Lett. **4**, 2513 (2004).

[28.14] R. Holm, Electrical Contacts: Theory and Applications (Springer, New York, 1967).

[28.15] www.cetr.com

[28.16] O. Yaglioglu, A.J. Hart, R.Martens and A.H. Slocum, Review of Sci. Instruments 77,095105 9 (2006)

Section - VII

Polymers

29 Experiments on Precision Measurements of Viscoelastic Properties of Industrial Polymers

29.1 A General Review

A novel experimental technology for studies of viscoelastic properties of polymers and composites has been developed, based on servo-control of either load or displacement and simultaneous real-time measurements of deformations and forces, as well as contact acoustical, electrical and temperature parameters. This technology has been effectively applied for parallel evaluation of elasticity, plasticity, creep, friction and wear of diverse industrial polymers, including gaskets, seals, and polishing pads.

29.2 Experimental Technology and Equipment

A universal materials tester mod. UMT has been designed for highly precision comprehensive measurements of mechanical and physical properties of polymers and composites, including with hard and soft coatings. It has two main characteristics: closed-loop servo-control mechanism, which allows for ultra-accurate control of either load (with displacement, or deformation, being monitored) or displacement (with load being monitored), and multi-sensor capabilities, with a number of additional parameters measured in-situ, at a total sample rate of 20 kHz, including: forces and torques in all X, Y and Z directions, high-frequency acoustic emission, contact or surface electrical resistance (or impedance), temperature, as well as digital video (or still) images of the surfaces.

The servo-control allows for dramatically improved data repeatability and reproducibility, as well as for precision measurements of friction, wear, elastic modulus, micro-hardness at various programmable compression and tension levels. The multi-sensing allows for greater sensitivity in detection of various microscopic phenomena of materials behaviour, as well as in comparison and ranking of mechanical properties of seemingly similar materials. All the additional parameters are mutually complimentary. For example, high-frequency contact acoustic emission is very sensitive to the localized micro-cracks, while contact electrical resistance allows for better quantification of thickness and breakthrough thresholds of coatings during testing.

29.3 Studies of Automotive Gaskets

The test setup is shown in Fig. 29.1, some experimental results for 3 gaskets of different polymers are presented in Fig. 29.2 (same gasket 1, different recovery times) and Fig. 29.3 (different gaskets, same recovery time).

Fig. 29.1 Setup for creep and recovery tests of automotive gaskets

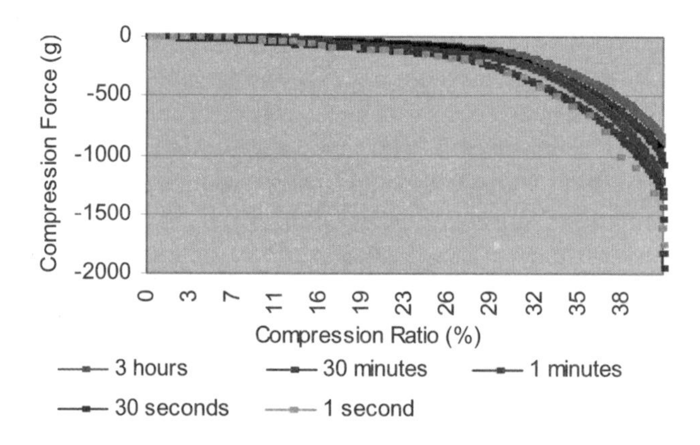

Fig. 29.2 Recovery at different test durations

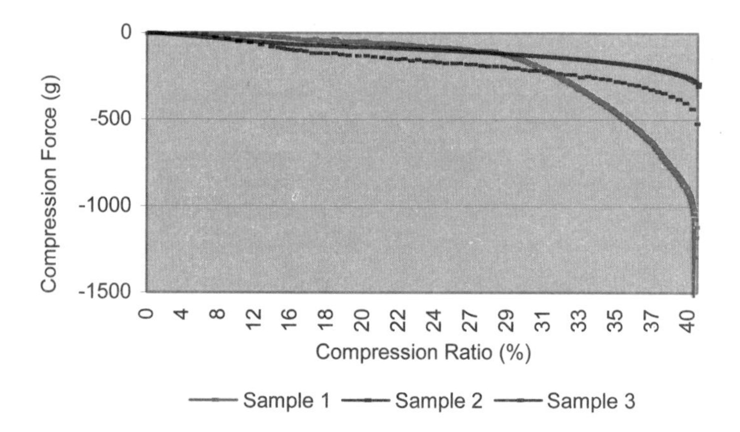

Fig. 29.3 Recovery for 3 gaskets, 30-min tests

29.4 Study of Pump Seals

A tester setup for elasticity, creep, friction and wear testing of pump seals is shown in Fig. 29.4, some experimental results are presented in Fig. 29.5.

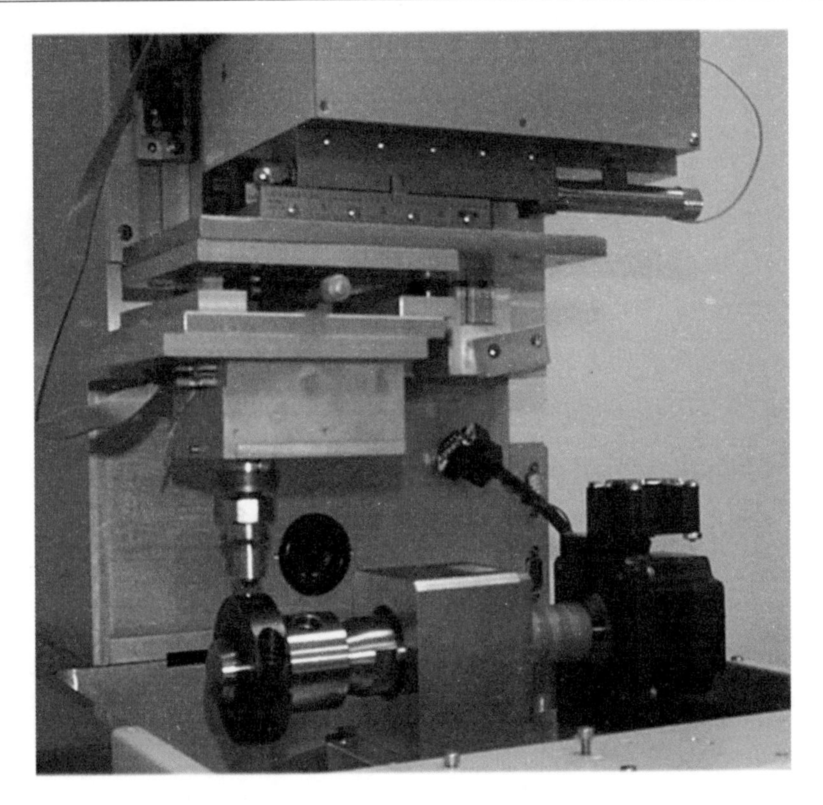

Fig. 29.4 Setup for seal testing

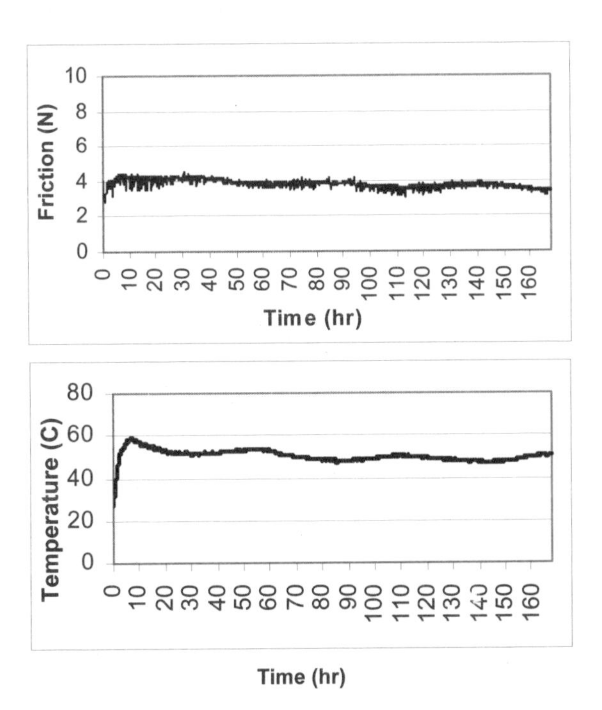

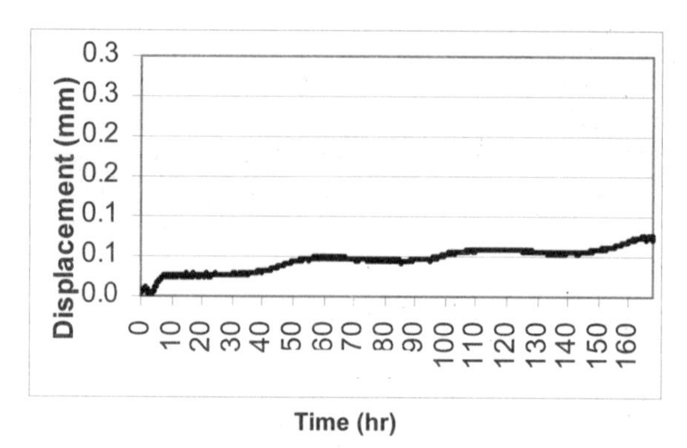

Fig. 29.5 Data for lip seal at 6000 rpm, load of 1N

29.5 Study of Polishing Pads for Wafer Planarization

Within one test, we measured elasticity (at loading), creep (under the load of 50 N) and plasticity (after load removal). In Fig 29.6., one can see large differences between pad 2 (stiffness of 0.12 micron/N, practically no creep, post-test plastic deformation of 2.2 micron) and pad 1 (stiffness of 0.24 micron/N, creep of 1.7 micron, post-test plastic deformation of 6.6 micron).

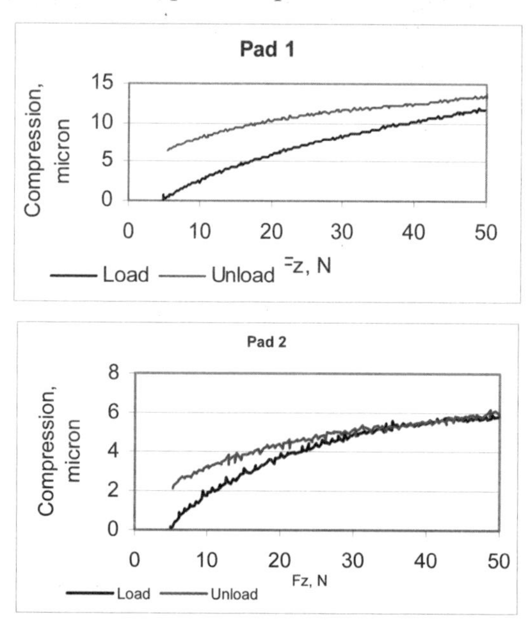

Fig. 29.6 Pad load-unload tests

An important question is which indenter to choose for pad deformation tests, a "2 disc (which averages over many pad grooves) or a 0.25" ball (between the grooves), as they produce different results not only quantitatively, but even in pad ranking (Fig. 29.7).

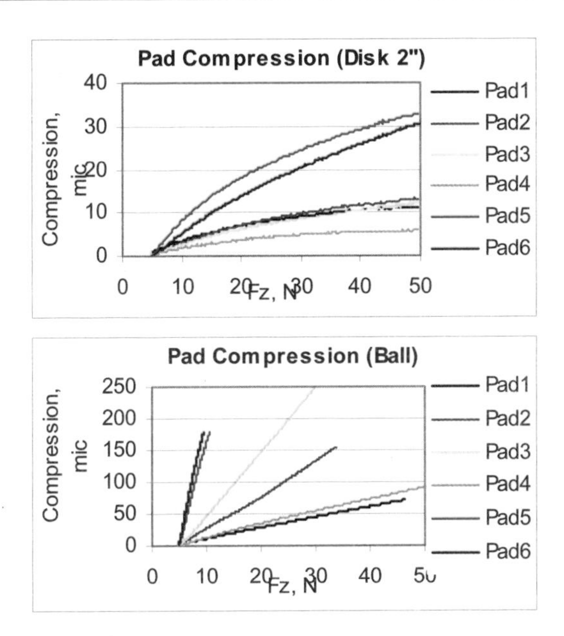

Fig. 29.7 Pad deformation summary

29.6 Conclusions

In depth lot of technical data need to be generated to chacterize the viscoelsatic properties of different industrial polymers.

30 Properties of Peek-Based Nano-composite Coatings Incorporating Inorganic Fullerene-Like Nano-particles

30.1 A General Review

The high strength, wear resistance and high operational temperature of polyetheretherketone (PEEK) have attracted increasing interests of this material for tribological applications. The addition of solid lubricant is an effective way to further improve the tribological properties of polymeric materials. In the present chapter, inorganic fullerene-like tungsten disulfide (IF-WS_2) nanoparticles were incorporated into PEEK coatings with the aim of reducing the coefficient of friction (COF) and improving the wear resistance of the coatings. The microstructures of IF-WS_2/PEEK nanocomposite coatings were characterized using a combination of SEM, XRD and FTIR measurements. The thermal behaviours of the coatings were determined using differential thermal analysis (DTA). Tribological tests had also been carried out to evaluate the friction and wear behaviours of IF-WS_2/PEEK nanocomposite coatings. The results showed that significant improvement can be achieved in the tribological properties of the nanocomposite coatings by incorporating lF-WS_2 nanoparticles.

30.2 Introduction

There is a trend to use self-lubricating solid materials and coatings to control friction and wear of components in severe applications [30.1]. Polyetheretherketone (PEEK) is one of the high performance engineering thermoplastic polymers, which has excellent mechanical properties, good chemical resistance and high long-term working temperatures (up to 260°C)[30.2, 30.3]. PEEK has attracted increasing interests as a bulk or coating material for bearing and sliding parts [30.4-30.6]. However, direct contact of PEEK and ferrous metal has relatively high coefficient of friction (COF) and the associated heat can have a deleterious effect on the PEEK [30.2]. There are some attempts on adding solid particles (e.g. AlN[30.7], Al_2O_3[30.8] and SiC [30.9], etc.) into PEEK matrix. Improvement on the wear resistance and hardness of these PEEK-based composites has been reported, but the COF of the composites has yet to be reduced.

The chapter reports the use of inorganic fullerene-like (IF) tungsten disulfide (WS_2) nanoparticles, as a source of self- lubricating for PEEK coatings to reduce COF. IP-WS_2 nanoparticles have a layered onion structure with a hollow core. They have been used as favourable solid lubricants under severe conditions [30.10, 30.11]. The incorporating IF-WS_2 nanoparticles into various matrices (e.g. Ni-P [30.12], epoxy [30.13] and ceramic [30.14], etc.) to form nanocomposites have also been reported. Considerable improvement has been achieved on the tribological properties of these nanocomposites. However, it seems that no work has been reported on the high performance thermoplastic like PEEK, due to the difficulty of processing. In this study, PEEK-based nanocomposite coating incorporating IF-WS_2 nanoparticles has been prepared, using aerosol-assisted deposition method. This chapter focuses on the microstructure and the tribological properties of the nanocomposite coatings, with the aim of obtaining a self-lubricating PEEK-based coating with lower COF and better wear resistance.

30.3 Experiment

Stainless steel plates ($L = 25$ mm, $W = 15$ mm) were selected as substrates. Prior to aerosol-assisted deposition, all substrates were cleaned and degreased in an ultrasonic bath with alcohol. IF-WS_2 nanoparticles ranging from 80 to 220 nm in diameter were supplied by NanoMaterials Ltd. The precursor dispersion was prepared by mixing IF-WS_2 nanoparticles with PEEK particles in an aqueous-based solution. The total solid content inside the dispersion was controlled at 0.01-5 wt.%.

Fine aerosol droplets were generated from the dispersion by a nebulizer. The droplets were subsequently directed towards a heated substrate where they underwent evaporation, melting and deposition during the aerosol-assisted deposition process. The substrate temperature was in the range of 280-350 °C. Post heat-treatment of the samples was carried out at 350-400 °C for 1 h to obtain smoother surface and enhance the interface strength between the coating and the substrate. The thickness of the coatings was controlled to 30 μm by adjusting the deposition time.

The microstructures of the IF-WS_2/PEEK nanocomposite coatings were characterized using a combination of X-ray diffraction (XRD), Fourier Transform Infrared Spectroscopy (FTIR), and Scanning Electron Microscopy (SEM). A Siemens D500 X-ray diffractometer with CuKα radiation was employed to detect phase and crystallinity of the coatings. The molecular structure of the deposited coatings was determined by a Perkin Elmer Spectrum - One Fourier Transform Infrared Spectrometer (FTIR, ATR mode). A Philips XL30 scanning electron microscope was used to observe the morphology of the coatings. The thickness of the coating was measured from the cross-section of the coating in SEM micrographs.

The thermal behaviours of the IF-WS_2 nanoparticles and the nanocomposite coatings were determined using Differential Thermal Analysis (DTA) (Setaram, Labsys 1600), from 30° C up to 700° C, at a heating rate of 5°C/min in air. This testing was performed by carefully scratching off the nanocomposite coatings from their substrates.

The nanohardness and modulus of the coatings were determined using a CETR Universal Micro-Tribometer (UMT-2) with a loading force up to 20 mN. The tribological tests were also carried out using this tribometer. Linearly reciprocating ball-on-flat sliding wear was performed under a constant loading of 1 N at a velocity of 1800 mm/min, using 1.6 mm diameter stainless steel balls. The loading and friction forces were simultaneously recorded during the testing in order to obtain the COF data. The duration of the sliding wear was 30 min and each sample was examined with three tests at different positions.

30.4 Results and Discussion

30.4.1 *Microstructure, phase and composition of the nanocomposite coatings*

Fig. 30.1 shows the SEM image of the cross-section of IF-WS_2/ PEEK nanocomposite coating with 2.5wt.% IF nanoparticles. The IF nanoparticles can be clearly observed in the cross-section of the coating. Most of the nanoparticles, sized from 80 nm to 220 nm, are distributed individually inside the matrix throughout the coating thickness, although a small degree of agglomeration of the nanoparticles may occur.

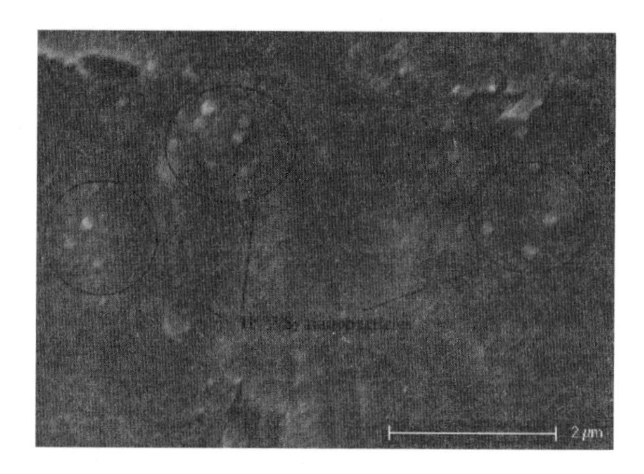

Fig. 30.1 SEM image of the cross-section of the nanocomposite coating with 2.5 wt.% IF nanoparticles.

Fig. 30.2 shows the XRD of the pure PEEK and the nanocomposite coating. PEEK is a semi-crystalline thermoplastic polymer [30.4]. Both coatings in Fig. 30.2 exhibit primarily orthorhombic crystalline form[30.7]. The three peaks around 2θ of 20° can be assigned to the (110), (113) and (200) planes of crystallized PEEK, respectively [30.15]. The peak around 29° is the (211) plane of PEEK. There is a small peak at 14.3° of 2θ in the diffraction pattern of the nanocomposite coating, which is attributed to the (002) plane of IF-WS_2. It indicates that the phase of IF-WS_2 nanoparticles remains unchanged inside the PEEK matrix.

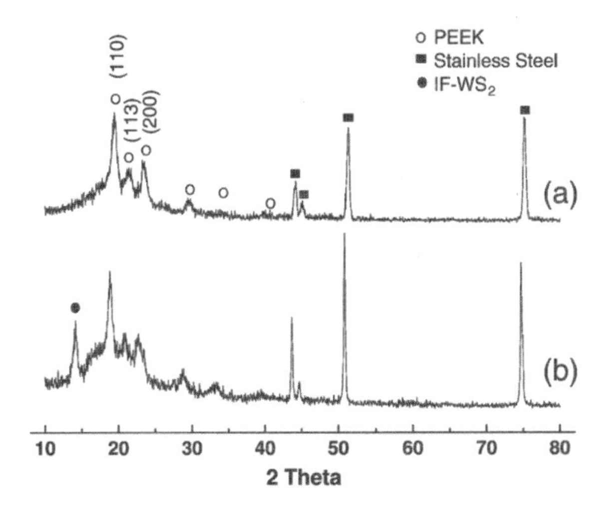

Fig. 30.2 XRD patterns of (a) pure PEEK coating and (b) IF-WS_2/PEEK costing.

Fig. 30.3 gives the FTIR spectra of the original PEEK powder and the IF-WS_2/PEEK coating. The IF/PEEK nanocomposite coating exhibits a similar IR response to the original PEEK powder, which indicates that the chemical composition of the PEEK matrix is preserved. The only difference is the peak of PEEK at 1159 cm^{-1} shifting to 1153 cm^{-1}, and becoming broader in the nanocomposite coating. This change might be attributed to the presence of sulphur-containing groups [30.16], at the interface of the IF nanoparticles and the PEEK matrix.

The results from SEM, XRD and FTIR confirm that IF-WS_2/ PEEK nanocomposite coatings have been successfully produced using aerosol-assisted deposition method.

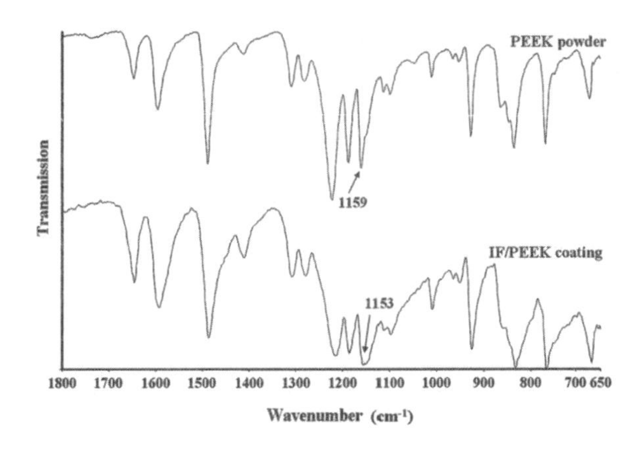

Fig. 30.3 FTIR spectra of PEEK powder and IF-WS_2/PEEK coating.

30.4.2 *Thermal Behaviour*

The thermal behaviour of the IF nanoparticles, the pure PEEK coating and the nanocomposite coatings were studied using DTA in air; the results are shown in Fig. 30.4. IF-WS_2 nanoparticles are readily oxidized to WO_3 at elevated temperatures in an oxidizing environment. According to the exothermic peak in Fig. 30.4 (a), the oxidation of the as-supplied IF-WS_2 nanoparticles started at 350 °C and reached its maximum rate at 400 °C. Pure PEEK started to decompose at 450 °C, and its decomposition rate peaked at 578 °C. It is interesting to observe that with the incorporation of IF nanoparticles, the temperature of maximum decomposition rate shifts 20–30 °C to the higher temperature region, even with 1 wt.% IF nanoparticles. The increase in thermal stability of PEEK may be due to the strong interaction and interfacial bonding between the matrix and the IF nanoparticles, which hinders the segmented movement of the PEEK molecules [30.7].

In the DTA curves of the nanocomposite coating, no exothermic peak around 400 °C can be observed. This shows that the IF-WS_2 nanoparticles are covered and protected by the PEEK matrix from oxidation at the elevated temperature, even in the nanocomposite coating incorporating 20 wt% IF nanoparticles. Only above 450 °C, does the PEEK matrix start to decompose and expose the IF nanoparticles to the oxidizing environment. Then the oxidation of both the matrix and the IF nanoparticles occurs together. The results indicate that the IF-WS_2 nanoparticles have better thermal stability towards air when inside the nanocomposite coating. Therefore, it can be concluded that the thermal stability of both PEEK matrix and IF-WS_2 nanoparticles is improved in the nanocomposite coating.

30.4.3 *Mechanical Properties*

The nanohardness and Young's modulus of the nanocomposite coating can be calculated from the loading-displacement curves derived from nano-indention testing. Fig. 30.5 shows the curve of nanohardness versus the content of IF nanoparticles in the coating. The hardness of the pure PEEK coating is around 0.33 GPa. With an increase of IF concentration, the hardness of the coating also increases, and reaches 0.55 (3Pa at 20 wt.% of IF nanoparticles). Fig. 30.6 shows the curve of Young's modulus of the nanocomposite coatings versus the concentration of IF nanoparticles. The Young's modulus of the nanocomposite coatings has a similar trend, and increases by as much as 60% with an increase of IF concentration in the coating to 20 wt.%.

The mechanical properties of composites are influenced by various factors, for instance, the particle size, particle concentration and the interface [30.7]. The increase in hardness and modulus may be due to the resistance of the deformation by IF nanoparticles. In the nanocomposite coating, IF nanoparticles exhibit a reinforcing effect on the PEEK matrix. Another possible reason is that the incorporation of IF nanoparticles may lead to better crystallization of the PEEK matrix, which improves the hardness and modulus of the nanocomposite coating. This hypothesis needs to be studied through future research.

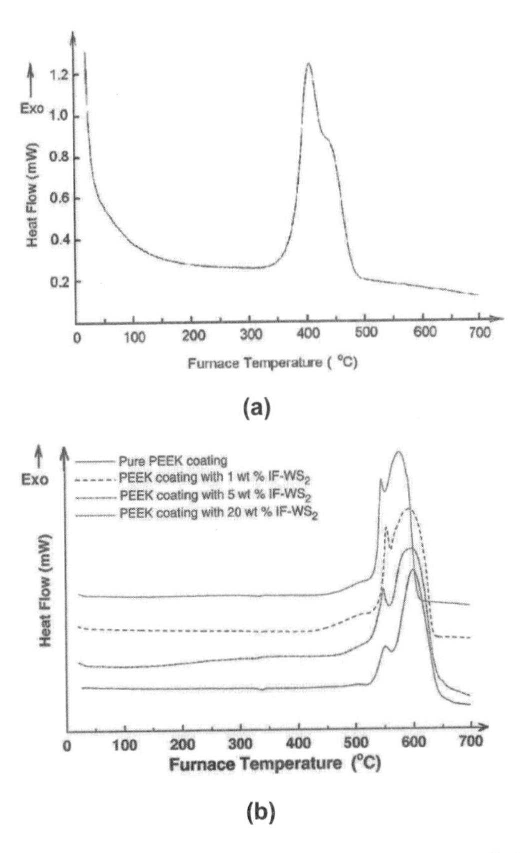

Fig. 30.4 DTA in air of (a) IF-WS_2 nanoparticles and (b) IF-WS_2/PEEK nanocomposite coatings.

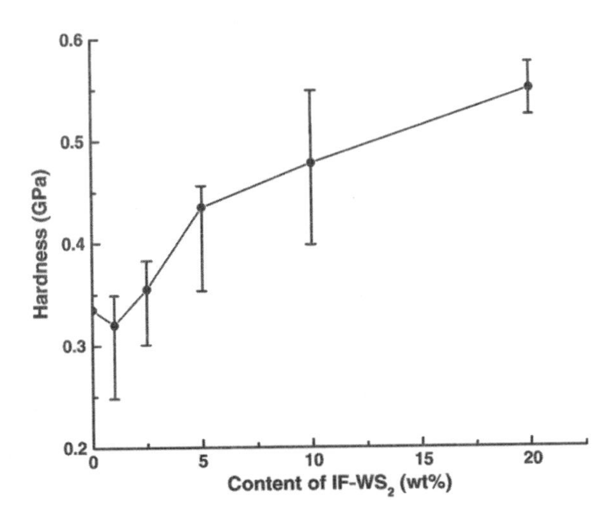

Fig. 30.5 Nanohardness of IF-WS_2/PEEK nanocomposite coatings as a function of the content of IF-WS_2.

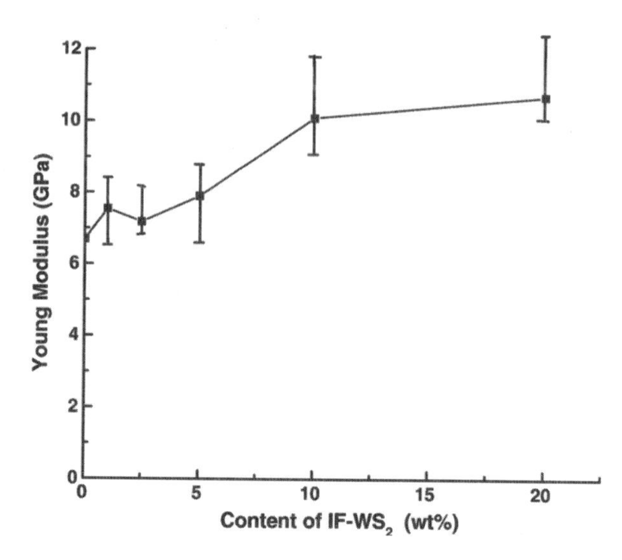

Fig. 30.6 Young's modulus of IF-WS_2/PEEK nanocomposite coatings with various IF contents.

30.4.4 *Tribological Properties*

Fig. 30.7 shows the coefficient of friction (COF) versus time in the ball-on-flat test. The results clearly show that with incorporation of IF nanoparticles (2.5 wt.%), the COF of the coating decreases significantly by up to 70%, from 0.4 to 0.15. The decrease in COF can be attributed to the lubricating capability of the IF nanoparticles. This is because the shearing force can be effectively decreased owing to the lower shearing strength of the lubricant particles, which in turn results in a lower friction coefficient. Such observation has also been made by Yu et al. in graphite/polyphenylene sulfide composites [30.17]. For polymers and their composites, the characteristics of the transfer film may play a dominant role in determining the tribological properties [30.9]. With the wearing of coating, the coating material may transfer to the counterpart of the ball surface, which leads to a self-lubricating effect. Further investigation on the transfer film will be carried out in future research for understanding the wearing of the IF-WS_2/ PEEK coatings.

Fig. 30.8 is the plot of COF versus the concentration of IF-WS_2 in the nanocomposite coatings. After incorporating IF-WS_2, the COF of the coating decreases to the lowest value of 0.15 at 2.5 wt.% IF-WS_2. However, further increase in the loading rate of IF-WS2 does not provide any further reduction, but causes a slight increase in COF of the nanocomposite coatings. There are two possible reasons: (i) over-amount of IF-WS_2 may affect the formation of the transfer film during wearing; (ii) agglomeration of IF nanoparticles tends to occur at higher IF concentration which may lose the best lubricating capability of IF nanoparticles. Hence, IF-WS_2 nanoparticles are effective solid lubricants at relatively low concentration (e.g. 2.5–5.0 wt.%).

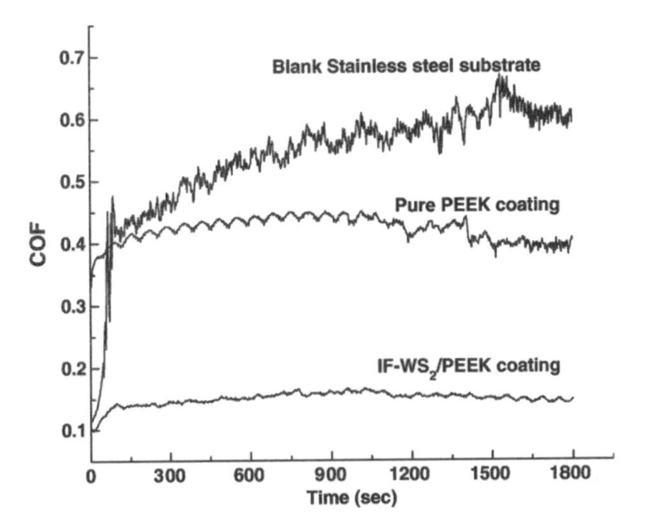

Fig. 30.7 Coefficient of friction (COF) of (a) stainless steel substrate, (b) pure PEEK coating and (c) 2.5 wt.% IF-WS_2/PEEK coating versus testing time in the ball-on-flat test

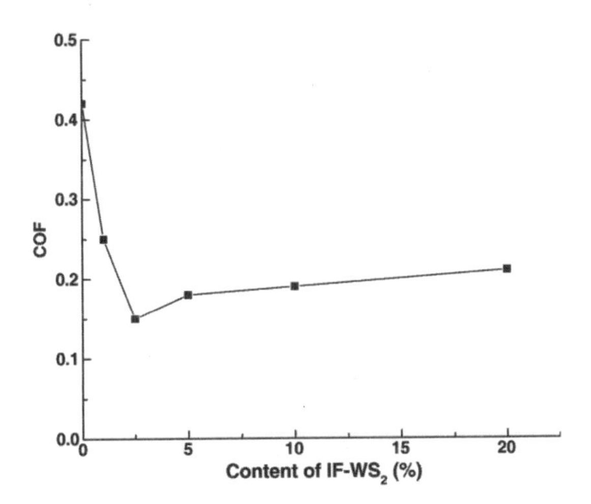

Fig. 30.8 COF versus the concentration of IF-WS_2 in the nanocomposite coatings.

To investigate the wear resistance of the nanocomposite coating, the friction tracks on the surface of the coating have been observed. Fig. 30.9(a) and (b) show the SEM images of friction tracks of pure PEEK and the nanocomposite coatings after the same frictional tests. On the pure PEEK coating surface, the PEEK coating has worn off and the substrate is exposed. As the thickness of the coatings is 30 μm, the depth of the worn area should be more than 30 μm. On the surface of the nanocomposite coating, only slight wearing occurs. It indicates that the wearing resistance has been improved by the incorporation of IF nanoparticles. The increase in wear resistance benefits from the increase of hardness and Young's modulus, as well as the decrease of COF. As a hard phase in the soft PEEK matrix, IF nanoparticles can reduce the coating deformation and true contact area. Then the nanocomposite coating exhibits an improved behaviour on reducing the plough and the adhesion between the relative sliding parts. Similar observation has been made by Zhang et al. in SiC/PEEK composite coatings [30.3].

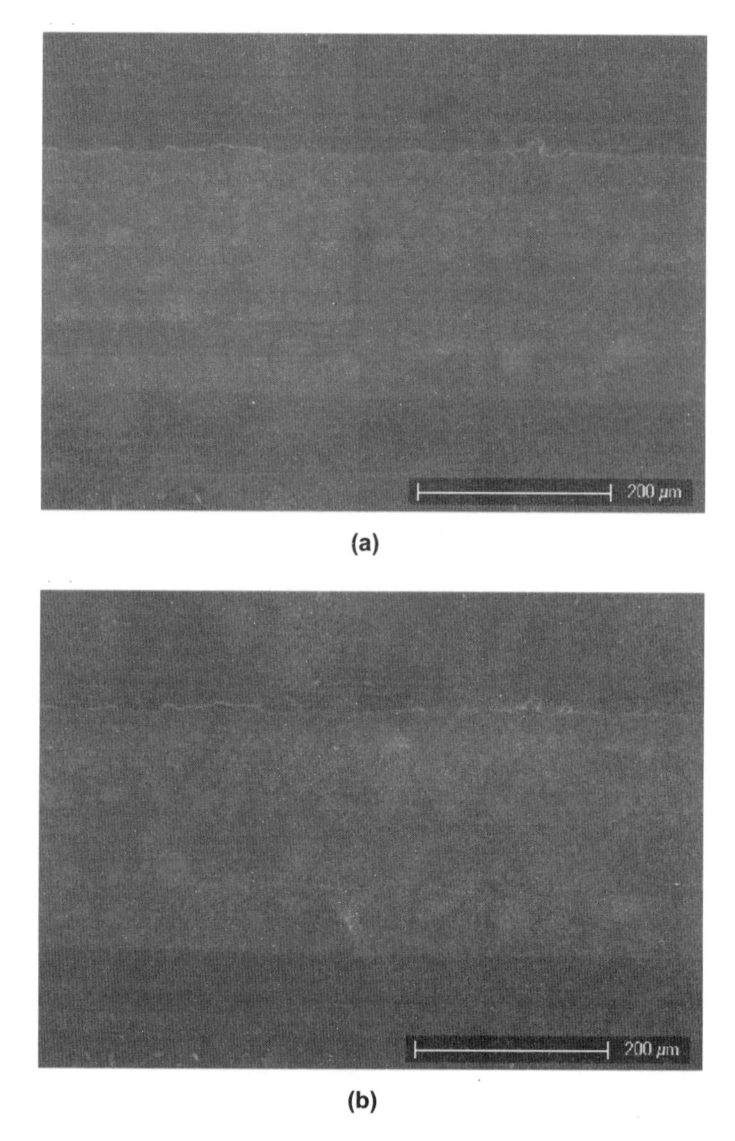

Fig. 30.9 SEM images of friction tracks of (a) pure PEEK coating and (b) IF-WS_2/ PEEK nanocomposite coating with 2.5 wt.% IF nanoparticles.

30.5 Conclusions

The following conclusions are drawn based on the above studies:[30.18]

(i) IF-WS_2/PEEK nanocomposite coatings have been successfully produced using our aerosol-assisted deposition process.

(ii) Thermal analysis shows that the thermal stability of both PEEK matrix and IF-WS_2 nanoparticles has been improved in the nanocomposite coating. Even with 1 wt.% IF nanoparticles, the temperature of maximum decomposition rate of nanocomposite coatings increases by 20-30 °C.

(iii) The hardness and modulus of the coating can be increased by as much as 60% (at 20 wt.% of IF concentration).

(iv) The COF of PEEK has been reduced significantly by up to 70% by incorporating 2.5 wt.% IF-WS2 nanoparticles. But a further increase in the loading rate of IF-WS_2 above 5% does not provide any further reduction in the COF of the nanocomposite coatings. The wear resistance of the nanocomposite coatings has also been significantly improved.

References

[30.1] C. Donnet, A. Erdemir Tribol. Lett. 17 (2004) 389.

[30.2] B.H. Stuart, Tribol. Int. 31 (1998) 687.

[30.3] G. Zhang, H. Liao, H. Li, C. Mateus, J.M. Bordes, C. Coddet, Wear 260 (2006) 594.

[30.4] B.H. Stuart, Tribol. Int. 31 (1998) 647.

[30.5] K. Friedrich, Z. Lu, A.M. Hager, Wear 190 (1995) 139.

[30.6] S.W. Zhang, Tribol. Int. 31 (1998) 49.

[30.7] R.K. Goyal, Y.S. Negi, A.N. Tiwari, Eur. Polym. J. 41(2005) 2034.

[30.8] H. B. Qiao, Q. Guo, A. G. Tian, G. L. Pan, L.B. Xu, Tribol. Int. 40 (2007) 105.

[30.9] Q.H. Wang, Q.J. Xue, W.M. Liu, J.M. Chen, Wear 243 (2000) 140.

[30.10] L. Rapoport, Y. Bilik, Y. Feldman, M. Homyonfer, SR. Cohen, R. Tenne, Nature 387 (1997) 791.

[30.11] L. Rapoport, N. Fleischer, R. Tenne, Adv. Mater. 15 (2003) 651.

[30.12] W.X. Chen, J.P. Tu, Z.D. Xu, R. Tenne, R. Rosenstveig, WL. Chen, H.Y. Gan, Adv. Eng. Mater. 4 (2002) 686.

[30.13] L. Rapoport O. Nepomnyashchy, A. Verdyan, R. Popovitz-Biro, Y. Volovik, B. Ittah, R. Tenne, Adv. Eng. Mater. 6 (2004) 44.

[30.14] X.H. Hou, K.L. Choy, Chem. Vap. Depos. 12 (2006) 583.

[30.15] G. Zhang, S. Leparoux, H. Liao, C. Coddet, Scr. Mater. 55 (2006) 621.

[30.16] B. Smith, Infrared Spectral Interpretation: A Systematic Approach, CRC Press, New York, 1998, p. 153.

[30.17] L.G. Yu, SR. Yang, W.M. Liu, Q.J. Xue, Polym. Eng. Sci. 40 (2000) 1825.

[30.18] Xianghui Hou, C.X.Shan, Kwang-Leong Choy. Surface & Coating Technology 202 (2008) 2287.

31

Synthesis of Cr_2O_3-Based Nano-composite Coatings with Incorporation of Inorganic Fullerene-Like Nano-particles

31.1 A General Review

The incorporation of inorganic fullerene-like nanoparticles into coating systems is considered an effective method to provide extra functionality and improve the specific properties of matrix coatings. In the studies reported in this chapter, inorganic fullerene-like tungsten disulfide (IF-WS_2) nanoparticles have been incorporated into Cr_2O_3 coatings via a modified aerosol-assisted chemical vapour deposition process (AACVD), for the purpose of reducing friction and improving wear resistance. The deposited nanocomposite coatings are characterized using a combination of X-ray diffraction, Raman spectroscopy and scanning electron microscopy. Ball-on-flat testing has been carried out to evaluate the friction and wear properties of the IF-WS_2/Cr_2O_3 nanocomposite coatings. The microstructure and tribological testing results show that the AACVD process is a promising method for the synthesis of nanocomposite coatings incorporating nanoparticles.

31.2 Introduction

With layered hollow structure and quasi-spherical shape, the inorganic fullerene-like (IF) tungsten disulfide (WS_2) and molybdenum disulfide (MoS_2) nanoparticles have demonstrated interesting tribological characteristics, which make them favorable solid lubricants under severe conditions, where fluids are unable to support a heavy load and are squeezed away from the contact area [31.1-31.5]. The IF-WS_2 and MoS_2 nanoparticles can be used as additives for lubrication fluids, oil and greases [31.6]. The sliding/rolling of IF is believed to be a dominant friction mechanism under loading conditions when the spherical shape of IF is preserved [31.7]. However, the slippery nature of these nanoparticles leads to their fast displacement from the contact area, and consequently the efficacy of their lubrication is maintained only so long as they can be replenished in the contact area [31.8]. Thus, the concepts of slow release and self-lubrication have been raised by confining the IF nanoparticles inside a porous and densified solid matrix [31.8, 31.9].

The incorporation of IF nanoparticles into nanocomposites and coatings has undergone preliminary investigations, for instance, by incorporating IF nanoparticles into polymeric and metallic matrices [31.7, 31.10, 31.11]. Considerable improvements have been reported on the tribological characteristics of nanocomposites with the incorporation of IF nanoparticles. However, there are not yet any reports on the incorporation of IF nanoparticles into ceramic matrices.

Chromium oxide (Cr_2O_3) is the hardest oxide, and exhibits high hardness values and low friction coefficients [31.12, 31.13]. Some applications of Cr_2O_3 coatings have been found in tribological and wearing environment, as protective coatings on digital recording system[31.14] and in gas-bearing applications [31.15, 31.16]. Chromium oxide coatings can be deposited with a variety of techniques such as plasma spraying [31.17], chemical vapour deposition [31.12], sputtering[31.13] and evaporation[31.18], etc., but a process for the incorporation of IF nanoparticles into the ceramic matrix has yet to be established.

This chapter reports the realization of the IF nanoparticles in ceramic coatings for tribological applications. Cr_2O_3 is selected as matrix material. With a modified aerosol-assisted chemical vapour deposition process (AACVD), IF-WS_2/Cr_2O_3 nanocomposite coatings have been produced. The AACVD method involves atomization of precursor liquid into fine aerosol droplets. The droplets are subsequently directed towards a heated zone, where the solvent is rapidly evaporated, and the chemical precursors undergo decomposition and/or chemical reaction near or on a heated substrate to form the desired films and coatings [31.19].

31.3 Experimental Details

Stainless steel plates and silicon wafers (L = 20 mm, W = 12 mm) were selected as substrates. Prior to deposition, all substrates were cleaned in an ultrasonic bath with alcohol to degrease. Chromium nitrate nonahydrate (Aldrich) was used as a precursor chemical of chromium oxide (Cr_2O_3), while IF-WS_2 nanoparticles ranging from 80 to 220 nm were supplied by NanoMaterials Ltd. The thermal behaviours of the precursors and IF nanoparticles were determined. Thermogravimetric analysis/differential thermal analysis (TG/DTA) was carried out, using a Setaram Labsys 1600, from 30 °C to 700 °C, at 5 K/mm in air. The basic precursor solution was prepared by dissolving chromium nitrate nonahydrate in alcoholic solvent to form solution of 0.05 M in concentration. IF-WS_2 was added into the basic precursor solution (0.23 g/L) to obtain a uniform suspension via an ultrasonic bath.

The basic precursor solution or the suspension containing IF-WS_2 nanoparticles was then atomized to generate fine aerosol droplets using an ultrasonic generator, at a frequency of 1.7 MHz, with nitrogen as a carrier gas [31.20]. The droplets were subsequently directed towards a heated substrate where they underwent evaporation, decomposition and chemical reactions. The chromium oxide coatings, with or without IF-WS_2 can be deposited onto the substrate. The deposition temperatures were set in the range of 280-300° C. Post heat treatment of the samples was carried out to obtain crystalline Cr_2O_3 and finalize the microstructure in the nanocomposite coatings. The as-deposited samples were treated at 500°C in argon for 1 h.

The chromium oxide coating and IF-WS_2/Cr_2O_3 coatings were characterized using a combination of X-ray diffraction (XRD), Raman spectra and scanning electron microscopy (SEM). A Siemens D500 X-ray diffractometer with Cu Kα radiation was employed to detect phase and crystallinity of the coatings. The measurement was operated in a step scan mode with a step of 0.01, in the range of $2\theta = 10\text{-}80°$, using 40 kV voltage and 25 mA current. Raman spectra were also recorded for the analysis of coating compositions on a LabRAM300 spectrometer ($\lambda = 633$ nm). A Philips XL30 scanning electron microscope equipped with an Oxford Instruments energy-dispersive X-ray spectrometry (EDX) was used to characterize the surface morphology and composition. The SEM operating voltage was 20 kV and the EDX spectra were collected during 100 s with calibrated standards under an acceleration voltage of 20 kV. The thickness of the coatings was measured by a MM-16 Ellipsometer from Horiba Jobin Yvon.

The tribological tests were carried out using CETR Universal Micro-Tribometer (UMT-2). Linearly reciprocating ball-on-flat sliding wear was performed under constant loading of 0.2 N at a moving rate of 100 mm/min, with 1.6 mm diameter alumina balls. The friction force and down force were simultaneously recorded during the testing, and the coefficient of friction (COF) was obtained.

31.4 Results and discussion

31.4.1 *Thermal behaviour of precursors*

The modified aerosol-assisted chemical vapour deposition process is carried out in open atmosphere, while decomposition and oxidation of chemical precursors occur during deposition. To achieve the best tribological characteristics of IF-WS_2/Cr_2O_3 nanocomposite coatings, the basic requirement for the deposition is to decompose chromium nitrate completely, but simultaneously to avoid oxidation of IF-WS_2. Fig. 31.1 shows the TG/DTA of chromium nitrate nonahydrate and IF-WS_2 nanoparticles in air. Fig. 31.1(a) shows that the crystalline state $Cr(NO_3)_3\ 9H_2O$ decomposes in air in a temperature range of 50 to 230 °C by several endothermic steps with rapid mass loss. The possible decomposition can be described by Eq. (31.1)[31.21]:

$$2Cr(NO_3)_3.9H_2O(s) \rightarrow 2Cr(NO_3)_3(s) + 18H_2O(g)$$
$$2Cr(NO_3)_3(s) \rightarrow Cr_2O_3(s) + 6NO_2(g) + 11/2O_2(g). \quad (31.1)$$

Thus, Cr_2O_3 composition can only be produced in air when the deposition temperature is higher than 230°C. Fig. 31.1(b) shows the TG/ DTA result of IF-WS_2 nanoparticles. IF-WS_2 nanoparticles are readily oxidized to WO_3 at elevated temperature in an oxidizing environment. The thermal stability of IF-WS_2 is highly depended on its particles size [31.22]. Larger particles have better stability in an oxidizing environment. According to the exothermic peak and the mass loss in Fig. 31.1(b), the oxidation of the supplied IF-WS_2 nanoparticles starts at 350 °C. As IF-WS_2 nanoparticles are involved in the form of suspension droplets in the AACVD processing, the maximum deposition temperature should be lower than 350 °C to avoid oxidation. Integrating the thermal behaviours of chromium nitrate and IF-WS_2, the deposition temperature of AACVD is set at 280-300 °C for the synthesis of IF-WS_2/Cr_2O_3 nanocomposite coatings.

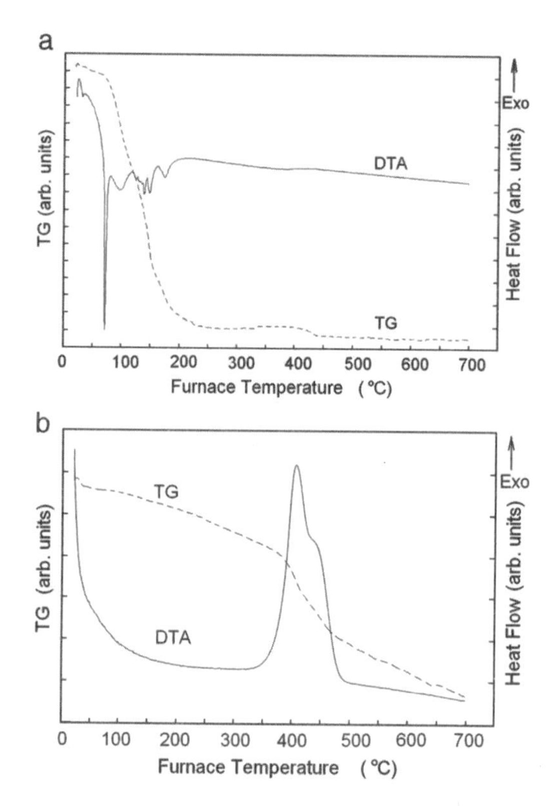

Fig. 31.1 TG/DTA of precursors: (a) chromium nitrate nonahydrate, and (b) IF-WS_2.

31.4.2 *Phase and composition of the nanocomposite coatings*

Fig. 31.2 shows the X-ray diffraction patterns of the nanocomposite coatings on stainless steel substrates. In both the as-deposited and the heat-treated samples, there is a small peak at $2\theta = 14.3^\circ$ which is attributed to the (0002) plane of IF-WS_2 nanoparticles[31.23]. It indicates that IF-WS_2 nanoparticles survive inside the coatings during the AACVD process and the subsequent heat treatment. No obvious oxidation and other chemical reactions have taken place with IF-WS_2. The XRD result also shows that the as-deposited coating matrix is mainly amorphous. After heat treatment, some diffraction peaks appear and all of them can be easily assigned to eskolaite, the hexagonal crystalline phase of Cr_2O_3 [31.24].

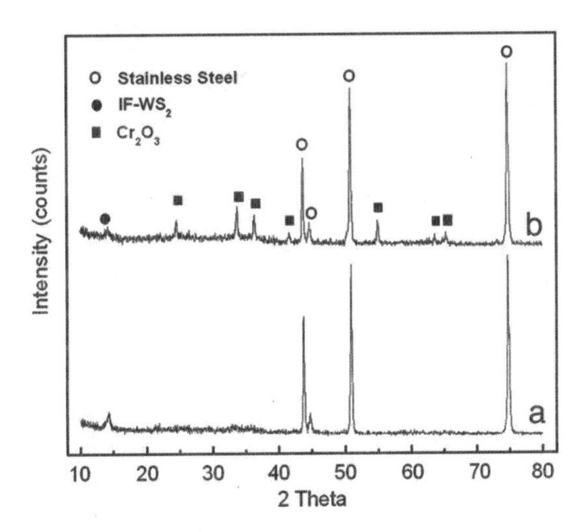

Fig. 31.2 X-ray diffraction patterns of the deposited IF-WS_2/Cr_2O_3 coatings on stainless steel substrates (a) as-deposited coating and (b) after heat treatment at 500 °C for 1 h.

Fig. 31.3 shows the Raman spectra of the IF-WS_2 nanoparticles and the composite coatings. IF-WS_2 nanoparticles present two strong peaks at 351 cm^{-1} and 421 cm^{-1} which correspond to the E_{2g} and A_{1g} modes, respectively[31.25]. Similar to the Raman spectrum of 2H-WS_2, these correspond to the first order Raman peaks. The band at 351 cm^{-1} is assigned as an E_{2g} mode for motion of W + S in the *x–y* plane and the band at 421 cm^{-1} assigned as an A_{1g} mode for the motion of two S atoms along the *z*-axis of the unit cell [31.26, 31.27]. The other smaller peaks observed in Fig. 31.3(a) are the second order Raman peaks that arise from phonon couplings with a non-zero momentum. The Raman spectrum of the as-deposited composite coating mainly reflects the spectrum of IF-WS_2 nanoparticles. The weak peak at 550 cm^{-1}, corresponding to the presence of Cr_2O_3, confirms that the as-deposited Cr_2O_3 matrix is amorphous or poorly crystallized [31.28]. In the heat-treated coating, the peak intensity related to IF-WS_2 nanoparticles decreases, while peaks centred at 549 cm^{-1} and 614 cm^{-1} become stronger, which correspond to the A_{1g} mode and E_g mode of Cr_2O_3, respectively [31.29-31.31]. It should be emphasized that the loading amount of IF-WS_2 nanoparticles inside the composite coating is 6 wt.%. However, the small loading of IF-WS_2 contributes to a high-intensity response in the Raman spectrums of the nanocomposite coating. Therefore, high sensitivity can be achieved with Raman spectroscopy for the characterization of IF-WS_2 nanoparticles.

The results of XRD and Raman spectroscopy suggest that IF-WS_2 nanoparticles have been incorporated into the Cr_2O_3 coatings by the AACVD process and the subsequent heat treatment

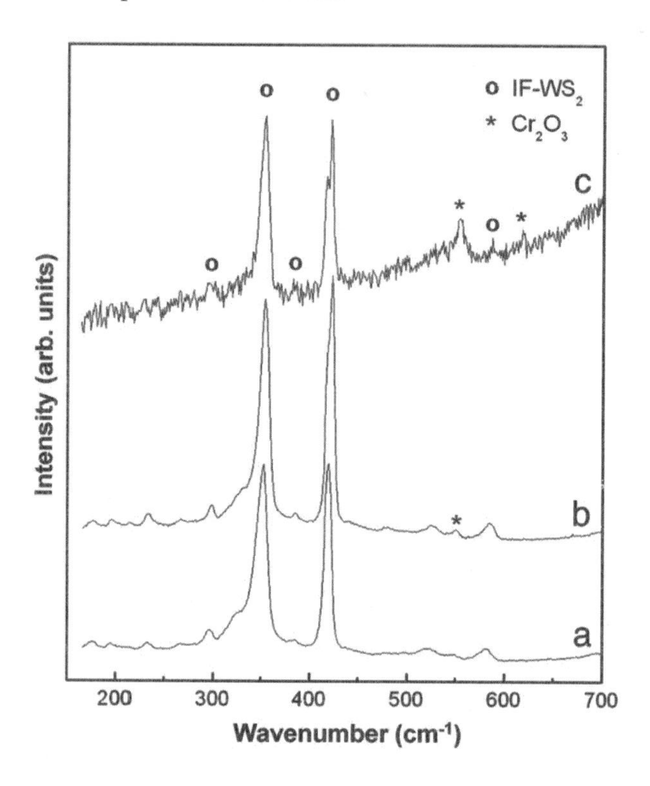

Fig. 31.3 Raman spectra of (a) IF-WS_2 nanoparticles and (b) as-deposited and (c) heat- treated IF-WS_2/Cr_2O_3 composite coatings.

31.4.3 *Surface morphology of the deposited coatings*

The surface morphology of the nanocomposite coatings is shown in Fig. 31.4. For comparison, a pure Cr_2O_3 coating is also provided as a reference sample. Both of them are dense coatings, with thickness around 1.5-2.0 μm. As seen in Fig. 31.4(a), the pure Cr_2O_3 coating is very smooth and there is no obvious aggregate observed on the surface of the coating. However, with the incorporation of IF-WS_2 nanoparticles, the nanocomposite coating presents a rougher surface consisting of numerous aggregates ranging from 100 nm to 2 μm in size.

Fig. 31.5 shows a SEM image and the corresponding EDX line scan of Cr and W elements across an embossing aggregate on the surface of the coating. The line scan of chromium indicates the uniformity of the chromium oxide (Cr_2O_3) matrix, while that of tungsten reflects the distribution of IF-WS_2 nanoparticles. According to the result in Fig. 31.5 (b), the concentration of tungsten inside the aggregate is much higher than that in the surrounding area, while the chromium is distributed uniformly. As the size of the individual IF-WS_2 nanoparticle is 80-220 nm, the dimension of the embossing aggregate in Fig. 31.5(a) indicates that it is a IF-rich region which may consist of tens or hundreds of IF nanoparticles. There are two possible reasons, which may cause the formation of these IF-rich aggregates using AACVD process.

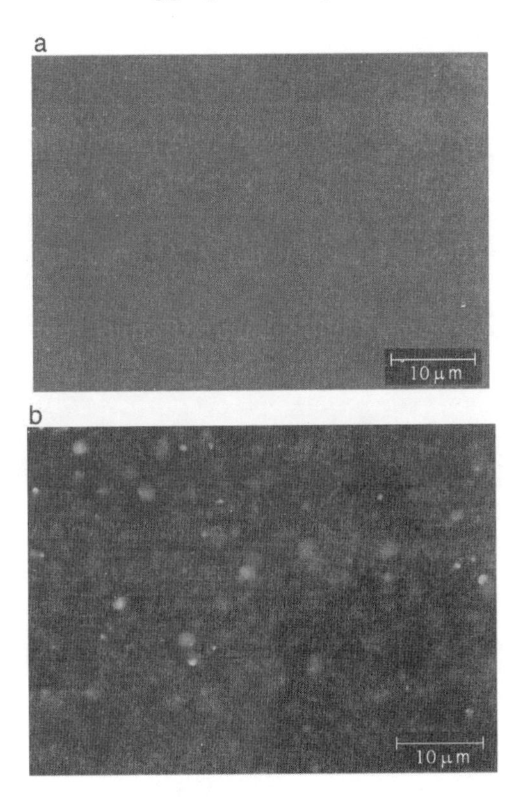

Fig. 31.4 SEM images of (a) Cr_2O_3 coating and (b) IF-WS_2/Cr_2O_3 composite coating.

1. The agglomeration of IF-WS_2 nanoparticles occurs before deposition and is directly transferred into the Cr_2O_3 matrix during the deposition. In this case, some nanoparticles have no direct contact with the Cr_2O_3 matrix. The agglomeration of IF nanoparticles will directly result in the aggregation inside the deposited coating. It would be helpful to reduce this kind of aggregation by synthesis of less agglomerated IF-WS_2 nanoparticles and preparation of a more stable suspension liquid for AACVD.
2. The agglomeration of IF-WS_2 nanoparticles takes place concurrently with the formation of the Cr_2O_3 matrix in the AACVD process. After evaporation of the solvent, the IF-WS_2 nanoparticles may serve as nucleation sites for the formation and growth of the chromium oxide

matrix, and thus more particles are attracted in the subsequent deposition, which leads to the formation of IF-rich aggregates in the coating. In this case, layers of Cr_2O_3 matrix will be formed between adjacent nanoparticles. The layers of the Cr_2O_3 matrix work as an adhesive and bind the nanoparticles together. Optimizing the processing parameters may thus be helpful for the reduction of this kind of aggregation.

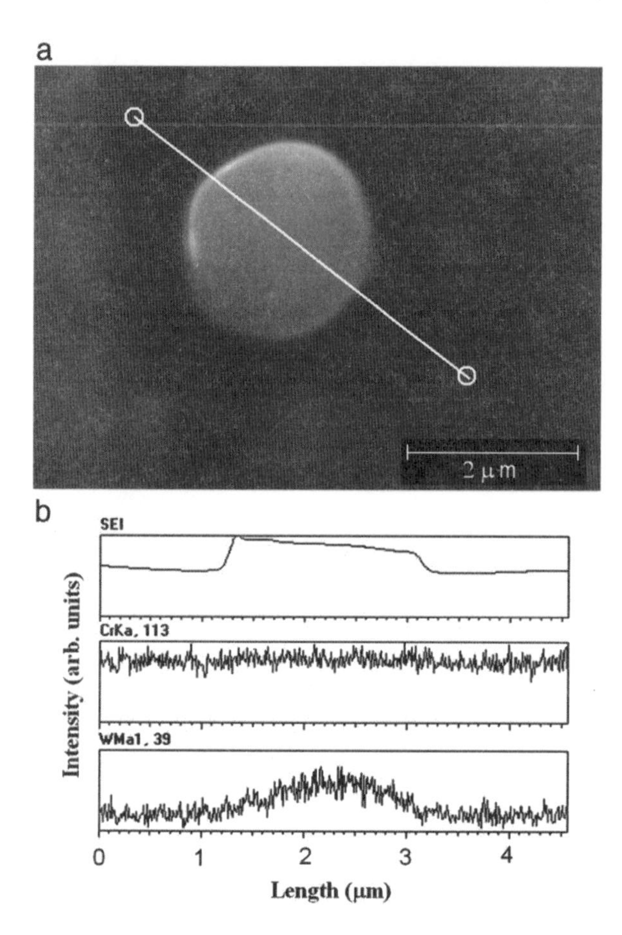

Fig. 31.5 SEM image of an aggregate on the surface of an IF-WS_2/Cr_2O_3 composite coating and the corresponding EDX line scans of Cr and w elements.

The aggregation observed in Fig. 31.4 may be a combination of the two above mechanisms. The effects of aggregation on the microstructures and tribological properties of the composite coatings are yet to be established.

31.4.4 *Friction and wearing test*

Preliminary ball-on-flat testing was performed to evaluate the friction properties of the deposited composite coatings. The duration of a test run is set at 30 mm.

Fig. 31.6 shows the coefficient of friction of both the deposited nanocomposite coating and the pure Cr_2O_3 coating. It is clear that the COF is significantly reduced by approximately 25%, with the incorporation of IF-WS_2 nanoparticles into the coating.

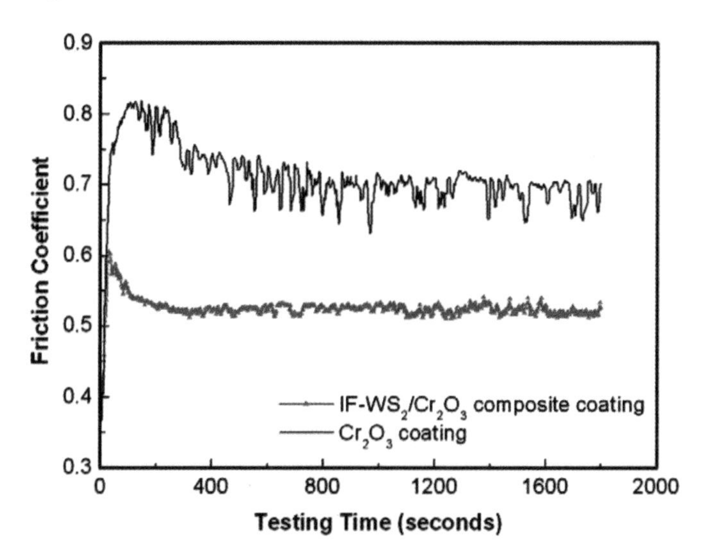

Fig. 31.6 COF of the deposited coatings under ball-on-flat test.

The agglomerates of IF-WS_2 are very soft and can split into individual nanoparticles under light loads and shear movement [31.9]. The IF-WS_2 nanoparticles can provide an effective rolling friction mechanism and serve as spacers, preventing asperity contact between the mating surfaces [31.5, 31.8]. Although the incorporated nanoparticles in the IF-WS_2/Cr_2O_3 composite coating are not free to offer rolling friction, some IF nanoparticles can be released to the mating interface under the testing condition in Fig. 31.6, which provides lubricating effect and reduces the friction. This result is similar to the phenomenon observed in the polymeric and metallic composite coatings [31.8, 31.9].

It should again be noted that the nanoparticles are not very well dispersed in the ceramic matrix. Thus, there is significant potential for the improvement of the friction properties of the nanocomposite coatings. This result demonstrates that the AACVD process has lots of potential for the synthesis of nanocomposite coatings incorporating nanoparticles.

31.5 Conclusions

The following conclusions are drawn based on the studies reported in this chapter[31.32]:

Inorganic fullerene-like tungsten disulfide (IF-WS_2) nanoparticles have been incorporated into Cr_2O_3 coatings via a modified aerosol-assisted chemical vapour deposition (AACVD) process. A Preliminary ball-on-flat test shows that the coefficient of friction is significantly reduced with the incorporation of IF-WS_2 nanoparticles in the nanocomposite coating.

It also demonstrates that aerosol-assisted chemical vapour deposition process is a suitable method for incorporating nanoparticles into ceramic matrices, which have huge potential for applications involving the incorporation of various nanoparticles or other nanoscale structures (nanotubes, nanorods, etc.).

References

[31.1] M. Chhowalla, A.J. Amaratunga, Nature 407 (2000) 164.

[31.2] L. Rapoport, Y. Bilik, Y. Feldman, M. Homyonfer, S.R. Cohen, R. Tenne, Nature 387 (1997) 791.

[31.3] L. Rapoport, N. Fleischer, R. Tenne, J. Mater. Chem. 15 (2005) 1782.

[31.4] Y. Feldman, A. Zak R. Popovitz-Biro, R. Tenne, Solid State Sci. 2 (2000) 663.

[31.5] R. Tenne, Colloids Surf, A 208 (2002) 83.

[31.6] L. Rapoport, N. Fleischer, R. Tenne, Adv. Mater 15 (2003) 651.

[31.7] L. Rapoport, O. Nepomnyashchy, A. Verdyan, R. Popovitz-Biro, Y. Volovik, B. Ittah, R. Tenne, Adv. Eng. Mater. 6 (2004) 44.

[31.8] L. Rapoport, M. Lvovsky, I. Lapsker, V. Leshinsky, Y. Volovik, Y. Feldman, A. Zak R. Tenne, Adv. Eng. Mater. 3 (2001) 71.

[31.9] L. Rapoport, M. Lvovsky, I. Lapsker, V. Leshchinsky, Y. Volovik, Y. Feldman, A. Margolin, R. Rosentsveig, R. Tenne, Nano Lett. 1 (2001) 137.

[31.10] X.H. Hou, C.X. Shan, K.L. Choy, Surf. Coat. Technol. 202 (2008) 2287.

[31.11] W.X. Chen, J.P. Tu, Z.D. Xu, R. Tenne, R. Rosenstveig, W.L. Chen, H.Y. Gan, Adv. Eng. Mater. 4 (2002) 686.

[31.12] G. Caro, M. Natali, G. Rossetto, P. Zanella, G. Salmaso, S. Restello, V. Rigato, S. Kaciulis, A Mezzi, Chem. Vapor Depos. 11 (2005) 375.

[31.13] P. Hones, F. Levy, N.X. Randall, J. Mater. Res. 14 (1999) 3623.

[31.14] M.D. Bijker, J.J.J. Bastiaens, E.A. Draaisma, L.A.M. de Jong, E. Sourty, S.O. Saied, J.L Sullivan, Tribol. Int. 36 (2003) 227.

[31.15] B. Bhushan, G.S. Theunissen, X. Li, Thin Solid Films 311 (1997) 67.

[31.16] P. Hones, M. Diserens, F. Levy, Surf. Coat. Technol. 120-121 (1999) 277.

[31.17] H.S. Ahn, O.K. Kwon, Wear 225-229 (1999) 814.

[31.18] C. Cantalini, J. Bin. Ceram. Soc 24 (2004) 1421.

[31.19] X.H. Hou, K.L. Choy, Chem. Vapor Depos, 12 (2006) 583.

[31.20] X.H. Hou, J. Williams, K.L. Choy, Thin Solid Films 495, (2006) 262.

[31.21] M. Garcia, M. Jergel, A. Conde-Gallardo, C. Falcony, G. Plesch, Mater. Chem. Phys. 56 (1998) 21.

[31.22] C. Schuffenhauer, G. Wildermuth, J. Felsche, R. Tenne, Phys. Chem. Chem. Phys. 6 (2004) 3991.

[31.23] Y. Feldman, G.L. Frey, M. Homyonfer, V. Lyakhovitskaya, L. Margulis, H. Cohen, G. Hodes, J.L. Hutchison, R. Tenne, J. Am. Chem. Soc. 118 (1996) 5362.

[31.24] L.W. Finger, R.M. Hazen, J. Appl. Phys. 51 (1980) 5362.

[31.25] L. Joly-Pottuz, F. Dassenoy. M. Belin, B. Vacher, J.M. Martin, N. Fleischer, Tribol Lett. 18 (2005) 477.

[31.26] C.J. Carmalt, I.P. Parkin, E.S. Peters, Polyhedron 22 (2003) 1499.

[31.27] J.W. Chung, Z.R. Dai, K Adib, E.S. Ohuchi, Thin Solid Films, 335 (1998) 106.

[31.28] P.M. Sousa, AJ. Silvestre, N. Popovici, O. Conde, Appl. Surf. Sci. 247 (2005) 423.

[31.29] J. Mougin, N. Rosman, G. Lucazeau, A. Calerie, J. Raman Spectrosc. 32 (2001) 739.

[31.30] D. Stanoi, G. Socol, C. Grigorescu, F. Cuinneton, O. Monnereau, L. Tortet, T. Zhang, I.N. Mihailescu, Mater. Sci. Eng. B 118 (2005) 74.

[31.31] T. Yu, Z.X. Shen, J. He, W.X. Sun, S.H. Tang, J.Y. Lin, J. Appl. Phys. 93 (2003) 3951.

[31.32] X. Hou, K.L. Choy, Thin Solid Films 516 (2008) 8620.

Section - VIII

Indentation, Scratch and Mapping

32

Comparison of Instrumented and Conventional Nano and Micro Indentation and Scratch Tests

32.1 General

Mechanical and tribological properties such as hardness, Young's modulus, friction, and scratch-adhesion strength of various coatings and thin films are reported in this chapter. These results for ultra-thin films, obtained using a Universal Nano and Micro Tester UNMT-1, indicate a substrate effect in static nano-indentation and its absence in dynamic nano-indentation.

32.2 Introduction

Evaluation of mechanical and tribological properties of bulk and coating materials is of great importance for design and development of engineering components[32.1-32.3].

In recent years, nano-indentation has become a popular technique, which allows for evaluation of hardness and Young's modulus of films and coatings. The stress distribution in the front of a nano-indenter tip makes this technique vulnerable to a substrate effect when evaluating ultra-thin films. Usually, the indentation depth should be restricted to 5-10% of the film thickness, but such ultra-shallow depths would require an extremely high accuracy of both depth monitoring and tip calibration. A novel Nano-analyzer enables quantitative characterization of ultra-thin films, including very hard ones, without a substrate effect; it can also perform ultra-shallow nano-scratches and subsequent imaging.

The objective of investigation discussed in this chapter is to compare different techniques of mechanical and tribological tests on micro and nano-levels using a single tester.

32.3 Experimental Details

The micro-indentation, micro-scratch, nano-indentation, nano-scratch, and nano-imaging techniques were employed for evaluation of mechanical and tribological properties of numerous bulk coating and film specimens using the same Universal Nano and Micro Tester model UNMT-1, designed and manufactured by CETR. The photograph of its nano-head is presented in Fig. 22.12 of chapter 22.

The UNMT-1 has several easily interchangeable modules for precision tests of hardness, Young's modulus, friction, and adhesion, including:

- traditional micro-indentation up to a load of 1.2 kN, with Rockwell, Vickers, and Knoop indenters according to ASTM E18-05, ASTM E92-82, and ASTM E384-99 standards, respectively,
- instrumented micro-hardness tests as per the ISO 14577-1/02 up to a load of 1.2 kN with the same Rockwell, Vickers, and Knoop indenters, in-situ monitoring of load and depth and automatic calculations of hardness and Young's modulus,
- instrumented static nano-indentation tests as per the ISO 14577-1/02 (loads from 0.1 μN to 0.5 N) with Berkovich, conical and cube-corner indenters, in-situ monitoring of load and displacement and automatic calculations of hardness and Young's modulus,
- instrumented dynamic Young's modulus tests with spherical, Berkovich, and other indenters, in-situ monitoring of tip frequency changes during surface scanning, calculations and maps of Young's modulus,
- micro-scratch-hardness tests as per the ASTM G171-03 (loads from centi-N to hecto-N, sliding distances from a few microns to many millimeters) with numerous indenters, under a constant load or any loading profile with simultaneous monitoring of friction, acoustic emission, electrical resistance, etc.
- nano-scratch-hardness tests as per the ASTM G171-03 (loads from 0.1 micro-N to hecto-N, sliding distances from 1 to 100 microns) with numerous indenters (Berkovich, spherical, etc.) and AFM-like imaging of scratches with the same tip.

The UNMT-1 allows for multi-scale measurements of the same sample without its removal.

32.4 Results & Discussion

The evaluation results of mechanical properties of numerous specimens are summarized in Table 32.1. Whereas the "+" indicates the technique being sufficient for measurements, the "–" indicates the failure to evaluate the film without the substrate effect.

Table 32.1 Summary of various tests on UNMT-1

Specimens	Micro-indentation		Instrumented		
	Traditional	Instrumented	Micro-scratch	Nano-indentation	Nano-scratch
Bulk Materials	+	+	+	+	+
20 μm metal film	–	+	+	+	+
2 μm metal film	–	–	+	+	+
4 nm DLC film	–	–	–	–	+

32.4.1 *Micro-indentation*

Instrumented indentation tests were performed on bulk metal specimens with a Rockwell diamond indenter. The maximum loads for aluminum, brass and steel specimens were 30, 70 and 150 N, respectively. Fig. 32.1 shows the representative load-displacement plots, the unloading portions of which were analyzed using the Oliver-Pharr methodology[32.4].

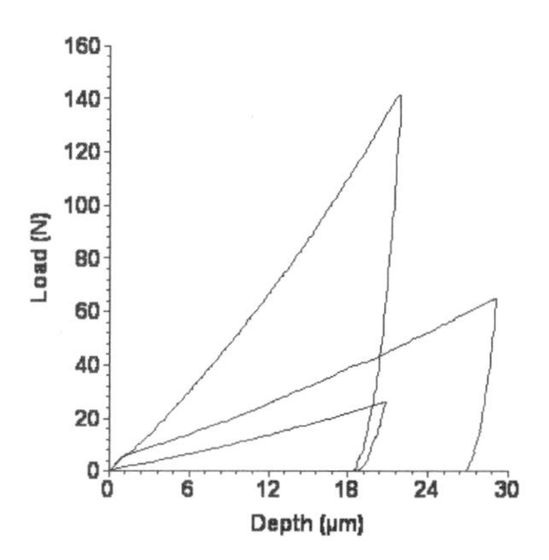

Fig. 32.1 Load-displacement plots for metals

Table 32.2 presents the hardness and Young's modulus values from 20 tests on each specimen. The mean results of the instrumented and traditional techniques were practically the same (the 10-time difference is due to the Vickers scale), while the deviations were smaller for instrumented indentation.

Table 32.2 Hardness and modulus data for metals

Specimen	Hardness, Mpa	Young's Modulus, GPa	Vickers hardness
Aluminum	968 ± 23	74 ± 2	97 ± 11
Brass	2004 ± 19	127 ± 2	201 ± 9
Steel	5202 ± 47	195 ± 8	519 ± 20

32.4.2 *Micro-Scratch*

The instrumented micro-scratch tests included sliding-scratching over a 5 mm length at a speed of 0.2 mm/s under a constant load, using a Rockwell diamond indenter, on test specimens and a reference material, usually fused silica with hardness of 9.5 GPa. The scratching loads for the test specimen and the reference sample should be such that the scratch width dimension on these materials should be similar. For example, Table 32.3 shows the scratch width and hardness values on a fused silica, pure and coated metal; the coating exhibited hardness three times higher than the metallic substrate.

Table 32.3 Scratch hardness data on fused silica, bare and coated specimens

Sample	Load, N	Scratch width, mm						Hardness, GPa
		1	2	3	4	5	Mean	
Reference Fused Silica	0.4	13.57	13.81	13.56	13.96	13.72	**13.72**	9.5
	0.2	9.96	9.78	9.68	10.23	10.03	**9.94**	9.5
Coating	0.1	7.74	7.79	7.83	8.23	7.98	**7.91**	7.5
Substrate	0.1	13.75	13.64	13.65	13.86	14.05	**13.79**	2.4

32.4.3 *Static and Dynamic Nano-indentation*

Instrumented nano-indentation was performed in both static and dynamic modes, using the same Berkovich indenter. Figures 32.2 and 32.3 show ten load-displacement curves, obtained in static nano-indentation tests on a fused silica and 2 μm thick polymer coating on Si, respectively. They show excellent repeatability of the nano-indentation data, which was then analyzed automatically to obtain hardness and Young's modulus values.

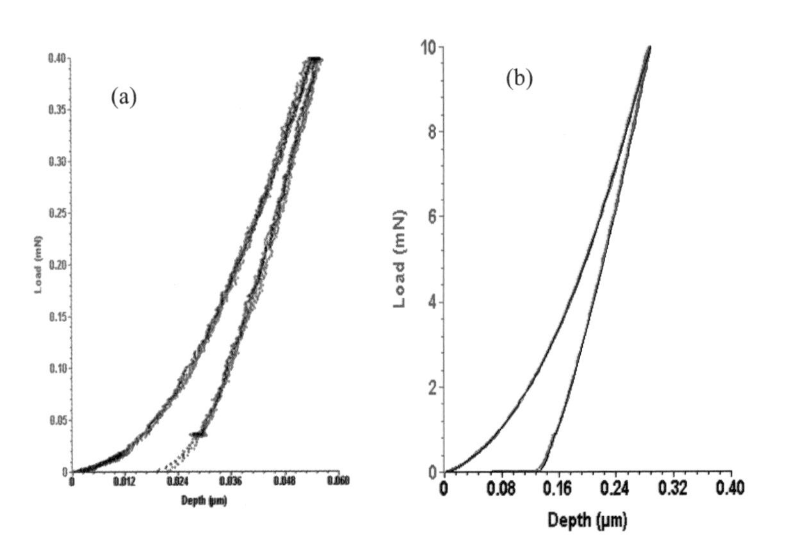

Fig. 32.2 Load-displacement curves from static nano-indentation on fused silica up to (a) 0.4 mN and (b) 10 mN.

Fig. 32.4 presents ten frequency-approach curves from dynamic nano-indentation on a polycarbonate and on 2 μm thick polymer film on a Si substrate. Again, one can see excellent data repeatability.

Table 32.4 shows the comparative Young's modulus data obtained from static and dynamic nano-indentation tests on various specimens. Each data are the average of 20 tests. The static nano-indentation showed good data for bulk materials and micron-thick coatings, but a significant substrate effect for nano-coatings. The dynamic nano-indentation showed good data for all the specimens, including ultra-thin films.

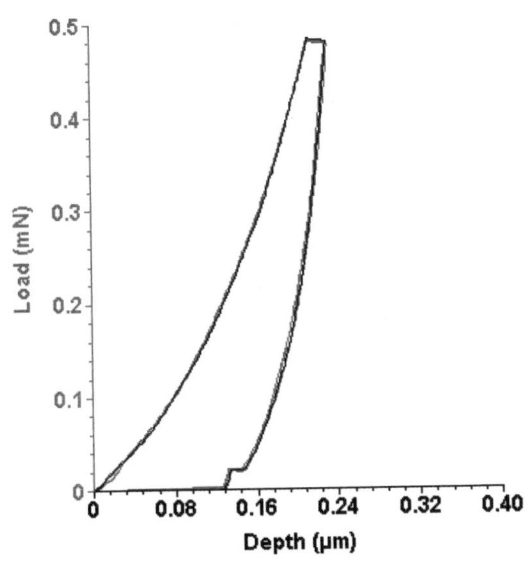

Fig. 32.3 Load-displacement curves from static nano-indentation on 2 μm polymer on Si substrate.

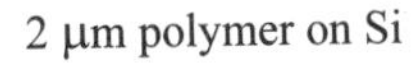

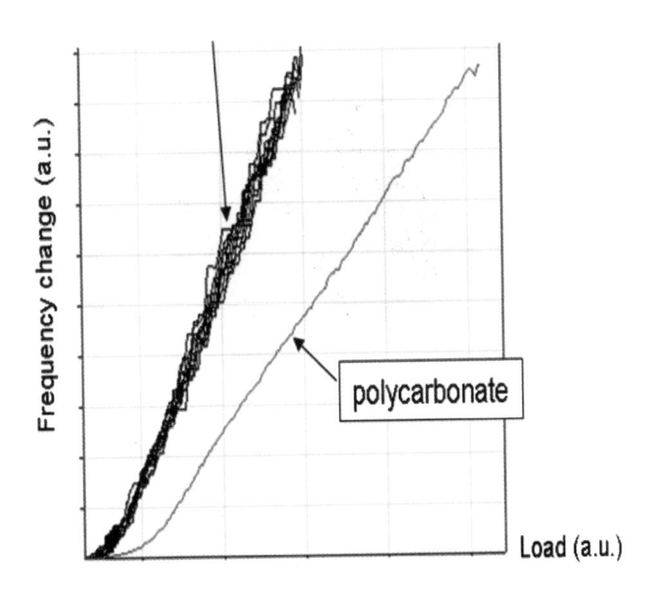

Fig. 32.4 Frequency-approach curves from dynamic nano-indentation.

Table 32.4 Young's modulus data from static and dynamic nano-indentation

Specimens	Young's Modulus, GPa	
	Nano-indenter	**Nano-analyzer**
Silicon 100	164 ± 18	164 ± 14
12 μm polymer on Si	6.85 ± 0.07	6.0 ± 0.5
2 μm polymer on Si	7.36 ± 0.31	6.0 ± 0.5
2 μm Ti on Si	98 ± 13	98 ± 10
2 μm DLC on Si	381 ± 24	382 ± 19
1 μm DLC on Si	372 ± 29	379 ± 21
100 nm DLC on Si	314 ± 32*	370 ± 25
4 nm DLC on Si	195 ± 39*	361 ± 27
Reference Polycarbonate	3.62 ± 0.06	3.50 ± 0.04
Reference Fused Silica	71.20 ± 0.65	72.9 ± 0.8
*Substrate effect		

32.4.4 *Nano-Scratch*

The Nano-analyzer module of the UNMT-1 was used to measure scratch-hardness by nano-scratching with a Berkovich tip, followed by nano-imaging with the same tip. Its software allows for image analysis to obtain scratch depth profile along any direction. The scratch width was then compared to a reference scratch on a material with known hardness.

Fig. 32.5 shows an image of 16 μm × 16 μm after nano-scratching with 500, 200 and 100 μN (left to right) loads on a 12-μm polymer coating on Si.

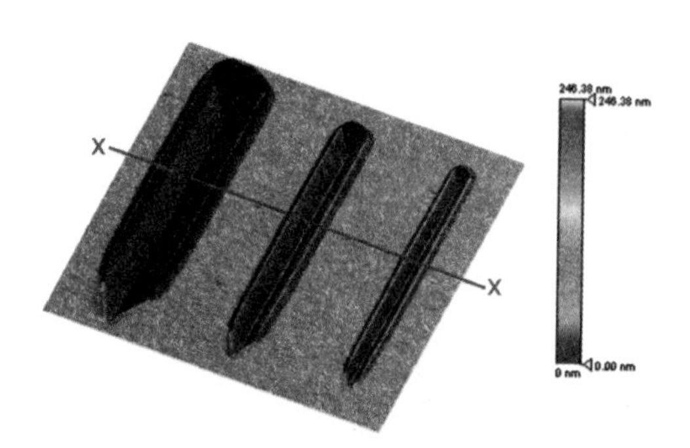

Fig. 32.5 Image of nano-scratches at 3 loads.

Table 32.5 shows hardness data obtained in 20 nano-indentation tests vs. 20 nano-scratch tests. The nano-indentation showed good data for micro-coatings, but a substrate effect for nano-coatings; such substrate effect was absent in the nano-scratch tests. Indeed, a stress distribution in the front of a moving indenter extends into the substrate during nano-indentation, but stay within the nano-coating in the nano-scratch mode.

Table 32.5 Nano-indentation and Nano-scratch hardness

Specimens	Hardness, Gpa	
	Nano-Indenter	Nano-Analyzer
Silicon 100	11.4 ± 1.8	11.6 ± 1.0
12 μm polymer on Si	0.29 ± 0.01	0.27 ± 0.03
2 μm polymer on Si	0.44 ± 0.01*	0.33 + 0.04
1 μm DLC on Si	30.5 ± 2.9	31.2 ± 1.7
100 nm DLC on Si	24.6 ± 3.6*	30.8 ± 1.7
4 nm DLC on Si	15.7 ± 3.2*	29.1 ± 1.9
Reference Polycarbonate	0.23 ± 0.004	0.21 ± 0.03
Reference Fused Silica	9.52 ± 0.07	9.57 ± 0.50

*Substrate effect

32.5 Conclusions

The following conclusions are drawn[32.5]:

In the micro- and nano-indentation tests, the indents under 5-10% of the film thickness have produced repeatable and apparently substrate-independent hardness and Young's modulus results.

In the micro- and nano-scratch tests, the scratches under 30-35% of the film thickness have produced repeatable substrate-independent hardness results.

The UNMT-1 provides a unique single platform for comparative studies of mechanical and tribological properties on micro- and nano-levels for thin films, thick coatings and bulk materials.

Future Study

Future studies shall be focused on indentation and scratch evaluation of coatings at extreme high and low temperatures, using UNMT-1 chamber modules.

References

[32.1] D. Tabor Phil. Mag., A74 (5), 1207 (1996)

[32.2] N. Gitis, J. Xiao, and M. Vinogradov, J. ASTM Int. STP 1463, 2 (9), 80(2005)

[32.3] N. Gitis, I. Hermann, S. Kuiry, V. Khosla, Proc. STLE/ASME Int. Joint Trib. Conf., October 22-24, San Diego, IJTC2007-44025 (2007)

[32.4] W.C. Oliver and G.M. Pharr, J. Mat. Res., 19(1), 1(2004); 7, 1564(1992)

[32.5] N.Gitis, S.Kuiry, I.Hermann, H.Prashad, ICIT conf., New Delhi, (2008)

33 Tribo-Mechanical Nano-Characterization of Thin Films

33.1 A General Review

Traditional techniques of tribological and mechanical characterization of coatings are limited to friction, wear and scratch tests on a micro-level with balls or pins having much larger sizes, and load-displacements much larger, than the film thickness. Lately popular nano-indentation allows for testing of thinner films, but its pressure distribution is concentrated under (in the front of) the indenter and thus makes it vulnerable to substrate effects for ultra-thin films, while ultra-shallow depths would require unachievable-yet resolution of tip and system calibration. Common AFM-based techniques use smaller tips and displacements, but limited to nano-dimensional and topographic characterization of surfaces.

A novel nano-analyzer enables effective quantitative tribo-mechanical characterization of ultra-thin films. It can perform repeatable nano-scratching at depths as shallow as 1 nm with both precision positioning and nano-imaging of the scratches, thus allowing for scratch-hardness and scratch-adhesion evaluation on a nano-level. It can measure Young's modulus of ultra-thin films with a negligible substrate effect. Combining topographical and mechanical surface mapping, it allows for detailed nano-morphology studies. Examples of the nano-analysis of various thin hard films are discussed in detail in this chapter.

33.2 Introduction

A growing demand for hand-held devices, such as cell phones, personal digital assistants, camcorders, etc., pushes technology to produce smaller, faster, lighter, and more functional electronic devices. They include a combination of several electronic and optical components, assembled in a tiny single package, each with precision wire bonding. This study involves non-destructive characterization of homogeneity and defects in a Cu/Ni interface in wire bonding.

During wire soldering, palladium and gold rapidly dissolve into the melted solder, which causes the nickel underlay to contact the solder and form inter-metallic compounds. Nickel from the pad, together with tin and copper, from the solder participate in the reaction to form a ternary inter-metallic $(Cu,Ni)_6Sn_5$ at the interface, which is observed on top of the Ni_3Sn_4 (formed on a nickel surface). There is high dislocation density from differences in crystal structure and lattice between Ni_3Sn_4 and $(Cu,Ni)_6Sn_5$. Voids will grow at the $Ni_3Sn_4/(Cu,Ni)_6Sn_5$ interface and create non-homogeneity in the wire bonding which is a major reason for failures. The analysis of the non-homogeneity was the objective of this study.

33.3 Experimental Technique and Equipment

Nano-indentation tests are the commonly applied means of testing nano-mechanical properties of thin films. In such tests, a hard diamond tip is pressed into the sample under known loads, and a nano-indentation machine collects load-displacement data in-situ.

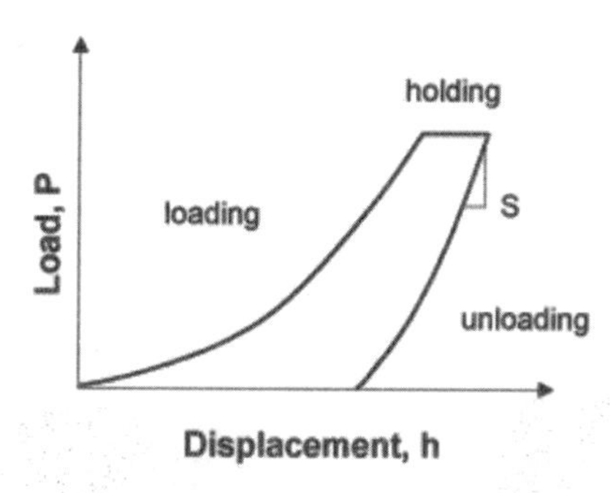

Fig. 33.1 Schematic plot of a typical load-displacement curve

In nano-indentation the pressure distribution is concentrated under (in the front of) the indenter and thus makes the results vulnerable to substrate effects for ultra-thin films. Moreover, ultra-shallow depths would require unachievable-yet resolution of tip and system calibration.

To detect the nano-non-homogeneities, a high-resolution surface mapping is required. Use of traditional nano-indentation for surface mapping is limited by the duration of a required series of numerous indents and their special resolution (the minimum space between indents is typically three times the diameter of the tip, or over 100 nm, to avoid the effect of the preceding indent). This makes nano-indentation unsuitable for studies of local nano-defects, or non-homogeneity.

A novel technique of nano-mapping has been used, when a diamond nano-tip is vibrating in a tapping mode, frequency and phase of its vibrations are monitored and analyzed, and simultaneous topographical and stiffness (Young's modulus) maps of surface are produced with nano-resolution in both vertical and horizontal directions.

The novel Nano-analyser model NA-1 can measure Young's modulus of films without leaving substantial residual imprints. Like a tapping (non-contact) AFM, it has resonating up (through unlike and AFM silicon cantilever, it is made from diamond), which changes frequency when it touches the sample surface. An ultra-low load is applied, so that the sample is deformed in the elastic regime. As the technique does not destruct the sample surface and makes continuous scans, the resultant nano-mechanical map generated by this technique has high resolution.

The nano-analyzer generates both topographical and nano-mechanical maps within the same scan. Its uniquely stiff design allows for measurements of hardness and Young's modulus of ultra-hard materials, up to 100 GPa and 1000 GPa, respectively.

33.4 Bond Characterization

Fig. 33.2 shows the optical and AFM-like topography images of the Cu/Ni interface under study. The NA-1 uses an optical microscope for micro-positioning and micro-imaging, then a user can click on the particular area of the image with a computer mouse, and perform SPM nano-positioning and nano-imaging there.

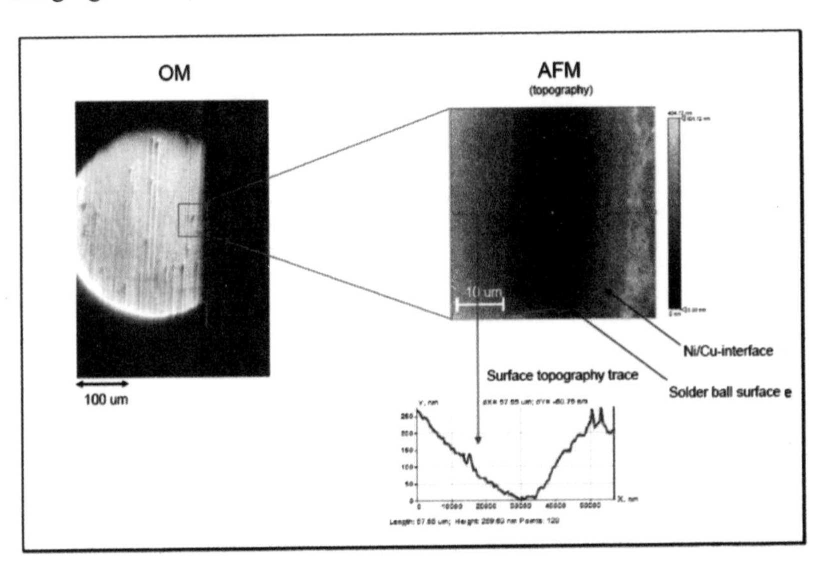

Fig. 33.2 Optical and AFM images of the bond

Fig. 33.3 shows detailed topographical and stiffness maps obtained with the nano-analyzer. Stiffness image here is the 2d distribution of compliance, which is proportional to the Young's modulus. The dark color corresponds to higher compliance as compared to the lightly-shaded area. One can observe a gradual change in the stiffness values from Cu to Ni surface.

Fig. 33.4 presents a topographical image of the inter-metallic phase of bond area. The scanned area looks uniform, with no defects or non-homogeneity visible.

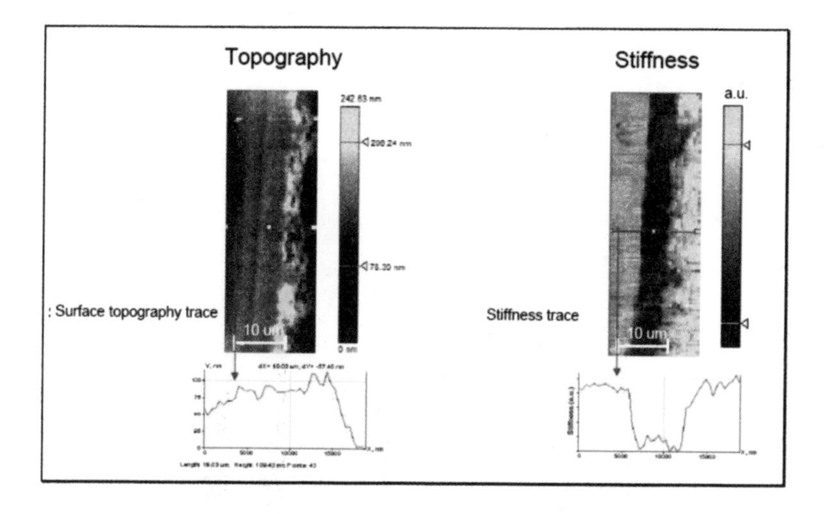

Fig. 33.3 Topographical and stiffness maps

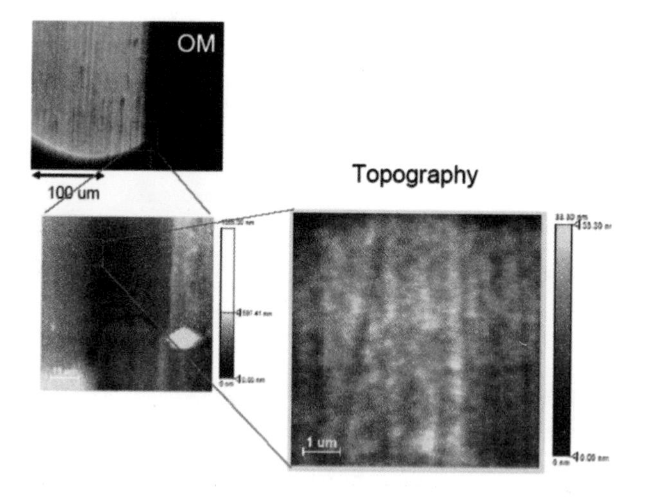

Fig. 33.4 Topographical map by Nano-analyzer NA-1

A stiffness map of the same area in Fig. 33.5 clearly shows multiple non-homogeneity defects. Some of the defects visible on the stiffness map are below the surface; hence, AFM or similar imaging techniques cannot detect them. These underlying defects in the lattice structure produce stress, which migrates to the surface and so revealed and measured on the stiffness map. Some of the other defects are on sample surface, but not picked up by AFM due to their shape and dimensions similar to the surface roughness.

To compare effects of destructive and non-destructive indentation, multiple indents with increasing load were done on the same ultra-thin copper film. Fig. 33.6 shows the residual indents. Young's modulus was measured at

150 GPa with the deeper destructive indent due to the effect of substrate, while it was correctly measured at 138 GPa with the shallower non-destructive indent with no substrate effect.

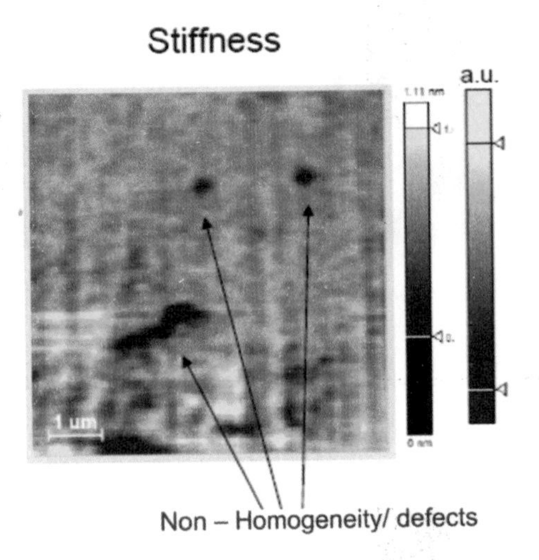

Fig. 33.5 Stiffness map by Nano-analyzer NA-1

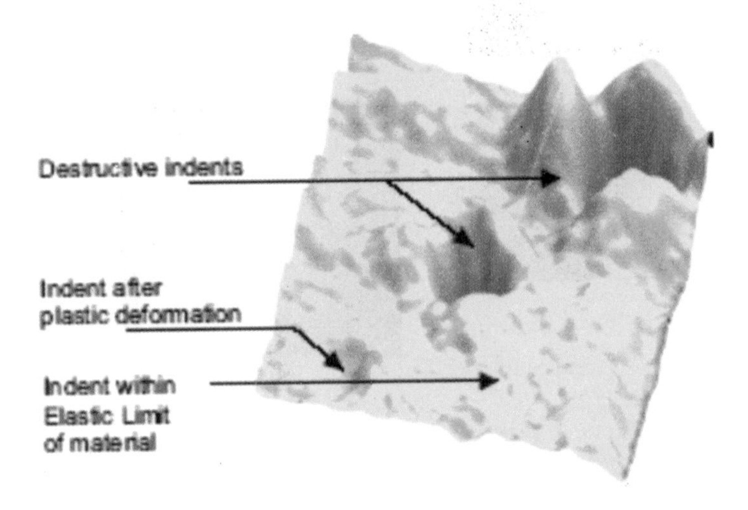

Fig. 33.6 Indents on ultra-thin metallic film under different loads

33.5 Conclusions

The above study brings the following conclusions[33.1]:

AFM-like surface analysis technique can miss defects of non-homogeneity in thin films when the defect size is of the order of surface roughness.

Nano-indentation techniques can miss defects or non-homogeneity in thin films due to insufficient special resolution and surface destruction.

A novel Nano-analyzer, combining AFM-like imaging with nano-indenter-like nano-mechanical mapping, may be perfect for ultra-sensitive non-destructive analysis of non-homogeneity in ultra-thin films.

Further studies are required to understand the defect formation and propagation. Further experiments shall involve study of bonding in diverse environmental conditions to study defect density and manufacturing conditions.

References

[33.1] N.Gitis, I.Hermann, M.Vinogradov, V.Khosla, Viennano 07, March 16, 200914-17, 2007, Vienna, Austria.

34

Non-Destructive High-Resolution Stiffness Mapping of Composite Engineered Surfaces

34.1 General

Experimental evaluation of hardness, adhesion and Young's modulus has been performed on polyimide polymeric coatings used in LCD displays and on composite polymer-based materials used in automobiles and aircraft.

A novel Universal Nano and Micro Tester UNMT-1 with a nano-analyzer module NA-2 has been utilized. It measures scratch-hardness of coatings and thin films, utilizing the same nano-tip for both scratching and nano-imaging under the constant load. It measures scratch-adhesion with the same diamond tip for both scratching and nano-imaging under the continuously increasing load. It evaluates homogeneity of films and composite materials by simultaneous Young's modulus and topography nano-mapping, with a diamond nano-tip in a tapping mode, while frequency and phase of its vibrations are analyzed. The Young's modulus maps allow to evaluate the distribution of SiO/SiO_2 particles embedded in araldite, with varying SiO/SiO_2 concentration. While the topography images could not distinguish between the particles and polymeric matrix, the nano-mechanical maps revealed the effects of particle concentration and agglomeration on the local modulus of the material and the relationship between the SiO/SiO_2 uniformity and uniformity in modulus.

The nano-scratches of 60 nm polyimide coatings at progressively increasing and constant loads generated adhesion and scratch-hardness data, respectively. Within the applied loads of 20 to 100 μN, critical load was observed and determined, at which the coating was delaminated from the glass substrate, and a corresponding lateral delamination force. The mutually complimentary nano-images and force graphs coincided nicely.

34.2 Introduction

A novel tool Nano-Analyzer on the Universal Nano and Micro Tester UNMT-1 (Fig. 22.1) was used to obtain non-destructive stiffness maps of composite materials, topographically-smooth surfaces and patterned sub-surfaces. Quantitative nano-mechanical characterization of a multi-component composite is a tedious chore. For example, nano-indentation involves indenting the material at many locations to obtain the mechanical property map, with the resolution limited by the need for sufficient spacing between the indents and the time of such mapping being unviably long.

34.3 New Technique – Nano-Analyser

To overcome the above shortcomings, a Nano-Analyser (NA) has been developed. The NA evaluates the surface by scanning it with a vibrating cantilever with a tiny (less than 100 nm) diamond nano-tip at very low loads (of micro-newtons). The amplitude of the tip vibrations is kept constant, while the vibrational frequency variations are monitored. The frequency shift depends on the tip-sample contact stiffness, which in turn, depends on the local elastic properties of the sample. The NA simultaneously and non-destructively generates AFM-like topographic images and stiffness maps of the scanned area.

34.3.1 *Stiffness Maps of Composite Material*

PMMA – Poly-methyl-meth-acrylate is a versatile polymeric material used in such industries as optics and electro-optics. TiO_2 addition to PMMA changes the linearity of its optical behavior, as well as of its mechanical properties.

The study was done to optimize the TiO_2 concentration to get optimal optical and mechanical properties. Fig. 34.1 shows the change in the response of the NA oscillating tip as it scans the pure PMMA and the one with TiO_2. These response curves are correlated to in-situ to generate a stiffness map as shown in Fig. 34.2.

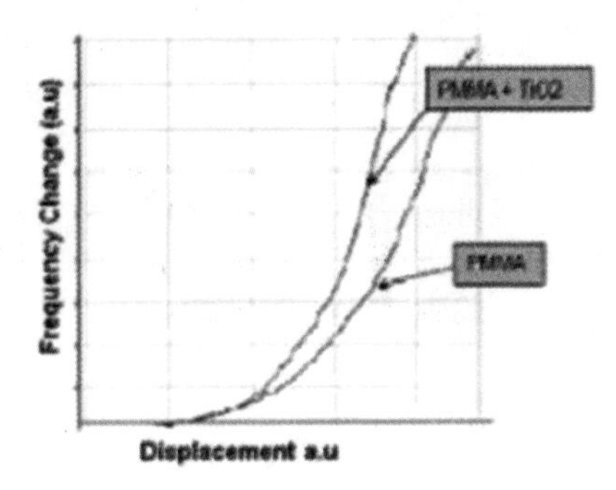

Fig. 34.1 Material response curves for TiO_2 in PMMA

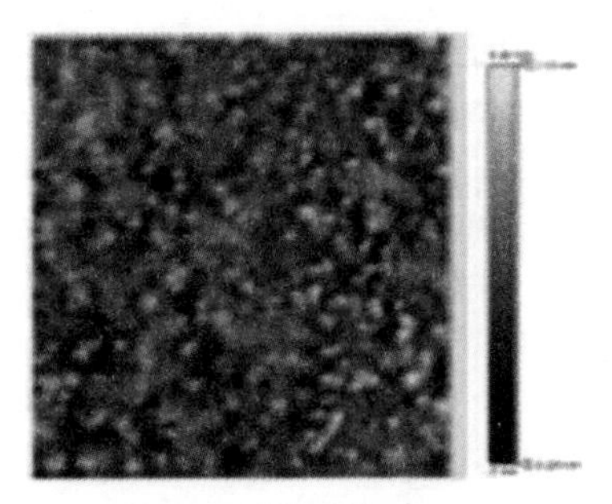

Fig. 34.2 Stiffness map of PMMA impregnated with TiO_2

The effects of particle concentration and agglomeration can be easily observed and evaluated with this map. Further tests were done with varying TiO_2 concentration.

34.3.2 *Stiffness Maps of Topographically Normal Material*

Ni is typically used for copper wire bonding in a semiconductor industry. Sometimes Ni diffuses in the copper wire-lines and thus increases electrical resistance considerably.

This study was undertaken to detect the Ni contamination in the copper surface. Using SEM for this purpose is laborious and requires cross-sectioning the specimen, while NA was used as it is on the sample. An AFM cannot distinguish between the matrix and the contaminants; the nano-indentation would require several days to locate Ni. Fig. 34.3 shows topographical and stiffness maps obtained with the NA. The stiffness map clearly showed the Ni contamination in copper due to a considerable difference in their mechanical properties.

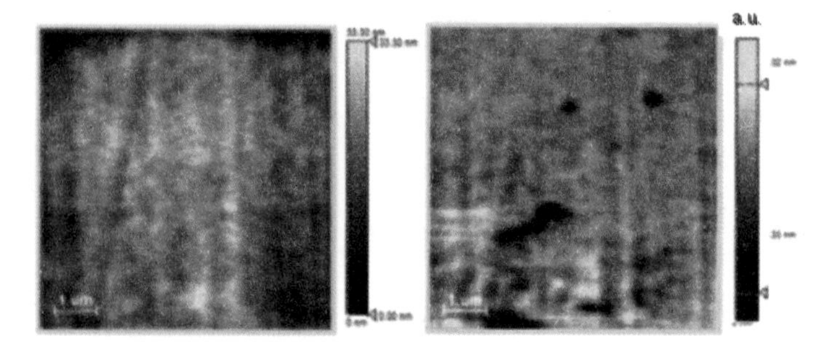

Fig. 34.3 Maps of copper with Ni contamination: (a) Topographical map (b) Stiffness Map – dark areas are Ni

34.3.3 *Stiffness Maps of Topographically Rough Material*

Lapping is a precision process used to provide flatness and surface finish to very demanding tolerances. It is widely used in a disk drive industry for head sliders. A lapping plate consists of a soft metal impregnated with hard diamond sub-micron particles. It is important to characterize the number of diamond particles remaining on the plate after lapping to estimate its lapping efficiency and life. AFM topographical maps cannot distinguish between the diamond particles and rough metal matrix. Using SEM for this purpose is laborious and requires destructive sectioning of the large plates. The NA is used to plot a stiffness map of the lapping plate as it is (Fig. 34.4), which clearly shows the diamond particles on the matrix.

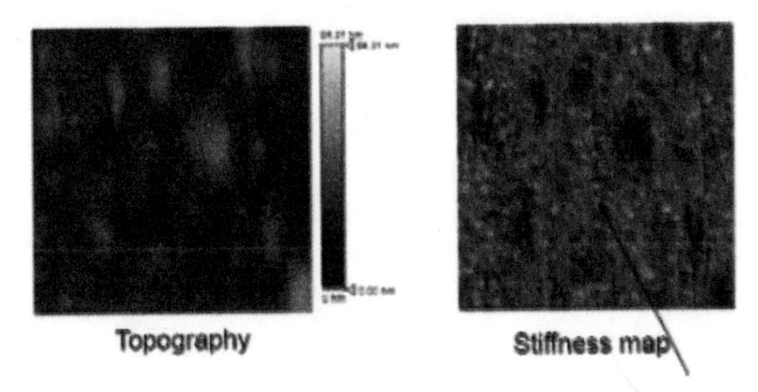

Fig. 34.4 Maps of lapping plate: (a) topographical map (b) stiffness map

34.3.4 *Stiffness Maps of Material with Subsurface Features*

The NA was used to generate the stiffness maps for determination of the effect of underlying patterns on the top surface. Fig. 34.5 shows a polymeric surface with underlying 3-D features. The underneath pattern imposes tensile stress on the top of the polished surface, which was revealed on the stiffness map.

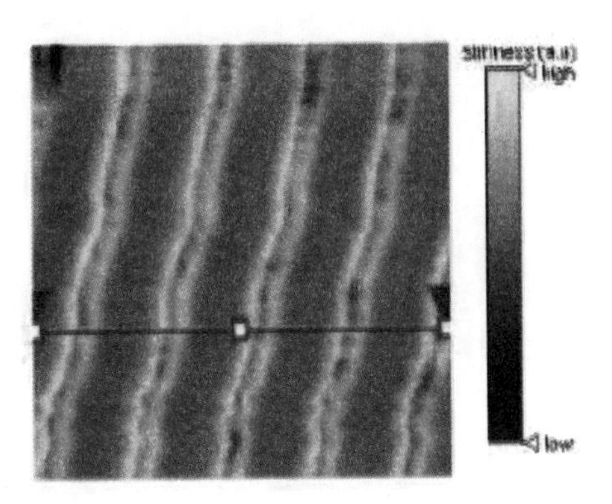

Fig. 34.5 Stiffness map of a polymer surface with underlying pattern

34.4 Conclusions

Based on the above studies the following conclusion is drawn[34.1]:

Non-destructive in-situ topography and stiffness mapping of various hard and soft materials with the novel Nano-Analyzer is effective for studies of multi-component composites and sub-micron particle distribution in the surface layers.

References

[34.1] N. Gitis, V.Khosla, I. Hermann, M.Vinogradov, Proc. IJTC, Oct.20-22 (2008)

Section - IX

CMP-Chemico Mechanical Processing

35

Post-CMP (Chemical Mechanical Polarisation) Cleaning Applications: Challenges and Opportunities

35.1 A General Review

CMP process defectivity and yield performance depends on effectiveness of the post-CMP (PCMP) cleaning process which should reduce roughness of the polished wafer and leave it defect-free, consistently removing particles (nanometer size in smaller feature next generation devices), organic residues, and ionic contamination. This chapter presents an overview of PCMP wafer cleaning attributes, trends, challenges and opportunities. PCMP cleaning effectiveness depends on the stability of brush PVA (Poly Vinyl Alcohol) properties, magnitude of brush wafer frictional force, and the adhesion forces between the particle and wafer, and the particle and brush. Evolution of current and next generation PVA roller brushes designs is discussed.

Tribological and PCMP cleaning performance characterization approaches for brushes are presented with some recent test data. Molded-through-the-core (MTTC) brush design with an integral/disposable core provides positive anchoring and adhesion of PVA with the core and excellent dimensional stability, eliminating any possibility of PVA slippage at the core interface unlike slip-on-the-core (SOTC) brushes. Stable behavior of brush-wafer contact-pressure, contact-area, and dynamic-friction could be useful indicators of post-CMP (PCMP) cleaning effectiveness and mechanical consistency of PVA brushes over brush lifetime. This chapter presents data from tribological studies using a new bench top tribology tester (specifically developed for PVA brushes) and a 200 mm wafer test tribometer (accelerated 48 hour tribological stress evaluation).

This study also reports results of comparative PCMP cleaning effectiveness fab evaluation of PVA brushes. This test involved MTTC and SOTC brushes, evaluated under brush break-in, scrub only and PCMP cleaning cycles, after Cu/Low-k barrier step CMP process in a 90 nm production fab using 200 mm (blanket and 180 nm feature M1T854 patterned) wafers on Mirra Mesa tool set. Present study highlights the importance of PCMP clean brush design and methods of tribological and PCMP cleaning evaluations to ensure consistent frictional characteristics and wafer cleaning performance over brush lifetime, and demonstrates the benefits of using MTTC design PVA brushes in the Cu/low-k PCMP cleaning applications.

35.2 Wafer Cleaning Attributes and Trends

Current PCMP wafer cleaning processes are contact cleaning techniques, which use chemical as well as mechanical action to effectively remove the particles from the wafer surface. Brush cleaning is a very effective PCMP cleaning technique and in an optimum mode, a contact between the particle and the brush is essential to the removal of submicron size particles from the wafer surface. In this operation, Rm>>1 for a 0.1 micron particle, for typical brush and wafer speed, based on a recent experimental study at the Northeastern University, where Rm is the ratio of removal moment to adhesive moment, and the particles are removed if Rm>1.

In the above reported test, 100 % particle removal could be achieved, employing intermediate brush pressure, speed and time. Studies show that brush cleaning is effective in removing particles down to 0.08 micron with different PCMP clean chemistries. Non-contact megasonic cleaning is also very effective in PCMP cleaning. It appears that the current and next generation PCMP cleaning process will continue to depend on PVA brush together with megasonic cleaning in the foreseeable future.

Based on recent published literature, the requirements of surface cleaning must be considered while designing future generations of ICs, since 60 % of fab-related (yield) problems are related to deans and another 12 % to etching steps, and the design dominates how wet processing is done and processing limitations influence the design process. To meet next generation challenges, many of the cleaning chemistries and approaches will have to change. There are plenty of potential solutions being considered for 45 nm and beyond, and the IC manufacturers will need to adopt new tech chemistries and cleaning regimens for the next generation devices.

Suggested non-damaging nanoparticle removal technologies include: Shock tube-enhanced laser-induced plasma (LIP) shockwaves for sub-50 nm nanoparticle removal, plasma-assisted cleaning by electrostatics (PACE), an ionized molecular-activated coherent solution, and parametric nanoscale cleaning by forming nanoscale bubbles to absorb the contaminants. On photoresist issues, several new or enhanced methods may be used for minimizing silicon and oxide loss during removal, including: photoreactive cleaning, a CO_2 cryogenic press and non-oxidizing chemistry, and methodologies for all-wet chemistries.

35.3 Post-CMP Cleaning Process and PVA Brush Design Evolution

Cleaning performance of PVA brush strongly depends on the chemical and mechanical properties and stability of the brush material, magnitude of wafer-brush frictional force, and adhesion forces between the particle and wafer as well as the particle and brush. Zeta potentials of the particle and the wafer in various cleaning solutions and pH of the CMP slurry are very important in the particle adhesion and removal in PCMP cleaning process. Common PCMP technologies include megasonic and double PVA brush based scrubbing (Fig. 35.1). MTTC design is a disposable PVA brush that reduces tool downtime and provides excellent dimensional stability over its lifetime.

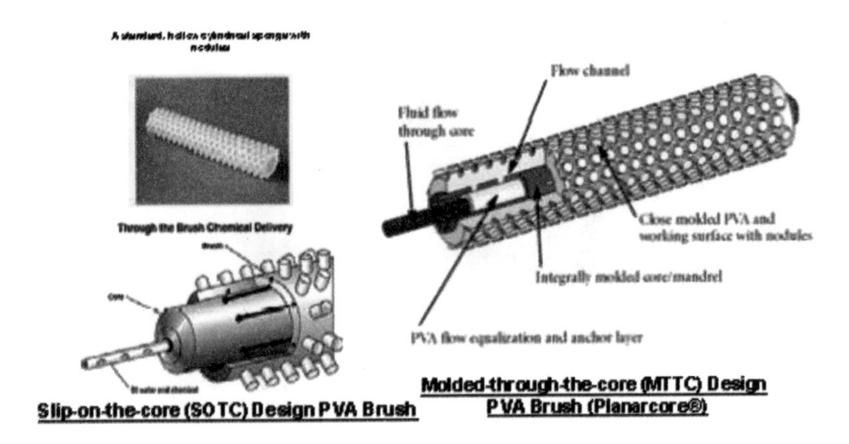

Fig. 35.1 PVA Roller Brush Designs.

This design eliminates possibility of any slippage at the PVA-core interface (possible in conventional SQTC, especially in the latter part of their lifetime due to possible swelling of PVA). PVA brushes based scrubbing, seems to be part of the next generation PCMP cleaning. New processes will require: low particle level PVA brushes for Cu/low-k applications and improved PVA with lower extractable levels for cleaner processes. Also, specific applications may require charged, IPA resistant or Cu/Low-k specific, and/or next generation CMP slurry compatible PVA brush technology.

More stringent cleaning process requirements over brush lifetime may require innovative products such as MTTC brushes. MTTC design with its positive adhesion/anchoring of PVA with the core provides ease of use and time saving, excellent dimensional stability, improved flow equalization in the core, and very uniform flow distribution (out-of-the-brush) along the brush length. Also, improved and cleaner PVA results in reduced particle counts/adders on the wafer, decreased defectivity, reduced tool downtime, and more consistent PCMP cleaning.

35.4 Characteristics of a New PVA Brush Tribological Evaluation Tool

A new PVA brush bench top tribological testing system (Center for Tribology (CETR modified PMT) was developed at the Center for Tribology (Fig. 35.2). Parameters measured by this tool include: coefficient of friction (COF), skin friction and normal force, creep and effect of chemistries, adhesion force, compressibility, Stribeck curves, temperature, acoustic properties, and material wear rate and lifetime. The PMT can accommodate full-size brushes or smaller brush coupons on semiconductor wafers, magnetic disks, flat displays and other to-be-cleaned specimens. Typical data from such measurements for different brushes are included in Figs. 35.3(a) – 35.3(h).

Fig. 35.2 Tribological Tester for PVA Brushes.

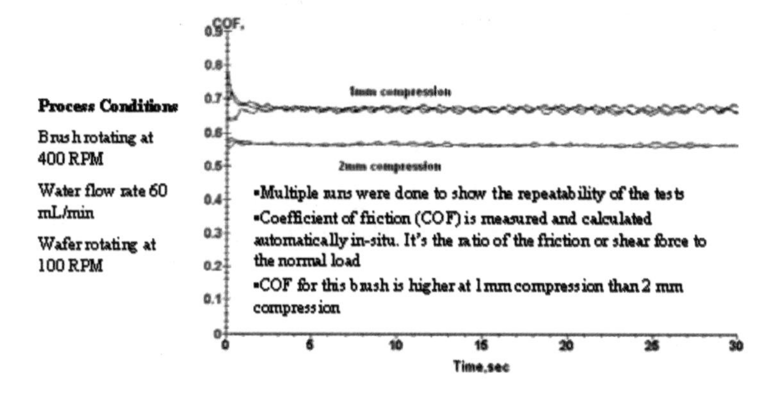

Fig. 35.3(a) Coeff. of Friction (COF) Results of a PVA Brush.

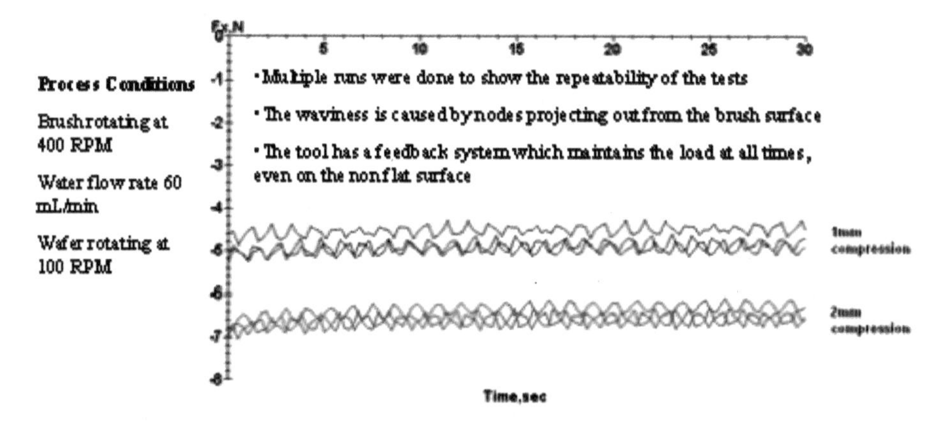

Fig. 35.3(b) Skin Friction Results of a PVA Brush.

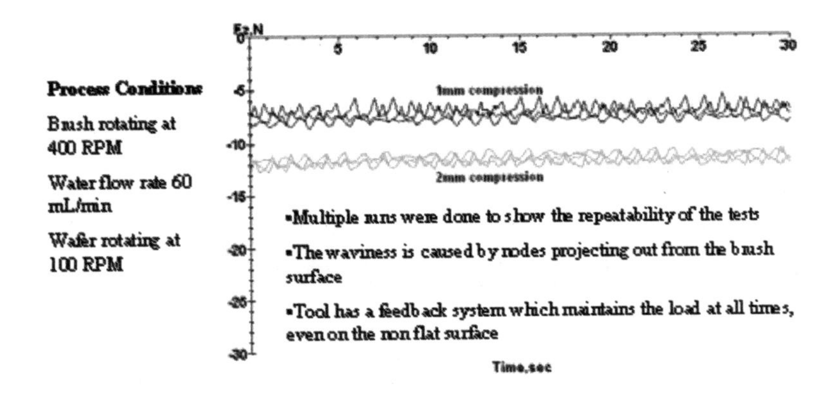

Fig. 35.3(c) Normal Load Results of a PVA Brush.

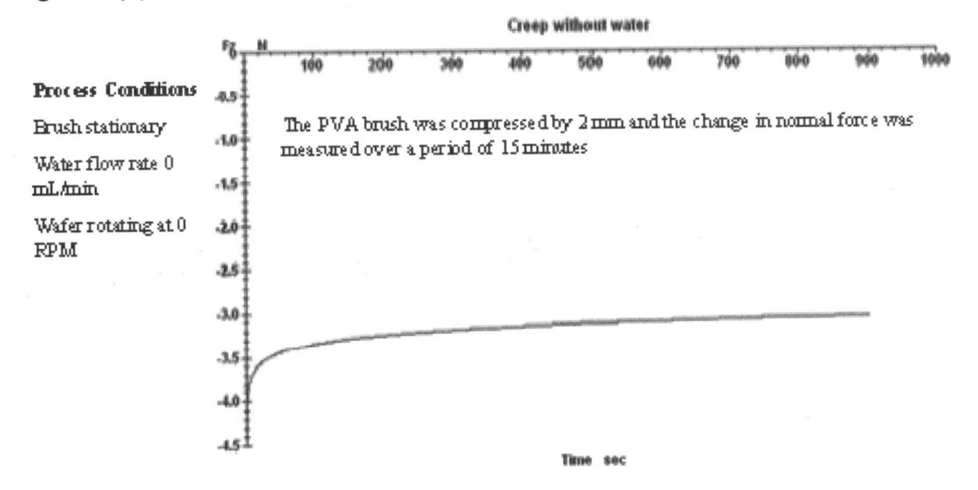

Fig. 35.3(d) Creep Measurement Data of a PVA Brush.

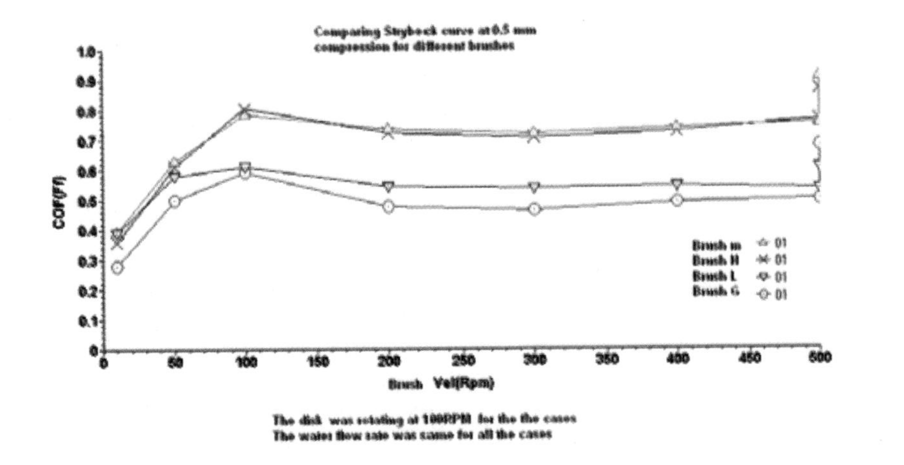

Fig. 35.3(e) Stribeck Curves for Different PVA Brushes.

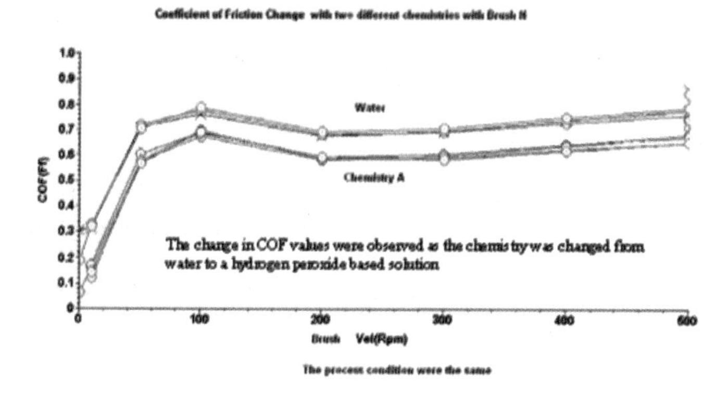

Fig. 35.3(f) Stribeck Curves for Brush with Different Chemistries.

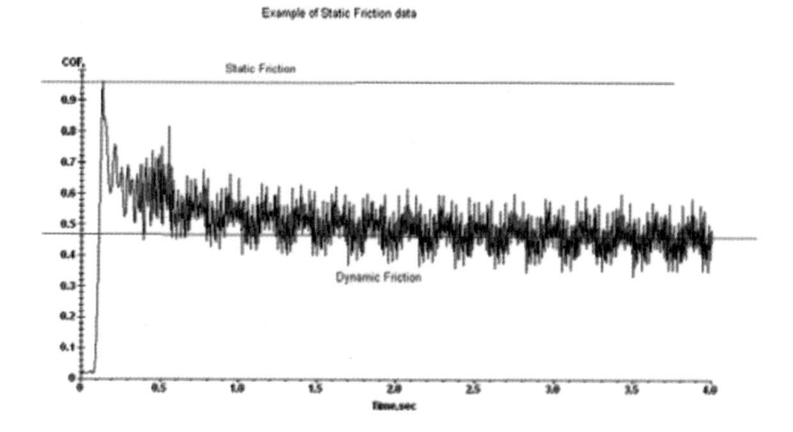

Fig. 35.3(g) Example of Static Friction Data for a Brush.

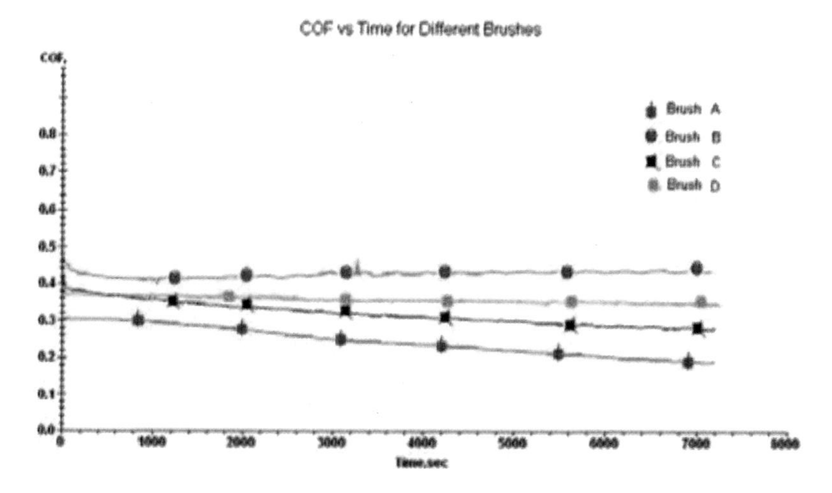

Fig. 35.3(h) COF Vs. Time Data for Brushes.

35.5 Results of PVA Brush Characterization Studies

(A) Case Study 1- PVA Brushes Comparative Tribological Performance

An accelerated tribological stress evaluation (48 hour marathon run) of PVA brushes was conducted employing two slip-on-the-core (SQTC) brushes (A and B) and one MTTC brush (C) at the University of Arizona. These tests demonstrate a very different behavior of wafer-liquid-brush contact–pressure, contact–area, and dynamic coefficient of friction (COF) for different brushes. Brushes A and C showed a more consistent behavior of mean COF, whereas design Brush B experienced catastrophic failure somewhere between 2 and 8 hours (Figs. 35.4(a), 35.4(c) and 35.5). Further, the total variation range of COF for MTTC Brush C was found to be minimum (Fig. 35.5). The experimental conditions and equipment details for this study are included in details.

These tests also demonstrate how the extent of brush deformation (as measured by the brush-pressure versus brush-wafer contact-area curves) and the magnitude of frictional forces (as measured by the brush – fluid – wafer coefficient of friction, COF) vary as a function of extended use for various types of brushes (Figs. 35.4(b) and 35.4(d)). This information is critical for brush performance consistency. Results further demonstrate that those brushes that experienced the least amount of deformation variability during the 48-hour marathon test also exhibited the least amount of variability in their frictional attributes.

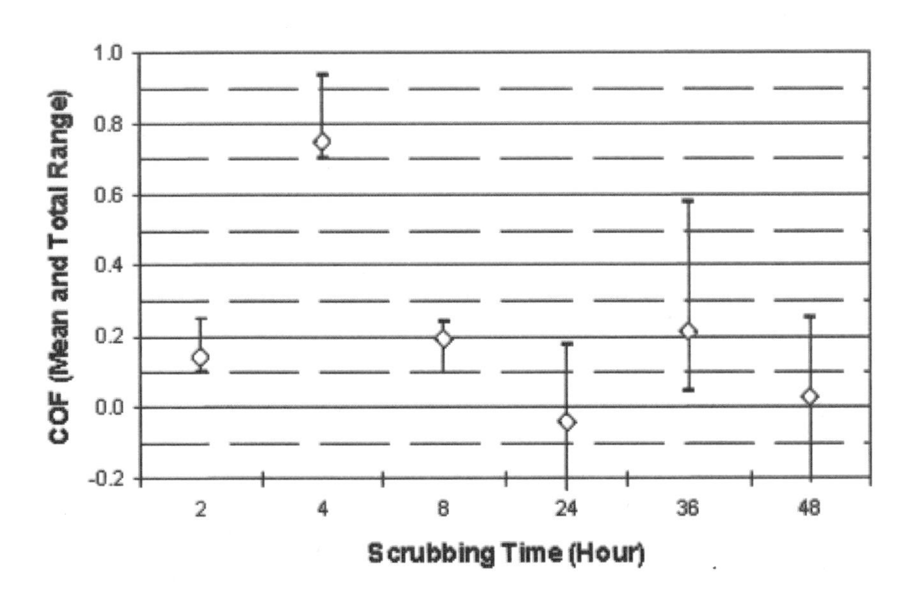

Fig. 35.4(a) COF Results for Brush B (SOTC Design).

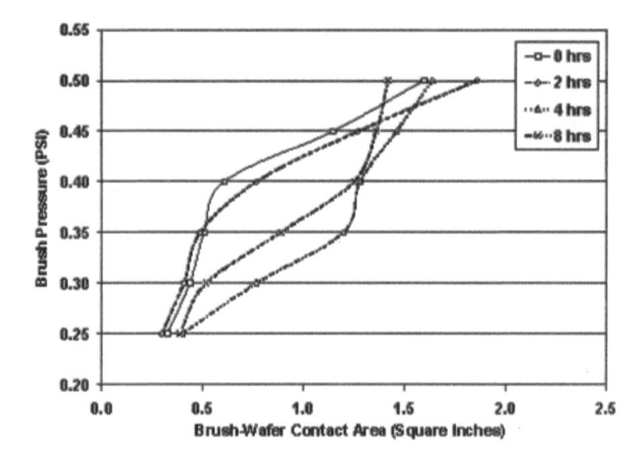

Fig. 35.4(b) Pressure Contact-Area Plot for Brush B. (The enveloped area bounded by the curves shows the extend of brush deformation)

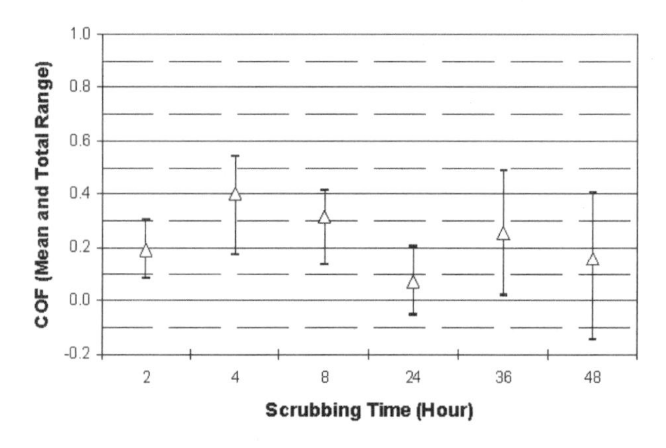

Fig. 35.4(c) COF Results for Brush C (MTTC Design).

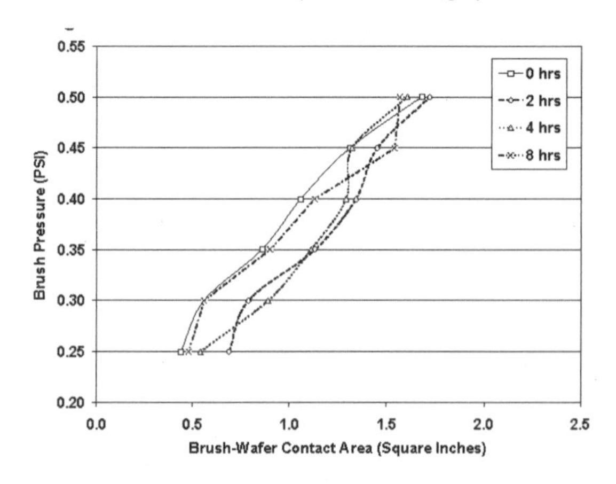

Fig. 35.4(d) Pressure Contact-Area Plot for Brush C.

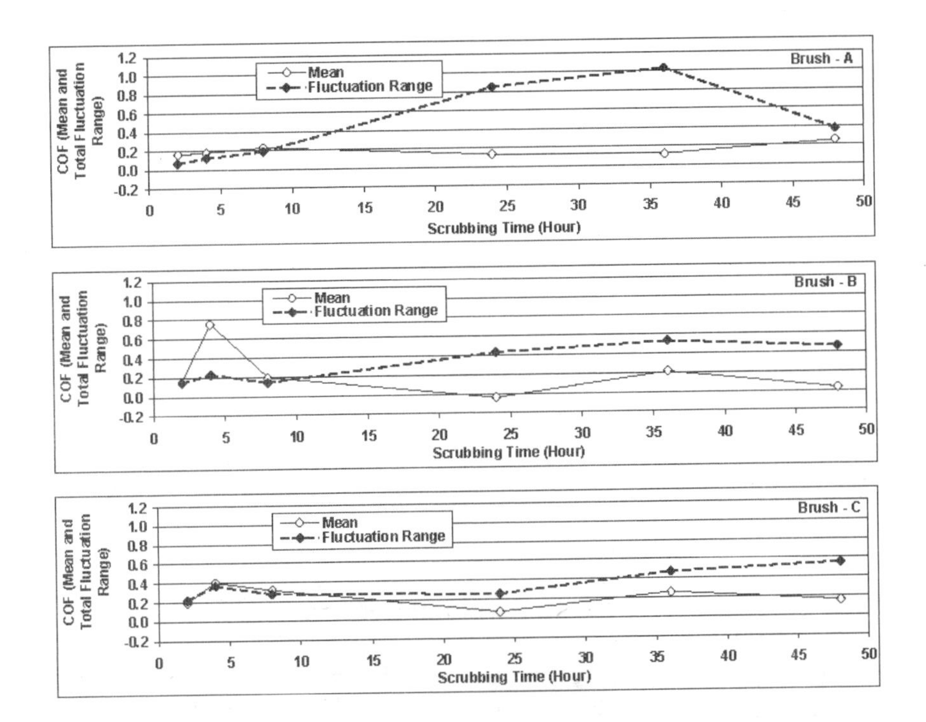

Fig. 35.5 COF Mean and Total Fluctuation Range during 48-Hour Accelerated Stress Test for Brushes A, B, and C.

Experimental Conditions and Equipment

Constants: Applied pressure = 0.5 PSI, Cleaning solution type and flow rate = Ashland CP – 70 at 120 cc/min, Brush and wafer rotational velocity = 60 and 40 RPM, respectively, Frictional force data acquisition frequency = 1,000 Hz (3.6 million samples / hour), Wafer type = 200 mm International Sematech 854 Copper wafer, Scrubbing time = 48 Hours marathon run (continuous). All tested PVA roller brushes were similar in dimension, commercially available and had cylindrical nodules.

Variables: PVA Brush Type; Brush A: Slip-on-the-core PVA sleeve design from Supplier A, Brush B: Slip-on-the-core PVA sleeve design from Supplier B, and Brush C: Molded-through-the-core PVA design from Supplier C (Entegris Planarcore®).

(B) Case Study 2 – PVA Brushes PCMP Cleaning Performance in Cu/Low-k Application

Objective: To generate comparative PVA brush PCMP cleaning data (defect maps/classification) for Entegris Planarcore® MTTC brushes and 3rd party fab POR SOTC design brushes in a 90 nm production fab, using 200 mm blanket and 180 nm feature M1T854 Cu/Low-k patterned wafers on a Mirra Mesa CMP toolset PCMP cleaner.

Tested Brushes and Equipment Set

MTTC technology brushes: Entegris PP core (enhanced cleanliness), thicker PVA (more tunable wider range down force), and advance PVA foam cleaning process (resulting in a shorter brush break-in cycle). POR brushes: PVA SOTC design brushes used as POR at the 3rd party site.

CMP Tool and Cleaner: AMAT Mirra Mesa. Wafer metrology: KLA-Tencor Surfscan 6420 and KLA-Tencor SPI (blanket wafer inspection), and KLA-Tencor 2139 Wafer Inspection System and KLA-Tencor AIT XP Wafer Inspection System (for patterned wafers).

The process conditions were optimized for current POR brush and were not specifically modified to ensure good comparative data for each brush type. Defectivity data classification for the above brushes is presented in Fig. 35.6. The PCMP cleaning defectivity performance of Entegris MTTC design Planarcore® brushes was found to be similar or better than the fab POR SOTC design brushes.

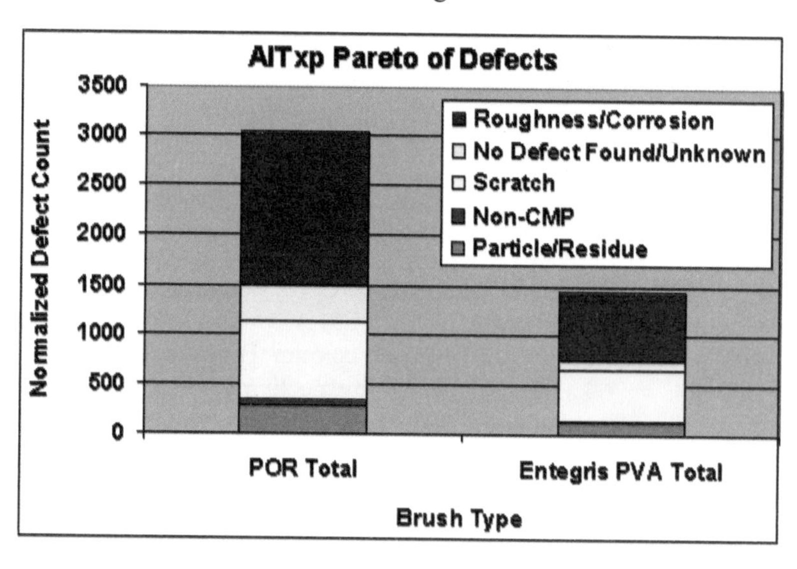

Fig. 35.6 Pareto of Defects for Two PVA Brushes.

35.6 Conclusions

Tribological and mechanical studies are very significant for brushes for hi-tech cleaning processes. Further investigations are required for these devices.

36

Bench Top Dual Mode eCMP Polisher with Multi Sensing Metrology

36.1 General

A traditional CMP (Chemical Mechanical Polarization) tool is designed to achieve the planarization process across the wafer via a down-force polishing effect. In doing so, the wafer is pressed onto a rotating polishing pad and a slurry material is applied. The traditional CMP process can cause delamination and fracture of low-k dielectrics that are soft and porous. As a new alternative, a low down-force process of electro-polishing or eCMP is being considered, in which a current is run from a conducting surface, through a conducting liquid and into an electrode. Under proper conditions, metal can be removed from the wafer surface into solution.

As the eCMP is not a mature technology, there are no standard consumables on the market. To develop slurries, pad, particles, conditioners, etc., a multifunctional bench-top eCMP polisher mod. eCP-4 has been developed.

The eCP-4 can be used to run both conductive and non-conductive pads. The tool can be used to do polishing in both constant current and constant voltage modes. It has a precise down-force servo-control from 0.05 to 10 psi. The eCP-4 polisher is equipped with multi-sensing metrology which measures in-situ friction force and coefficient in both wafer-pad and conditioner-pad interfaces, contact acoustic emission, applied voltage and current, temperature, pad wear at a 100 kHz data acquisition rate. The same tool can also be used for a regular CMP, which enables to use CMP and eCMP sequentially to completely clear the barrier after copper removal. This chapter covers the capabilities of the bench-top eCMP polisher with both conductive and non-conductive pads. The data presented is obtained on the eCMP tool and explain the benefits of multi-sensing metrology to develop novel processes and materials for eCMP.

36.2 Introduction

The performance of a miniaturized semiconductor device is governed by its signal processing speed, which in turn is determined by the gate and interconnect delay times. For devise features scaled down below 0.5 micron, the interconnect "RC delay" dominates the delay in signal processing, where R is the resistance of the wiring metal and C is the capacitance of the interlayer dielectric (ILD) used in the device. From simple considerations, it can be shown that RC = $[(rkl^2)/(yd)]$[36.1, 36.2]. Here *r*, *l* and *d* represent the resistivity, length and thickness of the wiring line, respectively; k and y represent the dielectric function and thickness of the ILD, respectively. Thus, decreasing the value of l and/or that of k can decrease the delay time. Most of such low-k ILD materials ($k < 2.5$), however, are often porous and mechanically fragile. Therefore, to avoid damages to the ILD during IC fabrication, planarization of the ILD materials and their overlying structures must be performed at a low applied down-pressure (usually at < 1 psi). Although low-P operation is difficult to incorporate in the currently available framework of CMP, it is possible to combine electrochemically controlled material removal with low-P mechanical polishing where the main role of the latter step is to provide uniform planarization across the sample surface.

Thus eCMP can be a technology of choice for future nodes. But as the eCMP is not a mature technology, there are no standard consumables on the market. To develop slurries, pad, particles, conditioners, etc., a multifunctional bench-top eCMP polisher mod. eCP-4 has been developed. This bench top polisher uses multi-sensing technology to generate data which can enable the researchers to learn and gain more about the process. The tool (shown in Fig. 36.1) can be used to polish low k wafers under 0.5 psi range and can accommodate both conductive and non conductive pads. The tool uses multi-sensing technology (explained later) and can be used to develop pads, pad conditioner, slurry, process etc.

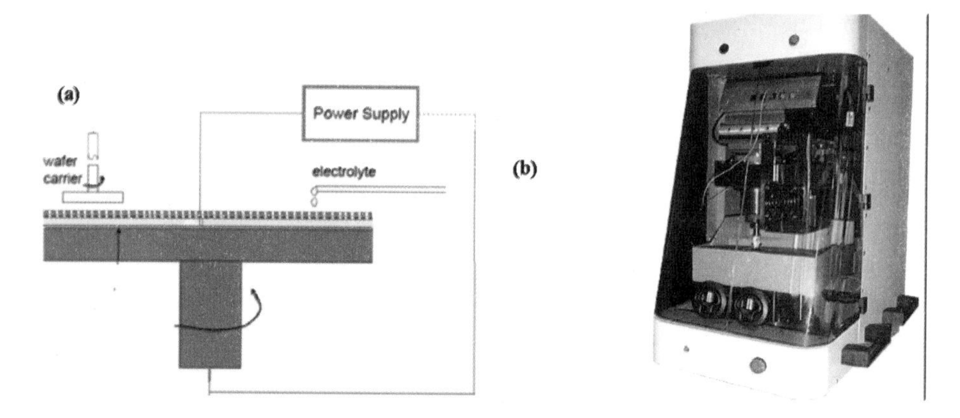

Fig. 36.1 (a) Schematic of eCMP (b) eCP4 tool by CETR

Multi-Sensing Technology: The technology involves measuring in-situ friction, coefficient of friction, down force, acoustic emission, current and voltages and pad wear at all times at 100 KHz frequency.

eCMP process can be run at constant voltage or at constant current. The tool is capable of running the process in either mode. Typically the modes are changed as copper thins down to prevent very high voltages at barrier and wafer contact interface. When running in constant voltage mode the tool measure in-situ current and vice versa.

The coefficient of friction (COF) is defined as a ratio of the tangential friction force, resisting relative motion of the surfaces, to the normal load pressing the surfaces together. To monitor friction, it is preferred to measure its coefficient instead of just measuring the force resisting the relative motion of rubbing surfaces. Indeed, there are cases when changes in down-force cause substantial changes in the friction force, while the coefficient of friction remains constant. Alternatively, sometimes important changes in the coefficient of friction cannot be observed by monitoring the friction force due to periodic fluctuations of the down-force.

The total wear of the interface, typically measured in the direction perpendicular to the rubbing surfaces, consists of wafer and pad linear wear. To monitor wear, it is preferred to measure the linear geometrical wear of each of the rubbing surfaces. As wafer material removal is of the order of nanometers, its measurements are very complicated during polishing, and are often performed after polishing. The pad wear is of the order of micrometers, and its measurements are straightforward.

Another important parameter of friction and wear is the acoustic emission (AE) from the contact of rubbing surfaces [36.3] [36.4]. Its spectrum may have numerous frequencies, corresponding to such different processes as plastic and elastic deformations of sub-surface material layers, micro-scratching and micro-fatigue, micro-corrosion and other electrochemical reactions, delamination of material layer. The mega-Hertz acoustics is more informative of the specific micro-tribo-processes on tiny micro-contacts, in comparison with a kilo-Hertz range reflective of integral characteristics of the interface and deci-Hertz range reflective of integral characteristics of the entire mechanical system. Among the main benefits of the acoustic measurements in CMP are monitoring the intensity of polishing processes and detecting polishing conditions when wafer layers, for example low-K polymers, delaminate.

Computerized real-time measurements and analysis of the coefficient of friction, contact high frequency acoustic emission and pad wear allow for the effective evaluation of dynamic characteristics of the polishing process. This Multi sensing technology can be used to develop new eCMP pads, slurry, conditioner, processes etc.

36.3 Results and Discussion

Fig. 36.2 shows real-time, in-situ monitoring of the wafer surface without optical observation. The transition in frictional values shows the material removal. Moreover, as the transition from an upper layer to a lower layer does not happen instantaneously and takes some time, this time is a direct characteristic of the non-uniformity of material removal. Parallel AE measurements compliment the friction ones (see Fig. 36.2) in detecting the defectivity rate and end point of polishing.

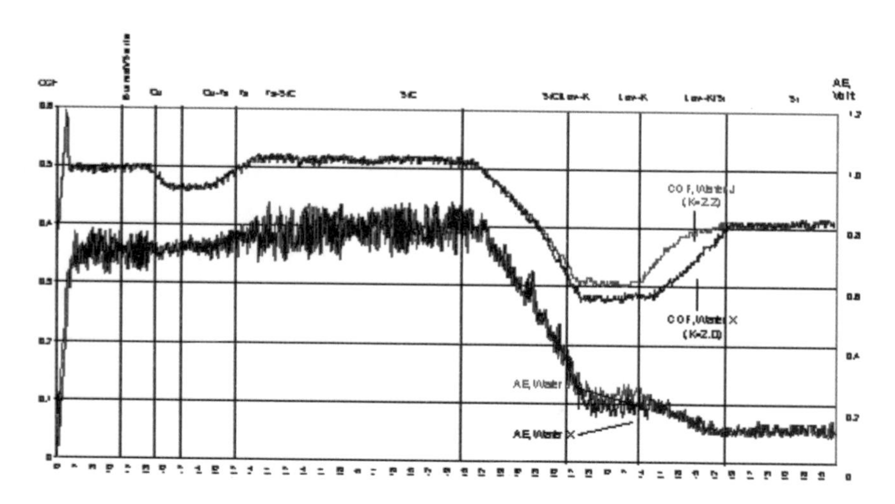

Fig. 36.2 COF and AE plot on low K wafer CMP

Fig. 36.3(a) shows the use of multi sensing technology on copper eCMP at constant current process. As copper thin down towards the end of process, the voltage increases. Fig. 36.3(b) shows the copper removal rate vs current density. These polishing experiments were done at sub (1psi) pressure.

Just like CMP, eCMP process will depend heavily on the pad design. The opening on the pad indicates the anode to cathode ratio.

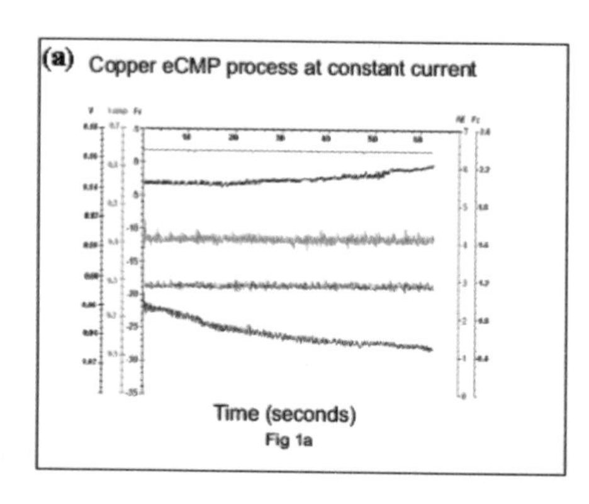

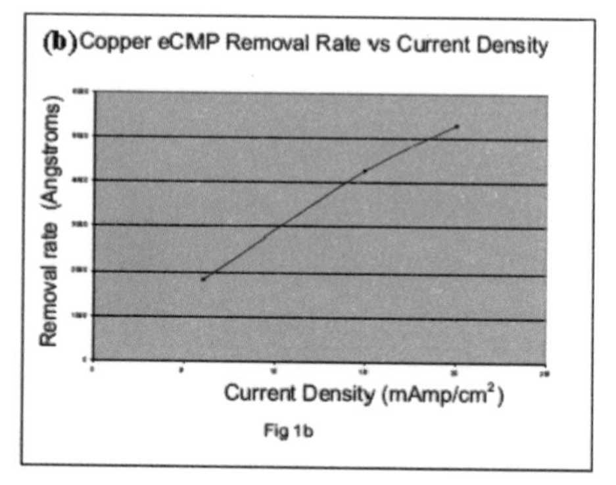

Fig. 36.3 (a) Graph showing Copper eCMP (b) Copper removal rate vs current Density

The groove patterns and the cell structure will dictate the level of electrolyte on top of the pad. The change in groove patterns or the pad material will change the material properties (hardness, young's modulus etc.). Fig. 36.4 shows the comparison between pad compressions of different pads using a disk. These slopes can then be used to calculate elastic modulus of the pads.

CMP or eCMP yield will depend on the efficiency of the conditioners. The newer pad conditioners need to be less aggressive on the pads (eCMP pads are typically softer in nature). The conditioner has to be more inert in nature to withstand eCMP electrolytes. Fig. 36.5(a) shows the in-situ pad wear rate after 2 hour of conditioning. Fig. 36.5(b) shows the conditioning efficiency, which indicates the time required to condition the pad to bring it to its original values.

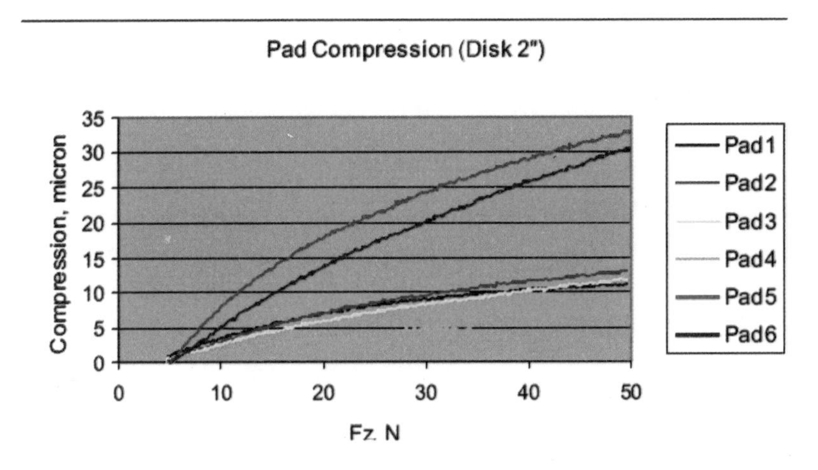

Fig. 36.4 Comparing Pad Compression of Different Pads

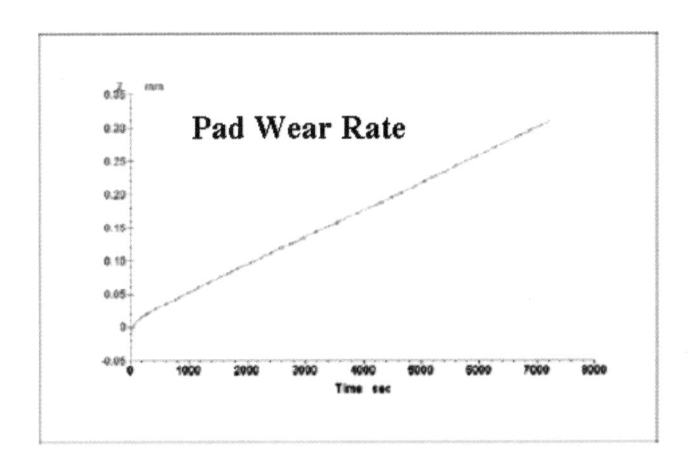

(a)

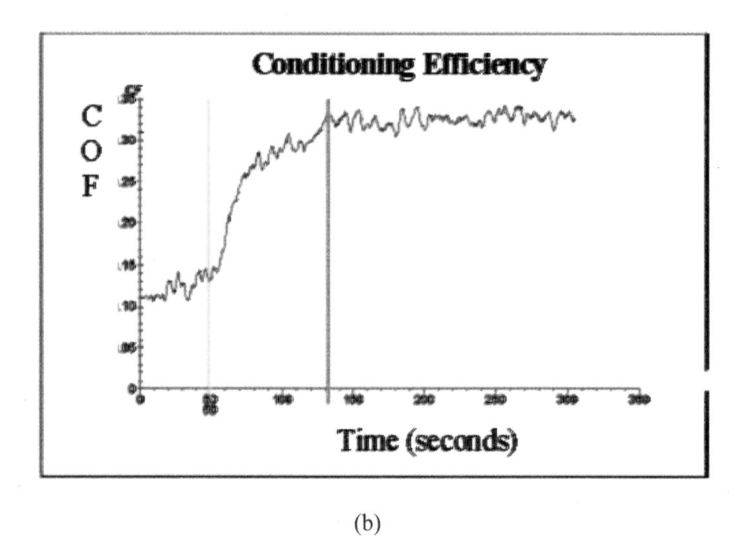

(b)

Fig. 36.5 (a) Insitu Pad Wear Rate during conditioning (b) Pad conditioning Efficiency

36.4 Conclusions

The CETR eCP4 is a tool, which can help the researchers to take eCMP to next level. The tool is capable to polish in sub (1 psi) range and can be used to develop pads, conditioners, slurries. The Multi sensing technology approach is a novel method to characterize the process completely. The bench top polisher is a cheaper viable option to do quick feasibility studies.

References

[36.1] S.V. Babu, K.C. Cadien, H. Yano, Eds. Chemical-Mechanical Polishing 2001: Advances and Future Challenges, Materials Research Society, Warrendale (2001).

[36.2] http://people.clarkson.edu/~samoy/ECMP.htm

[36.3] J. Fang, K. Davis, N. Gitis and M. Vinogradov, in Proceed. VLSI Multilevel Interconnection, 20th Annual Conference, Marina Del Rey, September 2003.

[36.4] N. Gitis and M. Vinogradov, in Proceed. VLSI Multilevel Interconnection, 20th Annual Conference, Marina del Rey, September 2003.

Section - X

Cosmetics

37
Comprehensive Nano-Mechanical and Tribological Characterization of Hair

37.1 General

This chapter presents test procedures and results of tribological and mechanical characterization of hair, treated with various shampoo and conditioners. They allow for a comprehensive functional comparison of hair care products.

37.2 Introduction

Mechanical and tribological analysis of human hair may provide dermatologists with several markers of considerable diagnostic importance. In cosmetic research, such analysis is important to understand how shampoo and conditioners affect the hair[37.1].

In this work, various tests developed to differentiate and optimize functional properties of common hair care products such as shampoos, conditioners and dyes have been reported.

A quantitative analysis of tribological and mechanical properties of hair has been performed on the Universal Nano and Micro Tester UNMT-1. Its nano-analyzer module allows for measurements of Young's modulus and hardness of hair in all directions, with precision positioning on different hair areas, such as cortex (melanocyte) and periphery (inner sheath and fibrous sheath)[37.2, 37.3]. Its tribology module allows for sensitive testing of hair tribology in both a cross-hair mode (between two perpendicular hair tresses) and a sledging mode (between a tress of hair and a counter-material), with a servo- controlled down-force. The tribology and nano-analyzer modules are easily interchangeable.

37.3 Nano-Mechanical Data

The study was performed on a European blonde hair, untreated and dyed, both impregnated in an epoxy matrix.

As shown in Fig. 37.1, the measurements were done in the longitudinal and transverse directions, the latter to distinguish the properties of the cortex and hair periphery. Fig. 37.2 shows topographical images of the smoother untreated and rougher dyed hair.

Fig. 37.3 shows stiffness maps of the untreated and dyed hair, where the former one was softer and more uniform.

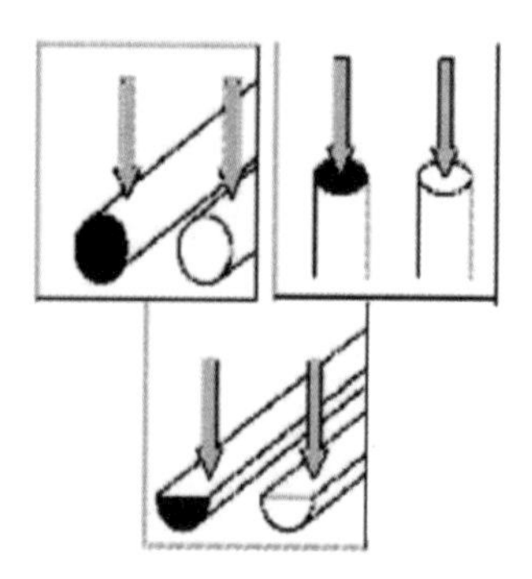

Fig. 37.1 Schematics of nano-mechanical hair analysis

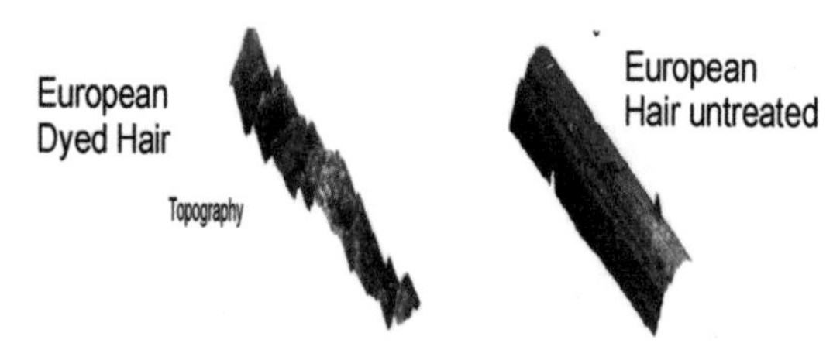

Fig. 37.2 Topographical images of hair

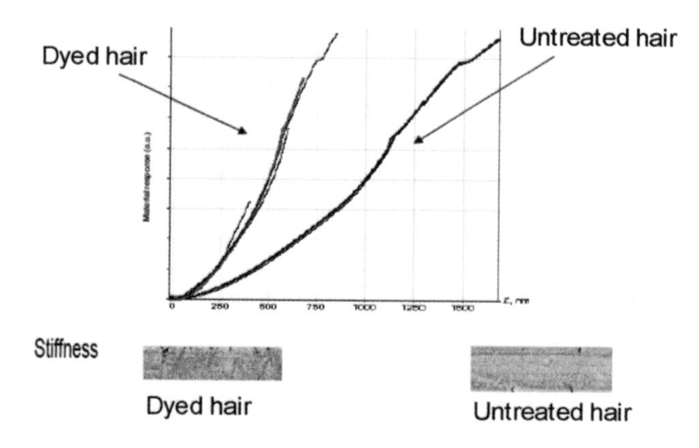

Fig. 37.3 Stiffness maps and raw data plots for hair

Fig. 37.4(a) shows an optical image of the epoxy matrix with impregnated hair. Fig. 37.4(b) shows the simultaneously obtained topographical and stiffness maps. A noticeable difference in stiffness between the core and the outer periphery of the hair is confirmed with the numerical data in Fig. 37.4(c).

37.4 Friction Data

Cross-hair tests were done using the UNMT setup in Fig. 37.5. A pair of cross-hair holders was used to mount two tresses of hairs, each of 10 hair strands. The applied load was 100 mN, sliding speed was 0.5 mm/s.

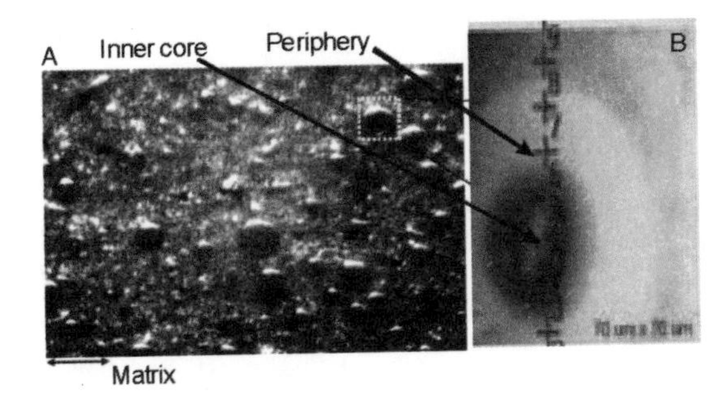

Fig. 37.4 (a) Optical image of matrix with hair; (b) Hair stiffness map with indentation positions

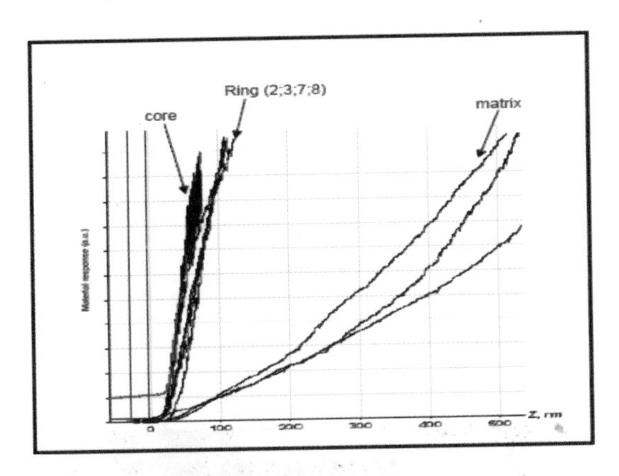

Fig. 37.4 (c) Young's modulus data for hair

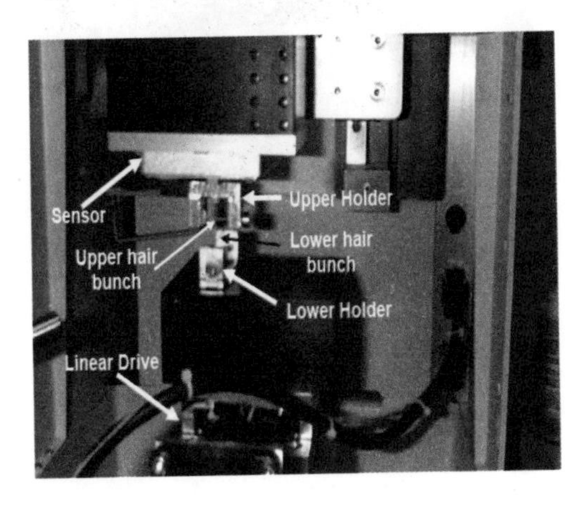

Fig. 37.5 Cross-hair friction test

Table 37.1 shows the coefficient of friction (COF) data on the hair treated with four different products. The coefficient of friction was the lowest after treatment with Pantene ProV and the highest after Pantene 2-in-i.

Table 37.1 Cross-hair friction data on hair pre-washed and treated with hair conditioners

Hair Coditioner	Sample #	COF						
		Test 1	Test 2	Test 3	Test 4	Test 5	Mean	SD
Pantene ProV *Sheer Volume*	#1	0.070	0.066	0.065	0.064	0.063	0.066	0.003
	#2	0.069	0.064	0.062	0.061	0.061	0.063	0.003
Dove *Extra Volume*	#1	0.088	0.080	0.079	0.079	0.078	0.081	0.004
	#2	0.078	0.079	0.074	0.073	0.072	0.075	0.003
White Rain *Extra Body*	#1	0.105	0.099	0.096	0.101	0.098	0.090	0.001
	#2	0.090	0.090	0.090	0.091	0.093	0.090	0.001
Pantene 2-in-1 *Shampoo*	#1	0.099	0.098	0.099	0.102	0.104	0.120	0.002
	#2	0.115	0.115	0.117	0.119	0.124	0.117	0.004

Sledge friction tests were done against a silicon counter- surface as presented in Fig. 37.6, under a constant load of 300 mN at the same sliding speed of 0.5 mm/s.

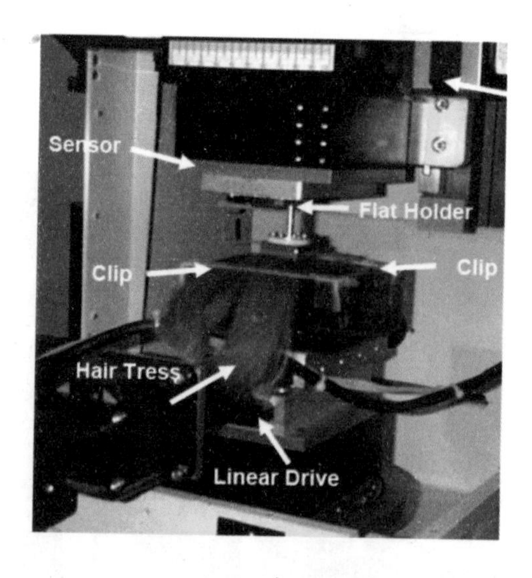

Fig. 37.6 Sledge friction test of hair

Table 37.2 summarizes the sledge friction test data. Though the friction levels in the sledge tests were different from those of the cross-hair tests, the effects of conditioners were similar, with the Pantene ProV leading to the lowest friction and the Pantene 2-in-1 shampoo revealing the highest friction

Table 37.2 - Sledge friction data on untreated hair pre-washed and treated with hair conditioners

Hair Coditioner	Hair Tress	COF						
		Test 1	Test 2	Test 3	Test 4	Test 5	Mean	SD
Pantene ProV *Sheer Volume*	#1	0.174	0.171	0.171	0.165	0.167	0.169	0.003
Dove *Extra Volume*	#1	0.191	0.184	0.181	0.180	0.181	0.183	0.004
White Rain *Extra Body*	#1	0.211	0.195	0.186	0.180	0.186	0.191	0.012
Pantene 2-in-1 *Shampoo*	#1	0.223	0.227	0.222	0.221	0.220	0.225	0.005

37.5 Conclusions

Based on above investigations, following conclusions are drawn[37.4].

A nano-mechanical analysis showed the untreated hair to be more homogeneous and more compliant than the dyed hair.

A tribological analysis showed the Pantene ProV to be more effective in reducing friction than the other hair care products tested.

The UNMT-1 is capable of comprehensive mechanical and tribological hair characterization.

References

[37.1] G. Wei, B. Bhushan, P Torgerson, *Ultramicroscopy*, 2005.

[37.2] G. Luengo, et. al., – IFSCC, 2005, v. 8, No 4.

[37.3] M. Sadaei, et. al., - Colloids and Surfaces B: Biointerfaces, v. 51, No. 2, 2006.

[37.4] N. Gitis, et. al., – IJTC 2007-44031

Index

A

B

C

D

E

F

G

H

O

P

R

S

T

U

V

W

X

Y